The Water of the Jordan Valley

SPONSORED BY THE

Federal Ministry
of Education
and Research

This book compiles results from projects funded by the Federal Ministry of Education and Research under the reference numbers 02WT9724, 02WT9725, KFK9701, 02WT0160, 02WT0161, 02WT0162, 02WT0250, and FZK0101. Only the authors are in responsibility of their contributions.

Heinz Hötzl (Project Leader) ·
Peter Möller (Ed.) · Eliahu Rosenthal (Ed.)

The Water of the Jordan Valley

Scarcity and Deterioration of Groundwater and its Impact on the Regional Development

Springer

Project Leader
Prof. Dr. Heinz Hötzl
Universität Karlsruhe
Inst. Angew. Geologie
Kaiserstr. 12
76131 Karlsruhe
Germany
heinz-hoetzl@agk.uka.de

Editors
Prof. Dr. Peter Möller
Helmholtz-Zentrum Potsdam
Deutsches GeoForschungsZentrum
Telegrafenberg
14473 Potsdam
Germany
pemoe@gfz-potsdam.de

Prof. Dr. Eliahu Rosenthal
Tel Aviv University
39040 P.O. Box
69978 Tel Aviv
Israel
elirose@netvision.net.il

Front Cover:
Sunset over the Dead Sea. View from Jordan towards the Judea Mts., Israel. © C. Siebert.

ISBN: 978-3-540-77756-4 e-ISBN: 978-3-540-77757-1

Library of Congress Control Number: 2008936030

Cover design: deblik, Berlin

Printed on acid-free paper

9 8 7 6 5 4 3 2 1

springer.com

THIS BOOK IS DEDICATED TO ALL PEOPLE LIVING ALONG THE JORDAN VALLEY IN HOPE THAT THE STUDIES CARRIED OUT IN THIS AREA OF SEVERE WATER SHORTAGE WILL CONTRIBUTE TO THEIR WELFARE, PROSPERITY AND TO THEIR PEACEFUL COEXISTANCE.

Preface

The long Jordan River, Dead Sea and the Arava/Araba chain of valleys compose together a segment of a global system termed The Syrian-African Rift which extends over 1/6 of the earth's circumference. Therefore, since the early 19th century this particular region became the target of multidisciplinary interest.

The Dead Sea-Jordan Valley Rift is not only an important geological and geographical feature but is of primary importance for the understanding of global zoological, botanical and anthropological migration processes. The prehistoric forefathers migrated 1.3 millions years ago through this Valley from Africa towards Europe and southern Asia. Among other places they settled south of Lake Tiberias and religious traditions, which evolved in this region which was once very prosperous, regarded this area as the mythical Paradise.

Unfortunately over long periods, the illusion of Paradise vanished and the valley became a pathway of armies - ancient Egyptians marching northwards and Assyrian and Babylonians going southwards. Later, the Valley was overrun by many others who wandered along this narrow border between green lands and desert.

During the present times, the political mosaic around the Jordan Valley is very complicated and - for a long time - did not facilitate free, easy and open dialogue neither between people nor between scholars and scientists. Nevertheless, the present book is the result of long and fruitful cooperation between scientists from Germany, Jordan, Israel and the Palestinian Authority. This may be the beginning of a new era of understanding and mutual help.

This is the first time when this complicate area has been studied by scientists from all riparian countries and without being hindered by territorial or by political limitations. This enabled us to create a complete, even 3-dimensional image of the area and solve hydrogeological issues which bothered all specialists in the area for many decades. One may say that following the studies described in the book, there is a clear picture as to all water bodies – fresh and saline - reaching the Jordan Valley and of their mutual interaction.

A most important achievement of the Project is the attenuation or even complete dissipation of suspicions between the riparians as to water resources. Moreover, its results indicate that in the study area, there is no negative transboundary influence which could have been caused by the utilization of local water resources. The joint and concerted scientific investigations are a tangible contribution to an atmosphere of cooperation not only in the domain of research but also in the ongoing exploitation of regional water resources. Throughout the duration of the investigations, all participants shared the feeling that they are contributing to the welfare of all inhabitants of the Valley and that the gradual progress of this concerted scientific effort, promotes the building of bridges joining the hitherto divided peoples of the area.

The Jordan Valley is in desperate need of additional water resources and we all hope that the results of our studies will enable to meet these acute demands. We all hope that the students and young researchers trained within the framework of the Project, will contribute to the development of additional water resources and to their rational management so as it should not be a Land of Lost– but a Land of Reborn Paradise.

The German-Israeli-Jordanian-Palestinian Joint Research Program on Water Resources Evaluation for a Sustainable Development in the Jordan Rift Basin was funded by the Federal Ministry of Education and Research, Germany, and by the Ministry of Science, Culture and Sport, Israel. Besides the financial support, the logistic assistance and the supply of information by Mekorot Co. and the Ministry of Water and Irrigation, Jordan, are greatly acknowledged. The Helmholtz-Centre of Environmental Research - UFZ, Germany, and the Geo-Research Center Potsdam, Germany, promoted the program by their man-power in field and laboratory work. The program was attended by special funds for PhD students from the four countries. All members of the international group acknowledge the financial support for publication of this volume received from the German Federal Ministry of Education and Research. Specifically, the editors thank Dr. H. Werz and Dr. C. Siebert for their assistance in editing the various contributions.

Wasim Ali, Akiva Flexer, Stefan Geyer, Joseph Guttman,

Marwan Haddad, Heinz Hötzl, Olaf Kolditz, Peter Möller,

Eliahu Rosenthal, Elias Salameh, Martin Sauter,

Abd el Rahman Tamimi, Annat Yellin-Dror

Contents

List of Contributing Authors

ALI, Wasim Dr.
Dept. Applied Geology
Karlsruhe University, Kaiserstr. 12
76128 Karlsruhe, Germany

ANKER, Yaakov
Tel Aviv University
P.O. Box 39040
Tel Aviv 69978, Israel

BEINHORN, Martin Dr.
Berghof, Lilli-Zapf-Str. 32
72076 Tübingen, Germany

BENSHABAT, Jacob Dr.
EWRE- Environmental & Water
Resources Engineering Ltd.
P.O.Box 6770
3106 Haifa, Israel

BERGER, Diego Dr.
Mekorot Co. Ltd.
P.O.Box 20128
Tel Aviv 61202, Israel

CHEN, Cui Dr.
Schickhart Str. 9/801
72076 Tübingen, Germany

DULSKI, Peter Dr.
Helmholtz-Zentrum Potsdam
Deutsches GeoForschungsZentrum
Telegrafenberg
14473 Potsdam, Germany

EL-NASER, Hazim Dr.
Dana Center
6 Al-Madeenah al.Munwarah St.
P.O.Box 3093,
Amman 11821, Jordan

FLEXER, Akiva Prof. Dr.
Tel Aviv University
P.O. Box 39040
Tel Aviv 69978, Israel

GEYER, Stefan Dr.
Helmhotz Centre for Environmental
Research-UFZ
Theodor-Lieser-Straße 4
06120 Halle/Saale, Germany

GLASER, Jens
Herkulesstrasse 101
34119 Kassel, Germany

GHANEM, Marwan Dr.
Palestinian Hydrology Group
Jerusalem-Beit Hanina
P.O. Box 25220
Jerusalem, Palestine

GRONEWOLD Jahn Dr.
Center of Applied Geosciences
University of Tübingen,
Sigwartstr. 10
72076 Tübingen, Germany

GUTTMAN, Joseph, Dr
Mekorot Water Company Ltd
9 Lincoln St, P.O.B. 20128
Tel Aviv 61201, Israel

HADDAD, Marwan Prof. Dr.
AN-NAJAH National University
P.O.Box. 7
Nablus, Palestine

HEINRICHS, Till Dr.
Dept. Applied Geology
Göttingen University
Goldschmidtstr. 3
37077 Göttingen, Germany

HÖTZL, Heinz Prof. Dr.
Dept. Applied Geology
Karlsruhe University, Kaiserstr. 12
76128 Karlsruhe, Germany

KHAYAT, Saed Dr.
Al Quds University
P.O.Box 20002
East Jerusalem, Palestine

KOLDITZ, Olaf, Prof. Dr.-Ing.
Helmholtz-Centre for
Environmental Research-UFZ
Permoserstr 15
04318 Leipzig, Germany

KUNTZ, David
Center for Applied Geoscience
Tübingen University, Sigwartstr. 17
72076 Tübingen, Germany

MESSERSCHMID, Clemens
Winzerstr. 121
80797 München, Germany

LENZ, Stefan
Igelbachstr. 16
D-76597 Loffenau, Germany

MÖLLER, Peter Prof. Dr.
Helmholtz-Zentrum Potsdam
Deutsches GeoForshungsZentrum
Telegrafenberg
14473 Potsdam, Germany

ROSENTHAL, Eliahu Prof. Dr.
Tel Aviv University,
P.O. Box 39040
Tel Aviv 69978, Israel

SALAMEH, Elias Prof. Dr.
University of Jordan
Faculty of Science
P.O. Box 13256
Amman 11942, Jordan

SAUTER, Martin, Prof. Dr.
Dept. Applied Geology
Göttingen University
Goldschmidtstr. 3
37077 Göttingen, Germany

SAWARIEH, Ali Dr.
Natural Resources Authority, Dept.
Geological Mapping Division
Amman, Jordan

SHULMAN, Heim Prof.
Tel Aviv University
P.O. Box 39040
Tel Aviv 69978, Israel

SIEBERT, Christian Dr.
Helmholtz-Centre for
Environmental Research-UFZ
Theodor-Lieser-Str. 4
06120 Halle/Saale, Germany

STORZ, Roman
Mainpfortstraße 1,
6914 Mainz-Kostheim, Germany

TAMIMI, Abed el Rahman Dr.
Palestinian Hydrology Group
Jerusalem-Beit Hanina
P.O. Box 25220
Jerusalem, Palestine

THIEL, Martin
Marienstr. 48
76137 Karlsruhe, Germany

TOLL, Mathias Dr.
Dept. Applied Geology
Göttingen University
Goldschmidtstrasse 3
37077 Göttingen, Germany

WERZ, Heike Dr.
Dept. Applied Geology
Karlsruhe University, Kaiserstr. 12
76128 Karlsruhe, Germany

WOLFER, Johannes
Dept. Appl. Geology
Karlsruhe University, Kaiserstr. 12
76128 Karlsruhe, Germany

YELLIN-DROR, Annat Dr.
Tel Aviv University
P.O. Box 39040
Tel Aviv 69978, Israel

1 Introduction

1.1 Scope and goals

H. Hötzl, A. Flexer, M. Haddad, E. Salameh

The lower Jordan Rift Valley is a large geotectonic feature bordered by a deep-seated fault system that creates an elongated and narrow basin. Three nations are neighbouring and sharing this basin and its margins: Jordan, the Palestinian Authority and Israel. The area has great potential for agricultural development being the natural greenhouse for a widespread region. The scarcity of water due to arid conditions is, however, the limiting factor for profitable growth. With the exception of Turkey and of Lebanon, the Middle East is an area deficient in water with yearly average natural recharge which is much lower than in Europe.

There are several major drainage basins in the Middle East, i.e., of the Euphrates, the Tigris, the Nile, Orontes, Litani, and of the River Jordan. The basin of the latter extends over Lebanon, Syria, Jordan, Israel and the Palestinian Authority. As it is a transboundary basin, neither country may independently dispose of its waters and is usually at the mercy of other riparians even if their political relations are not the best. Moreover, west of the River Jordan, groundwater basins are also transboundary and have to be shared by different political entities such as Israel and the Palestinian Authority

In the Jordan Valley, in which the studies presented in this book were carried out, rainfall is low, irregular, and often torrential. Most of it runs off without penetrating into the subsurface and replenishing groundwater. This delicate hydrological situation occurring in a complicate hydrogeological setting is often taken out of equilibrium due to overexploitation. Other results of the disturbance of the hydrological equilibrium are acute processes of natural salinization conditioned by the occurrence of brines and of saline groundwater bodies. Anthropogenic pollution is an additional factor for the deterioration of water resources and for their increasing deficiency. During the last two decades, the countries of the Jordan River basin are facing one of the most severe environmental problems which is the result of the retreat of the Dead Sea. This is manifested by diminishing groundwater resources, decreases in the discharges of springs, and formation of sinkholes along the shoreline causing the collapse of roads and of tourism facilities and the loss of agricultural areas

During the last century this area was ruled by different authorities such as the Ottoman Empire, the British Mandate, the Jordanian Kingdom, and by Israel and by the Palestinian Authority. In view of the fact that water flows across political boundaries, the present study investigates the water resources occurring to the east and west of the river Jordan disregarding any administrative or political boundaries. We all believe that water in general and the waters of the Jordan basin in particular, can play a significant role in building peace and ensuring that riparian neighbors work together to achieve together beneficial goals. Prince Hassan bin Talal from Jordan defined our strivings in a very precise manner …"Water, so essential to life, is a means to extinguish fires, not to ignite them…".

In order to improve the knowledge on available water resources a multidisciplinary research project with German, Israeli, Jordanian, and Palestinian counterparts was carried out for the sustainable utilization of aquifer systems in the Jordan Rift in the years 1997–2004. Scientists and engineers from universities, research institutions, and commercial companies of all enlisted countries joined their skill and experience. It has allowed close continuous and collegial cooperation between partners of all parties, even under the most difficult political circumstances. The main fields of activity included: hydrogeology and database management, water quality, salinity and salinization processes, vulnerability studies, quantitative modeling of groundwater flow and solute transport processeses.

This project supported by the German Ministry of Education and Research seems to be one of the most important catalysts for future cooperation among the riparian countries of the Dead Sea and of the Jordan River. This free choice of scientists to cooperate in such a project, should find the strong support it deserves. Jordan, Israel, and Palestine with the assistance and help of German scientific institutes and the support of the Federal Government of Germany would strengthen their commitment to peace, when they cooperate to find the solution for the upcoming regional deficiency in water resources and its socio-economic implication. Such an exercise in regional cooperation could encourage these countries also to overcome other, more ideologically based conflicts.

1.2 Regional setting

A. Flexer, J. Guttman, M. Haddad, H. Hötzl, E. Rosenthal

1.2.1 Topography

The project domain is located within the Jordan Rift Valley and encompasses an area of approximately 5,000 km^2. It extends from the Lake Tiberias in the north to the Dead Sea in the south. It is limited on the east by the mountainous crest along the axis Irbid Amman-Madaba (Jordan) and in the west by the mountainous crest along the axis Bardala-Faria-Ramallah-Jerusalem and Hebron (Israel and Palestine).

The lower Jordan Valley is 8–15 km wide and extents over a length of 100 km. Its elevation decreases from -200 m msl in the north, to -415 m msl at the Dead Sea. The Jordan River flows all along the valley, from Lake Tiberias in the north to the Dead Sea in the south (Fig 1.2-1). In the past the river carried a considerable amount of freshwater to the Dead Sea. However, at present its flow consists of only a small amount of high salinity water. To the east, the valley is outlined by the Jordanian Rift escarpment. Its elevation goes up to +1200 m msl and the slopes, extending in the cross-section over a distance of 20 km, decline flexure-like with increasing gradient to the west. The slopes are deeply incised by numerous wadis. The valley-floor is limited to the west by the fault-escarpments of the Hebron and Judea Mountains. The elevation increases to +900 m msl and generally, the slopes are milder as compared with the Jordanian side. They too are deeply cut by several important wadis.

Due to the rather inhospitable dry and hot conditions, there are only few larger urbanised areas down in the valley. The largest city is Jericho, situated close to the Dead Sea. In the west, the larger settlements such as the cities of Hebron, Jerusalem and Ramallah, are located along the bordering mountain crest. In the east the cities of Madaba, Amman and Salt are in a comparable situation.

Fig. 1.2-1. Satellite-Mosaic of the Near East region with the Dead Sea and Jordan Rift Valley as a part of the Aqaba-Syrian transform fault. Modified after image source: http://parstimes.com/MODIS/. (Abreviations of city names: A: Amman, Aq: Aqaba, E: Elat, J: Jerusalem)

1.2.2 Climate

The lower Jordan Valley is mostly arid (precipitation rates of 50–150 mm/yr and potential evaporation rates of up to 2600 mm/yr). Replenishment from rainfall occurs mainly along the western and eastern margins, with precipitation rates of 600–800 mm/yr and potential evaporation rates of up to 1900 mm/yr.

In the Jordan Rift Valley there is a close correlation, both on regional and local scale, between physiographic features, geological structures and climate. Three different climatic zones prevail in the study area:

- *Mediterranean climate zone*: This type of climate prevails along the highlands with precipitation rates of 600–800 mm/yr, and potential evaporation rates of about 1600–1900 mm/yr.
- *Arid climate zone*: This type of climate prevails in the Jordan Valley. Precipitation rates range from 50–150 mm/yr, and the potential evaporation rates reach 2,600 mm/yr.
- *Semi-arid to Mediterranean climate zone*: Between the arid zone of the Jordan Valley floor and the Mediterranean zone along the highlands, a transitional zone prevails along the slopes extending from the highlands to the Jordan Valley. This zone receives average precipitation rates ranging from 200–300 mm/yr, with average potential evaporation rates of 1,900–2,400 mm/yr.

The amount of groundwater recharge into the aquifer and surface runoff depends on factors such as rainfall intensity, duration, evaporation and on the sizes and rock types of the catchments areas.

- *Precipitation*: Water resource in the Jordan Rift Valley depends directly on the amount of precipitation (rain and occasionally snow) in the mountainous areas. Rain falls between October and April. The wettest months are January and February.
- *Temperature*: The average annual temperature in the mountainous region of the West Bank is 17°C and 23–28°C in the Jordan Valley. In the mountainous region of Jordan, the average annual temperature decreases to 18°C.
- *Relative humidity*: The average relative humidity in the West Bank varies from 50–70%. Down in the Rift Valley it varies between 20–30% and in the mountain region of Jordan it range from 35–40%. The minimum humidity is recorded in June, while the maximum occurs in January.

- *Evapotranspiration*: No direct evapotranspiration measurements are available in the Jordan Valley. Annual evaporation rates of 1,900 mm/yr are typical in the highlands, while down in the valley they reach 2,600 mm/yr.

1.2.3 Introductory remarks on geology and hydrogeology

The study area is part of the Dead Sea Rift system (a segment of the East African-Red Sea Rift System). The Rift Valley faults are the major structures that affect the geology and hydrology of the Rift Valley.

The regional lithologic column contains carbonates, cherts, chalk, gravels, sandstones and evaporites which range in age from Triassic to Holocene. The ancient formations of Triassic to Turonian age are composed mainly of limestone, sandstones and marl layers and crop out in the highland areas on both sides of the Jordan Rift. As a result of the creation of the Rift Valley, the oldest formations of Triassic to Lower Cretaceous age are exposed on the eastern side of the Rift. Within the Jordan Rift Valley the sediments consist mainly of Quaternary gravel, sand, clay and evaporites. A more detailed description of the geology is given in chapter 2.1.

The Jordan Rift Valley is a topographic low acting as the drainage basin for groundwater originating both from the eastern and the western mountain aquifers bordering the Rift basin as well as from ephemeral streams and floods generated in the surrounding mountains.

Due to the irregular pattern of replenishment in many parts of the arid zone, most water resources are groundwater from local aquifers. Their storage capacity is of crucial importance. Two groundwater systems can be identified:

- The regional mountain aquifers, including the Judea Group aquifer on the western side of the rift and the Ram, Zerka, Kurnub and Ajlun Groups on the eastern side of the Rift.
- The Jordan Valley aquifer system, where the water is pumped from gravels and from sand beds.

The replenishment of the mountain aquifers occurs essentially on the outcrops of their formations. The replenishment of the Jordan Valley system is manifold: precipitation, base flows, flash floods, return-flows from irrigation and upward seepage from deeper aquifers (mainly on the eastern side of the Rift).

The predominant elements governing the flow regime are the anticlinal and synclinal structures crossing the entire area and the intensive fault system that form graben, horsts and step structures. They dictate the groundwater flow patterns, the distribution of salinity and the size and shape of the replenishment areas. The dense faulting system allows hydraulic connections between the upper and lower aquifers.

The hydrogeology of the Jordan Valley itself is still ambiguous, first because of the complexity of the Rift system and associated faults and second, because the entire rock mass is covered under young sedimentary fill. Interpretations of available seismic sections reveal several conspicuous structural and stratigraphic features, which are of considerable hydrological importance. A preliminary seismic interpretation indicates the occurrence of several deep-seated salt bodies. The interpretation of salinity distribution and of the chemical composition of brines show that these salt bodies are one of the factors causing salinization of groundwater in the Rift Valley and along the foothills. Several wells located along the mountain edges are pumping brackish water which is the mixing product of freshwater and of the saline end-members.

Due to the ever-increasing population in the main cities located on the hills on both sides of the Rift, substantial amounts of raw sewage are created. These flow through the wadis to the Jordan Valley (Fig.1.2-2). The wadis are incised into the aquifer rocks and allow hydraulic connections between the sewage and the groundwater causing contamination events in many wells located along the foothills and particularly at the outlets of the wadis to the Jordan Valley. When wells are contaminated by sewage, they have to be shut down until the contamination disappears. This means that such a well cannot supply water to customers and therefore water shortage increases. It becomes crucial particularly if this happens during the peaks of seasonal water demand.

The springs that are outlets of the phreatic aquifers are traditionally an important water source of many villages in the region. Most of the springs emerge along the banks of the wadis in proximity to the sewage flowing in the wadis. The springs are highly sensitive to contamination and to the deterioration of their water quality. This occurs also in the shallow phreatic aquifers of the Rift Valley as a result of leakage of sewage, irrigation return flow and leaching of salt crusts that exist in the section. These deteriorations cause numerous water supply problems. Water that only one decade ago could be used directly for drinking and for agriculture, has to be mixed with less saline water (if such is available) nowadays. Reservoirs such as the Kafrain Dam contain water deteriorated by eutrophication processes with all the resulting impacts on downstream irrigation projects.

Therefore, there is a high demand for waste-water treatment plants and for responsible reuse of treated effluents.

1.2.4 Environment

On both sides of the Jordan Valley natural forests and afforested lands cover parts of the upper reaches of the catchments area. Irrigated agriculture is practiced along the wadi courses, whereas the high rainfall areas are used for rain-fed agriculture. Natural rain-fed plants are used for grazing.

Desertification is increasing in the lower parts of the escarpment slopes which are covered by hard rocks and rock debris and are barren of any vegetation. The floor of the Valley is covered by desert pavements with reg and sand sheets. At the outlets of wadis to the Valley as well as around springs, natural vegetation cover is common. Further eastwards and mainly due to irrigation, the valley plain becomes progressively covered by agricultural land.

The environmental scenery is governed by the presence of urban centres and of industrial areas. The cities are located along the mountain crests (the axis of the major anticlines that build the mountains).

The population growth in the area puts real threats on the development of agriculture. For instance, the Jordan Valley with its tributary wadis is the most important agricultural area in the West Bank. This is due to water allocation from groundwater wells and springs. The area has a good potential for fertile soil and is suitable for agricultural communities. However, the development of this sector is affected by water quality. The shallow aquifers and springs such as Ein Sultan in Jericho become more and more deteriorated due to cesspits and waste-water disposal system such as open sewage channels which facilitate infiltration of pollutants into the shallow aquifers.

The main activity of the people in the Valley is agriculture. Fertilizers and pesticides are applied rather generously; for example, 15,000 metric tons are applied every year in Israel and the West Bank. When applied from the air, the effect of pesticides on natural ecosystems adjacent to agricultural land is particularly evident. Fertilizers are also applied through irrigation water. Fertilizers reach aquatic ecosystems causing eutrophication and also contaminate ground water. In some of the aquifers the concentration of nitrate in the water reaches 130 mg/l.

Agriculture and urban development requires water-resource development. The impact caused by increase in agriculture has a stronger negative effect on biodiversity than the urban development. The development of agriculture and urban areas and the infrastructure connecting them (e.g., roads and pipelines) adversely affect biodiversity. Therefore, changes in the agricultural land use are necessary. The recent reduction in cotton production in Israel, for example, not only saves water but also reduces the pesticide damage to aquatic and other ecosystems. Thus, water drawn from wadis and aquifers for agriculture, contaminates and alters ecosystems. Because dry-land ecosystems are limited not just by water but also by nutrients, the enrichment of fertilizers "escaping" from desert agriculture may dramatically change the functioning and structure of these ecosystems.

Human activities that affect water quality in the Jordan Rift Valley include the followings (Fig.1.2-2):

- Discharge of inadequately treated effluent from municipal treatment plants,
- Discharge of untreated domestic and agricultural wastes,
- Extraction, use, and disposal of low-quality groundwater,
- Leachates from solid waste landfills,
- Run off from urban drainage,
- Fertilizer and pesticide residues, and
- Drainage of wetlands.

1.2.5 Socio-economic aspects

The main economic activity both in the Palestinian, Jordanian and Israeli communities at present and in the foreseeable future, is agriculture. The Jordan Valley has great potential for agricultural development, being a natural greenhouse. In the coming years, Palestinian and Jordanian towns and villages expect fast growth rates. This will result in additional demand for freshwater, which can only be supplied by reducing the amount of available water for agriculture in the Jordan Valley. On the other hand, the volumes of sewage will increase and, after adequate treatment, could be used for the agriculture in the Valley. Waste-water treatment is almost inexistent and most of the sewage flows in the wadis and rivers of the area, thus causing serious environmental problems and high risks of aquifer pollution.

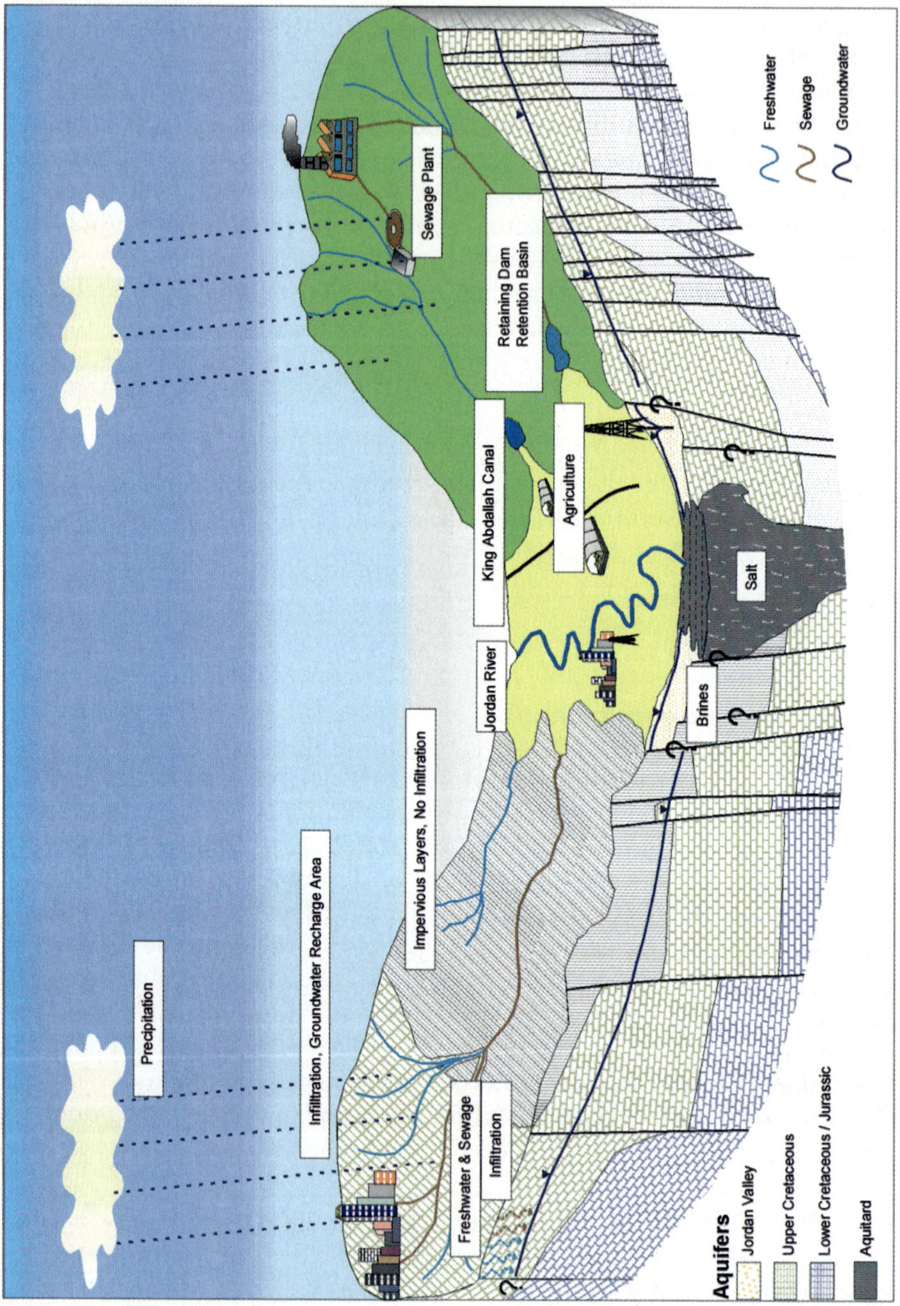

Fig. 1.2-2. Main components of the hydrogeological system in the lower Jordan Valley (after Sauter and Toll, unpubl.)

Joint industrial ventures are being planned in the region. All these activities depend on the availability of relatively large quantities, of high-quality water which, as of yet, are unavailable. Large portions of groundwater resources, which constitute the major source of water in the area, are of a relatively high salinity, which makes them unsuitable for urban- and for most of traditional agricultural uses. The water demand is expected to rise significantly but the expected supply is far from satisfying it. New prospect for future economic development is tourism. Here too, freshwater is a precondition for its large scale development. It is clear that additional freshwater resources are needed. Import of freshwater from other regions in the following 10–20 years is clearly unfeasible, since the neighboring areas also suffer from severe hydrological deficits. Due to their prohibitive costs (of both desalination and transport) so are also the prospects of import of desalinated seawater. These are the reasons which encourage the need to explore possibilities and options for increasing the sustainable exploitation of groundwater in the area - fresh, saline residual waters alike - for the benefit of all the populations in the region.

1.2.6 Current status of water exploitation

The total amount of water produced in the western part of the area amounts to 100 Mm^3/yr: 70 Mm^3/yr from wells of both the Judea Group and alluvial aquifers and 30 Mm^3/yr from springs. Production wells are located in two well fields: (i) in the hills, at the vicinity of the urban agglomerations, along the eastern edge of the mountain slopes, and (ii) in the Jordan Valley (Jericho, Marj Naja). There is considerable advantage of drilling in the mountain area in the proximity to the main urban concentrations. However, the disadvantages are deep water levels, low discharge and quick influences of drought periods on the water level and on exploitation possibilities.

In the mountain areas east of the Jordan River the aquifer conditions are much more inappropriate and restricted due to the outcrop of the main aquifer (Ajlun Group) in the high altitude and due to the close watershed to the eastern Basin. The lower aquifers lack recharge areas and the waters are more or less brackish. The intercalated mighty and marly aquitard abets the fast discharge of surface flow into the deeply carved wadis. Therefore, the water production is mainly based on (i) seasonal surface water from the wadis partially stored in reservoirs (120 Mm^3/yr), (ii) treated and untreated waste water of about 65 Mm^3/yr discharged into the wadis and reservoirs and thereby indirectly used for irrigation, (iii) spring and well water from the upper aquifer of about 80 Mm^3/yr, and (iv) few deep wells tapping the lower aquifers with a still small amount of brackish water.

In the lower Jordan Valley and along the margins of the mountain ranges usage of shallow wells is limited by salinization. In order to avoid the rise in salinity, pumping (hourly and yearly) had to be reduced. This reduction has caused the loss in supply of large amounts of freshwater. In the shallow aquifers there is a constant degradation of water quality as a result of the migration of saline water into the freshwater bodies, of the irrigation return flow and leakage of raw waste water.

On the Jordanian side about 130 Mm^3/yr of the Yarmouk River are deviated by the King Abdullah Canal and distributed along the lower Jordan Valley for municipal and irrigation use.

1.2.7 Geographic nomenclature

The study area extends over a region populated by a diversity of people, cultures, religions, languages, dialects and customs. During the last 100 years the area was administered by different political authorities, each of them left behind its imprints on the geographic nomenclature of the area.

The papers assembled in this book were written by authors from different national entities in the area and Germany. Each such entity preserves zealously their geographic terms. This produces a long series of local names which at first sight seem to be different but are actually synonymous. Therefore, a reader which is not familiar with these regional characteristics may suspect editorial inconsistencies or even negligence, which is not the case. Moreover, the rules of transliteration of local Hebrew and Arabic names into English are not identical in the three national entities residing in the area.

The editors took in consideration the sensitivities and decided not to impose one system of geographical terms and to respect each national tradition. We read critically all the papers and in those cases in which divergent nomenclature could cause confusion, necessary corrections and changes were made while preserving the original national terminology.

We would like to present several commonly encountered examples (A = Arabic name; H = Hebrew name; E = English translation):

Geographic terms

En (H)	Ein (A)	Spring (E)
Nahal (H)	Wadi (A)	Ephemeral stream (E)

Different names for the same locations

Lake Tiberias, Lake Kinneret, Sea of Galilee

Different transliterations of same location names
Qilt, Qelt
Feschkha, Faschkha Fashkha, Fescha,
Zerka, Zarka, Zerqa, Zarqa
Mitspe, Mitzpe
Nezer, Netzer
Salt, E-Salt
Fasail, Fatzael, Fazael, Pezael
Yorqeam,Yorkeam, Yorqe'am
Arava Valley (H), Araba Valley (A)

2 State of the Art

2.1 Geology

A. Flexer, A. Yellin-Dror

2.1.1 Geology and tectonic setting

The geologic setting of the Eastern Mediterranean and the Levant consists of the triple junction of the African, Arabian, and Anatolian plates, bounded by the Cyprian Arc, the Peri-Arabian Ophiolitic crescent and the Dead Sea Transform.

Israel, Palestine and Jordan are located along the northern and northwestern margins of the Arabian-Nubian continental craton that is a part of the African-Arabian Plates. It comprises a series of rigid blocks composed of a complex assemblage of Precambrian igneous, metamorphic, and sedimentary rocks. During Cambrian to Holocene time, regional geologic conditions were dominated by the interaction between this rigid continental mass and the oceanic areas extending to the north and northwest which produced the stratigraphic sequence that has so far been observed in the surface and subsurface geological record of the study area. This interaction between the Arabian Craton and adjacent oceanic areas is controlled by:

- Alternating advance and retreat of shallow epicontinental seas on the Arabian Craton in response to cratonic subsidence/emergence and eustatic sea level changes
- Opening and closing of the Neotethys, accompanied by faulting and volcanism
- Convergence and development of an Alpidic folding system: the Levantid folds or the (Senonian-Eocene) Syrian Arc fold System
- Regional uplift and fracturing of the Arabian-African Plate, isostatic readjustments and Dead Sea stress field activity since Oligocene, the opening of the Red Sea, the Dead Sea Transform, and Palmyride thrusting and folding (Flexer et al. 2005)

The area subdivides into a number of tectonic provinces. These provinces are defined in the geological sense, i.e., as zones with common geophysical, seismological, structural, and sedimentary characteristics. In most cases a typical morphological character expresses a tectonic province. The area can largely be divided in the sense of the five classical morphotectonic units of the Middle East as suggested by Picard (1937), (Fig. 2.1-1):

1. The Arabo-Nubian Massif, consisting of Precambrian terrain devoid of sedimentary cover.

2. The Tabular Zone, consisting of a thin sedimentary cover on a relatively thick continental crust that has been relatively stable during and since the Paleozoic. It has been subjected to slow, epeirogenic, long-wave time-wise oscillations which brought about rhythmic advances and retreats of the sea along its margins. During the Paleozoic and the greater part of the Mesozoic, these generated a typical platform sedimentary cover of mature clastics (primarily sandstones) on a flat continental landscape, or

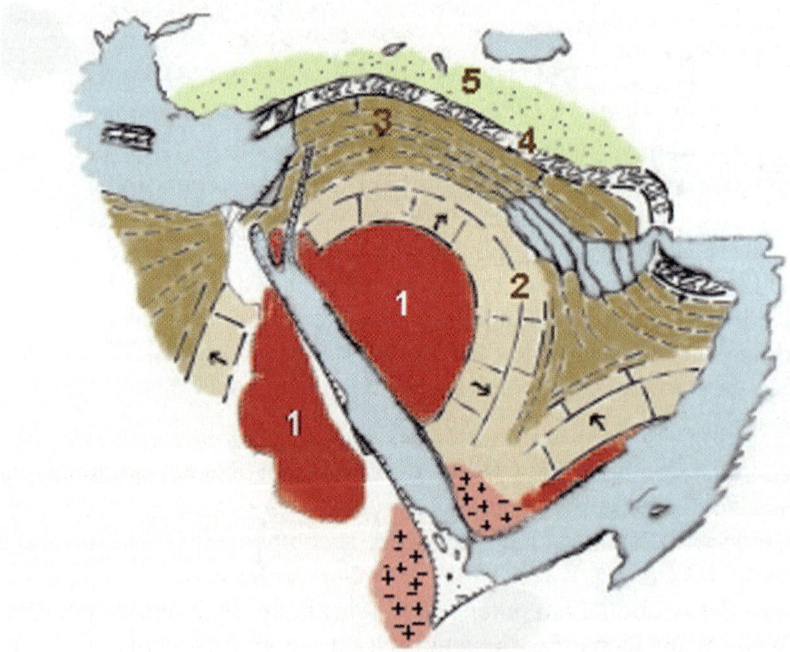

Fig. 2.1-1. The legendary morphotectonic scheme of the Middle East originally plotted by Picard (1937). The numbers present structural units with an unique relief expression: (1)The ancient Arabo-Nubian Massif with an uneven topography; (2) the Tabular Zone, a sedimentary flat highland; (3) the valleys and range topography of the Folding Belt; (4) the high mountainous view of the Alpine Orogenic Belt; (5) the Flat Median Plateau extending between the mountain belts.

carbonate-shale sequences in shallow epeiric seas. An accentuation of earth movements took place in Late Triassic and in Late Jurassic-Early Cretaceous times.

3. The valley-and-range topography of the Folding Belt consists of the Levantid folds of the Syrian Arc. The simple folding zone is characterized by a flat S-shaped band of folds that extends from northwestern Sinai, through Israel and Lebanon into Syria. The common characteristic of this tectonic province is the presence of generally normal concentric folds which were developed in the sedimentary cover. Most of the folds are strongly asymmetric, a feature generally attributed to the presence of basement-controlled, deep-seated reverse faults. Starting in the Coniacian, the Levantid compression decomposes into transversal faulting with strike-slip components and vertical displacements, and in reverse faulting with asymmetric folding. These compressional pulses re-occurred until the Early Miocene, post-Oligocene deformation, decomposed into folding and sinistral strike-slip movements, possibly in both combinations. In Mt. Hermon, the compressional deformation, though present from the Senonian, reached its peak only after the Oligocene, as in the Palmyrides (Chaimov et. al. 1990).

4. The Alpine Orogenic Belt thrusted over the African-Arabian Plates. The belt, characterized by a chain of high mountains, starts in the Kyrinia Range in Cyprus and continues eastwards to the Taurus chain in Turkey and to the Zagros Mountains in Iran.

5. Median blocks (Anatolian and Ionian masses). These are Paleozoic blocks that were consolidated after the Hercynian Orogeny and behaved as massifs during the Alpine Orogeny.

2.1.1.1 Tectonics

Plates and plate boundaries

The area of the Eastern Mediterranean countries can be subdivided geologically into three plates: the African Plate, the Arabian Plate and the Eurasian Plate (Fig. 2.1-2).

The Arava-Dead Sea-Jordan Rift Valley is the plate boundary which separates the Arabian Plate from the African Plate. It should be pointed out that since the Miocene this boundary has developed as a major left-lateral transform fault zone (Fig. 2.1-2) in response to the culmination of the major collision of the African-Arabian and Eurasian Plates. During most of the Phanerozoic era, the African-Arabian Plates were united and consisted of the Arabian-Nubian Massif or Craton.

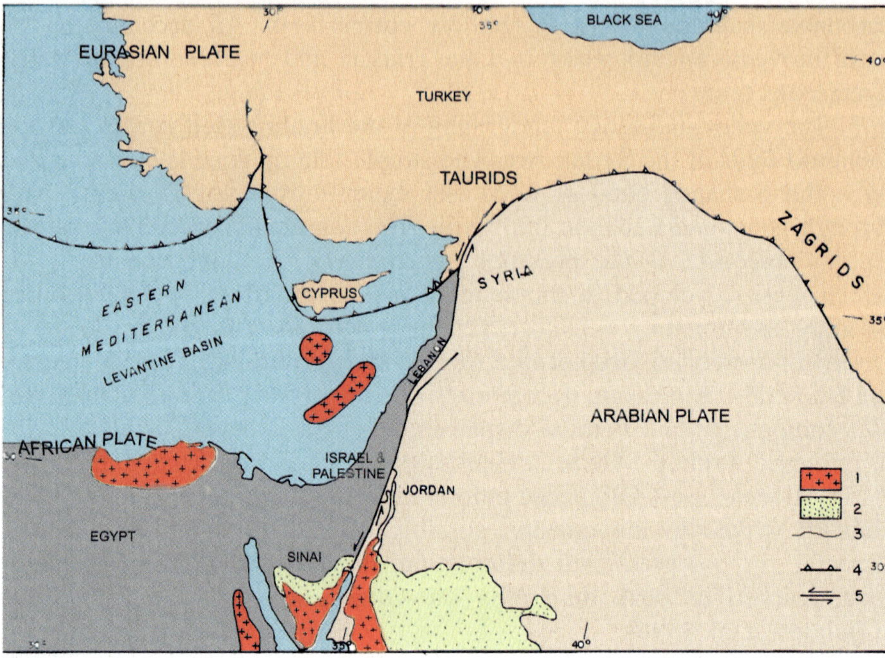

Fig. 2.1-2. Plate-tectonic pattern of the Levant highs (modified after Hirsch et al. 1995):

1. African-Arabian Precambrian crystalline
2. African-Arabian Paleozoic sediments
3. Present day coastline
4. Eurasian Plate thrust front
5. Dead Sea transform

Interplay and collision of plates

The strongly interwoven geological and tectonic history of the Levant is punctuated by periods of subsidence and uplift which are subdivided into phases (Hirsch et al. 1995):

The early geological history of the Levant as part of the ancient Gondwana continent was a build-up of a basement composed of accreted terranes, consisting of plutonics, volcanics, ophiolites and metasediments. It was consolidated by Late Precambrian Pan-African orogeny, including the Menderes, Eratosthenes and Arabian-Nubian massifs, prior to regional peneplanation.

The Paleozoic history of the Levant is one of domal uplifts during the Late Devonian, Permian and post-Permian, which resulted in the erosion of much of the Paleozoic deposits. Hardly anything is left of the Ordovician-Devonian sedimentation and the Earliest Triassic is missing (Bender 1974).

The Early Triassic opening of the Neotethys across Gondwana was echoed by Mesozoic epicontinental sedimentary cycles (Druckman 1974). Echoing the Alpidic Eurasic-Gondwana collision, the Levantid deformation of the Negev-Palmyride belts is the result of the inversion of the Early Mesozoic faults, which produced the asymmetrical folds of the S-shaped Syrian Arc. These extend from Northern Sinai to the Palmyrides. The Syrian Arc shaped a paleogeography of basins and heights, sometimes islands which controlled the carbonatic and siliceous sedimentation of the Senonian-Paleocene Mount Scopus Group and the Eocene Avedat Group (Braun 1967; Flexer 1968). The Oligocene lower part of the Saqiye Group represents the last period of this phase. Sediments of this age were widely destroyed by the regional uplift that started the next phase, leaving only a few marine outcrops in the northern Negev, the Coastal Plain and Golan Heights.

A possible mid-Oligocene tectonic rebound coincides with a sea-level drop that enhanced regional regression, emersion and erosion. Since the Late Oligocene, a Red Sea opening has split the Gondwanian foreland into separate African and Arabian Plates which continue to move northwards toward the Eurasian-Anatolian Plates (Flexer et al. 1984).

2.1.1.2 The Syrian Arc folding system

The following text is mostly based on Flexer (2001). The term "Syrian Arc folding system" is applied to a tectonic province in the Levant, distinguished by its huge "S"-shaped belt of folds. The term, as introduced by Krenkel (1924), has recently been revived, but "Levantides Fold Belt" is also in use. The Syrian Arc is only a segment of a larger arc of foreland folds, encircling the Arabian Craton from Egypt via Israel, Palestine, Jordan, Lebanon, Syria, Iraq, the Emirates and Saudi Arabia. The Syrian Arc folds have many common geological characteristics such as the northeast trend, the asymmetry due to the presence of basement controlled reverse faults, and the multiphase history of deformation. The folding occurred from Turonian to Neogene (Quaternary?), whereas the deformation peaks are discerned during the Late Turonian, Middle Campanian, post-Middle Eocene and Late Neogene-Early Quaternary (Shahar 1994).

Israel, Palestine, Northern Sinai, Northern Jordan, Lebanon and southern Syria are crossed by a series of surface and subsurface anticlines and synclines which strike mostly in east-northeast, northeast and north-northeast directions (Fig. 2.1-3). This folding system comprises a segment of a broader belt of folds which continue eastward to Iraq and the north-eastern parts of Saudi Arabia (Fig. 2.1-1). The belt is mainly located on the unstable shelf between the Arabian Craton and the Zagros-Taurus thrust zone.

Different tectonic interpretations and descriptive terms were proposed for the belt of folded mountains of the lands bordering the eastern Mediterranean. The common characteristic of this tectonic province is the presence of generally simple folds, developed in the sedimentary cover. However, the belt is not uniform in its structural details, as some of the folds are basement-controlled through deep-seated reverse faults while others are rootless, more superficial, epidermal folds. Moreover, the tectonic activity is not synchronous, as some of the folds exhibit a multiphase tectonic history, whereas others underwent simple deformation at different periods of time.

The Syrian Arc has many common tectonic characteristics and quite a definite time for its peak of activity. The general understanding is that this tectonic stage was initiated in the Turonian, roughly coinciding with the main Alpine stage, associated with the collision of the African-Arabian and Eurasian Plates. The folds are well preserved in the Negev, in the mountains of Ramallah in the Judean and Jordanian mountains. Due to the Plio-Pleistocene uplift, they form broad structural highs of the anticlinorial style.

The Syrian Arc system is characterized by a "Jura"- type fold morphology, with erosion cirques, cluses, ruzes and hogbacks on the frequently steep flanks of the asymmetric anticlines, with mesas and tabular features in the synclines. The fold morphology is less noticeable in western Judea, Ramallah, Galilee and Northern Jordan, most likely due to the effect of subsequent faulting and the influence of the more humid Mediterranean climate, forming a caliche crust and terra rossa soils which cover most of the rocks.

The folds are strongly asymmetric, with a steep flank to the southeast, a feature generally attributed to the presence of reverse faults at depth. Other structures, such as the Hebron anticline, are strongly asymmetric to the west (Fig. 2.1-4). Still other structures are roughly symmetric or of the box-fold type, with steeply dropping flanks on both sides. The folds are often described as concentric or parallel, i.e., not changing the thickness of their beds as a result of the compressive stress (changes in thickness inherited from the sedimentological pattern are common, but irrelevant to this topic). Folding started in the Late Turonian and the folds acquired their final shape between the Eocene and Miocene. They became uplifted and, especially to the north, block-faulted during the Neogene and still more so during the Pleistocene.

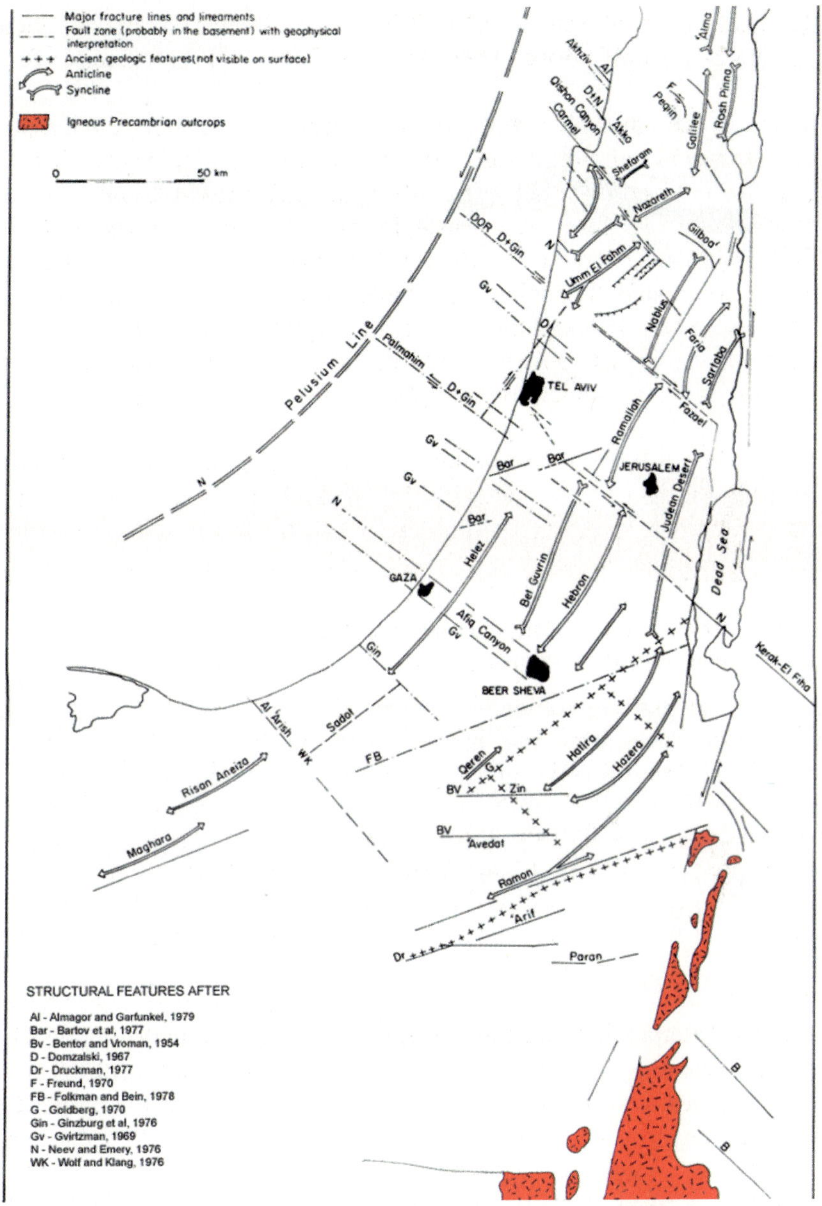

Fig. 2.1-3. Folds and morphotectonic lineaments of the southern Levant (modified after Flexer et al. 1984)

The location of the Syrian Arc fold belt, together with the Zagros-Taurus, Cyprean and Hellenic Arc thrust zones, between Eurasia to the north and Arabia-Africa to the south, has interested geologists for decades.

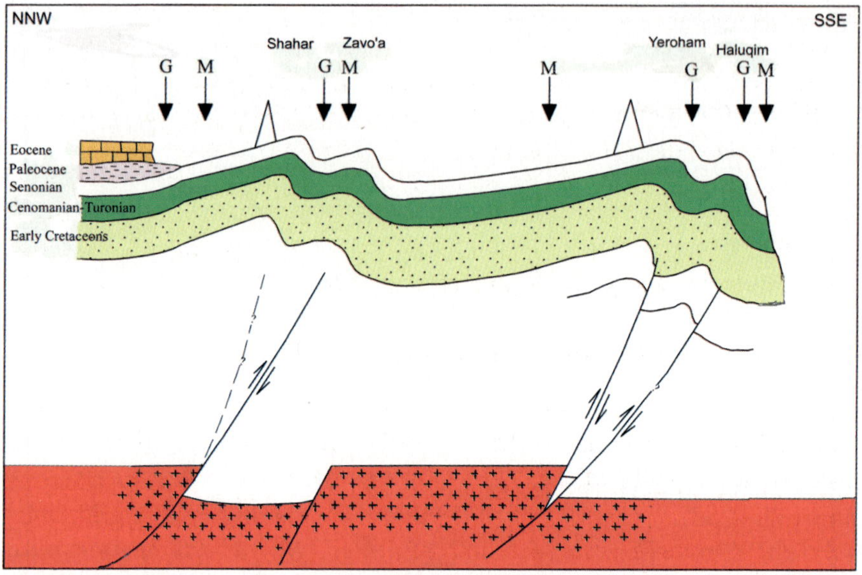

Fig. 2.1-4. A typical cross section through the Syrian Arc folds, showing the association between structure, gravimetric (G) and magnetic (M) anomalies

The advent of plate tectonics led to new explanations of the relationship between these two continental masses, with the basic concept that the African-Arabian Plate is moving relatively northward and underthrusting the the Eurasian Plate (Scott 1981). The situation is even more complex, since the Arabian and African Plates, now separated, were united ever since the Precambrian and up to Neogene times, forming the Arabo-Nubian Massif, eventually surrounded by the Syrian Arc fold system.

2.1.1.3 The Syrian-African Rift System

The Jordan-Dead Sea depression is part of the Syrian-African Rift System (Fig. 2.1-5) running all the way from southern Africa through East Africa and the Red Sea to southern Turkey (Horowitz 2001 and a long list of references therein).

The Dead Sea-Jordan Rift Valley

The approximately 1000 km-long Dead Sea Transform (DST) (Fig. 2.1-5) is an active plate boundary that connects the sea floor center of the Red Sea with the Alpine convergence zone in Turkey. It was formed during the Late Cenozoic by the break-up of the once continuous Arabian-African continent. A remarkable feature of the DST is that several basins, in places

≥10 km in depths, are present along much of its trace, especially along its southern half. These are interpreted as pull-apart basins, some being among the largest known structures of this kind (Garfunkel and Ben-Avraham 2000; Ben-Avraham and Lyakhovsky 1990). The deep pull-apart basins are: Gulf of Elat (Aqaba), Dead Sea, Sea of Galilee, Hula and Ghab basins. Basin lengths range from 15–150 km, widths from 5–20 km, and depths reach 5 km to almost 15 km. These pull-apart basins formed between left-stepping master strike-slip faults, but the deep troughs are commonly bordered by marginal normal faults.

History of the Jordan Rift Valley geological research

The Jordan Rift Valley had attracted the attention of geologists since the middle of the 19th century. In the beginning, the structure of the linear valley, which extended northward from the Gulf of Aqaba (Elat) to the Sea of Galilee, was attributed to tectonic motion (Anderson 1852; Hull 1886; Lartet 1869; Blanckenhorn 1912, 1914).

Since then, various aspects of the Rift Valley and its fill have been studied extensively. Suess (1909) and Gregory (1921) recognized that the Jordan Rift is the northern branch of the continental-scale rift system which extends through the Red Sea to East Africa.

These early studies have mainly led to the view that the Jordan Rift is an extensional feature dominated by normal faulting. Some early researchers suggested, however, that important lateral motion took place along the Rift. This view became widely accepted when the role of large strike-slip faults was recognized, and especially following the advent of plate tectonic theory. Picard (1943) called the depression "the Jordan-Aqaba graben or Rift Valley" because it was formed by downfaulting along border faults. Today it is obvious that the terms Rift Valley and Rift Zone are somewhat incorrect. Although a large vertical motion is associated with linear depression, there is overwhelming evidence that the “rift” is a continental transform plate boundary. In the current book we use the term "The Jordan-Dead Sea Rift System" in order to emphasize the tectonic aspects of the plate boundaries. However, other authors use the term "Dead Sea Transform" or more simply “the Rift”.

An estimated 105 km of cumulative left-lateral, strike-slip offset (e.g., Quennell 1959, 1996; Freund et al. 1970) has been attributed to the transform that connects the spreading center of the Red Sea with the continental collision of the Taurus-Zagros mountain portion of the Alpine Orogenic belt in Turkey (Fig. 2.1-5). The accumulative lateral displacement of 105 km has been estimated since the time of the Rift initiation near the

Oligocene-Miocene boundary or since Mid-Miocene as is postulated by Bayer et al. (1988) from studies in the Midyan Peninsula east of the Gulf of Aqaba.

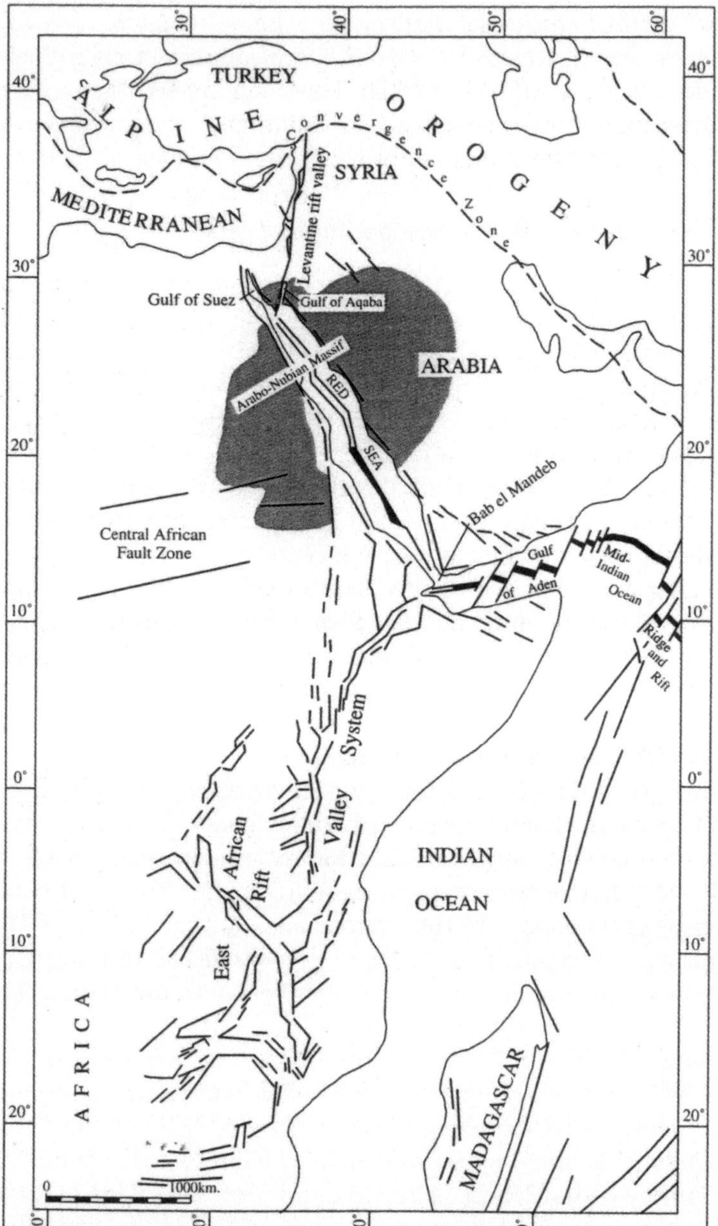

Fig. 2.1-5. The distribution and configuration of the Syrian-African Rift system (modified after Horowitz 2001)

Uplifted Rift shoulders bordering deep depressions and filled with young sediments are common along the Rift. On the Highlands west of the basin floor, Middle Cretaceous carbonates and Upper Cretaceous clays, marls, chalks and cherts are exposed. The eastern highlands are composed mostly of Paleozoic to Lower Cretaceous Nubian sandstone. The different rock units on the two opposite sides of the basin reflect both relative motion and differential uplift (Figs. 2.1-6; 2.1-8 and 2.1-9).

The most significant syntheses and interpretations of the Jordan Rift Valley history, structure, tectonic setting and geophysical characteristics were given by Neev and Hall (1979), Zak and Freund (1966; 1981), Garfunkel (1981), Kashai and Crocker (1987), ten Brink et al. (1993), and Niemi et al. (1997).

2.1.1.4 Stratigraphy

The stratigraphic succession exposed on both sides of the lower Jordan Valley and along the margins of the Dead Sea starts with top Precambrian molasses, sediments and volcanics, and terminates with young Pleistocene and Holocene Rift Valley fill. Valuable stratigraphic information is obtained from deep oil wells drilled in the valley itself and on the shoulders.

Structurally and stratigraphicaly, the western and eastern banks of the Rift Valley are rather different. The eastern side is structurally higher and exposes deeper stratigraphic levels. Precambrian, Paleozoic and Triassic rocks crop out along the eastern shore of the Dead Sea, whereas in the west only Cretaceous sediments occur. In the north, along the eastern Jordanian side, Triassic to Neogene successions prevail. On the western side only Late Jurassic to Neogene sediments crop out.

The geological map (Fig. 2.1-6) and the stratigraphic table (Fig. 2.1-7) clearly show this asymmetry. In the "Central Israel" and "Palestinian Authority" columns, the stratigraphic table adds subsurface details which were collected from deep drillings. The following text is a brief description of the stratigraphy of the four eras from bottom to top.

The Precambrian

The sole outcrop of Precambrian rocks in the study area is the Saramuj Conglomerate which is exposed in the SE corner of the Dead Sea (Bender, 1974). This Late Precambrian arkose and volcanic series is exposed in Jordan and Israel (Figs. 2.1-7–8) and has been penetrated by many deep drill holes in Jordan, Egypt and Israel. The following description is taken and modified from Weissbrod and Sneh (2002) who studied the sedimentary petrography of the Late Precambrian–Early Cambrian clastic facies in the

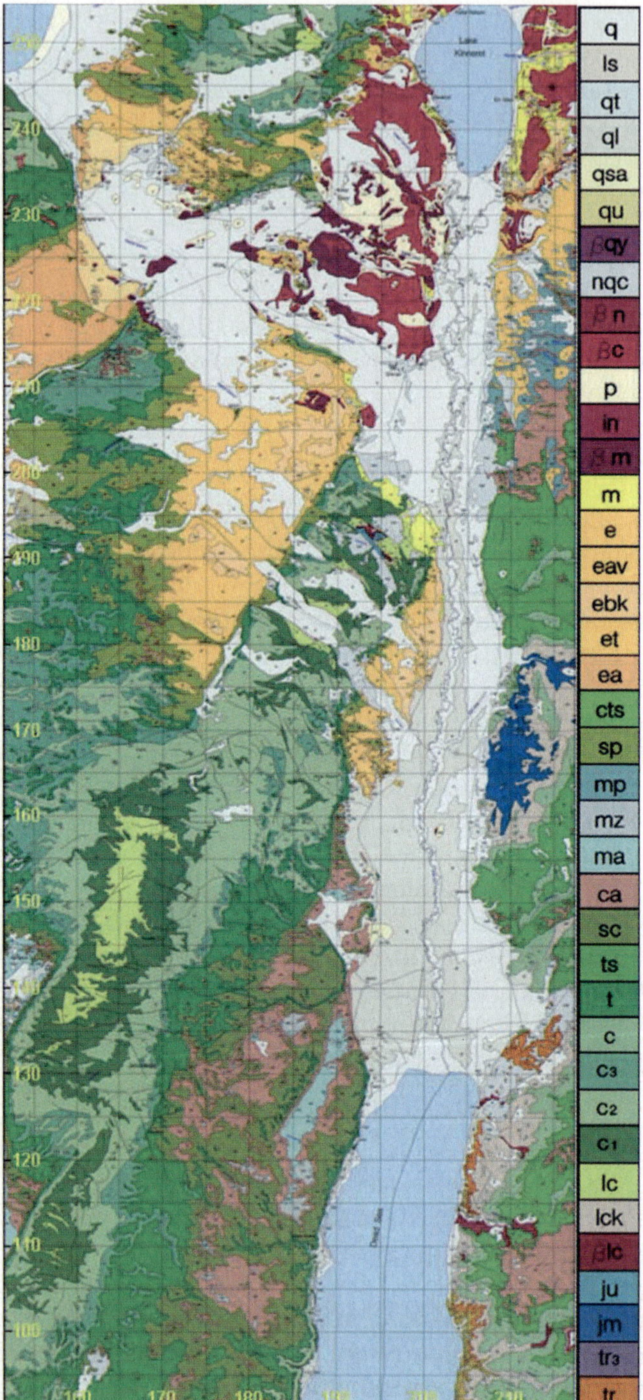
q
ls
qt
ql
qsa
qu
βqy
nqc
βn
βc
p
in
βm
m
e
eav
ebk
et
ea
cts
sp
mp
mz
ma
ca
sc
ts
t
c
c_3
c_2
c_1
lc
lck
βlc
ju
jm
tr_3
tr

Fig. 2.1-6. Geological map on both sides of the Jordan Rift Valley and surrounding areas (coordinates 125-230N; 180-220E; mapped area 105×40km). From: Sneh et al. (1998) The geological map of Israel, 1:200,000. Jerusalem: Israel Geological Survey (by permission of the Geological Survey of Israel). (q) Alluvium, Faza'el and Damiya Fms, Late Quaternary; (ls) landside, Quaternary; (qt) Bet Sh'an and other travertines, Quaternary; (ql) Lisan Fm, Late Quaternary; (qsa) Samra and Oolitic Fms, Pliocene; (qu) Erk el Ahmar and Ubeidiya Fms, Quaternary; (βqy) Yarmouk Basalt, Quaternary; (nqc) undivided conglomerate units, Neogene-Quaternary; (βn) undivided Neogene volcanics; (βc) Cover and Intermediate basalts, Neogene-Quaternary); (p) Bira and Gesher Fms, Pliocene; (in) intrusive and pyroclastic rocks, Miocene; (βm) Lower Basalt and Late Miocene Volcanics, Miocene; (m) Hazeva and Dana Fms, Miocene; (e, eav, ebk, et, ea) various Eocene Fms; (cts) undivided Cenomanian-Turonian-Senonian (Jordan); (sp, mp, mz, ma, ca, sc) Mount Scopus Group, Senonian-Paleocene; (ts) undivided Turonian-Santonian (Jordan); (t) Turonian Fm (Israel); (c) undivided Albian-Cenomanian (Jordan); (c3, c2, cl) Cenomanian Fm; (βlc) Igneous rocks, Mesozoic; (ju, jm) Jurassic (Israel); (tr) Permian and Triassic Fms (Jordan); solid lines represent faults (after Horowitz 2001)

northern margins of the Arabo-Nubian shield. They support the existence of a presumed longitudinal rift (geosuture) which developed in the Late Precambrian to Early Cambrian along the later Tertiary Dead Sea Transform, where relicts of the conglomeratic facies have been protected from erosion in the downfaulted depression.

The final stages of cratonization of the Arabo-Nubian shield during the Late Precambrian-Early Cambrian (~620–525 Ma) are characterized by uplift, extensional tectonics, alkaline volcanism, plutonic granite intrusions and erosion. The denudation was extensive and molasses-type sediments rapidly accumulated as gravel aprons and conglomerates in intramontane valleys and in subsiding grabens or as alluvial fan systems in the piedmont zone. The conglomerates, either pebble or matrix-supported, are dominated by clasts derived mostly from gneisses, granitoid rocks, their dike suite and rhyolitic-andesitic volcanics. Deposition was accompanied by volcanism. Today, relicts of these coarse clastics are exposed in fault- bound depressions in the massif terrain or have been locally encountered in the subsurface. Along the northern and northeastern margins of the shield, a broad pericratonic basin developed in which a thick section of immature arkose-dominated clastics and conglomeratic tongues associated with alkaline volcanics accumulated (Zenifim Formation). These rocks have been encountered in the subsurface.

The presence of the Saramuj and the Zenifim Formations at the bottom of eleven deep boreholes in southern Israel, their reinterpretation in seven boreholes in Jordan and one in central Sinai (Egypt) and the limited exposures of the Late Precambrian-Early Cambrian greywacke-volcanics series

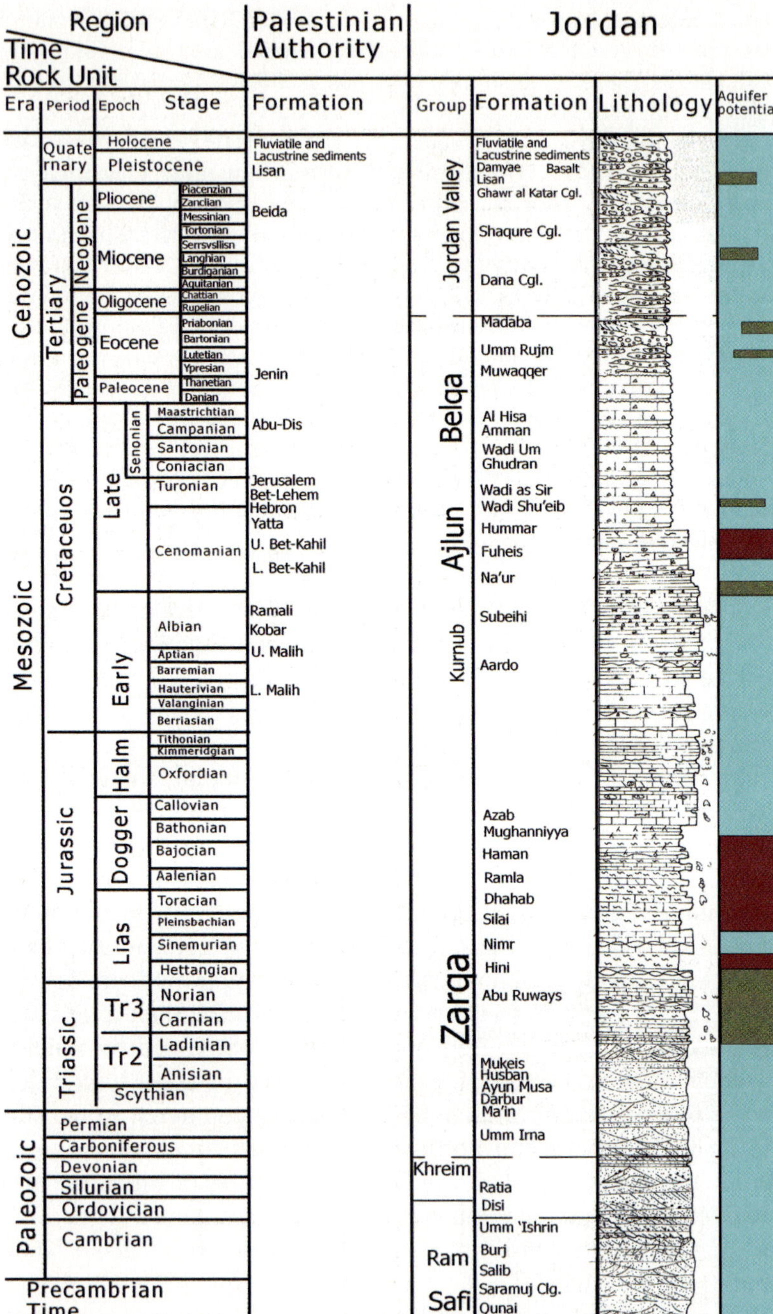

Edited by Yaakov Anker (Anker 2007)

Fig. 2.1-7. Correlative stratigraphic table for the study area

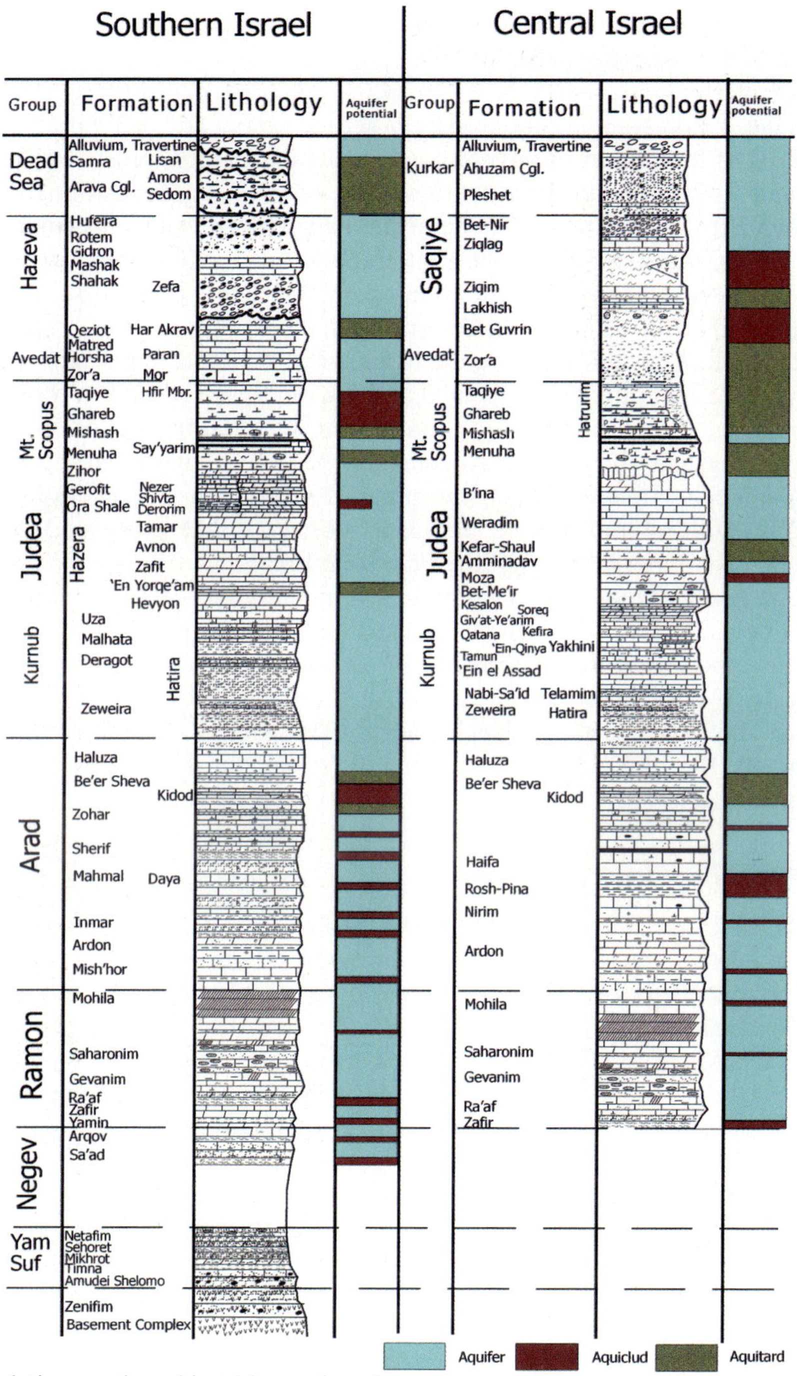

Correlative stratigraphic table continued

in three regions in SE Turkey, enable the demarcation of the wide regional extent of the pericratonic basin.

The pericratonic basin is an immense zone of deposition extending from the Arabo-Nubian Shield margin across Arabia and NE Africa to the foothills of the Taurus in the north and along the Zagros thrust zone in the NE and east. It displays an entire system of evolving depositional environments from the piedmont zone of the Pan-African chains in Saudi Arabia, Jordan, Israel, Sinai and the Eastern Desert of Egypt all the way across the Levant and Mesopotamian countries to the base level in the passive margin and open sea in SE Turkey and Iran.

By the late Early Cambrian, the extensional regime and the igneous activity in the region were terminated and the landscape eroded to form vast peneplains. The base level subsided leading to a resumption of an active drainage system and sediment deposition. Due to a change in the paleogeographical setting and different mode of weathering and alteration, the detritus evolved into mature platform sediments that proceeded almost continuously through much of the Paleozoic Era.

Late Precambrian-Early Cambrian sedimentary or volcano-sedimentary strata occur in restricted depressions within the igneous basement and are locally exposed on the eastern margin of the Dead Sea Rift (Fig. 2.1-8). They have been described by several authors since the mid-1800's (e.g., Lartet 1869; Hull 1886; Blanckenhorn 1912; Blake 1935; Picard 1941; Quennell 1951; Wetzel and Morton 1959; Bender 1974; Powell 1988; Jarrar et al. 1991).

Saramuj Conglomerate (Jordan)

The most extensive and continuous exposure of the Late Precambrian coarse clastic facies occurs at the southeastern shore of the Dead Sea in a wedge-shaped, 2x9 km, NNE-SSW trending zone between Wadi Numeria and Wadi Hasa (Fig. 2.1-8).

It consists of reddish brown to dark green-grey, hard, thick-bedded or massive, clast-supported, poorly-sorted polymictic conglomerate and alternating bands of well-stratified, coarse- to medium-grained arkosic or lithic sandstone which may be up to several meters thick. The clasts are well- to sub-rounded, ranging in size from pebbles to boulders set in a consolidated arkose matrix. They are derived mostly from the igneous basement, rarely with a reworked lithified matrix. Granite and diorite are the common clasts, although Picard (1941) noted that the majority are prophyritic dike components. The conglomerate beds vary from 1 to 30 m in thickness and the sandy layers are 10 cm to several meters thick. The proportion of thick conglomerate beds decreases upwards. A monzo-dioritic

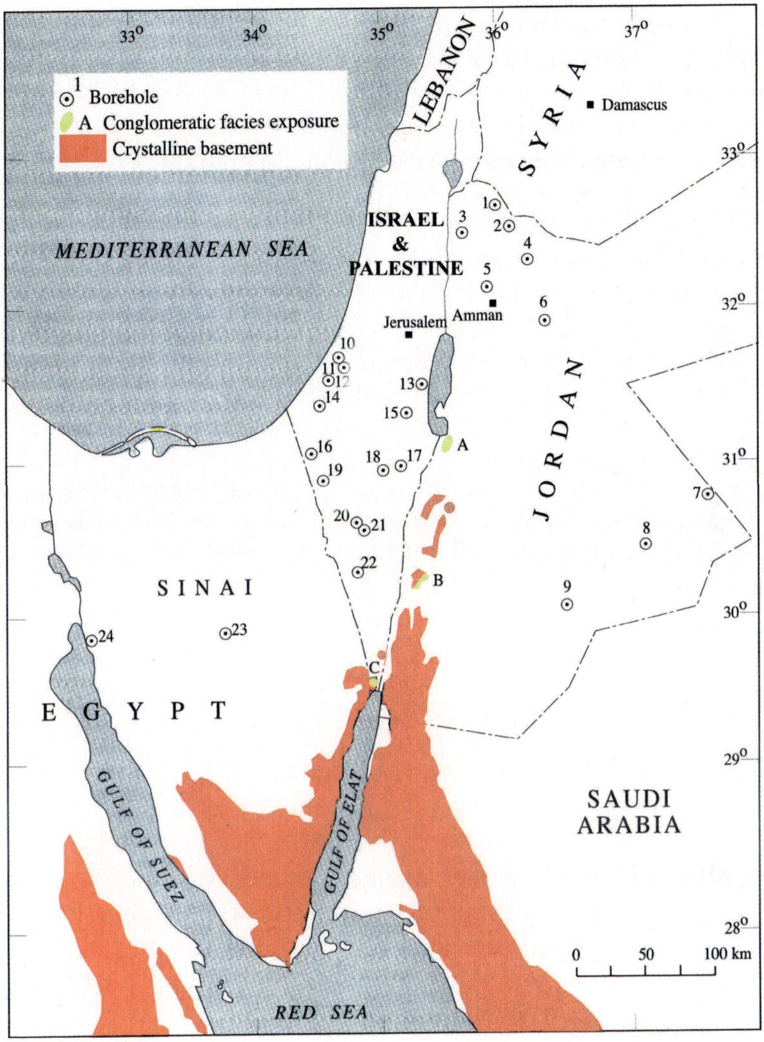

Fig. 2.1-8. Location map of boreholes that penetrated the arkosic facies (Zenifim Formation) and/or igneous basement in Israel, Sinai and Jordan and exposures of the conglomeratic facies (Weissbrod and Sneh 2002). Boreholes (a) Jordan: 1=Northern Highlands-2; 2 = Er Ramtha-1A; 3 = Ajlun-1; 4 = Northern Highlands-1; 5 = Suweileh-1; 6 = Safra-1; 7 = Wadi Sirhan-10; 8 = Wadi Sirhan-3; 9 = El Jafr-1. (b) Israel: 10 = Helez Deep-1A; 11 = Pleshet-1; 12 = Gevim-1; 13 = Emuna-1; 14 = Besor-1; 15 = Zohar-8; 16 = Agur-1; 17 = Makhtesh Qatan-2; 18 = Kurnub-1; 19 = Shezaf-1; 20 = Ramon-1; 21 = Hameishar-1; 22 = Sinaf-1; (c) Sinai: 23 = Abu Hamth-1; 24 = Ayun Musa-2. Exposures: A = Wadi Saramuj; B = Wadi Abu Barqa; C = Elat area

stock and numerous dikes and sills of dolerite, and quartz-porphyry intrude the Saramuj Conglomerate. The formation is about 200 m thick and its base is not exposed.

Zenifim Formation (Israel and Sinai)

The arkosic facies of the Zenifim Formation were formed at the same time as the conglomerate facies of the Saramuj. A more detailed description can be found in Weissbrod and Sneh (2002). The Zenifim Formation consists mostly of well cemented, impervious, red to dark brown, massive, horizontally bedded or cross-stratified matrix-rich arkose, often with thin interlayers of siltstone and mudstone, and in certain places, of conglomerate. The formation is well exposed in Jordan, Israel, Saudi Arabia and Egypt (Fig. 2.1-8 after Weissbrod and Sneh (2002). The Zenifim Formation was encountered in many drill-holes in Egypt (Sinai), Southern Israel (Negev) and Jordan (Fig. 2.1-8). It comprises of the molasse-type Volcano-Conglomeratic complex as well as of the Elat Conglomerate and their finer clastic series and overlies rocks of previous metamorphic and intrusive complexes along a very pronounced unconformity. Rocks of the Volcano-Conglomeratic Complex constitute the entire Amram and Neshef massifs and occur in restricted areas in Elat and Timna, located in southern Israel.

The Infracambrian

The term Infracambrian (or "Lipalian Interval", see Picard, 1943, p. 43) was coined (Wolfart, 1967) to describe a succession of non-metamorphosed or just barely altered sedimentary, mostly clastic sequences which clearly overlie metamorphosed or crystalline Precambrian rocks. The Volcano-Conglomeratic Complex, the Saranuj in Jordan and the Elat Conglomerate and Zenifim Formation in Israel comprise this Infracambrian succession. According to geophysical studies (Ginzburg and Folkman 1981) the Zenifim Formation increases in thickness from a few tens of meters in the south up to 3 km in central Israel. The Infracambrian is widely distributed in the adjacent countries attaining a thickness of 2–3 km. The same holds true for the conglomerate and greywacke series of the Saramuj Formation (Picard 1939) in Transjordan, the Derikand Sadan Redbeds and volcanics in Turkey, as well as numerous clastic shallow marine formations in Saudi Arabia, Oman, Yemen and other countries (Al-shahran and Nairn 1997, p 84).

Paleozoic

During the Paleozoic, the Middle East underwent a series of epeirogeneic movements which resulted in regional structural developments and caused pronounced stratigraphic breaks. The area acted as a relatively stable, passive continental margin, over which terrestrial and shallow marine sediments have accumulated (Alsharhan and Nairn 1997).

The Early Paleozoic of the Middle East was a period of gentle, epeirogenic movements of large, generally rigid crustal units. Widespread transgressions and regressions crossed a shallow, epicontinental shelf, covered by clastic sediments.

In the Middle East, evidence of Caledonian deformation is totally lacking, but the time when it occurred elsewhere maybe represented by sea level changes. Evidence of the Late Ordovician glaciation is more complete, as compared with North Africa, and thereby provides a better understanding of the details of that event (Alsharhan and Nairn 1997).

The effect of the Late Paleozoic Hercynian Orogeny in the Middle East is reflected by uplift which was accompanied by erosion and diversification of the topography such as in North Africa (Gvirtzman and Weissbrod 1984). This was brought to an end by the deposition of the younger Permian Khuff carbonates during the transgression following the clastic deposits of the Late Carboniferous-Early Permian Unayzah Formation.

The Paleozoic column is incompletely represented in southern Israel (Cambrian and Permian), in southern Transjordan (Cambrian-Sillurian, Bender 1974, pp 39–59) and in the western Sinai Gulf of Suez area (Cambrian and Early Carboniferous (Said 1962, p 129). The reconstruction of the Paleozoic paleogeography is rendered difficult by the incomplete sections in sparse outcrops or drill holes. The synthesis that follows is primarily based on data presented in Weissbrod (1981) and Gvirtzman and Weissbrod (1984).

As visualized by Wolfart (1967, pp 6–31) and many others, the generalized regional framework upon which Paleozoic deposition began to take place was the continental crystalline core (Arabo-Nubian Massif), accurately surrounded by a mobile shelf that included Israel and northern Sinai. The sedimentary record of the Paleozoic is rather fragmentary for the eastern Mediterranean and Persian Gulf areas. All of the Paleozoic periods have been identified in the whole region, but in no locality has a complete representation of the whole Paleozoic sequence been found. There were marine transgressions and regressions, but a slight angularity of the unconformities has also been reported from field observations.

Weissbrod (1981) noted that the Nubian Sandstone Formation marks the beginning of the Paleozoic era. It unconformably rests on a flat pene-plain surface of the crystalline basement in southwestern Sinai which resembles the basement of southern Transjordan, which is also flat. On the other hand, in southern Israel, the surface is uneven. This difference may be due to block movements at the end of the Lipalian Interval, which caused dislocation and rejuvenation of the relief in southern Israel. Weissbrod identifies Cambrian sediments and "sandstone, dolomite and several shale sequences of Precambrian to Lower Carboniferous age". Isopachs of the Paleozoic by Weissbrod show a general northeast-southwest depositional strike, with thickens towards the northwest.

The cumulative thickness of the Paleozoic section in the Levant area amounts to only 2,200 m. This rather thin column suggests either low rates of sedimentation, and/or extensive erosional events. The discontinuous distribution pattern of thin, epicontinental marine intercalations of Early Cambrian and Early Devonian age and of fluvio-deltaic complexes of Late Ordovician age on the northern margins of the Arabian Craton, can be interpreted to be either due to differential tectonic movements of a block located at the northern margins of the Gondwana (Turkey and Cyprus), akin to the behavior of other cratonic blocks in North Africa (Garfunkel 1970, Garfunkel and Derin 1984). Alternatively, the missing section(s) (e.g., Permian epicontinental deposits overlaying infra-Cambrian arkoses in most of the drill holes in southern Israel) can be attributed to a Carboniferous epeirogenic uplift, related to the Hercynian Orogeny in Europe (Weissbrod 1981; Gvirtzman and Weissbrod 1984).

Early Permian deposits in southern Israel and central Transjordan (part of the Negev Group) are of fluviatile to deltaic facies (Weissbrod 1981; Eshet 1983), very much like the older Paleozoic deposits in the area. The Late Permian signals the onset of the fast, localized subsidence of the margins of the Arabian Craton, where prisms of epicontinental marine shales, sandstones and limestones accumulated (Garfunkel and Derin 1984). The sedimentation pattern of the Late Permian is the precursor of the Mesozoic style (Weissbrod 1976). This may mean that the transient formation of the Mesogean Sea or the Paleotethys (Argyriadis et al. 1980) is a shallow predecessor of the Tethys.

Mesozoic

Triassic

The Triassic succession in the Levant is characterized by shallow marine, lagoonal to continental sedimentation. This rock sequence is typical for the Peri-Tethys countries and the Germanic Facies (so-called by Gignoux

1955) or the Sephardic (meaning Spanish in Hebrew) Facies, as defined by Hirsch et al. (1995).

The discoveries of the Triassic in Transjordan by the Turkish Petroleum Company were published by Cox in 1924. However, the main body of information concerning the Triassic from other parts of the Levant and of Saudi Arabia has become available only after World War II, when oil companies and governments published the results of their field surveys and wildcat drillings. This information appeared first in a series of reports but was finally collected in the "Lexique Stratigraphique International" as well as in internal reports. Shaw's synopsis of 1947 on the work of the British Petroleum Companies in the Negev during the Second World War (Awad, 1945) on the Arayif en Naqa (Sinai-Negev) should also be mentioned.

Landmarks for the Triassic in Cis- and Transjordan were published by Druckman (1974); Druckman and Kashai (1981), Druckman et al. (1982); Druckman (1984); Bandel and Khoury (1981); Picard and Flexer (1974); and Salameh (2001).

The Triassic section is more than 1,000 m thick in southern Israel, a few hundred meters in Jordan and over 2,000 m in northern Israel (Druckman, 1984). Fig. 2.1-9 shows the outcrops in Egypt (Sinai mainly), Israel, Palestine, Jordan, Syria and Iraq. The Triassic succession is penetrated by several deep boreholes that were drilled for oil exploration in the area.

Israel and Palestine: The Triassic in southern Israel consists of alternating clastic, carbonate and evaporitic sections, reflecting transgressions and regressions of the Tethys over the platform. The Triassic sequence is exposed in two anticlinal structures in the Negev but is penetrated by many deep oil wells all over the area. The average thickness of the stratigraphic column is approximately 1000 m.

An almost continuous shallow marine sequence of carbonates and evaporites was discovered in the north (Druckman and Kashai, 1981). Druckman (1984) indicated the northward thickening of the formations and the increase of marine influence. The Triassic is not exposed in the study area and is penetrated by one well: the Ramallah borehole.

The Late Triassic sequence is distinguished by evaporite facies (gypsum, anhydrite and traces of halite) typical for the end of the period throughout the Near East. The Late Triassic anhydrite is exposed in Wadi Zerka in Jordan (see the next paragraph) and it would be plausible to asume a possible influence on groundwater salinization and on the formation of deep brines.

Jordan: In the Nubian Sandstone formations, which built the eastern cliffs of the Dead Sea Rift up to the plateau of Humrat Main, shaly-marly and slightly calcareous beds appear over a length of 20 km between the

Zerka River and the NE corner of the Dead Sea. According to Bender (1968), these Lower Triassic beds wedged out south of the Arnon River.

Large Pleistocene alluvial fans hide other outcrops of the Triassic near the NE corner of the Dead Sea, but to the north, in Wadi Hisban and Ayun Musa, Middle Triassic is exposed. Still farther north, in the valley floor of the Zerqa (Yabok) river, Triassic beds are also exposed, but in spite of detailed sectioning, exact Triassic dating is rendered difficult due to the lack of fossils (Wetzel and Morton, 1959). An assumed Triassic age for some undeterminable corals and gastropods collected near Fenan in southern Jordan (Cox 1924) have never been verified.

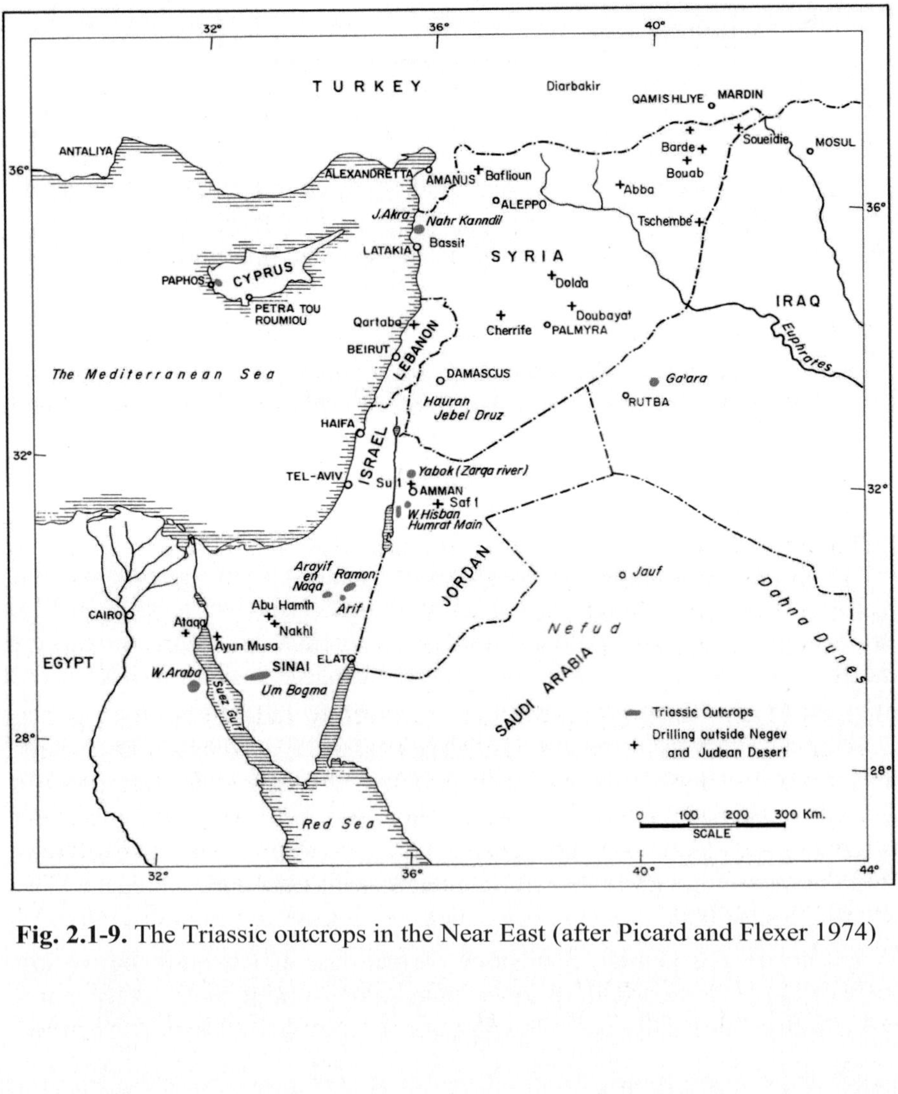

Fig. 2.1-9. The Triassic outcrops in the Near East (after Picard and Flexer 1974)

Fig. 2.1-10, compiled by Picard and Flexer (1974) after Wetzel and Morton (1959) shows the lithology and age of the main reference section of Triassic outcrops in Jordan

Salameh et al. (2003) describe the stratigraphic section along the NE corner of the Dead Sea as follows:

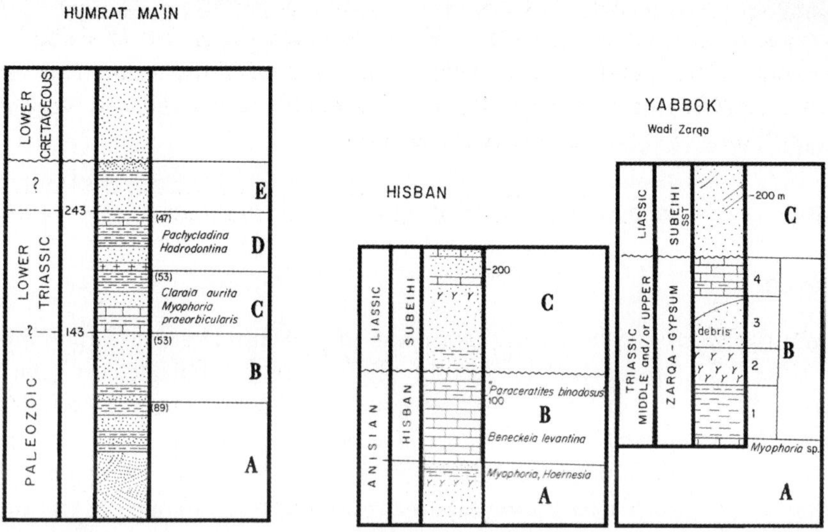

Fig. 2.1-10. Jordan main reference sections (compiled after Wetzel and Morton, 1959). Stratigraphic divisions and members A–E and 1–4 added by Picard and Flexer (1974)

The Dardur Member at the bottom is 60 m thick and consists of dolomitic limestone and laminated marl changing upwards to cross-bedded sandstone with dolomitic sandstone and marl. The uppermost part consists mainly of dolomitic limestone. Shale, marl, clay, and fine to medium white sandstone are found intercalated with green to yellow siltstone and clay. The Dardur Member was deposited in near-coast environment (Bandel 1981).

Ayun Musa Member (Scythian) is 80 m thick and can be divided into three units: the lower part which consists of white medium-grained massive sandstone with siltstone, clay and marl. The middle part consists mainly of siltstone intercalated with clay, and sandstone and limestone near the base. The upper part consists of clay and marl with fossiliferous bioturbated calcareous beds and glauconite with marl and sandstone. The Ayun Musa Member was deposited in a marine environment (Bandel 1981).

The overlying Hisban Limestone Formation (Anisian) is 35 m thick and consists of massive limestone and dolomitic limestone with marly intercalations including shell fragments and burrows. This formation was deposited in a shallow marine environment.

The Mukheiris Formation (Anisian) is the youngest unit of the Triassic rocks exposed in the study area (Wadi Na'ur). The formation is 70 m thick and in the lower part consists of cross-bedded sandstone, silt, clay and calcareous sandstone intercalalated with sand and clay. The middle part consists of clay and green silty shale. The upper part consists of sandstone, intercalated with claystone and siltstone with cross-bedded sandstone. The Mukheiris Formation was deposited in a fluvial to intertidal environment (Bandel 1981; Bandel and Khoury 1981).

The Triassic rocks in the study area are intruded by dikes and sills of chlorite, plagioclase, olivine, diabase, gabbro and iron oxides.

Jurassic

During the Jurassic, the Levant was part of the Gondwanian-Tethys platform-shelf. The paleotectonic setting of the Levant consisted in relative highs and lows, controlled by differential rates of subsidence and sedimentation: the shallow Negev high in southern Israel (900–2,000 m), the Maghara basin in northern Sinai (3,000 m), the central Israel-Jordan embayment (3,000 m), which extends to the southern Antilebanon (Hermon) range and the northern platform of Galilee (1,500–2,000 m), which itself extends to Lebanon and northern Syria. Clastics are the result of the wearing down of the Arabian Massif in the south and southeast. Thickening of the Early Oxfordian shales in offshore wells points to a source which could be a land mass in the west, now hidden below the Mediterranean (Hirsch et al. 1998). Volcanics mark the beginning and the end of the Jurassic.

Since the monograph on the Jurassic of Israel and adjacent countries (Picard and Hirsch 1987; Hirsch and Picard 1988), contributions have been added in the domain of paleontology and biostratigraphy and have been summed up by Hirsch et al. (1998). In the more general field of paleogeography, the interpretation of seismic profiles in the east Mediterranean Levantine basin and the drilling activity offshore the Sinai coast (Hirsch et al., 1995) have shed new light on the existence of a thick sedimentary succession, including Jurassic.

Israel and Palestine: At its base, the Mishhor Formation (Goldberg and Friedman 1974) consists of bauxitic-lateritic palaeosoil and flint-clays which rest on an eroded Triassic land surface. Its average thickness doesnot exceed 50 m. In the north of the country, it is replaced by the Asher volcanics. Normally, the Mishhor Formation overlies a karstic surface, leaching the Carnian Mohilla at Makhtesh Ramon or Noro-Rhaetian Shefayim Formation at the coastal plain of Israel in the west.

From the northern coastal plain northward (Carmel, Galilee, Hermon and Lebanon), over 1000 m of dolomicrites, occasionally anhydritic build up an undifferentiated Aalenian to Oxfordian succession of the carbonatic Haifa Formation (Derin 1974). The uppermost part of the uniform sedimentary regime of the Haifa Formation comprises micrites, which yield the algae Pseudoclypeina and the foraminifer Levantinella Egyptiensis of possibly the Late Callovian–Early Oxfordian age (Hirsch et al. 1998).

A reefoidal facies of biolithites of stromatoporoidea, corals, porifers and bryozoans occurs under the coastal plain of Israel, identified as the Nir-Am Formation (Derin 1974).

Upper Jurassic limestone (Haluza Fm.), attaining a thickness of almost 200 m, crops out along Wadi el Malih in the NE corner of the Faria anticline south of Bet-Shean Valley. Upper Jurassic shales, marls and limestones with a thickness of about 350–400 m are exposed on the southern slopes of Mt. Hermon (Kidod, Beer Sheva and Haluza Formations).

Jordan: According to fossil determinations, the lower contact of Jurassic strata in the gorge of the Zerqa River is preliminarily placed where 50–70 m thick-bedded limestones overlie predominantly clastic sediments of Triassic age. The limestones in turn are overlain by about 70 m of sandstones with plant fossils, calcareous sandstones and shales, which Wetzel and Morton (1959, p 126) dated as Rhaetic-Liassic (Bender 1974).

Bender (1974) indicated that further Jurassic sediments, attaining a total thickness of 220 m, are exposed above the limestones in the Wadi Huni (northern slope of the Zerqa River). The lower part of this sequence consists of alternating calcareous sandstones, oolitic marls, blue-grey shales, grey and yellow limestones, crystalline limestones, marly limestones and partly dolomitic limestones (about 80 m). Above, clastic components increase and coarse sandstones with sandy limestone beds, sandy marls, calcareous sandstones and marly limestones are developed (approx. 65 m). In the upper part, clastic components decrease again and dense limestones, marls, crystalline dolomites, detritial limestones, shales and marls form this part of the Jurassic sequence (about 75 m).

In the oil wells of Suweileh 1 and Safra-1 no subdivision was established between the Jurassic and the Triassic. However, Bender (1974) indicates that a distinct decrease in thickness of the Jurassic sequence towards the E and SE is recorded. The general character of these Middle Jurassic sediments in Jordan is of shallow marine depositional environment.

Cretaceous

Three major transgressive-regressive sedimentation cycles occur on the northwestern margins of the Arabian Craton. These cycles are the Neocomian-Aptian which is mostly terrigenous sediments, the Albian-Turonian, which is characterized by basinal marls and platform carbonates, and the Senonian which consists of marly chalks. The cycles are separated by two major regional unconformities, the Aptian-Albian and Turonian-Coniacian boundaries. The sedimentary cycles are related to regional tectonic and to volcanic events and eustatic fluctuations. The paleo-depth curve illustrates the gradual sea-level rise, reaching its maximum during the Late Cretaceous, with conspicuous advances during the Late Aptian, Late Albian-Cenomanian, Early Turonian, Santonian and Early Campanian. Major lower stands occur at the Aptian-Albian, Cenomanian-Turonian, Turonian-Coniacian, and Campanian-Maastrichtian boundaries (Flexer et al. 1986).

During the Early and Middle Cretaceous two main sedimentary provinces can be distinguished in Israel and in the adjacent areas of the Levant (Lebanon, Syria, and northeast Egypt): a platform in the southeast and a basin in the northwest which more or less coincides with the present-day Mediterranean Sea. A large carbonate platform extended between the open Tethys Sea in the northwest and the Arabo-Nubian massif in the southeast. The edge of the platform was located close to the present coastline of Israel and Sinai (Sass and Bein 1982), which has rotated counterclockwise from its Cretaceous position (Dixon and Robertson, 1984). The Cretaceous sediments over the platform of Israel, Palestine and Jordan consist of thick sequences, averaging 1,500 m of high- and low-energy carbonates, sandstones, and shales containing abundant microfossils (planktonic and benthonic foraminifera, ostracodes, and calcareous nannofossils) and megafossils (e.g., ammonites, rudists and other bivalves, and gastropods).

The basin's sedimentary province of the Eastern Mediterranean is characterized by thicker homogeneous accumulations (2,000–4,000 m) of detritial carbonates, i.e., calcilutites and fine calcarenite laminae. Microfossils are the most abundant fauna group in these sediments.

This environmental pattern changed completely during the Late Cretaceous (Senonian). The platform was tilted, the rudist reefs disappeared, and pelagic chalk was deposited over large parts of the platform.

Cretaceous rocks in the study area are economically important. The Cenomanian-Turonian hard carbonates, such as dolomites and limestones, comprise the main aquifer of the region. The Lower Cretaceous shales and sandstones serve as source rocks and reservoirs for oil and water. A thick layer of oil shale was found in the Senonian and the entire sequence is used as a source for building stones and as raw material for cement. Many wells have penetrated the Cretaceous succession. Material from more than 300 boreholes, supplemented by many field exposures, is available for lithostratigraphic and biostratigraphic study.

The Cretaceous of the Levant has been extensively studied. Basic studies include Arambourg et al. (1959), Picard (1959), Daniel (1963), and Bender (1974). In addition, various aspects of the lithostratigraphy and biostratigraphy have been examined in detail (e.g., Bartov et al. 1972; Kafri 1972; Lewy and Raab 1976; Flexer 1968, 1971; Flexer and Honigstein 1984; Flexer et al.1986).

Jordan: The following text is mainly based on Salameh et al. (2003). It describes the Cretaceous sequence east of the Jordan River. Cretaceous sediments unconformably overlie Triassic and Jurassic rocks.

The Kurnub Sandstone Group (Neocomian-Early Albian) unconformably overlies the Azab Group in Wadi el Malih. The Group is 220-320 m thick. Its lower part consists of white, medium- to coarse-grained, pebbly, planar cross-bedded sandstone, changing upwards into cross-bedded sandstone. The upper part is varicolored, medium grained, cross-bedded sandstone with burrow-mottled horizons and common plant fragments. This group was deposited in braided to meandering streams, possibly interbedded with sediments of shallow marine ingressions (Wetzel and Morton 1959).

The overlying Na'ur Limestone Formation (Cenomanian) is 100–125 m thick, includes the Juhra Member, and marks the basal part of the Ajlun Group. The basal part of this formation consists of 40 m fine to medium-grained sandstone, rich in glauconite, and interbedded with siltstone, mudstone, marl, with secondary gypsum and horizontal burrows. The upper part of this formation consists of alternating limestone and marl overlain by yellowish-grey marl, interbedded with thin layers of marly limestone in the middle part. The top of the formation is marked by a 10–15 m thick, light to dark grey and pink dolomitic limestone bed with chert nodules. The formation was deposited in tidal to lagoonal and open marine environments (Bender 1968).

The overlying Fuheis Formation (Cenomanian) is 50 m thick and consists mainly of yellow gray calcareous silt, marl and marly limestone, and nodular and fossilifereous limestone. The formation was deposited in very shallow subtidal environments (Wetzel and Morton 1959; Bender 1968).

The overlying Hummar Formation (Cenomanian) is 65 m thick and consists of grey limestone, dolomitic limestone and dolomite. The formation was deposited in an intertidal marine environment.

The overlying Shueib Formation (Cenomanian to Early Turonian) is 50 m thick and consists of red-green and buff calcareous siltstone with secondary gypsum and thin limestone beds with dark grey calcareous siltstone, gypsum nodules, algal laminated dolomite or dolomitic limestone and mudstone. The formation was deposited mainly in a moderate to shallow subtidal marine environment (Bandel 1981).

The overlying Wadi es Sir Limestone Formation (Turonian) is 100 m thick and is marked at the base by massive, hard, buff dolomitic limestone. The formation consists of bedded massive dolomitic limestone and dolomite with chert bands, chalky limestone and marl. The depositional environment was fully marine to subtidal (Bender 1968).

The overlying Wadi Umm Ghudran Formation (Coniacian-Santonian) is 0–20 m thick and represents the basal part of Belqa Group. The formation consists of white to buff massive chalk and marl. The depositional environment was moderately to deep-water pelagic.

The overlying Amman Silicified Limestone Formation (Santonian - Campanian) is 50 m thick and consists of dark grey thick-bedded chert, grey limestone, dolomite, oyster-rich coquinal limestone and silicified phosphate. The phosphate percentage increases upwards.

Israel and Palestine: The Kurnub Group (Quennell 1951). The Kurnub Group (Early Cretaceous) was deposited and accumulated while most of the area of Israel and Palestine was a landmass, covered by a considerable thickness of a sedimentary sequence of eolian-fluviatile sands, silts and shales which were interbedded with an increasing northward marine intercalation (Flexer 2001). The base of the Kurnub Group in the Jericho-1 borehole is at a depth of 1,458 m. The group consists of sandstone, shale and limestone with a total thickness of about 440 m. The thicknesses of the formations are (from bottom to top): Zweira Formation - 160 m; Dragot Formation - 63 m; Malhata Formation - 110 m; Uza Formation - 105 m. These thicknesses are comparable with the thicknesses known in the northern Negev (Begin 1975).

The Judea Group (Arkin and Hamaoui, 1967) The total thickness of the Judea Group which is penetrated by the Jericho-1 borehole is 700 m. The total thickness of the Judea Group in the research area is between 800-850 m. Most lithostratigraphic units have similar thicknesses throughout the research area. The Jericho-1 borehole reaches the base of the Judea Group at a depth of 1020 m. A rather shallow sea covered the region during the Cenomanian-Touronian, giving rise to the deposition of a thick limestone and dolomite section. In general, the environment of deposition changed slightly from an open sea of normal salinity which caused the deposition of mainly limestones to restricted shallow environments during the deposition of the dolomite (Soreq Formation).

The Kesalon Formation (the limestone facies) is the lowest exposed unit in the mapped area. It consists mainly of limestone and some dolomite. The age is Cenomanian and the base of the formation is not exposed (Arkin and Braun 1965).

The Ein Yorqe'am Formation consists of limestone chalk and marl. Its thickness ranges between 40–72 m in the exposed area. Its age is Cenomanian (Arkin and Braun 1965).

The Beit Me'ir Formation (Tongue) in the Jerusalem-Beit Shemesh region occurs as a tongue within the Ein Yorqe'am Formation in the Jericho area. It consists mainly of dolomite with some chert nodules. The thickness is between 42–49 m and the age is Cenomanian (Arkin et al. 1965).

The Aminadav Formation overlies the Ein Yorqe'am Formation and is overlain by the Avnon Formation. It consists of hard dolomite and limestone. Its thickness is between 67–120 m and the age of the formation is Upper Cenomanian (Arkin et al. 1965).

The Avnon Formation is identical to the Kefar Sha'ul Formation in the Jerusalem-Beit Shemesh area. The Avnon Formation consists of well-bedded dolomite. Its thickness ranges between 28–41 m and its age is considered to be Upper Cenomanian.

The Weradim Formation consists of dolomites and limestones; its thickness ranges from 19–40 m and the age of the formation is Upper Cenomanian. It is identical to the Tamar Mbr. of the Hazera Formation of the Negev (Arkin and Braun 1965).

The Derorim Formation consists of limestone and chalk with some chert concretions. Its thickness ranges between 8–25 m and the age of the formation is Lower Turonian (Bentor and Vroman 1963; Type Section: Ecker 1962).

The Shivta Formation consists of limey dolomite and forms typical cliff morphology with many caves. The thickness of the formation ranges from 25–40 m in the area and its age is Lower Turonian. The formation is identical to the Detrital Limestone Member of the Bi'na Formation in the Jerusalem area (Arkin et al. 1965).

The Nezer Formation consists of limestone, dolomite, marl and some chert. Two facies are discerned in the formation: a 70 m thick limestone facies and a dolomite and limestone facies of 16–65 m. In both cases appears a marl bed of 2–4 m in thickness in their lower parts. The age of the Nezer Formation is Turonian-Cenomanian. The upper contact of the Nezer Formation with the Senonian Menuha Formation usually is conformable.

The Mount Scopus Group (Flexer 1968): Starting at the Santonian, all the way through the Senonian, a better circulation with the open sea (deepening) was established. This is indicated by the deposition of the pelagic chalk as it occurs in the Menuha Formation and in the upper part of the

Mishash Formation. The environment of deposition was a shallow basin with constant connection to the open sea. During the Maastrichtian, open basin conditions resulted in deposition of chalk of the Ghareb Formation.

The Menuha Formation consists of chalk and limey chalk, sometimes bituminous. The thickness of the formation ranges from 25–82 m. The lower contact with the Nezer Formation is usually conformable. The upper contact with the Mishash Formation is defined by the first appearance of chert beds. The age of the formation is Santonian-Campanian.

The Mishash Formation is composed mainly of chalk and chert, with some limestone and phosphorite. The thickness of the Mishash Formation ranges from 20–115 m and the age of the formation is Santonian-Campanian.

The Ghareb Formation consists of chalk, sometimes bituminous. It reaches a thickness of 35 m and its age is Maastrichtian. In the Jericho area the top of the formation is missing due to erosion.

Cenozoic

Paleogene

This paragraph is based mostly on Flexer (2001). At the early stage of the Paleogene (Paleocene, Eocene and Oligocene), a transgression re-established extensive marine conditions over most of the northern part of the Arabian shelf, persisting until the Late Eocene. A widespread transgression occurred between the Paleocene and Early Eocene. More restricted conditions developed as the sea level fell during the Oligocene, culminating in a Late Paleogene hiatus, during which almost all the Oligocene and Late Eocene beds were removed by erosion from much of Arabia and the Levant. The Paleogene of the northern margins of the Arabian Craton is characterized by its homogeneity and uniformity of sediments in space, and by the transition in time from a transgressive regime at the beginning of the period (e.g., Paleocene to Middle Eocene) to regressive regime at the end (Late Eocene and Oligocene).

Extensive deposition of marl, shale and chalk across the Middle East marks the Paleocene transgression. These soft sediments are part of the Late Cretaceous sedimentary sequence, resemble them in the field and together comprise the Mount Scopus Group (Flexer 1968). No lithologic break is discerned in the field at the Cretaceous-Tertiary boundary. Sedimentation was controlled by the tectonic relief of the incipient folds (Arkin et al. 1972; Salem 1976). The Paleocene marls (Taqiye Formation in Israel and Esna Shales in Egypt) accumulated mainly in the synclines, whereas on the flanks of anticlines they are thinner or wedge out completely.

An extrusive Early Eocene transgression which mainly deposited chalk and flint of the Mor Formation (Braun 1967) or the Adulam Member of the Zor'a Formation (Buchbinder 1969), is terminated by the occurrence of conglomerates, extensive intraformational slumping and disharmonic folding (in Jordan: Wetzel and Morton 1959; in Lebanon: Beydoum 1977; in Israel: Benjamini, 1979, 1980).

These data suggest some sort of tectonic activity over the Middle East at the end of the Early Eocene. The combined evidence suggests a peak in the folding activity, an event of regional extent, well expressed by orogenic movements in the Alpine-Tauride system. This orogenic phase, along the northern margins of the Syrian Arc, probably reflects the resumption of compression between the African and European Plates (Early Eocene), after a period of relative transcurrent movement between them, during the Late Senonian-Paleocene (Dewey et al. 1973).

A new transgressive cycle began during the Middle Eocene, depositing chalk with a high content of planktonic fossils in the deeper areas to the west, while benthonic nummulitic and reef limestones were laid down over shallow shelf areas. The southward and eastward extension of this Middle Eocene rise in sea level was probably surpassed only by the Late Cretaceous transgression. A new tectonic and sedimentary regime started after the Middle Eocene (Picard 1943, p 55).

Relics of Late Eocene chalk, marl and some coral bioherms occur in the coastal plain, and the foothill areas of Israel, in the western and southern Negev and in the Arava (Bentor and Vroman 1963; Sakal et al. 1966; Gvirtzman 1970; Benjamini 1980). These rocks represent the last major transgression which reached into the interior of the Levant area from the west. Rocks from the Oligocene age (Saqiye Group) are confined only to the area of the coastal plain foothills in Israel, to a depression in the Lake Kinneret-Wadi Taiyiba area (Wetzel and Morton, 1959; Gvirtzman 1970; Michelson et al. 1987) and to the deeper basin of the southern Dead Sea (Horowitz 2001).

The present-day morphotectonic landscape of the Middle East began to form during the Late Eocene and Oligocene. Fission track dating of apatites established that the domal uplift of the Sinai area started in the Late Oligocene, 26.6 Ma ago (Kohn and Eyal 1981). However, there is extensive evidence that the accumulation of alkaline volcanics and coarse clastics in the Suez Graben, presumably post-updoming, was already well under way in the Oligocene (Garfunkel and Bartov 1977), concomitantly with the opening of the Red Sea and the Gulf of Aden (Azzaroli 1968). Whether the updoming of what would become the Jordan Rift Valley area was the decisive factor behind the newly established sedimentation pattern cannot be determined at present. An alternative explanation may have to do with the establishment of the ancestral Nile in the Early Eocene, as a major supplier of clastic sediments to the southeastern corner of the Levant basin (Salem 1976). Not only did it affect the lithologic composition of the Saqiye Group in the coastal plain of Israel, it also elicited the accumulation of a thick sedimentary prism, which may have induced a fast subsidence of the continental margin.

To sum up, the Paleogene period commenced with a wide, extensive transgression (continued from the Cretaceous time) and terminated with folding, uplift, faulting and volcanism. Generally, during the Paleogene there are great similarities between the global eustatic record and the inferred record of the northern end of the Arabian Craton. The major regressive cycle of the Paleogene in the study area starts in the Late Eocene and precedes the major global eustatic fall of the Early Chattian.

Neogene and Quaternary

The Neogene and Quaternary successions are built mainly of marine and continental clastic formations, marine and limnic chalk, evaporites and magmatic rocks (Schulman 1959). These formations are limited mostly to the Jordan Rift Valley itself. The thickness varies between a few hundred meters on the rift shoulders to a few thousand meters in the deep depocenters. At the southern tip of Lake Kinneret (Zemah-1 borehole), the thickness is probably more than 4,200 m as the base of the Neogene was not penetrated by the well.

The sequence is divided into two groups: the Tiberias Group of Neogene age at the base and the Dead Sea Group (or the Jordan Valley Group in Jordan) of Plio-Pleistocene - Holocene age at the top (Fig. 2.1-7).

The geological history and the rock sequences of an evolving and subsiding Rift basin system are very complicated to understand. Conglomerates and sandstone bodies most frequently bear a local signature as well as limnic sedimentation that could also be of individual local nature. Accordingly, too many formational names are coined which makes the stratigraphic correlation so complicated. A palynological study by Horowitz (2001) chronologically positioned the Neogene-Quaternary stratigraphic succession in correlation tables. Tables 2.1-1 and 2.1-2 are slight modifications of those published by Horowitz (2001) and present the most relevant formations of the Neogene-Quaternary succession of the study area.

Table 2.1-1. Oligocene to earliest Quaternary formations, subdivided according to their geographic distributions. Some formations occur in both regions (modified according to Horowitz 2001)

Age	Northern sector: The upper and lower Jordan Valley	Southern sector: The Dead Sea and the Arava
Earliest Quaternary	Mahanayim Marl, Amud Conglomerate, Mechki Basalt, Ruman Basalt, Dalton Basalt, Amud Basalt, Cover Basalt, Bethlehem Conglomerate, Dhuleil Fm.	Melekh Sedom Sands HaMeshar Fm., Ar Risha Gravel Fm., Ghor ei Qatar Series
Pliocene	Zahle' Beds Tel Hai Fm., Tanur Conglomerate Gesher Fm., Bira Fm., Intermediate Basalt, Fajjas Tuff, Jabal Bakiya Fm., Grain Sabt Basalt	Oolitic Fm., Samra Fm., Shagur Fm., En Feshha Conglomerate, Amora Fm., Sedom Fm., Mazzar Fm., Arava Fm., En Yahav dike
Late Miocene	Umm Sabune Conglomerate Hordos Conglomerate? Lowest part of Bira Fm., Late Miocene Volcanics Uppermost Herod Fm.?	Hufeira Fm., Uppermost part of Ashalon Mbr., Dhira Ibn Salih Mbr.? (Upper part of Dana Fm.)
Early and Middle Miocene	Herod Fm., Herods Conglomerate, Lower Basalt, Barda Fm.	Hazeva Fm., Ghor Qatar Series? Wadi Bustan Mbr. (Upper part of Dana Fm.) Ashosh and Karak volcanic plugs, Red Sea Dike System

Table 2.1-2. Correlation of the Quaternary formations in the Jordan Valley and Dead Sea region (modified according to Horowitz 2001)

Age (Palynozone) Ma	Jordan Group	Upper Dead Sea Group	Wadi terraces	Sites and cultures
0.011 to present day	Tabigha Fm.; Rehov Fm.	Ze'elim Fm.; Sub-Recent Travertines	"Lower Terrace"; Sukhne Fm.	Neolithic through Present day
0.07 – 0.011	Tabigha Fm.; Lisan Fm.; She'an Fm.; "Knob Limestone"; Travertine	Damiya Fm.; Fatza'el Mbr.; Iddan Mbr.; Lisan Fm.; Fasayil Travertine; Mo'a Travertine	Wadi Hammeh Congl.; Nahshon Congl. Khirbet Samra Fm.	Epipaleolithic; Upper Paleolithic; Middle Paleolithic
0.13 – 0.07	Raqqad Basalt		Lower beds of Zuttiyeh Cave	Middle Paleolithic; Acheulian Yabrudian
0.425 – 0.07	Naharayim Fm.; Tabaqat Fahl Fm.; Kufrinja Gravel	Sayif Travertine; Lower Dome Country Terrace	Baq'a Congl. Bira Fm.	Jisr Banat Yaqub; Umm Qatafa Cave; Middle and Upper Acheulian
0.79 – 0.0425	Yarmouk Basalt			Gesher Benot Ya'aqob
1.25 – 0.79	Ubediya Fm.; Upper Abu Habil Fm.	Upper Dome Country Terrace	Dauqara Fm.	Dauqara; Ubediya
1.5 – 1.25				
1.8 – 1.5	Erk el Ahmar Fm.; Lower Abu Habil Fm.	Shezaf Hills Congl.; Zehiha Fm.	"Uppermost Terrace"	Erk el Ahmar; Lower Abu Habil
2 – 1.8		"Red Unit"?		

REFERENCES

Almagor G, Garfunkel Z (1979) Submarine slumping in continental margin of Israel and Northern Sinai. Am Assoc Petrol Geol Bull 63:324–340

Alsharhan AS, Nairn AE M (1997) Sedimentary basins and petroleum geology of the Middle East. Elsevier, Amsterdam, pp 1–843

Anderson HJ (1852) Geological reconnaissance of part of the Holy Land. In W.F. Lynch (ed.), Official report of the United States' expedition to explore the Dead Sea and the River Jordan. Murphy, Baltimore, pp 75–206

Arambourg C, Dubertret L, Signeux J, Sornay J (1959) Contributions a la stratigraphie et a la paleontologie du Cretace et du Nummultique de la marge NW Peninsule Arabique: Notes et Memoires sur le Moyen Orient 7:193–252

Argyriadis I, de Graciansky PC, Marcoux L, Ricou LE (1980) The opening of the Mesozoic Tethys between Eurasia and Arabia-Africa. Mem BRGM 115:199–214

Arkin Y, Braun M (1965) Type section of Upper Cretaceous Hazera Formation in the Northern Negev (Southern Israel). Stratigraphic Section, No. 2a, Geol Surv, Israel

Arkin Y, Braun M, Starinsky A (1965) Type sections of Cretaceous formations in the Jerusalem-Beit Shemesh Area. Stratigraphic Sections, No. 1, Geol Surv Israel, pp 1–42

Arkin Y, Hamaoui M (1967) The Judea Group (Upper Cretaceous) in Central and Southern Israel. Bull Geol Surv Israel 63:1–17

Arkin Y, Nathan Y, Starinsky A (1972) Paleocene-early Eocene environments of deposition in the northern Negev (southern Israel). Bull Geol Surv Israel 56: 1–17

Awad GH (1945) On the occurrence of marine Triassic (Muschelkalk) deposits in Sinai. Bull Inst Egypt 27:397–427

Azzaroli A (1968) On the evolution of the Gulf of Aden. Intern Geol Cong 23, Prague, Sect. 1:125–134

Bandel K (1981) New stratigraphical and structural evidence for lateral dislocation in the Jordan Rift Valley connected with a description of a Jurassic rock column in Jordan. N Jb Geol Palaont Abh 161:271–308

Bandel K, Khoury H (1981) Lithostratigraphy of the Triassic in Jordan. Facies 4: 1–26.

Bartov J, Eyal Y, Garfunkel Z, Steinitz G (1972) Late Cretaceous and Tertiary stratigraphy and paleogeography of southern Israel. Isr J Earth Sci 21: 69–97

Bartov Y, Mimran Y, Karcz I (1977) Lineaments in the Coastal Plain of Israel. Isr J Earth Sci 26:1–14

Bayer HJ, Hötzl H, Roscher B, Voggenreiter W (1988) Sedimentary and structural evolution of the northwest Arabian Sea margin. Tectonophysics 153:137-151

Begin ZB (1975) The geology of the Jerico sheet (Geological map series 1:50,000). Bull Geol Surv 67:1–35

Ben-Avraham Z, Lyakhovsky V (1990) Faulting processes along the northern Dead Sea transform and the Levant margin. Geology 20:1139–1141

Bender F (1968) Geologie von Jordanien. Beitrage zur Regionalen Geologie der Erde. Berlin, Gebrüder Borntraeger, pp 1–230

Bender F (1974) Geology of Jordan: Contribution to the Regional Geology of the Earth, Supplementary Edition of Volume 7, Gebrüder Borntaeger Berlin Stuttgart, pp 1–196

Benjamini C (1979) The geological history of central and southern Israel during the Eocene. PhD dissertation, Dept. Geol., Hebrew Univ. Jerusalem (in Hebrew, English summery, unpubl.) pp 1–41

Benjamini C (1980) Stratigraphy and foraminifera of the Qezi'ot and Har Akrav formations (latest middle to late Eocene) of the western Negev, Israel. Isr J Earth Sci 29:227–244

Bentor YK, Vroman A (1954) A structural contour map of Israel (1:250,000) with remarks on its dynamical interpretation. Bull Res Counc Isr G4:124–35

Bentor YK, Vroman AJ (1963) The Geological map of Israel, 1:100,000. Series A, The Negev, Sheet 17, Nizana, Geol Surv Israel

Beydoun ZR (1977) Petroleum prospects of Lebanon: re-evaluation. AAPG Bull 61:43–64

Blackenhorn M (1912) Naturwissenschaftliche Studien am Totem Meer und im Jordanthal: Berlin, Friedlander and Sohn, pp 1–478

Blackenhorn M (1914) Syrien, Arabien und Mesopotamien. In: G Steinmann, O Wilckens (eds), Handbuch der Regionalen Geology, Winter, Heidelberg, pp 1–159

Blake GC (1935) The staratigraphy of Palestine and its building stonesBraun M (1967) Type sections of Avedat Group Eocene formations in the Negev (southern Israel). Jerusalem: Geol Surv Isr Stratig. Sec. 4, pp 1–14

Buchbinder B (1969) Geological map of Hashfela region, Israel, with explanatory notes. Geol Surv Isr Rep OD/1/68, pp1-13

Chaimov T A, Barazangi M, Al-Saad D, Sawaf T, Gebran A (1990) Crustal shortening in the Palmyride fold belt, Syria, and implications for movement along the Dead Sea fault system. Tectonics 9:1369–1386

Cox LR (1924) A Triassic fauna from the Jordan Valley. Ann Mag Nat Hist Ser 9, 14:52–96

Daniel EJ (1963) Lexique Stratigraphique Internationale. Vol. 3, Fasc. 10C, Jordan, Center Natl Res Sci, Paris, pp 293–399

Derin B (1974) The Jurassic of central and northern Israel. Ph.D. dissertation, Dep. Geol., Hebrew Uni. Jerusalem (in Hebrew, English summary, unpubl) pp 1–152

Dewey JF, Pitman WC, Ryan WBF, Borin J (1973) Plate Tectonics and the evolution of the Alpine System. Geol Soc Am Bull 84:3137–3180

Dixon JF, Robertson AHF (1984) The geological evolution of the Eastern Mediterranean. Geol. Soc. Spec Pub, London, No.17, pp 1–824

Domzalski W (1967) Aeromagnetic survey of Israel. Inst Pet. Res Geophys Isr Open File Rep SMA/482/67

Druckman Y (1974) The startigraphy of the Triassic sequence in southern Israel. Bull Geol Surv Isr 64:1–92

Druckman Y (1977) Differential subsidence during the deposition of the Lower Jurassic Ardon Formation in western Jordan, southern Israel and northern Sinai. J Earth Sci 26:45–54

Druckman Y (1984) Evidence for Early–Middle Triassic faulting and possible rifting from the Helez Deep Borehole in the Coastal Plain of Israel. In: JE Dixon and AHF Robertson (eds), The geological evolution of the eastern Mediterranean: Geol Soc London, pp 203–212

Druckman YY, Hirsch F, Weissbrod T (1982) The Triassic of the southern margin of the Tethys in the Levant and its correlation across the Jordan Rift Valley. Geol Rundsch 71:919–936

Ecker A (1962) The microfauna and microfacies in the type section of the Derorim Formation. Geol Surv Isr Intern Rep 2:1–62

Eshet Y (1983) A palynostratigraphyc study of the Permian-Triassic succession in the Makhtesh Qatan-2 well, Negev, Israel. Geol Surv Israel Curr Res 1982: 15–20

Flexer A (1968) Stratigraphic and facies development of Mount Scopus Group (Senonian-Paleocene) in Israel and adjacent countries. Isr J Earth Sci 17:85–114

Flexer A (1971) Late Cretaceous paleogeography of northern Israel and its significance for the Levant geology, Paleogeog Paleoclimatol Paleocol 10:293–316

Flexer A (2001) The pre-Neogene geology of the Near East. In: A Horowitz (ed) The Jordan Rift Valley, Balkema, Abington, pp 123–171

Flexer A, Dimant A, Polishook B, Livnat A (1984) Relation of joints in the Avedat Group (Eocene) to the tectonic pattern of Israel. Isr J Earth Sci 33:12–25

Flexer A, Honigstein A (1984) The Senonian succession in Israel – lithostratigraphy, biostratigraphy and sea level changes. Cretaceous Res. 5:303–312

Flexer A, Rosenfeld A, Lipson-Benita S, Honigstein A (1986) Relative seal level changes during the Cretaceous in Israel. AAPG Bull 70:1685–1699

Flexer A, Hirsch F, Hall JK (2005. Introduction. In: JK Hall, VA Krashennikov, F. Hirsch, C Benjamini, A Flexer (eds). Geological Framework of the Levant, Vol II, pp 215–256

Folkman Y, Bein A (1978) Geophysical evidence for a pre-Late Jurassic fossil continental margin oriented east-west under central Israel. Earth Planet Sci Lett 39:335–340

Freund R (1970) The geometry of faulting in the Galilee. Isr J Earth Sci 19:117–140

Freund R, Garfunkel Z, Zak I, Goldberg M, Weissbord T, Derin B (1970) The shear along the Dead Sea Rift. R. Soc. London Phil Trans A 267:107–130

Garfunkel Z (1970) Tectonics of the western margins of the southern Arava. PhD dissertation, Dept. Geol., Hebrew Univ. Jerusalem, (in Hebrew, English summary, unpubl.), pp 1–204

Garfunkel Z (1981) Internal structure of the Dead Sea leaky transform (rift) in relation to plate kinematics. Tectonophysics 80:81–108

Garfunkel Z, Bartov Y (1977) The tectonics of the Suez Rift. Bull Geol Surv Israel 71:1–44

Garfunkel Z, Ben-Avraham Z (2000) The architecture and history of the Dead Sea Transform. Dead Sea: Abstracts, The first Stephan Mueller Conf Eur Geophys Soc, pp 1–27

Garfunkel Z, Derin B (1984) Permian–Early Mesozoic tectonism and continental margin formation in Israel and its implicationfor the history of the Eastern

Mediterranean. In J.E. Dixon and A.H.F. Robertson (eds), The geological evolution of the eastern Mediterranean. Geol Soc London, pp 187–201

Gignoux M (1955) Stratigraphic Geology. Freedman and Co., San Francisco, pp 1–682

Ginzburg A, Cohen SS, Hay-Roe H, Rosenzweig A (1976) Geology of the Mediterranean shelf of Israel. Am Assoc Pet Geol Bull 59:2142–2160

Ginzburg A, Folkman Y (1981) Geophysical investigation of crystalline basement between Dead Sea Rift and the Mediterranean Sea. AAPG Bull 65:490–500

Goldberg M (1970) The lithostratigraphy of Arad Group (Jurassic). Geol Surv Isr, Isr Open File Rep MM/3/70

Goldberg M, Friedman GM (1974) Paleoenvironments and paleogeographic evolution of the Jurassic system in southern Israel. Bull Geol Surv Isr 61:1–44

Gregory JW (1921) The rift valleys and geology of East Africa: Seeley, Service and Co., London, pp 1–479

Gvirtzman G (1969) The Saqiye Group (late Eocene to Early Pleistocene) in the coastal plain Hashfela regions, Israel (Vol 2, maps). Isr Geol Surv Bull 51. Scale 1:100,000, 18 sheets

Gvirtzman G (1970) The Saqiye group in the coastal plain and HaShfela regions, Israel. PhD thesis, Dept Geol, Hebrew Univ Jerusalem, (in Hebrew, English summary, unpubl), pp 1–170

Gvirtzman G, Weissbroad T (1984) The Hercynian geanticline of Helez and the Late Paleozoic history of the Levant. In: JE Dixon and AHF Robertson (eds), The geological evolution of the Eastern Mediterranean. Geol Soc London, pp 177–186

Hirsch F, Bassoullet JP, Cariou E, Conway B, Feldman HR, Grossowitch L, Honigstein A, Owen EF, Rosenfeld A (1998) The Jurassic of southern Levant. Biostratigraphy, palaeogeography and cyclic events. In: S. Crasquin-Soleau and E. Barrier (eds), Peri-Tethys Memoir 4: epicratonic basins of Peri-Tethyan platforms. Mem Mus Hist Nat 179:213–235

Hirsch F, Flexer A, Rosenfeld A, Yellin-Dror A (1995) Palinspatic and crustal setting of the Eastern Mediterranean. J Petr Geol, 18:149–170

Hirsch F, Picard L (1988) The Jurassic facies in the Levant. J Petr Geol 11:277–308

Horowitz A (2001) The Jordan Rift Valley. A.A. Balkema publishers, Lisse, Abingdon, Exton (PA), Tokyo, pp 1–730

Hull E (1886) The survey of Western Palestine, memoir on the geology and geography of Arabia Petraea, Palestine and adjoining districts, with special reference to the mode of formation of the Jordan-Araba Depression and the Dead Sea. R. Bentley and Sons, London, pp 1–145

Jarrar G, Wachendorf H, Zellmer H (1991) The Saramuj Conglomerate: Evolution of a Pan-African molasses sequence from southwest Jordan. N Jb Geol Palaontol Mh 6:335–356

Kafri U (1972) Lithostratigraphy and environment of deposition, Judean Group, western and central Galilee, Israel. Isr Geol Surv Bull 54:1–56

Kashai EL, Croker PF (1987) Structural geometry and evolution of the Dead Sea-Jordan rift system as deduced from new subsurface data: Tectonophysics 141: 33–60

Kohn BP, Eyal M (1981) History of uplift of the crystalline basement of Sinai and its relations to opening of the Rad Sea as revealed by fission track dating of apatites. Earth Planet Sci Lett 52:19–141

Krenkel E (1924) Der Syrische Bogen. Zentralbl Miner. 9:274–281; 10:310–313

Lartet L (1869) Essai sur la gèologie de la Palestine et des contreès avoisinates, telles que l'Egypte et l'Arabie, comprenant les observasions recueillies dans le cour de l'Expedition du Duc de Luynes á la Mer Morte. Masson, Paris, pp 1–292

Lewy Z, Raab M (1976) Mid Cretaceous stratigraphy of the Middle East: Annales du Musee d'Histoire Naturelle de Nice, 4:32.1–32.20

Michelson H, Flexer A, Erez Z (1987) A comparison of the eastern and western sides of the Sea of Galilee and its implication on the tectonics of the northern Jordan Rift Valley. Tectonophysics 141:125–134

Neev D, Hall JK (1979) Geophysical investigations in the Dead Sea. Sediment Geol 23: 209–238.

Niemi TM, Ben-Avraham Z (1994) Evidence for Jericho earthquakes from slumped sediments of the Jordan River delta in the Dead Sea. Geology 22: 395–398

Niemi TM, Ben-Avraham Z, Gat JR (1997) The Dead Sea, the lake and its setting. University Press, Oxford, pp 1–286

Picard L (1937) On the structure of the Arabian Peninsula. Bull Geol Dept Hebr Univ. Jerusalem, No. 3, pp 1–12

Picard L (1939) Outline of the tectonics of the Earth with special emphasis upon Africa. Bull. Geol. Dept. Hebrew Univ. Jerusalem 2(3-4), pp 1–66

Picard L (1941) The Precambrian of the North Arabian-Nubian Massif. Bull Geol. Dept Hebr Univ, Jerusalem, 3:1–30

Picard L (1943) Structure and evolution of Palestine, with comparative notes on neighboring countries. Bull Geol Dept Hebrew Univ, Jerusalem 4:1–134

Picard L (1959) Geology and oil exploration of Israel. Bull Res Counc Israel 8G(1):1-30. Am Spec Pap 84:337–366

Picard L, Flexer A (1974) Studies in the stratigraphy of Israel: the Triassic. Isr Inst Petrol Ener, Tel Aviv, pp 1–62

Picard L, Hirsch F (1987) The Jurassic stratigraphy in Israel and the adjacent countries. Israel Acad Sci Hum, Jerusalem, pp 1–107

Powell JH (1988) The geology of the Karak area. Map sheet (1:50,000) No. 3152 III (1987). NRA Geol Bull 8:1–172

Quennell AM (1951) The Geology and Mineral Resources of (former) Trans-Jordan. - Colonial Geology and Mineral Resources, London, 2:85–115

Quennell AM (1959) The structural and geomorphic evolution of the Dead Sea Rift. Quart J Geol Soc, London 114:1–24; 432–433

Quennell AM (1996) The western Arabia rift system. In: JE Dixon and AHF Robertson (eds.): The geological evolution of the eastern Mediterranean (2 ed), Geol Soc, London, pp 775–788

Said R (1962) The geology of Egypt. Elsevier, Amsterdam, pp 1–377
Sakal E, Raab M, Reiss Z (1966) Upper Eocene in the southern Negev. Isr J Earth Sci 15:42–45
Salameh E (2001) Source of water salinities in the Jordan Valley. Acta Hydrochim Hydrobiol 29:6–7
Salameh E, El-Naser H, Al-Zoubi A (2003) German-Israeli-Jordanian-Palestinian Joint Research Program for Sustainable Utilization of aquifer systems. Final Report 2001-2003, submitted to the BMBF, pp 1–21
Salem R (1976) Evolution of Eocene-Miocene sedimentation patterns in parts of northern Egypt. AAPG Bull 60:34–64
Sass E, Bein A (1982) The Cretaceous carbonate platform in Israel. Cretaceous Res 3:135–144
Schulman N (1959) The geology of the central Jordan Valley. Bull Res Coun Isr 8G(2-3):63–90
Scott B (1981) The Eurasian-Arabian and African continental margin from Iran to Greece. J Geol Soc, London 138:719–733
Shahar J (1994) The Syrian Arc system: an overview. Paleogeogr Paleoclimatol Paleoecol 112:125–142
Shaw SH (1947) Southern Palestine, Geological map on a scale 1:250,000, with explanatory notes. Govt of Palestine Jerusalem, pp 1–43
Sneh A, Bartov Y, Weissbrod T, Rosensaft M (1998) The Geological map of Israel, 1:200,000. , Jerusalem: Israel Geol Surv, 4 sheets
Suess E (1909) Das Antlitz der Erde. Tempsky, Vienna Vol. I, (1892) pp 1–778; Vol. II (1888) pp 1–703; Vol. III (1901) pp 1–508; Vol. IV (1909) pp 1–789
Ten Brink US, Ben-Avraham Z, Bell RE, Hassouneh M, Coleman DF, Andreaen G, Tibor, G, Coakley B (1993) Structure of the Dead Sea pull-apart basin from gravity analysis. J Geophys Res 98:21,877–21,894
Weissbrod T (1976) The Permian in the Near East: In H. Falke (ed.), The Continental Permian in Central, West and South Europe. Reidel, Dordrecht, pp 200–214
Weissbrod T (1981) The Paleozoic of Israel and adjacent countries (lithostratigraphic study). Jerusalem: Geol Surv Israel Rep MP/600/81, (in Hebrew; English summary), pp 1–276
Weissbrod T, Sneh A (2002) Sedimentology and paleogeography of the late Precambrian–Early Cambrian arkosic and conglomeratic facies in the northern margins of the Arabo-Nubian Shield. Geol Surv Israel Bull 87:1–44
Wetzel R, Morton DM (1959) Contribution a la geologie de la Transjordanie Notes et Mem Moyen Orient VII:95–191
Wolf O, Klang A (1976) The ancient El-Arish Channel (Haruvit Prospect). Oil Exploration (investments) Ltd., Tel Aviv, Rep. 76/3
Wolfart R (1967) Geologie von Syrien und dem Lebanon: Berlin: Gebrüder Borntraeger, pp 1–326
Zak I, Freund R (1966) Recent strike-slip movements along Dead Sea Rift. Isr J Earth Sci 15:13–37
Zak I, Freund R (1981) Asymmetry and basin migration in the Dead Sea Rift: Tectonophysics 80:27–38

2.2 Hydrogeology

Joseph Guttman

2.2.1 Introduction

This chapter reviews the behavior of water levels, salinity and flow patterns in the main aquifers on both sides of the Rift Valley. Additional detailed information on local cases is dealt with in Chapt. 3 (Research studies).

The understanding of flow patterns from the regional aquifers on both sides of the Rift into the valley itself is important for building a conceptual model and for understanding groundwater salinization processes. From a practical point of view, the understanding of the flow mechanisms is an important tool for the sustainable exploitation of groundwater resources for the benefit of the population in the region.

The study area covers a surface of approximately 2,800 km^2. Its southern boundary is a line running from Suweimeh to the northern shores of the Dead Sea and further towards the northern Judean Desert. The northern margin of the area extends from the city of Salt westwards to the Fari'a Valley and from the water divide running along the Ramallah anticlinal axis in the west to Madaba-Suweilih line in the east (Fig. 2.2-1). This area is one of the few locations, where water resources are utilized by capturing freshwater flows before they reach the Rift Valley and become saline.

2.2.2 Surface water

The initiation of the Jordan Graben (Jordan Rift Valley) created a base level for the surrounding surface- and groundwater.

The Jordan River which flows in the middle of the Valley is the main tributary of the Dead Sea. Streams and wadis flow from the mountains to the Jordan River. Due to the large elevation differences of more than 1000 meters between the mountain crests and the Jordan Valley the steeply incised wadis cut into the rocks. Some of the wadis in the area (Wadi Qilt, Wadis Auja, Shueib and Kafrain) receive return flow from springs that discharge along their flow-paths. During dry seasons, these flows are not big enough to reach the Jordan River. During the winter season, in response to heavy rainfalls, surface water flows in wadis for variable periods of time.

During the last 20 years, waste-water treatment plants of the main cities have discharged their effluents into the wadis thereby causing contamination of surface and groundwater bodies.

2.2.3 Aquifers

The stratigraphic profile exposed in the study area consists of carbonates, chert, chalk, gravels, and sandstones and evaporite layers, which range in age from Triassic to Holocene (see Chap. 2.1 and Fig. 2.1-7 for details). The limestone and sandstone layers of Cambrian to Turonian Formation

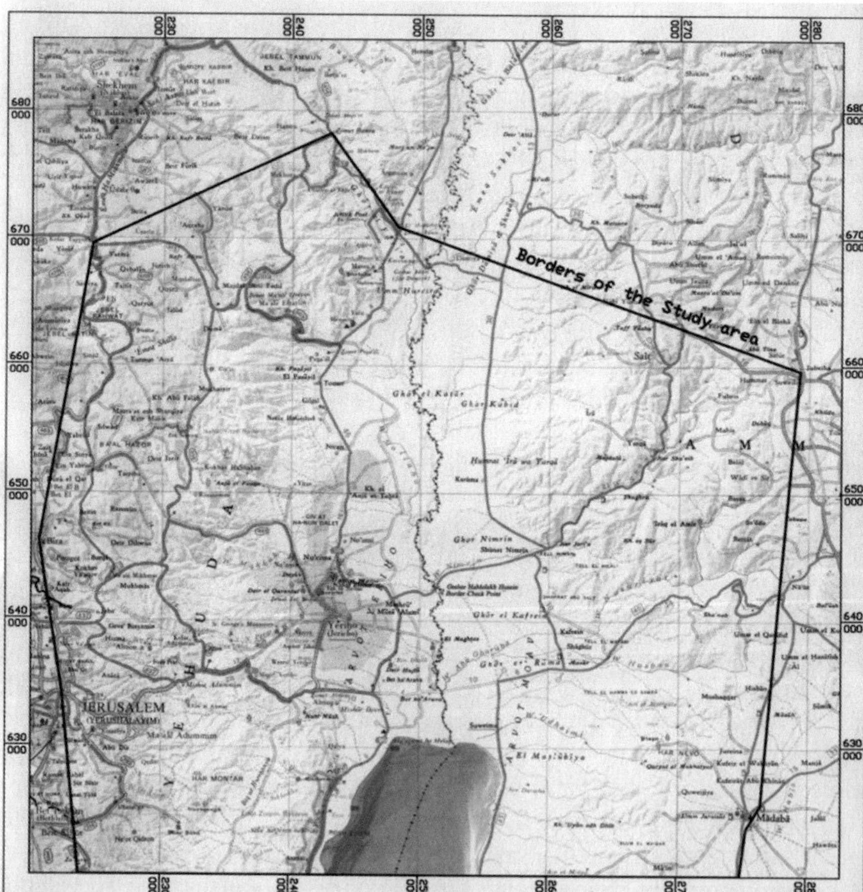

Fig. 2.2-1. Location map showing the lower Jordan Valley and its escarpments

comprise the regional aquifers in the highland areas on both sides of the Rift. As a result of the creation of the Rift Valley, the aquifers of Triassic to Lower Cretaceous are only exposed on the eastern side (the Jordanian side) of the Rift. Inside the valley, young formations built of gravels, sand, clay and evaporites comprise the aquifer.

The main aquifers are (from old to young age):

Ram aquifer (Cambrian age)

The Ram Group aquifer is built of Cambrian sandstone, limestone and siltstone. The sandstone layers make a good aquifer. There are no outcrops of this aquifer in the study area itself. The closest outcrops are located south of the study area in the vicinity of Wadi Zerqa Ma'in in Jordan.

Zerqa aquifer (Permo-Triassic age)

In Jordan, the Zerqa Group aquifer overlies the Ram Group aquifer. The sequence consists of limestone, argillaceous rocks with shales, dolomites and siltstone. The thickness of the group ranges from 300 m in the Zerqa Ma'in area to more than 350 m in the Hisban area.

The rocks are highly fractured and jointed with low primary, but high secondary permeability. Together with the overlying Lower Cretaceous sandstones and the underlying Ram Group they form an interconnected aquifer system.

The Kurnub sandstone aquifer (Lower Cretaceous age)

In Jordan, the Kurnub sandstone aquifer overlies the Zerqa Group aquifer. The thickness of the aquifer is around 220 m in the southern parts of the study area increasing to about 320 m in the Hisban and South Shuna areas. On the eastern side of the Rift, the Ram, Zerqa and Kurnub Groups form an interconnected aquifer complex called "Lower Aquifer Sandstone Complex". Locally, parts of this complex are separated from each other by marl or silt. Direct recharge into this aquifer complex is restricted to its outcrops extending along the slopes of the escarpment. Nevertheless, this complex receives direct and indirect recharge in the southern and eastern parts of the country.

The Upper Cretaceous aquifer

Most outcrops in the highland on both sides of the Rift belong to the Upper Cretaceous Formations. Because of their importance to the water balance, flow regime and salinization process, these aquifers are described in greater detail below.

The western side (West Bank or Judea-Samaria area)

In the western side of the area the permeable units in the Judea Group constitute the regional aquifer. The total thickness of the Judea Group is of 800–850 m. The aquifer is composed of limestone and dolomite and is interbedded by marls and clays (Begin 1976; Guttman 1979; 2001). A detailed lithological description of the formations is given in Chapt. 2.1.

In most of the study area, the aquifer is divided into two sub-units. The lower sub-aquifer includes the Kefira and the Givat Yearim Formations (Lower Beit Kahil Fm.) and the upper sub-aquifer includes the Aminadav (Hebron Fm.), Weradim (Bet-Lehem Fm.) and Bina Formations (Jerusalem Fm.). The main outlets of the lower and upper aquifers are the springs emerging along the western shores of the Dead Sea (Feshkha, Qane-Samar springs). Other outlets are local springs (such as Uja, Fazael, Samiya) and lateral flow to the Jordan Valley in the Fazael area. In the vicinity of Jericho and Wadi Qilt, there is a local aquifer which is known as the uppermost sub-aquifer in the Bina Formation. The outlets of this sub-aquifer are the Jericho springs (see also in Chapt. 3.3.1).

The eastern side (Jordanian side)

The aquiferous formations make part of the Ajlun Group the total thickness of which ranges from 360 to 480 m. Like on the western side, the Ajlun Group aquifer is also divided into two sub-units, i.e., a lower sub-aquifer in the Hummar Formation and an upper sub-aquifer in the Wadi Sir and Amman Formations. The Hummar Formation consists of dolomite and limestone and the Wadi es Sir and Amman Formations consist of limestone and chert beds.

The Shueib Formation is composed of calcareous siltstone, mudstones and shales which act as aquicludes separating between the two sub-aquifers. The lower part of the Ajlun Group, the Na'ur-Fuheis Formations, forms an aquiclude which separates this aquiferous sequence from the underlying sandstone aquifer.

The Jordan Valley deposits (Dead Sea Group)

Young formations of Pleistocene-Holocene age are exposed in the Jordan Rift Valley and in inland wadis and valleys. On both sides of the Rift there are differences in the names of the formations. The gravel and the sandstone layers create local aquifers.

The western side of the Jordan Valley floor area

The Samra Formation crops out along the western part of the Jordan Valley floor, at the base of the marly cliffs lining the Jordan River. They were deposited as marginal sediments along the Jordan Valley. The formation consists of conglomerates, sandstones and silts and is subdivided into two members: the silt member and the clastic member. The two members show interfingering relationships and both interfinger with the Lisan Formation. The clastic member is a local aquifer. It is laterally connected to the regional aquifers or receives water from the wadis.

The Lisan Formation which acts as an aquiclude occurs over wide areas of the Jordan Valley. It consists mainly of laminated aragonite-chalk, gypsum and clay with some sandstone and pebble beds. The section contains hypersaline and brackish fauna.

The eastern side of the Jordan Valley floor

The Shaqure Formation has a very low primary permeability and is composed of well-indurated gravels with clays and marls. In areas in which it is jointed and fractured, its permeability increases but due to its small thickness and limited distribution it does not form an appreciable aquifer. As on the western side, the Lisan Formation consists of two alternating lithofacies: laminated evaporites and massive mudstone and sandstone. The formation acts as an aquiclude.

The thickness of the Damya Formation ranges between 0–14 m. It consists of red-brown, thin to medium bedded silty limestone and calcareous mudstone. The formation acts as an aquiclude.

Good aquifers that contain freshwater exist in alluvial deposits and rock debris which is deposited along the foothills of the escarpment bordering the rift. Their distribution, thickness and permeabilities vary widely. These aquifers extend over areas of few kilometers in width and lengths, their maximum thickness may exceed locally 100 m and their permeabilities are between $5 \cdot 10^{-4}$ m/s and $5 \cdot 10^{-5}$ m/s (Salameh et al. 2003).

2.2.4 Springs

In the study area, springs are an important but secondary source for drinking water and irrigation. Some cities and villages (Jericho, Uja, Wadi Sir,

Naur, and Fuheis) depend to a much greater extent on spring water than on pumped groundwater.

The western side

The springs emerge from the layers of the Judea Group and are the main outlets of this aquifer. They can be divided into two groups.

The first group emerges north of the Jerusalem-Jericho road. It includes the springs of Wadi Qilt, the Uja and Fazael springs and formerly also the spring of En Samiya. These springs are the outlets of the upper and the uppermost sub-aquifers (Rosenthal 1979; Shaliv 1980; Hoetzl et al. 2000) and are characterized by quick response to precipitation changes (such as droughts or very rainy periods). Guttman (2004) investigated the behavior of the Uja Spring as an example of a typical karst spring in the study area (see Chapt. 3.2.2). In general, the maximum discharge of the spring appears shortly after the rainiest month and dies out towards the end of the winter season. A similar hydraulic behavior was found in others springs such as En Samiya and En Gihon (located in East Jerusalem). The Jericho springs behave differently. Their discharges are more stable and constant and do not show a high response to precipitation changes as is the case of the Uja spring. The reason for this is that the recharge area of these springs is located far away from the outlets of the springs and their storage is much greater than that of the Uja and En Samiya springs.

The second group includes the springs which emerge along the western shore of the Dead Sea such as Feshkha and Qane-Samar. These springs are the outlets of the regional sub-aquifers. Their average annual discharge is between 80–95 Mm^3/year (Guttman and Simon 1984; Rofe 2003). Their outcrops extend from the vicinity of Ramallah in the north to the vicinity of Hebron in the south. The recharge area is located far away from their outlets. The springs have a large storage volume and are characterized by stable discharges with only small annual fluctuations.

The eastern side

The springs on the eastern side of the study area discharge from the Upper Cretaceous aquifer system, from the Lower Cretaceous sandstones and from the Triassic siltstones. Springs which flows out from the Triassic rocks are restricted to the Suweimeh area northeast of the Dead Sea where Triassic rocks are tilted northwards facilitating the occurrence of strata springs. Over time, the discharge of these springs has been rather constant which indicates the large extent of the catchments and the large reserves of groundwater. Water from these springs is brackish with salinities of

3,000–5,000 mg/l of dissolved salts, mostly sodium chloride (Salameh 2001). Only small discharge finds its way from the Lower Cretaceous sandstone aquifer to the surface where it appears as springs and seepages. Their locations are concentrated along the intermediate reaches of the Kafrain, Shueib (Mahis) and Hisban wadis. Discharge is normally less than 1 l/s, but the water is fresh and of bicarbonate type with salinities of 400–700 mg/l. The groundwater in this aquifer does not form a continuous water body but perched water lenses which extend eastwards into the highlands of Jordan.

The Upper Cretaceous limestone springs are generally karst springs with appreciable discharges, such as Wadi es Sir, Naur, Hisban, Jadour, Hizzir, Fuheis and Mahis. The discharge of these springs ranges from a few cubic meters per hour (Hummar spring) to a few hundred cubic meters per hour (Fuheis or Azraq spring). The springs immediately react to precipitation events. The Cl^- concentration ranges from as low as 270 mg/l in the Fuheis spring to 650 mg/l in the Hizzir spring. The groundwater, which feeds these springs in their respective catchments, has been affected (to different degrees) by human activities and urbanization. Not only has the composition of the water changed and deteriorated, but also their discharges decreased. In the catchments of some of these springs, wells were drilled which lead to reduction in discharge. In some cases, leakage from water supplies and sewage systems, irrigation return flows and other activities has lead to induced infiltration of water into the aquifers which feed the springs. Accordingly, the discharge rates and the chemistry of the water have also changed.

2.2.5 Wells and groundwater abstraction

A large number of active boreholes have been drilled in the study area, mostly at two locations. The first group includes wells that are located in the Jordan Valley and in the adjacent foothills. The second group includes wells that are located on the mountain tops (Jerusalem–Herodion–En Samiya well-field in the western side and in the Salt–Wadi es Sir area on the eastern side).

The wells located along the foothills and on the mountain tops exploit water from the “Regional Mountain Aquifers". The wells located in the Jordan Valley abstract water from the Jordan Valley deposits. The depth of the wells located on the mountain tops, range from 250–900 m, depending on their geological location and the depth of the aquifer. The wells tap water from the calcareous aquifers (Judea and Ajlun Groups) and are characterized by low to moderate discharges (100–300 m^3/h) and by deep static water levels. These wells are located close to the recharge area and suffer from large water level fluctuation that sometimes (especially during drought periods) influence their discharge. The water is characterized by low salinity. Some wells that are located close to municipal areas are sensitive to contaminations from sewage. In the Salt area, some wells were abandoned as municipal water supply sources because of contamination.

The wells located in the Jordan Valley (on both sides) pump groundwater from the young sediments in the subsurface of the Jordan Valley. The wells are shallower than the first group. The average depth is around 100 m. Their discharge is smaller than that of the first group and is only some tenth m^3/h. The young sediments in the Jordan Valley are replenished by water from different sources such as lateral flows from the surrounding mountain aquifers, infiltration of direct precipitation and of flood water, return flows from irrigation, and by ascending of saline water from deep aquifers. Therefore, the groundwater in the young sediments of the Jordan Valley contains saline water. As the result of pumping, the water level dropped and the saline waters flow towards the pumping wells thereby raising the salinity of the water.

2.2.6 Groundwater levels

The western side

The water levels in the wells that are located in the recharge area (Jerusalem and Herodion well fields) are between 300–450 m above mean sea level (msl) (Fig. 2.2.-2). These wells react to fluctuation of replenishment. Further away from the recharge area, the influence of the natural recharge dies out and fluctuations are small. In wells located close to the foothills (Gitit, Fazael, Uja) the water levels are between -200– -300 m msl (Fig. 2.2.-3). The water levels rose for 2–2.5 years after the very rainy year of 1992 (see Chap. 3.2.2. for details). This large delay is a result of the long distance between the recharge area and the wells, the gradient and the permeability, which, as described above, is different from springs like En Uja (see Chapt. 3.2.2).

There are hundreds of wells located in a few clusters in the Jordan Valley (Jericho, Uja, Fazael-Jiftlik and Marj Naja) which pump from the Jordan Valley deposits. There is a hydrological separation between the clusters of wells. The water level declines from north to south. In the northern cluster of Marj Naja, the water level elevation is around -290 m msl, whereas in the southern cluster, the Jericho area, it varies between -300 and -340 m msl. There is also a decline trend from west to east. The wells located in the western part have a higher water level than the wells located in the eastern part of the cluster.

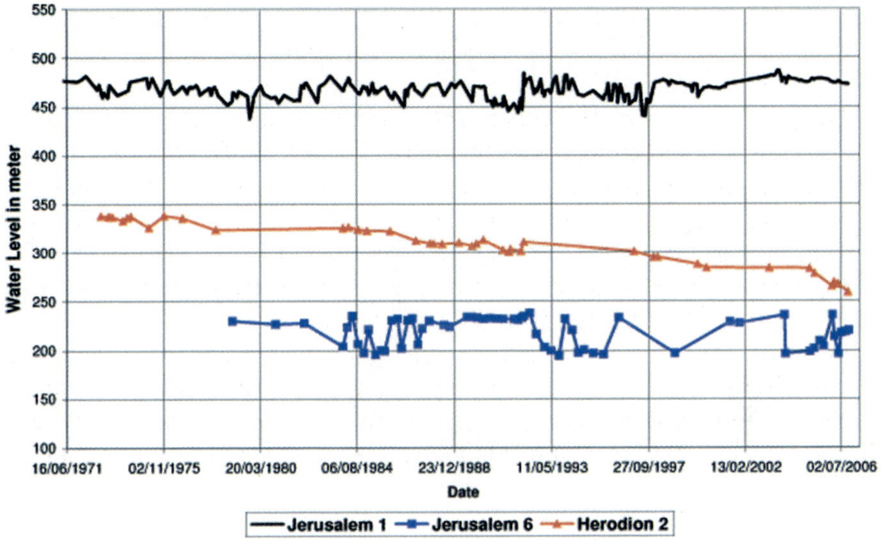

Fig 2.2-2. Water levels (msl) in Herodion and Jerusalem wells

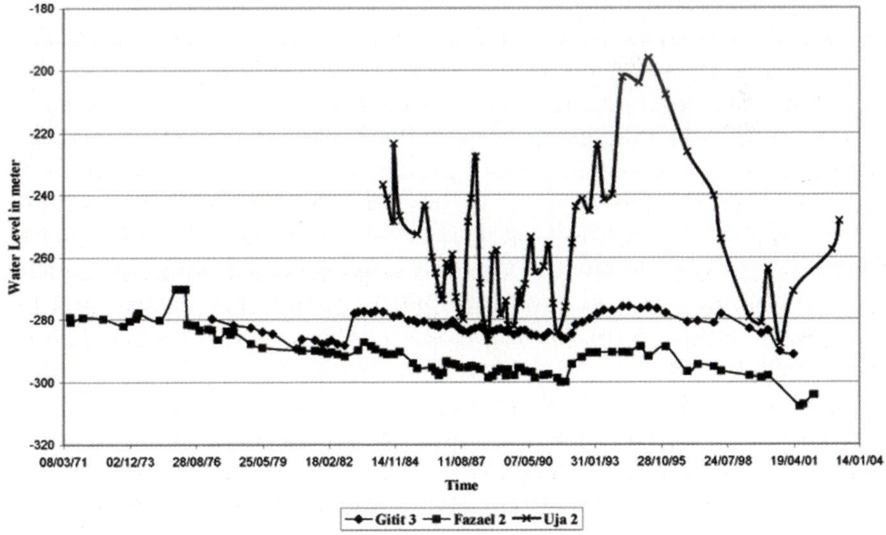

Fig 2.2-3. Water levels (msl) in Fazael, Gitit and Uja wells

The eastern side

Groundwater levels in the Jordan Valley area fluctuate seasonally as a direct reaction to precipitation and flood flows. Water levels rise during and after the rainy season and drop gradually during summer and autumn. For example, the water level in observation well Shuna South No. AB 1340, rose 2 to 5 m after the rainy season and drops back during the dry season. In 1992, a very rainy year, the level rose by about 15 m. It returned gradually to its original level after 10 years (Fig. 2.2-4).

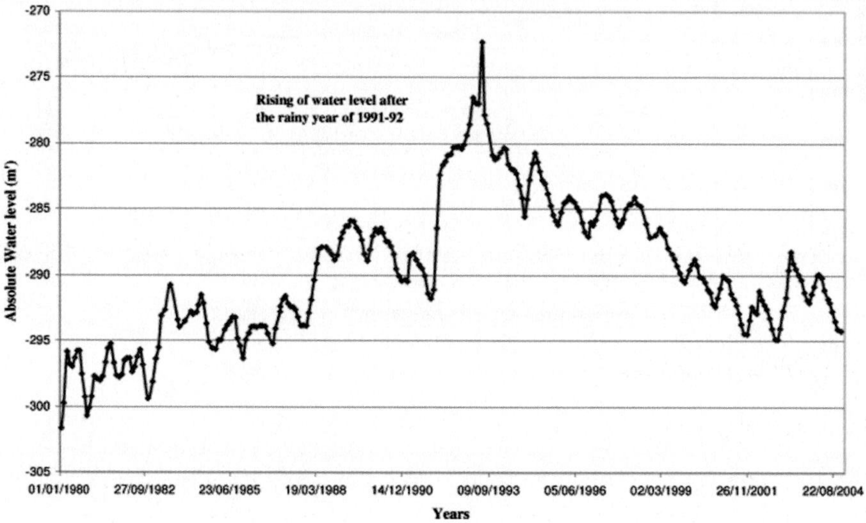

Fig. 2.2-4. Water level (msl) in well Shuna North No. AB 1340 (located on the eastern margin of the Rift) during the years 1980–2004

Generally, groundwater levels in the Jordan Valley fluctuate around the same level which means that the aquifer is in state of equilibrium and that there is no overexploitation. In the past, water levels of wells in the vicinity of the Dead Sea, e.g., well Suweimeh Ca 1100 seasonally rose and dropped by about 50 cm. However, since 1995, the water level continuously declined, possibly as a reaction to the drop of the Dead Sea level (Fig. 2.2-5).

There are no observation wells along the upper reaches of the wadis. The water levels in producing wells fluctuate and react rapidly to rain events. The same happens to springs emerging nearby. Similar occurrences were recorded west of the Jordan Valley. Major drops in the groundwater levels have not been registered because the water produced from these wells is taken from the discharge of the neighboring springs.

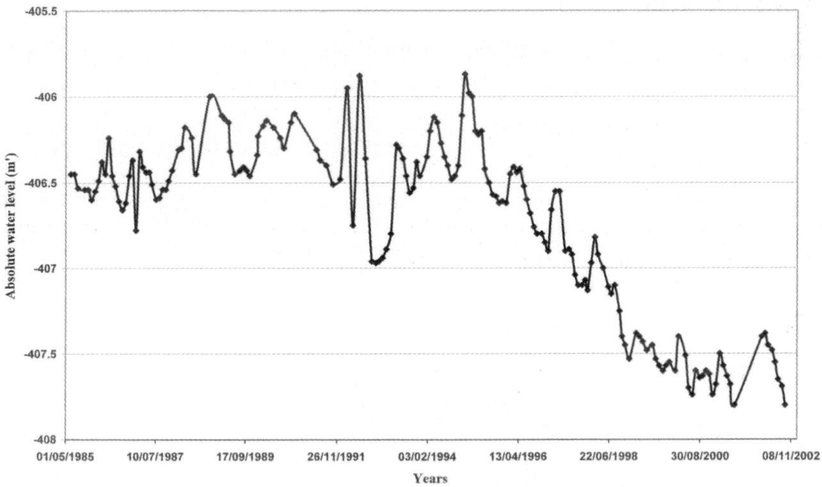

Fig. 2.2-5. Water level (msl) in well Suweimeh Ca 1100 (located on the eastern margin of the Rift) during the years 1985–2002

2.2.7 Water salinity

The western side

Cl^- concentration values in the natural replenishment zones ranged from 30–50 mg/l and remained constant throughout the years. A wide range of Cl^- concentration values was observed in wells located along the eastern foothill, ranging from 30–600 mg/l (Kronfeld et al. 1990). Two types of salinity behavior were observed in the same area (Fig. 2.2-6). In certain wells, such as Fazael 2, the salinity remains constant with time. In other wells, such as Fazael 6, Cl^- concentration seasonally fluctuates by tens to hundreds of mg/l. The chemical composition of the water indicates mixing of fresh and saline water. The data points related to the pumped wells, plot in the middle of the mixing line (see Chapt. 3.2.2). When water production from the wells began, salinity was low. It started to rise with the decline of the water levels. The increase in salinity was accompanied by seasonal fluctuations. During high demand periods in which the monthly pumpage rate was very high, there was an increase in salinity and vice versa. Such a behavior indicates that there is saline water very close to the radius of influence of the well and that the wells are pumping mixtures of fresh and saline water (see Chapt. 3.2.2).

Salinity in the Pleistocene aquifer is highly variable. This variability is due to the hydrological and geological situation in the well field. The natural replenishment of the Pleistocene aquifer is limited and most of the recharge derives from infiltration of runoff in the wadis and from irrigation return flow. Therefore, the lowest salinity occurs in wells located close to wadi banks. Local overexploitation caused a local drop of water levels. The saline water that occurs along the eastern banks of the wadi and in the deeper beds of the well-field, and flows towards the pumping wells causing an increase in salinity. During the last 30 years, irrigation return flow has become an important source of salinization. For more details refer to Chapts. 3.3.3 to 3.3.5.

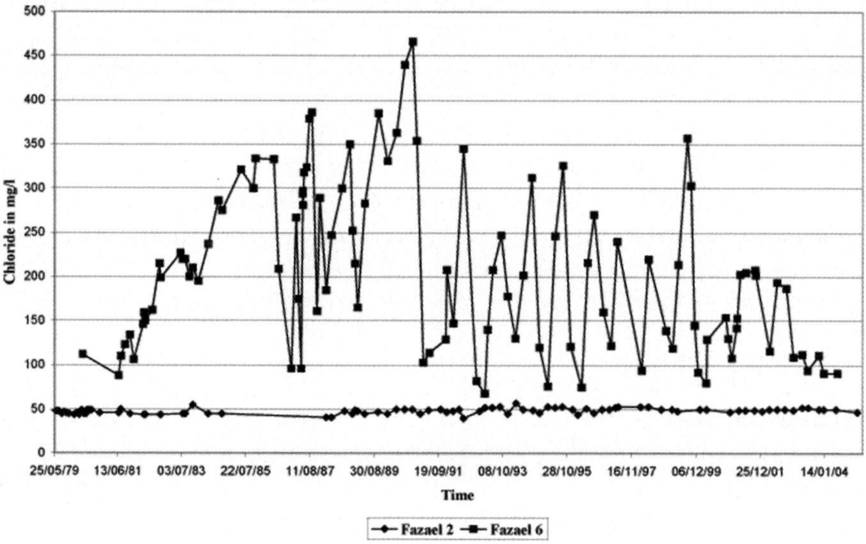

Fig 2.2-6. Two types of Cl^- concentration (mg/l) behavior in the Fazael well field

The eastern side

The hydrochemistry of the various aquifers in the area differs according to the history of the water, the type of aquifer and the impacts by human activities. The water in the Triassic aquifer is brackish and of Na-Cl type, whereas that of the Lower Cretaceous aquifer is fresh with relatively low amounts of dissolved salts. The water in the Upper Cretaceous limestone aquifer is of bicarbonate type and is directly exposed to pollution sources. Its Cl^- concentration is less than 1000 mg/l (Salameh and Banayan 1993; Salameh 1996; 2001).

The chemistry of groundwater in the Jordan Valley area differs from one well to another but evolves in a westerly direction, towards the River Jordan. The water is affected by irrigation return flows and by upward leakages from the deep saline groundwater. Cl^- contents may be as high as 5000 mg/l, but only water of less than 2000 mg/l is used for irrigation.

2.2.8 Conceptual flow model

The predominant elements governing the geological structure and flow regime in the area are the anticlinal and synclinal structures crossing the entire area and the dense fault system that create graben, horsts and fault-step structures (see Chapt. 3.2.2). These elements influence the sizes of the recharge areas, the flow regime and the water quality. The faults allow hydraulic connections between the aquifers, and the groundwater can flow from one aquifer to the other. In some places, faulting elevates the base of the aquifer above the regional water table and creates local barriers to groundwater flow. Such a hydraulic barrier exists in the En Samiya well field in which there is a difference of more than 100 m between the water levels in the up-thrown and down-faulted sides of the En Samiya fault strips.

There are two main groundwater systems in the study area:

- The "Regional Mountain Aquifers" including the Judea Group aquifer on the western side of the Rift and the Ram, Zerqa, Kurnub and Ajlun Groups on the eastern side of the Rift. Their natural replenishment is mainly from direct precipitation on their outcrops.
- The Jordan Valley system, in which the water is extracted from gravel and sand beds. The Jordan Valley system is recharged from various sources such as direct precipitation, flood- and base flows in wadis draining to the valley, irrigation return-flows and upward-seeping water from deep aquifers (particularly on the eastern side of the Rift).

The water levels in recharge areas (Hebron and Ramallah anticline axes in the west and Salt-Amman structure in the east) are around +450 m msl (Jerusalem 1, En Siniya 5 in the west) and around +500–600 m msl (Salt area in the east). The water levels descend to the Jordan Valley and to the Dead Sea to levels of about -300 m msl in the Jordan Valley and about -400 m msl on the shore of the Dead Sea. Consequently, the difference in the water levels between the recharge area and the aquifer outlets in the Jordan Valley and the Dead Sea are between 500–850 m over a distance of only about 20–30 km (Guttman and Zuckerman 1995; Guttmann 2000). This signifies a very steep flow gradient assuming that the gradient is uniform. A huge amount of rainfall and recharge is required to maintain such a high flow gradient. Available data on average rainfall on the recharge areas as well as on recharge coefficients show that such huge volumes of recharge do not exist in this area. This means that the volume of groundwater in the aquifers cannot account for such a high and constant flow gradient. The only mechanism that yields a satisfactory hydrological explanation for such a steep gradient is a combination of structural elements and of the flow pattern. To account for the high water levels in the higher areas of the aquifer and the low water levels in the lowermost area, a series of hydrological barriers is required. These structural elements create a cascade on the anticline flanks and at places, in which the base of the aquifer is located above the regional water level. These hydraulic barriers force the groundwater to bypass them and to flow along individual paths towards the outlets in the Jordan Valley and along the Dead Sea. In such cases, the flow line is extended and the gradient decreases.

The strike-slip movement of the Rift transform fault is the major tectonic element that defines the location of the outlets of the regional aquifers and the flow pattern in the Jordan Valley floor.

The Uja-Sartaba syncline is located close to the foothills, along the western side of the Rift Valley (between the Gitit and Uja wells). In this syncline, the Judea aquifer is at a depth of several hundred meters and is covered by thick aquiclude layers of the Mt. Scopus and Avedat Groups and by Pleistocene layers (for more details, refer to Chapt. 2.1). These thick layers confine the Judea Group aquifer and prevent any hydraulic connection between the regional aquifer (Judea Group aquifer), the Jordan Valley aquifers and the River Jordan, all of which cannot be the outlets of the Judea Group aquifer in this area. Under these structural conditions the aquifer is blocked eastwards and groundwater flow is diverted southwards to the lowest point in the region, i.e., the Dead Sea (see Chapt. 3.2.2 for more details).

The elongated monoclinal flexure which extends from the Maon area in the south (Um Daraj) through the Marsaba anticline in the north (towards Jericho) is of high hydrological importance for the understanding of the flow regime towards the springs emerging along the western shores of the Dead Sea. The Marsaba structure is an asymmetric anticline plunging northeastwards. Along the anticlinal axis, the bottom of the fresh upper aquifer is above the expected water level. Because of the geological structure, the lower part of the upper aquifer is only saturated in the Mizpeh Jericho area. This geological situation creates an individual groundwater flow path which drains to the Feshkha Springs.

The western major fault of the Rift is the dominant tectonic element which acts as a hydrological border between the Judea Group aquifer strata and the young formations of the Rift Valley. The vertical movement of the fault in the Feshkha springs area creates an offset of hundreds to thousands of meters. Between Jericho and the Feshkha springs, this fault is a relatively sealed boundary through which the few Mm^3 leak annually. This sealed fault forces groundwater to flow southward to the Feshkha Springs instead of flowing directly to the Jordan Valley fill in the east, which would be the shortest flow path (for more details refer to Chapts. 3.3.2 and 3.1.3).

Unlike the western side of the Jordan Rift, east of it, the geological conditions bring the Pleistocene Formations in contact with the regional aquifers thereby creating direct hydraulic connection between them. Groundwater which flows unhindered from the regional aquifers into the Pleistocene layers in the Rift, also reaches the Jordan River and the Dead Sea (for more details refer to Chapt. 3.2.2). The aquifers in the fill of the Jordan Valley which covers the entire area between the eastern foothills and the Jordan River are the major water resources for the Jordanian agriculture. The natural recharge of the aquifers in the fill of the valley is almost negligible. Most of it derives from inflow and leakage of surface water, leakage of spring water, from flow of irrigation water in open channels and irrigation return-flow, from lateral flow from the regional aquifers and by ascending saline water from deep-seated layers (see Chapt. 3.4.3 for more details).

The quality of water in the aquifers is variable, with low salinity near the recharge areas and high salinity in the areas close to salinization sources. The salinization sources are salt layers located in the Lisan Formation, salt domes, brines and irrigation return-flow. Overpumping creates local depression which allows the saline water to flow into the freshwater body which consequently increases its salinity. Due to its higher piezometric head, groundwater in the deep layers of the Jordan Valley fill moves upwards and flows into the shallower groundwater body. Due to the high salinity of the deep-aquifer water and that of the Lisan marls underlying the area, the upward leakages negatively affect the fresh surface-water aquifer in the Jordan Valley area.

2.2.9 Numerical modeling of the flow system - requirements and strategies

As previously stated, the hydrogeological situation in the lower Jordan Valley is rather complex. Therefore, when attempting to model regional-groundwater flow, a variety of different aspects have to be taken into consideration. Such aspects are the complexity of the aquifer geometry, the contact between the unconsolidated fill in the Jordan Valley and the consolidated mountain aquifer system, and - as in the case of the Jordan Valley aquifer system - the effects of density in groundwater flow.

The modeling of an extremely heterogeneous aquifer such as the mountain karst aquifer system requires a precise analysis of various model assumptions such as Darcy flow (Kolditz et al. 2004; Kalbacher et al. 2005). It has to be examined whether the flow system complies with the assumptions or whether only parts of the aquifer system can be represented by a given model. In any case, the first step is a classification of the complex aquifer systems. Each category can then be treated individually, depending on the investigated problem.

In order to understand the groundwater flow mechanisms, i.e., the interaction between different aquifer sub-systems, the different hydrogeological situations have to be studied in detail. Small-scale models (sub-models) have to be constructed in order to investigate specific problems and processes that are important for large-scale catchment-type modeling. Such sub-models allow the testing of certain processes and provide information on whether these specific processes are important within the context of the complex large-scale flow model which comprises the entire area, or whether they can be neglected. This accounts particularly for the variable density flow, e.g., between the regional mountain aquifer system and the Jordan Valley alluvial aquifer system and salinization processes (e.g., sources of salinity, movement of saltwater bodies, influence of Dead Sea level changes on salinity and groundwater resources). Another reason for the division into sub-areas is the lack of sufficient spatial data which is essential for the reliable modeling of the dynamics of movement of salt water fronts.

Detailed modeling investigations were focused on two different aspects, which address the most imminent concerns of water exploitation in the region. These aspects are further associated with specific areas where sufficient data is available:

- Flow and transport in the Jericho area, an area which partly suffers from high salinities in the alluvial and lacustrine sediments.
- The flow in the Marsaba-Feshkha system and the flow in the consolidated mountain aquifer system with its complex geology, steep hydraulic gradients and sparse data. The principal scenarios are listed below. They illustrate general concepts of how the conceptual hydrogeological model, which was summarized in the previous chapter, can be made into a numerical model for the prediction of flow and transport.

Jericho: groundwater salinization problems in an alluvial aquifer

Due to the high water demand in the area of Jericho and the consequently high groundwater abstraction rates, the groundwater quality (i.e., high salinity) has deteriorated considerably.

The proposed model intends to assist the management of aquifer exploitation. It also helps to develop appropriate pumping strategies in order to keep the salinity of the pumped water at a minimum. The research area comprises the alluvial and lacustrine valley fill around Jericho. The information from measured groundwater levels suggests that parts of the groundwater in the alluvium is connected to the carbonate aquifers (Chapt. 3.3.4).

A flexible model code is needed for the modeling of density-dependent flow in aquifers with small saturated depths and large variations in the saturated depth such as is the case of the Jericho alluvial aquifer. The same is true for the interaction between the Dead Sea and the adjacent aquifer system. Both should fulfill the following requirements:

- Model adaptation for coupled processes: Model adaptation is necessary for large-scale simulation of complex processes such as density-dependent flow and transport in confined/unconfined systems. The idea is to simulate complex processes only where necessary, i.e., density-driven convection only in areas of large salinity gradients. Anywhere else, the groundwater composition is treated as homogeneous, without a locally varying density. This way, the computational effort can be reduced to a reasonable measure.
- Modeling systems with variable saturated depths (unconfined aquifers): Salt water which comes up to the surface results in the flow of highly density-variable fluids in the unsaturated zone. This is a superposition of two non-linear phenomena, i.e., variable saturated depths in shallow groundwater systems (change in transmissivity with time) and variable density (change in hydraulic conductivity and gravity flow with time).

Once the geometry of the aquifer is altered, this type of model can be transferred to the alluvial aquifer on the eastern side of the Jordan River.

Marsaba-Feshkha groundwater flow in a karst aquifer

The Feshkha spring system is one of the few locations where water resources can be utilized by capturing freshwater flows before they reach the Dead Sea. This area can be regarded as a case scenario for flow in the mountain karst aquifer system. It addresses the problems which are encountered when modeling flow in a complex fractured, karst aquifer system: a very steep hydraulic gradient, complex aquifer geometry and sparse and unevenly distributed measurable information.

The modeling approach of such a difficult system should fulfill the following requirements:

- A well defined conceptual model should consider the aquifer geometry and extension. It should also include an approach for modeling the very steep hydraulic gradient observed in the area (e.g., cascade flow due to structural fragmentation of the study area).
- High resolution of the aquifer computational mesh: the complex geological features which are characteristic for the area need to be included in the model. The model code should be able to simulate groundwater flow in a three dimensional anisotropic, heterogeneous and phreatic aquifer. A module which takes care of dynamic boundaries using a moving mesh technique is desirable.

Another important feature to be included is the hydraulic connection between the consolidated mountain aquifer and the unconsolidated alluvial aquifer of the Jordan Valley system. Highly saline brines are encountered along this boundary. A representation of the processes at this boundary requires the consideration of both salt transport and density-dependent flow. At first however, the source area and the distribution of the saline water have to be characterized and investigated in detail in order to be able to build a flow and transport model for this particular region.

References

Begin ZB (1976) Geological Map - Jericho Sheet (1:50,000), published by the Geological Survey (in Hebrew).

Guttman J (1979) Jordan Valley - Hydrogeological Conditions in the Cenomanian Aquifer on the Margins of the Jordan Valley, TAHAL Consulting Engineers Ltd. 01/79/26 (in Hebrew).

Guttman J (2000) Hydrogeology of the Eastern Aquifer in the Judea Hills and Jordan Valley- Final Report. Mekorot No. 468

Guttman J (2001) The hydrogeology of Fazael-Salt Cross Section. Ann Rep 2001. Mekorot No. 787

Guttman J, Simon E (1984) Hydrological study of the western Dead Sea shore, stage 1. TAHAL Consulting Engeneers Ltd. Pp1–61 (in Hebrew)

Guttman J, Zuckerman H (1995) Flow Model in the Eastern Basin of the Judea and Samaria Hills, TAHAL Consulting Engineers Ltd. 01/95/66 (in Hebrew).

Hötzl H, Ali W, Wolfer J, Lenz S, Rother M (2000) Hydrologic, hydrogeological and hydraulic interrelationship of the different aquifer of the Rift escarpment. Final Activity Report in the Multi-Lateral Project, University of Karlsruhe, pp 1–37

Kalbacher T, Wang W, McDermott C, Kolditz O, Taniguchi T (2005) Development and application of a CAD interface for fractured rock. J Environ Geol 47:1017–1027

Kolditz O, Bauer S (2004) A process-oriented approach to compute multi-field problems in porous media. J Hydroinform 6:225–244

Kronfeld J, Vogel JC, Rosenthal A (1990) Natural Isotopes and Water Stratification in the Judea Group Aquifer (Judean Desert), Isr J Earth Sci 39:71–76

Rofe B (2003) Measurement campaign in Feshkah Spring - Summer 2003 (in Hebrew, unpubl)

Rosenthal A (1979) Non-Equilibrium of $^{234}U/^{238}U$ Isotopes in the Water of the Judea Group Aquifer on the Eastern Slopes of the Judea and Samaria Hills. M.Sc. thesis, Tel Aviv University (in Hebrew).

Salameh E (1996) Water Quality Degradation in Jordan. Friedrich Ebert Stiftung, pp 1 179

Salameh E (2001) Sources of water salinities in the Jordan Valley area. Acta Hydrochim Hydrobiol 29:329–362

Salameh E, Bannayan H (1993) Water resources of Jordan present status and future potentials. Friedrich Ebert Stiftung, pp 1–183

Salameh E, El-Naser H, Al-Zoubi A (2003) German-Israeli-Jordanian-Palestinian Joint Research Program for Sustainable Utilization of Aquifer System. 2nd year report 2001–2003, unpubl

Shaliv G (1980) Bet Shean and Eastern Samaria Basins - Updating of Hydrogeological Model, TAHAL Consulting Engineers Ltd. 01/80/51 (in Hebrew)

2.3 The hydrochemical history of the Rift

E. Rosenthal, A. Flexer, P. Möller

2.3.1 Introduction

The Jordan Valley and the area of the Dead Sea are a closed drainage base for numerous aquifers and water bodies of diverse ages and of different chemical and isotopic compositions. The elucidation of the relationships between these water bodies, of interaquifer flow and of salinization processes is a necessary step for the rational management of these waters and of primary importance for the further development of water resources in an area with an ever increasing population and expanding agriculture.

2.3.2 The mineralization of water: general considerations

From the perspective of a solute, catchments and aquifers can be viewed as "reaction vessels" through which meteoric waters are processed stepwise. In the absence of major lithologic variability, it is tempting to think that the weathering contribution of a catchment might have a unique, characteristic signal, determined primarily by the particular major and minor mineralogy. However, even in catchments characterized by uniform mineralogy, numerous lithology-controlled factors such as dissolution rates, cation exchange capacity, fluid mobility and reaction kinetics at mineral surfaces determine together the net solute inventory added to an evolving water parcel along a flow path. If the lithology along the flow path varies or if the infiltrating water mixes with saline formation water enclosed in sedimentary rocks, the solute inventory is altered by water-rock interaction.

The gross chemistry of groundwaters depends on the leachability of minerals and of enclosed residual fluids, on the behavior of dissolved ionic species in solution and surface complexation and on mineral alteration reactions. For instance, halite is easily soluble in the molar range and Na^+ and Cl^- are insignificantly adsorbed onto mineral surfaces. Gypsum and anhydrite are less soluble than halite but still more soluble than carbonates (calcite, aragonite, dolomite). Sulfates are soluble in the mM range and the carbonates even below and their resulting dissolved species are also insignificantly adsorbed. However, mineral alteration reactions such as dolomitization, albitization, chloritization, adularization etc. significantly change the chemical composition of groundwaters. Contrasting the group of major elements, trace elements behave differently, particularly if present in the pM range. Such low concentrations result from adsorption onto mineral surfaces and co-precipitation with alteration minerals. Most trace elements are constituents of major minerals but due to dominant surface complexation, their solubility behavior differs widely from that of their carriers.

2.3.3 The history of the formation of saline waters in the study area

In the study area, during several major phases of the geological history, brines were formed which endanger at present all groundwater reservoirs of the Dead Sea-Jordan Rift system. Between the various phases, the rock sequences were flushed of previously formed brines and evaporites and were "made ready" for new generations of liquids (Rosenthal et al. 2006).

A major evaporative phase occurred in the Middle East (Stoecklin 1968) during the Late Proterozoic-Early Cambrian. Later, the Triassic was characterized by large-scale marine oscillations and due to alternations of normal shelf and intertidal hypersaline to brackish environments thick layers of Upper Triassic gypsum (and locally of halite) were laid down both west and east of the present-day Jordan River (Zak 1963; Bandel and Khoury 1981). According to Möller et al. (2007), the brines encountered in the Triassic beds (Cl^- concentrations in the 55–140 g/l range) were formed by massive ablation of halite and gypsum with significant influence of igneous processes the evidence of which derives from numerous boreholes in northern Israel.

The transition from the Triassic to the Jurassic was marked by intensive surface and groundwater flow which dispersed and flushed the previously formed saline water bodies. The Jurassic and the Cretaceous were periods of predominantly marine regimes during which thick sequences of carbonate sediments were laid down. This was also a period of volcanic activity. Such conditions prevailed until the end of the Turonian during which regional large-scale folding processes began. As the result of subsequent folding and emergence during the Oligocene, the mainly marine fluids drained and were flushed out from the Jurassic-Oligocene rock sequence.

During the Mio-Pliocene, brines were mainly generated by the post-Messinian ingression of seawater which dissolved evaporites previously deposited on the dried-up Mediterranean Sea bed and in the erosive channels incised into the adjoining coastal areas (Rosenthal et al. 1999). In addition, brines reacted with the invaded rock sequence and with contemporaneous basic volcanics. East of the emerging water divide, between the Rift and the Mediterranean Sea, the dominant brine was the final product of the evaporation of an inland marine lagoon, the Sdom Sea, which infiltrated into a previously formed environment mainly built of carbonate rocks and particularly of clastic beds, filling at that time the nascent Rift.

Evaporites precipitated in this lagoon and the subsequent dissolution of the resultant salts contributed to the generation of brines. During the Pleistocene, fluvio-lacustrine conditions prevailed in the Jordan Valley and the freshwater Samra Lake developed. This lake became progressively saline, probably as the result of dissolution and flushing of salts from the previous hypersaline Sdom Sea (Picard 1943) which resulted in the saline Lisan Lake. From the Pleistocene onwards, the Lisan Lake became more and more saline by evaporation and by subsequent flushing of evaporites. At present, the remnant of this period is the Dead Sea. Thus, during four major periods, the rock sequences were flushed of previously formed brines and evaporites and were made ready for the following generations of liquids. The current phase in the Rift is characterized by an ongoing process of flushing-out of residual brines and dissolution of evaporites by continuously recharged freshwater. The sequence of geological events in the Rift is portrayed in Fig. 2.3-1

Lake Tiberias, which currently serves as a reservoir of freshwater, is constantly endangered by saline springs emerging along its shores and on the bottom of the lake. The Cl^- concentrations of the saline waters, which are mostly thermal up to 60°C, vary within a wide range of 400–18,000 mg/l. Considering Mg/Ca molar ratios, two hydrochemical types of saline waters are identified (Goldschmidt et al. 1967; Rosenthal 1988): (i) predominantly phreatic and non-thermal waters characterized by Mg/Ca > 1 which occur along the eastern shore of Lake Tiberias and in the north-eastern parts of the Jordan Valley (Newe Ur type), and (ii) confined, thermal and saline groundwaters with Mg/Ca < 1, which occur mostly along the western margins of the Rift, in the area of Lake Tiberias, in the Jordan Valley and further south along the shore of the Dead Sea and in the Arava Valley (Devora type). In all cases, the phreatic Newe Ur type overlies the confined Devora type.

The following major scenarios of brine formation and of salinization of groundwaters are considered:

- Seawater penetrated from the Mediterranean Sea into the Rift, and was subsequently confined (Mazor and Mero 1969; Kafri and Arad 1980; Gat et al. 1969; Magaritz and Nadler 1980). The later upflow of brines was assumed to be controlled by the movement of blocks as a result of

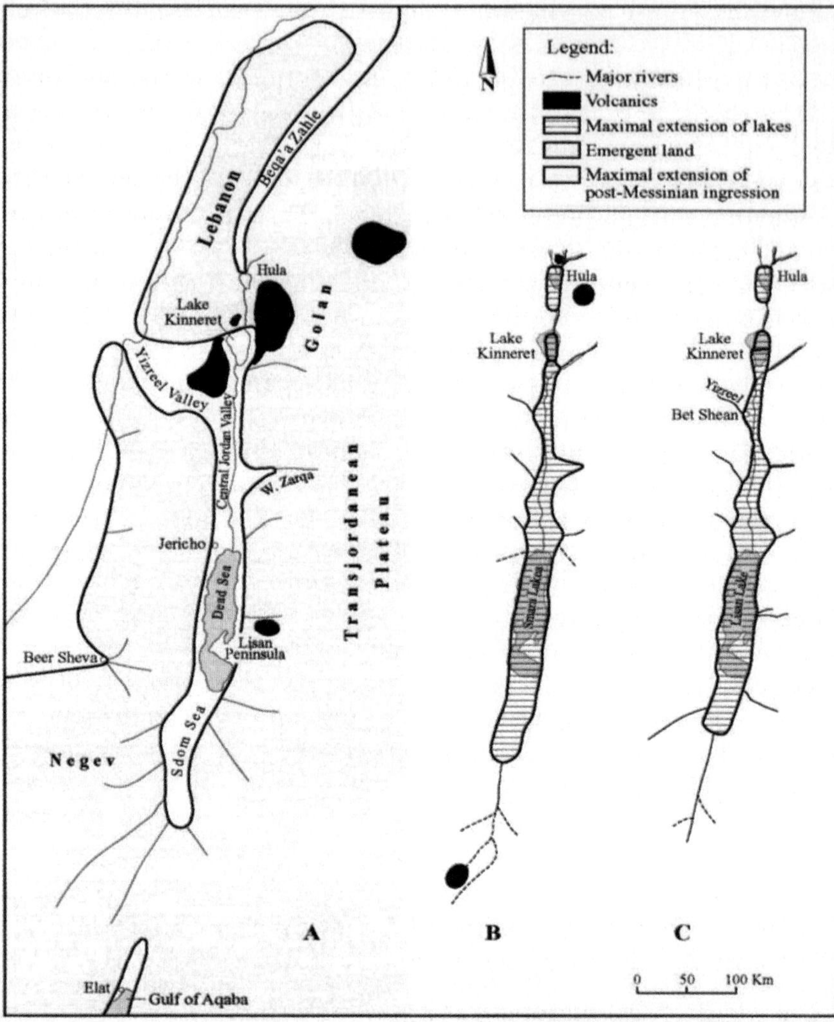

Fig. 2.3-1. Schematic maps portraying the expansion of various water bodies in the Rift during the Plio-Pleistocene (acc. to Horowitz 2001). A- Maximum expansion of the post-Messinian ingression and the formation of the Sdom Sea. B- The mid-Pleistocene Samra Lake (790–1250 ka). C- The Lisan Lake (11–70 ka)

tectonic processes. Bergelson et al. (1999) outlined a schematic hydrochemical evolution of the saline springs in the region of Lake Tiberias according to which seawater was evaporated to generate "original brine", which was later diluted to yield the "Sea of Galilee brine". It was suggested that the "original brine" is represented by the thermosaline springs of Tiberias. The groundwaters emerging in the springs of Fuliya and Tabgha were assumed to be dilution products of the “Sea of Galilee brine”.

- The Mg-rich brines, emerging from the springs along the western shore of the Dead Sea were thought to represent residual brines derived from evaporated Pliocene seawater which was trapped in the primordial Sdom depression within the Rift (Rosenthal 1988; Starinsky 1974). The Ca-chloride composition of these brines was attained by subsequent dolomitization of limestone and became confined and pressurized in an unexplained manner. Ca-chloride brines are known from a thick sequence of evaporites in the Zemah 1 wildcat well located close to the southern shores of the lake (Simon and Mero 1992). Based on $\delta^{11}B$ studies, Vengosh et al. (1994) considered these Rift brines to be the main source of salts in the contemporary Dead Sea water.
- Ablation of post-halite evaporites which were identified in the rock sequence penetrated by borehole Zemah 1 (Flexer et al. 2000). There is geological and geophysical evidence for the existence of similar evaporite bodies at other places along the Rift. Considering evidence of the presence of bittern salts (Inbar, pers. com.; Bender 1974; Raab 1998), ablation of such bodies creates Mg-rich brines such as is encountered along the eastern and southern shore of Lake Tiberias. The Ca-rich brines encountered along the western littoral of Lake Tiberias and elsewhere in the Rift could be the result of Mg removal by dolomitization.
- Mixing of two different evaporation brines of seawater was proposed by Klein-BenDavid et al. (2004). In their dual-mode evaporation scenario, the salinity of one end member was controlled by the rates of inflow and outflow of seawater. The second end member was generated, when evaporative loss exceeded the inflow of seawater. When the total evaporation rate from the surface of the shrinking water body equalled seawater supply, the mixing process came to a halt.

The mixing of saline and freshwaters is nowadays controlled by the head and the flow of fresh groundwater leaching brine pockets trapped in the subsurface (Kolodny et al. 1999) and possibly by thermal convective flow (Gvirtzman et al. 1997). However, the very high hydraulic pressures (several hundred bar) which characterize Ca-chloride brines in the deep boreholes around Lake Tiberias such as Devora 2A, Jordan 1 and Rosh Pina-1 (Rosenthal 1988) and Zemah-1 (Baida and Nevo, pers. com.) clearly indicate deep confinement of some of these brines and contradicts the model of Starinsky (1974) for brine formation. Moreover, during the Pliocene, no geological processes are known that could have generated such deep confinement (Rosenthal 1988; Illani et al. 1988). The divergence between the high pressures and the history of brine replacement in the Rift indicate a possible existence of several generations of Ca-chloride brines which were formed and infiltrated at different times and under different conditions.

2.3.4 The history of Neogene formation waters in the Rift

The marine estuary that formed the Sdom Sea may have split into a series of ever shrinking evaporative pans which contained the concentrated residual liquor “floating” on a substratum of halite, gypsum and carbonates which were precipitated during a long-lasting process of evaporation. According to Valyashko (HJ Voigt, pers.com.) halite beds have a certain degree of porosity, particularly in their upper parts and may therefore facilitate downward percolation of heavy brines. Gypsum layers undergo dehydration, thereby setting free water which creates solution channels which in turn facilitate gravitational flow of the brines. In the first place, the recipient of these brines would be the strata underlying the evaporation pans within the Rift and, probably to a lesser extent, the pervious strata along the margins of the Rift. These could be the beds laid down in the Rift during the Oligo-Miocene as well as the Upper Cretaceous and Paleogene formations. The intensive faulting in the Rift and along its margins (particularly along the western margins), probably enhanced percolation and penetration of brines into the underlying and adjacent rock units in which they are under phreatic conditions. Today’s average Cl^- concentrations increase from 5 g/kg within the Eocene Avedat Group to 140 g/kg in the Triassic rocks of the Ramon group which may indicate that the heavier brines consequently replaced anything lighter or were mixing with less saline water in the younger formations. Many of the deep brines are pressurized and of enhanced temperatures above the normal gradient, which cannot be related to any events of the Plio-Pleistocene sequence of geological processes.

During later phases of post-Messinian inland penetration, the liquid became less and less loaded with dissolved salt and finally regained the composition of Mediterranean seawater. During the late phases of the post-Messinian, the penetrating seawater covered extensive parts of the country reaching deeply into the foothills. This seawater probably penetrated into any then exposed permeable formations, i.e., mainly into the outcrops of the Judea and Avedat Groups. This is the reason for the present occurrence of remnant "seawater brines" in these groups. To make things precise, the hosting rocks are of Upper Cretaceous and Eocene age but the marine brines which penetrated them are of Pliocene age.

In the Rift, the hydrological system hosts differently composed Ca-chloride brines which may represent the chemical evolution of one and the same "fluid of origin", i.e., the Mg>Ca brines which were generated during the drying up of the Sdom Sea and even during earlier events, possibly during the Triassic.

The isotope data of groundwaters from the Rift (which will be discussed in the following chapters) suggest that contemporary freshwater mixes with water which was formed during the Pleistocene. This occurs in the groundwater hosted in the Lower Cretaceous (Kurnub Group) sandstones along the western margins of the Arava Valley (Rosenthal et al. 1998; Möller et al. 2007). Thus, the oldest so far known groundwater in the Rift occurs in the sandy beds of the Lower Cretaceous Kurnub Group. The recharge areas of this groundwater are in the Sinai Peninsula (Rosenthal et al. 1998; 2007). The waters from all other aquifers turn out to be more recent ones. The water from basalts and from the Avedat aquifers are the youngest without any contribution of older water. The thermosaline waters around Lake Tiberias are mixtures of Avedat groundwater with local brines that are derived from a common Mg>Ca type brine underlying Lake Tiberias and which is known from wells along the eastern and southern shore. All brines are diluted by recent recharge water from the Avedat- and Judea Groups (Chapt. 3.2.3).

References

Bandel K, Khoury HN (1981) Lithostratigraphy of the Triassic in Jordan. Facies 4:1–26

Bender F (1974) Geology of Jordan. Gebrüder Bornträger, Berlin, pp 1–196

Bergelson G, Nativ R, Bein A (1999) Salinization and dilution history of groundwater discharging into the Sea of Galilee, the Dead Sea transform Israel. Appl Geochem 14:91–118

Flexer A, Yellin-Dror A, Kronfeld E, Rosenthal E, Ben-Avraham Z, Artzstein PP, Davidson L (2000) A Neogene salt body as the primary source of the salinity in Lake Kinneret. Archiv Hydrobiol Spec Issue Adv Limnol 55:69–85

Gat JR, Mazor E, Tsur Y (1969) The stable isotope composition of mineral waters in the Jordan Rift Valley. J Hydrol 7:334–352

Goldschmid MJ, Arad A, Neev D (1967) The mechanism of saline springs in the Lake Kinneret depression. Geol Surv Isr Bull 45:1–19

Gvirtzman H, Garven G, Gvirtzman G (1997) Hydrogeological modelling of the saline hot springs in the Sea of Galilee, Israel. Water Resour Res 33:913– 926

Illani S, Rosenthal E, Kronfeld J, Flexer A (1988) Epigenetic dolomitization and iron mineralization along faults and their possible relation to the paleohydrology of southern Israel. Appl Geochem 3:487–498

Horowitz A (2001) The Jordan Rift Valley. Balkema, Rotterdam, pp 1–730

Kafri U, Arad A (1980) Current subsurface intrusion of Mediterranean Sea water, possible source of groundwater salinity in the Rift Valley system, Isr J Hydrol 44:267–287

Klein-BenDavid O, Sass E, Katz A (2004) The evolution of marine evaporitic brines in inland basins: The Jordan Dead Sea Rift Valley. Geochim Cosmomochim Acta 68:1763–1775

Kolodny Y, Katz A, Starinsky A, Moise T (1999) Chemical tracing of salinity in lake Kinneret (Sea of Galilee). Isr Limnol Oceanogr 44:1033–1044

Magaritz M, Nadler A (1980) Re-interpretation of ^{18}O and D isotopic composition of the Tiberias subgroup hot waters. Proc Isr Geol Soc Ann Meeting Southern Sinai and the Gulf of Elat. Ophira, Sinai Feb 4th, pp1–26

Mazor E, Mero F (1969) The origin of the Kinneret-Noit water association in the Kinneret-Dead Sea Valley. Isr J Hydrol 7:318–333

Möller P, Rosenthal E, Geyer S, Guttman J, Dulski P, Rybakov M, Zilberbrand M., Jahnke C, Flexer A (2007) Hydrochemical processes in the lower Jordan valley and in the Dead Sea area. Chem Geol 239:27–49

Picard L (1943) Structure and evolution of Palestine with comparative notes on neighbouring countries. Bull Geol Dept Hebrew Univ Jerusalem 4:1–143

Raab M (1998) The origin of the evaporates in the Jordan-Dead Sea Valley in view of the evolution of brines and evaporates during seawater evaporation. Geol.Surv.Israel Rep.No. GSI/1/98. Jerusalem, Israel, pp 1–138.

Rosenthal E (1988) Hydrochemistry of groundwater at unique outlets of the Bet-Shean-Harod multiple aquifer system. J Hydrol 97:75–87

Rosenthal E, Flexer A, Möller P (2006) The paleoenvironment and the evolution of brines in the Jordan-Dead Sea Transform and in adjoining areas. Intern J. Earth Sci 95:72–740

Rosenthal E, Jones BF, Weinberger G (1998) The chemical evolution of Kurnub Group paleowater in the Sinai-Negev Province – a mass balance approach. Appl Geochem 27:1–17

Rosenthal E, Weinberger G, Kronfeld J (1999) Ground water salinization caused by residual Neogene and Pliocene seawater - an example from the Judea Group aquifer, southern Israel. Ground Water 37:261–270

Rosenthal E, Zilberbrand M, Livshitz Y (2007) The hydrochemical evolution of brackish groundwater in central and northern Sinai (Egypt) and in western Negev (Israel). J Hydrol 337:294-314

Simon E, Mero F (1992) The salinization mechanism of Lake Kinneret. J Hydrol 138:327–343

Starinsky A (1974) Relationships between Ca-chloride brines and sedimentary rocks in Israel. PhD thesis, Dept. of Geology, Hebrew University, Jerusalem

Stoecklin J (1968). Salt deposits of the Middle East. In: R.B. Mattox (ed). Inten Conf of Saline Deposits, 1962. Geol Soc Amer Spec Pap 88:157–181

Vengosh A, Starinsky A, Kolodny Y, Chivas AR (1994) Boron isotope geochemistry of thermal springs from the northern Rift Valley. Isr J Hydrol 162:155–169

Zak I (1963) Remarks on the stratigraphy and tectonics of the Triassic of Makhtesh Ramon. Isr J Earth Sci 12:87–89

3 Research Studies

3.1 Hydrochemical aspects

3.1.1 Characterization of aquifer environments by major and minor elements and stable isotopes of sulfate

P. Möller, E. Rosenthal, S. Geyer

3.1.1.1 Scope of study

The divergence of opinions reported in Chap. 2.3 emphasizes the essential questions on the origin of the brines, their present distribution, and the processes of increasing salinization of groundwater reservoirs in the Jordan-Dead Sea Rift. Hence, it is of outmost importance to elucidate whether the brines generated in the Rift (i) evolved as the result of sea- or lake water evaporation (primary), (ii) resulted from the leaching of evaporite minerals (secondary), or whether they are altered primary or secondary brines (tertiary). To find out, it is necessary to define hydrochemical end-members and to elucidate the processes of their formation, distribution, and their probable alteration by reactions with minerals of their environment. The processes which cause the salinization of groundwater are examined on both sides of the Rift system and the probability of exchange of groundwater between both sides is discussed.

3.1.1.2 Evaluation of chemical data

The discussion in the following papers is based on the chemical analyses of fresh and saline groundwater which were recently compiled by Möller et al. (2003, 2006a, 2006b, 2007), Farber et al. (2004), and Klein-BenDavid et al. (2004). The data set is complemented by unpublished analyses of brines encountered in deep boreholes (Table 3.1.1-1). Many of these deep wells were wildcat drillings with only poor and incomplete documentation. Although highly desirable, a re-sampling of these wildcat drillings is impossible. The reliability of the brine analyses has repeatedly

Table 3.1.1-1. Compilation of analyses of groundwaters from the Jordan-Dead Sea Rift system. Sources of analyses: 1 = Möller et al. 2003; 2007; 2 = compiled by Rosenthal; 3 = Klein BenDavid, 2004; 4 = Salameh, 2001 with corrected concentration of bivalent ion species; W = well; S = spring; L = lake; Fm = formation

Sample ID	Location	Sample origin	Data Source	Hosting Fm.	Spider type	T °C	Na, mg/l	K, mg/l	Mg, mg/l	Ca, mg/l	Cl, mg/l	Br, mg/l	SO_4, mg/l	HCO_3, mg/l	B mg/l	SiO_2 mg/l	TDS, mg/l	$\delta^{34}S$ (SO_4) ‰	$\delta^{18}O$ (SO_4) ‰
	LAKE KINNERET AREA																		
1	Rosh Pina 1	W	2	Jur.			33415	1157	2918	26946	105349	1742	347	171			172044		
1	Rosh Pina 1	W	2	Jur.			8004	358	724	4794	22819		1161	370			38229		
2	Qalanit 2	W	1	U. Cret.		25.5	17.6	1.8	32.6	79.3	29.6	0.1	8	464	0.04	13.1	646		
3	Huqoq well	W	1		*B1*	18.6	42.5	2.7	28.5	76.6	89.6	0.2	12.8	357	0.05	22	632	18.9	9.60
	Tabgha cluster																		
4	Tabigha-Druzi Spring	S	1	U. Cret.	*B2*	18.3	670	29.1	84.7	305.2	1437	15.3	140	219	0.41	13.7	2915	23.2	10.1
4	Tabigha-Druzi Spring	S	3	U. Cret.			579	21.7	69.9	249	1280	12.5	107	269			2588		
5	Tabigha-En Sheva	S	1	Tert.	*B2*	23.9	293	12.1	42	170	690.4	7.24	56.5	259	0.04	15.4	1545	20.1	10.0
5	Tabigha-En Sheva	S	3	Tert.			493	19.1	62.3	249	1110	10.6	86.8	291			2322		
6	Tabigha-Sartan spring (saline)	S	1	U. Cret.	*B2*	28.8	1274	51.4	149	518	3000	28.1	245	325	0.59	13.1	5604	21.6	12.9
6	Tabigha-Sartan spring (saline)	S	3	U. Cret.			1420	51.8	146	501	3200	26.5	259	304			5908		
7	Tabigha-Sartan spring (fresh)	S	1	Tert.	*B2*	27.8	812	31.2	91.7	330	1930	17.6	143	287	0.40	15.4	3659	20.9	12.4
7	Tabigha-Sartan spring (fresh)	S	3	Tert.			741	29.2	84.8	316	1770	15.9	140	266			3363		
8	Tabigha-well Kinneret 7	W	1	U. Cret.	*B2*	23.2	139	6.6	24.4	123	320	2.55	29.4	265	0.25	15.7	926	33.3	15.7
8	Tabigha-well Kinneret 7	W	3	U. Cret.			317	12.7	42.1	189	741	6.93	55.1	286			1650		
9	Tabigha- well Kinneret 8	W	1	U. Cret.	*B4*	27.8	7296	260	466	1742	15490	120	706	17	1.79	0.26	26099	39.6	8.5
10	Barbutim well	W+L	1	Quat.		23	1103	73.9	132	448	2662	22.3	203	346	0.00	13.4	5004	21.5	16.2
	Fuliya cluster																		
11	Fuliya-spring A	S	1	U. Cret.	*B4*	28.4	374	12.1	65.3	167	755	4.49	98	313	0.18	20.6	1810		
12	Fuliya- well D 906	W	1	U. Cret.	*B4*	25	186	4	45.6	111	431.7	2.6	58.3	312	0.09	14.9	1166	14.5	6.1
13	Fuliya-well A	W	1	U. Cret.	*B4*	28	1111	40	85.5	129	2169	15.2	3.9	48	0.25	0.49	3603		
14	Fuliya-spring B	S	1	U. Cret.	*B4*	28.1	475	20.3	81.2	212	986.9	6.3	191	326	0.18	21.8	2321	19.6	9.6
15	Fuliya- well Kinneret 5	W	1	U. Cret.	*B4*	25	1869	67.4	181	426	4271	28.1	295	35	0.68	2.1	7175	19.6	10.3
16	Fuliya- 6	W	3	U. Cret.			352	11.2	64.1	151	720	3.91	92.6	314			1709		
16	Fuliya- well 6/2	W	3	U. Cret.			450	15	73	178	939	5.71	121	310			2092		
17	Fuliya-well Kinneret 10 B	W	1	U. Cret.	*B4*	27	8044	306	748	1856	17213	124	1606	21	1.96		29919	26.8	9.9

	Tiberias Hot Spring cluster																		
18	Tiberias, Qiqar well	W	1	U. Cret.	*B4*	40.8	10540	491	1093	5157	29746	484	474	40	2.64	12.2	48040		
19	Tiberias-Main Spring	S	1	U. Cret.	*B4*	59	7197	333	655	3418	18810	241	796	197	2.85	28.2	31678	7.3	4.6
19	Tiberias-Main Spring	S	3	U. Cret.			6940	325	662	3400	18300	238	812	143			30820		
19	Tiberias-Main Spring	S	3	U. Cret.			6800	316	651	3330	18000	230	788	140			30255		
20	Jordan-2	W	2	L. Cret.			1436	19.9	24	40	1584		227	912			4242		
20	Jordan-1	W	2	L. Cret.			6080	257	377	1407	11413		2188	250			21971		
21	Tiberias-Roman Spring	S	1	U. Cret.	*B4*	57.5	6335	336	639	3403	19018	252	755	135		35.0	30908	23.37	6.2
21	Tiberias-Roman Spring	S	3	U. Cret.			7050	326	666	3480	18400	240	808	138			31108		
	Eastern Shore of Lake Tiberias																		
22	Bet Zaida, well	W	3	Quat.			3580	35.3	857	724	9680	123	10.9	185			15195		
23	Gofra spring	S	1	Quat.	*D*	25.5	1187	71.5	213	230.4	2522	36.4	97.9	436	1.26	32.9	4827	20.6	9.3
24	Haon beach, well	W	1	Quat.	*D*	24.2	5212	442	1657	1078	15800	202	59.8	504	1.76	53.3	25010	17.9	9.7
	YARMOUK GORGE																		
	Northerrn side																		
26	Yarmouk Gorge-En Reah	S	1	Tert.	*B2*	38	143	12.2	39.5	142	309	4.2	115	362	0.18	20.3	1148	21	8.1
27	Yarmouk Gorge-En Makla	S	1	Tert.	*B2*	47.3	212	18	44.2	178	488	6.2	158	340	0.25	26.2	1472	24.7	6.2
28	Yarmouk Gorge-Mezar well 2	W	1	U. Cret.	*B3*	60	178	21.8	34.8	142	317	3.7	278	244	0.28	37.8	1258	26.1	11.7
29	Yarmouk Gorge-Mezar well 3	W	1	U. Cret.	*B3*	41.8	48.5	4.7	20.9	65.3	62	0.3	4	315	0.07	23.0	544		
	Southern side																		
30	Yarmouk Gorge-Bir Mukheiba 8	W	1	U. Cret.	*B1*	29.1	39	3.1	30	91.3	57	0.3	54.3	357	0.06	15.6	648	5	7.1
31	Yarmouk Gorge-Bir Mukheiba	W	4	U. Cret.		44.3	42.6	5.3	46.2	86	62		77.8	400			720		
32	Himma	S	1	U. Cret.	*B3*	41.5	124	13.9	34.9	108	208	2.3	148	337	0.46	25.0	1001	29.1	2.8
	LOWER JORDAN VALLEY																		
	Western side																		
25	Zemah 1	W	3	Quat.			79274	140	63	1200	119812	54.1	7978	110			208630		
33	Devora 1	W	2	U. Cret.			759		146	216	1633		211	354			3319		
33	Devora-2A	W	2	Tr.			40204	402	1944	34820	129540	2102	192	427			209630		
34	Bira 2	W	1	Neog.	*C*	25.2	99.5	2	30.8	37.4	111	0.40	35.9	233	0.15	40.3	590	14.1	6.5
35	Newe Ur	W	1	Quat.	*C*	29.2	258	14	22.1	20.6	178.7	0.80	94.9	364	0.64	31.7	985	11.95	11.3
36	Bet Yosef	W	1	Neog.	*C*	31.1	569	51.8	63.5	47.8	665.4	4.4	484	260	0.99	28.5	2175	13	17.0
37	En Huga	S	1	Tert.	*B1*	25	764	24.8	169	256	1716	9.4	303	316	0.48	25.5	3584	14.6	12.3
38	En Amal	S	1	U. Cret.	*B1*	28.2	498	11.8	93.6	175.7	1076	6.5	136	310	0.20	19.8	2328	17.1	16.4
39	En Moda	S	1	U. Cret.	*B1*	26.3	97.5	3.2	42.5	102	250	1.0	44.7	361	0.07	20.4	921	9.4	18.0

40	Hammam el Malih	S	1	Jur.	*B2*	35.1	755	44.5	103	380	1729.0	23.8	343	257	0.73	19.9	3656	22.9	12.5
41	Beqaot 2	W	1	U. Cret.	*B1*	31.1	85.1	4	34.3	58.9	135	1.8	36.5	245	0.17	20.2	621	7.6	8.5
42	Beqaot 1	W	1	U. Cret.	*B1*	28.9	55.4	3.1	33.3	59.4	103.1	1.2	31.2	267	0.09	18.9	572	9.3	10.0
43	Well 20-17/11	W	1	Cret.-Tert.	*B1*	29.2	495	36.7	171	202	1248	21.8	313	249	0.83	26.6	2764	1.8	10.6
44	Argaman 14	W	1	Cret.-Tert.	*B1*	30	156	13.4	45.7	73.2	304	4.6	60.7	245	0.33	19.3	922	11.1	12.1
45	Argaman 29	W	1	Cret.-Tert.	*B1*	29.7	225	16.4	71.2	97.3	472	8.4	108	259	0.43	20.6	1279	5.5	12.8
46	Pezael 2	W	1	U. Cret.	*B1*	25.3	26.3	1.9	26.2	56.9	49	0.2	14.3	267	0.05	15.5	457	3.1	12.2
47	Pezael 6	W	1	U. Cret.	*B1*	26.1	56.5	4.6	38	68.4	154	3.7	16.2	228	0.07	15.0	584	3.6	10.6
48	Pezael 8	W	1	U. Cret.	*B1*	28.5	88.5	6.8	49.7	78.5	238	6.6	22.1	230	0.10	15.5	735	-0.1	8.7
49	Mizpe Jericho 2	W	1	U. Cret.	*B1*	24.3	38.5	2.8	31.1	74.9	71.9	0.4	31.1	264	0.04	13.3	528	6.4	12.6
	Eastern side																		
50	Adasiya	W	4	Quat.		28	271.2	9.44	58	102	380		229	217			1267		
52	Shuna-North	W	4	U. Cret.		23.8	127	7	69	152	181		258	400			1194		
52	Shuna-North	S	1	U. Cret.	*B1*	54.8	83.6	4.2	50	69.5	120	0.6	69.7	406	0.19	20.0	824	15	-0.1
54	JICA well 3	W	4	Quat.		27.5	1186.8	117	158.0	670	1928		1822	1003			6885		
55	Manshiya 2	W	1	L. Cret.	*B1*	48.8	28.6	1.9	37.2	64.3	34.9		39.8	348	0.06	16.0	570	3	1.6
	Manshiya well	w	4	L. Cret.		52	30.4	1.6	41.3	76.2	47.2		45.6	372			614		
56	Waqqas	W	1	U. Cret.	*B1*	49.4	31.7	3.1	36.8	64.1	43.8	0.1	26.7	363	0.08	17.7	587	10.8	-7.0
	Waqqas well	w	4	U. Cret.		52	40.5	3.12	38.5	88	54.7		60	396	0.06		681		
57	Wadi Ziglab	W	4	Quat.		24.3	51	4	50	88	99		74	393			760		
59	Teibeh	W	4	Quat.			41	3.9	41	56	78		115	192			528		
61	Rajib	W	4	U. Cret.			102	9	62	53.8	108		142.6	309			786		
62	Abu Zigan	S	1	L. Cret.	*A*	34.8	1054	76.9	140	546	1318	4.5	1309	1223	2.71	17.5	5691	18.7	14.3
66	Muallaha sp.	W	4	Quat.		23.3	4600	361.9	331.7	2408	10877		3479	397			22454		
68	Bassat el-Faras	W	4	Quat.		28.7	23.1	2.5	30	42	51.4	0.31	27.8	248			425		
	Sweima Dead Sea well	w	4	Quat.		32.5	706	71.0	65	328	1193		648	370			3380		
	DEAD SEA AREA																		
	Western side																		
71	Yam-7C	S	3	Quat.			1230	155	782	410	4840	98.6	103	320			7939		
71	Yam 7N	S	3	Quat.			954	117	579	342	3620	72.2	93.3	325			6103		
71	Yam-1	S	3	Quat.			462	45.9	216	141	1490	25.7	59.2	270			2710		
71	Yam-2	S	3	Quat.			414	41.4	193	139	1310	22.5	56.6	290			2467		
71	Yam-5	S	3	Quat.			451	45.2	207	138	1440	24.9	57.1	290			2653		
72	Ein Feshkha creek	S	1	Quat.	*B4*	28.2	446	45.1	250	187	1383	28.9	76.4	262	1.00	14.0	2693	3.4	11.7
72	Ein Feshkha spring	S	1	Quat.	*B4*	26.6	436	41.4	215	182	1415	28.7	61.2	276	0.47	13.6	2669	5.8	12.6
72	Ein Feshkha spring	S	3	Quat.	*B4*		5130	815	4000	2030	23800	541	147	227			36690		
72	En Feshkha spring	S	3	Quat.			8560	1350	6900	3550	40200	886	225	298			61969		

73	Qane A (spring)	S	1	Quat.	*B4*	25.8	138	17.9	90.1	85.2	400	12	50	302	0.25	15.3	1111	0.9	10.2
73	Qane B (north)	S	1	Quat.	*B4*	26.9	178	21.4	114	112	597	11.6	66.2	300	0.26	14.7	1415	-1.1	8.4
73	Qane C (south)	S	1	Quat.	*B4*	26.5	193	25.9	130	124	697	14.3	72.8	284	0.29	13.6	1555	-2.3	11.3
74	Mizpe Shalem	W	1	Quat.	*B4*	43.6	25501	3769	20169	10790	120048	3148	1061	177	25.00	11.7	184700	25.4	16.0
74	Mizpe Shalem	W	3	Quat.		43.6	26200	4020	21000	11500	123000	2690	1110	100			189620		
75	En Gedi spring	S	1	U. Cret.	*B1*	27.9	47.4	5.0	27.9	55.3	81.9	1.07	33.9	208	0.09	13.3	474	9.8	12.3
76	Dead Sea, 31° 28.664' N; 35° 24.980' E; Teufe: -60 m	L	1	Quart.	*B4*	25.8	33800	8050	46012	18000	222680	5700	343	140	28.00	5.1	334758	14.6	
77	En Gedi Spa	W	1	Quat.	*B4*	39.4	23990	2380	11705	9136	88161	1673	466	163	11.60	15.6	137701	24.9	14.0
77	En Gedi-3	W	3	Quat.			25800	2520	12700	8980	92600	1770	1580	150			146100		
77	En Gedi-2	W	3	Quat.			54500	4780	21800	15000	176000	3100	400	230			275810		
78	Yesha Springs	S	2	Quat.			27824	0	18685	10290	117663	2209	1060	142			177873		
79	En Kedem	S	3	Quat.			27000	4110	21500	12500	126000	2770	1240	86			195206		
80	En Zeruya	S	3	Quat.			26900	2870	14500	9730	102000	2000	1730	160			159890		
81	Massada 1	W	2	Jur.			2990	316	1081	1840	9017		2640	61			17945		
81	Massada 1	W	2	Jur.			8395	924	2742	6360	33974		2496	122			55013		
81	Massada 1	W	2	L. Cret.			1357		106	844	1916		2436	317			6976		
82	Zohar Springs	S	2	Quat.			20518	0	22481	11389	120035	2954	1551	130			179058		
83	En Boqeq, Gofrit well	W	1	Quat.	*B4*	27.3	24223	4551	34844	15113	159474	4101.5	2312	63	33.80	4.0	244719	23.5	14.7
84	En Boqeq, PB 26	W	1	Quat.	*B4*	31.1	6830	304	1229	202.8	13720	291.68	295	199	2.35	13.2	23088	5.1	-9.8
85	Hammei Zohar 15	W	1	Quat.	*B4*	34	902	51	246	235	2048	24.6	395	188	0.89	13.9	4104	-0.3	9.1
86	Lot-1	W	2	Tr.			9890	799.5	9185	42840	119209	2468	816				185207		
86	Lot-1	W	2	Tr.			15111	1677	13936	34600	131137	3050	1008	61			200580		
87	Sdom-1	W	2	Quat.			23110	30023	21083	85030	275109	3103					437459		
88	Sdom-2	W	2	Quat.			33200	999	8510	17300	107000	1790	905	121			169825		
89	Sdom Deep-1	W	2	Tr.			32904	4778	5456	77345	216444	4305	42.0	150			341424		
	Eastern side																		
90	Rama	W	1	Pal.-L.Cret.	*A*	33.5	515	63.2	85.8	240	771	2.5	319	966	1.11	14.9	2978	14.20	10.4
91	JICA well 5	W	4	Quat.		35.3	1328	175.5	145.557	458.8	2272		536.16	1263.92			6180		
92	Hisban artesian deep well	W	1	L. Cret.	*A*	32	347	46.2	77.4	197	560	1.8	251	767	0.96	11.4	2259	13.9	11.6
93	Hisban deep	W	4	L. Cret.		32	689	94	82.62	266	1015		349.44	873			3369		
94	JICA well 1	W	1	Pal.-L.Cret.	*A*	33.7	582	73.5	121.7	284	814	2.9	450	1134	1.76	13.7	3477	21.7	10.9
96	Dead Sea Research well	W	4	Quat.		32.5	41.1	18.3	69.7	69	113		120	326			758		
97	Zarqa Ma'in Waterfall	S	1	Pal..-L.Cret.	*A*	57	452	49.1	34.1	177	727	3.2	258	371	0.59	22.8	2093	20.8	9.3
98	Zara Princess Basma Farm	S	1	L. Cret.	*A*	51.7	268	31.4	21.1	122	494	2.4	155	212	0.40	20.4	1328	19	13.5
100	Ibn Hammat 2	S	1	L. Cret.	*A*	38.8	78.6	6.1	18.2	53.2	106	0.5	65.3	206	0.11	13.1	547	10.4	13.6

	ARAVA/ARABA VALLEY																		
	Western side																		
101	Arava 1	W	2	Quat.			50508	1030	14081	50636	202457	3674	144	61			322590		
101	Arava 1		3	Quat.			50500	1030	12000	50600	202000	3670	122	61			319983		
102	En Ofarim 5	W	1	L. Cret.	*A*	25	367	19.8	77	193	663	1.4	454	243	0.52	12.9	2032	1.1	11.6
103	En Yahav 6	W	1	L. Cret.	*A*	44.3	313	19.8	100	217	607	1.9	545	265	0.35	16.7	2085	1.5	9.6
104	Zofar 220	W	1	U. Cret.	*A*	39.4	179	26.7	96.1	177	312	0.7	570	364	0.35	18.0	1743	12.5	13.6
105	Zofar 20	W	1	L. Cret.	*B4*	35.3	1176	64.8	191	895	3636	35.6	729	14	1.11	21.2	6764	14.9	12.5
106	Zofar 4a	W	1	U. Cret.	*A*	26.5	196	8.7	79.5	212	421	0.9	448	204	0.23	17.5	1587	0.92	12.2
107	Paran 20	W	1	L. Cret.	*A*	59.2	323	28	95.1	286	748	3.8	609	207	0.46	25.2	2325	8.3	11.1
108	Paran 18	W	1	L. Cret.	*A*	36.3	200	29.3	80.6	208	408	0.9	627	307	0.37	22.0	1883	7.63	12.8
109	Yaalon 116	W	1	U. Cret.	*A*	32.6	255	12.1	151	335	746	1.5	653	322	0.38	15.5	2491	3.9	14.5
110	Yaalon 6A	W	1	L. Cret.	*A*	42.2	106	6.6	51.3	80.8	257	0.8	165	234	0.17	17.0	918	0.2	16.5
111	Yaalon 7	W	1	L. Cret.	*A*	41.6	168	5.7	74	100	340	1	295	264	0.18	17.0	1266	2	18.2
112	Yaalon 117	W	1	Quat.	*A*	39.6	215	16.2	120	211	368	1.1	656	264	0.44	16.8	1869	8.7	15.6
113	Yotvata 13	W	1	U. Cret.	*A*	29.1	285	10	141	203	479	1.5	738	213	0.45	19.1	2089	3.1	15.6
114	Yotvata 12	W	1	U. Cret.	*A*	29	476	14.2	221	326	873	2.1	1113	241	0.67	19.9	3288	3.2	13.0
115	Gerofit 4	W	1	U. Cret.	*A*	33.3	312	15.8	110	272	765	3.9	574	220	0.36	14.2	2288	7.4	17.0
116	Samar 4	W	1	U. Cret.	*A*	31.8	337	10.9	106	175	631	3.3	678	131	0.51	19.1	2092	0.8	12.3
117	Samar 2	W	1	U. Cret.	*A*	28.3	681	45.9	248	619	1346	9.4	1832	195	1.33	15.1	4993	5.4	11.7
119	Elat 103	W	1	Quat.	*D*	32	1139	22.4	312	679	3151	30.2	967	66	88	18.2	6473	4.4	11.6
120	Elat 108	W	1	Quat.	*A*	31.8	431	12.8	160	233	967	7.7	636	140	0.52	20.0	2607	2.4	17.8
	Eastern side																		
121	Wadi Araba 1	W	1	Jur.	*B3*	29.9	79.1	2.2	32.2	129	202	2.4	125	235	0.10	12.1	819	-4.1	14.2
122	Wadi Araba 5	W	1	Jur.	*C*	27.3	783	9.6	111	676	2723	43.6	66.1	8	0.37	6.9	4427	23.5	14.3
	SOUTHERN JORDAN																		
123	Disi 1	W	1	Paleoz.	*B1*	28.5	25.7	1.5	7	34.4	33.6	0.1	25.3	97		14.4	239	8.3	7.0
124	Wadi Yutum	S	1	Quat.	*B3*	29.7	114	2	20.2	78.6	194	1.5	93	109	0.13	20.8	633	7.8	8.0

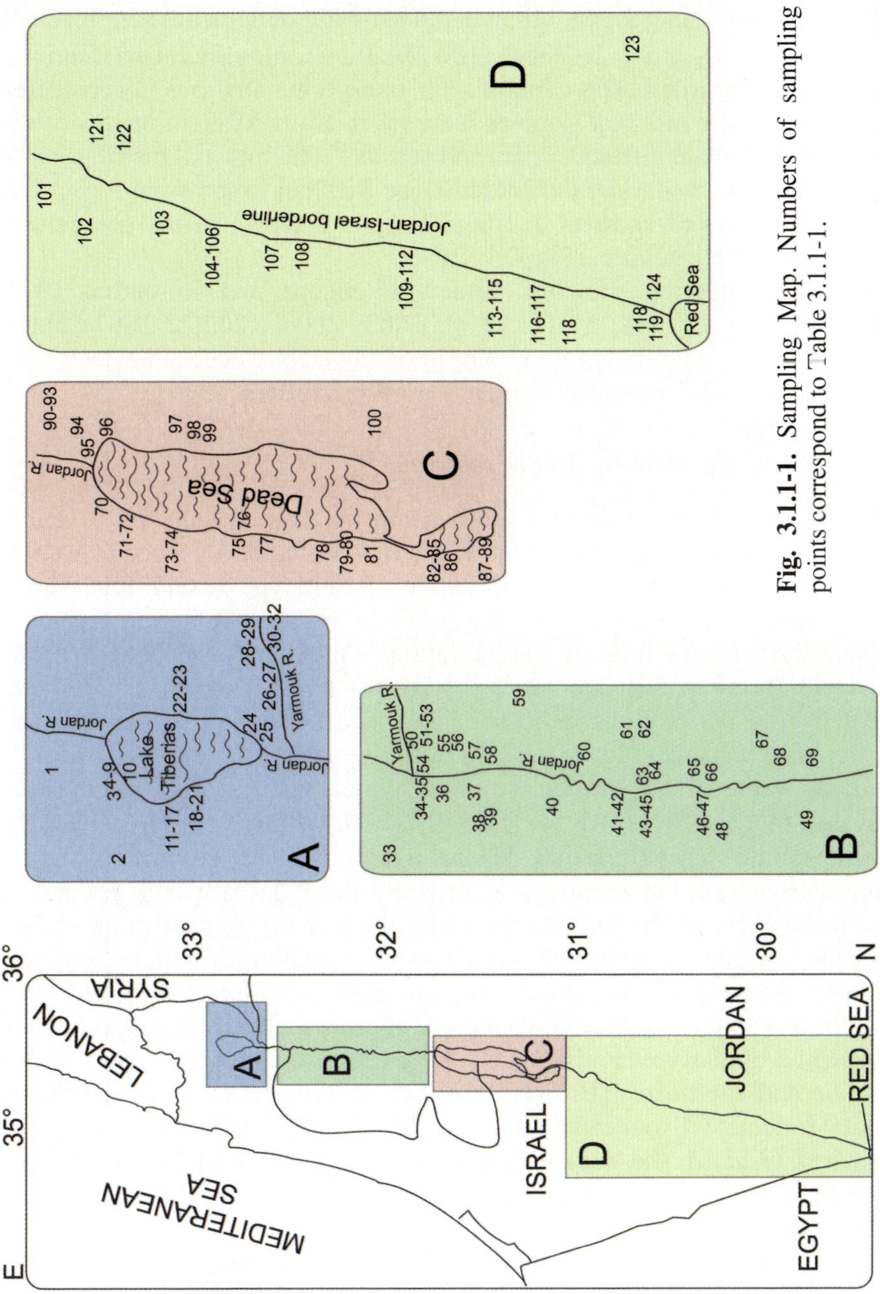

Fig. 3.1.1-1. Sampling Map. Numbers of sampling points correspond to Table 3.1.1-1.

and critically been checked by Bentor (1969), Starinsky (1974), Fleischer et al. (1977), Greitzer (1980), Menashe (1991), and Fleischer and Varshavsky (2002). In general, all analyses showing deficits of more than 7% in their anion-cation balance $(100\cdot(\Sigma\text{cation}-\Sigma\text{anion})/(\Sigma\text{cation}+\Sigma\text{anion}))$ have been discarded. The variability of these brine analyses is considered to be due to the fact that samples were taken from different depths of the wells as well as in different years and seasons. The regional distribution of well and spring waters in the Jordan-Dead Sea Rift is shown in Fig. 3.1.1-1. The position of aquifers and aquicludes are shown in the generalized lithostratigraphic column (Fig. 2.1-7).

As indicated by previous authors (Vengosh and Rosenthal 1994; Rosenthal et al. 1998; Möller et al. 2003; 2006a; 2006c; 2007), inter-aquifer flow is significant and many groundwaters listed in Table 3.1.1-1 represent mixtures of waters from at least two aquifers.

Visualization of data by spider patterns

The chemical composition of groundwater is easily visualized in spider patterns, which show the concentrations of selected dissolved species normalized to seawater (Möller et al. 2006c). In this type of semi-logarithmic plot, seawater is represented by the zero line. Hence, any species below or above the zero line indicate concentrations which are lower or higher than concentrations in seawater. Any horizontal trend between species presented in spider patterns indicates element ratios similar to those in seawater. For instance, Fig. 3.1.1-2 shows the patterns of seawater brines at various stages of evaporation with their typical element distribution (Fontes and Matray 1993). $Rb_{sw} \approx K_{sw} \approx Br_{sw} \approx B_{sw} \approx Mg_{sw}$ are typical for the pre-epsomite stages (index sw denotes normalization to seawater). Any dilution of seawater or its brines by groundwater reduces mostly Na and Cl in the spider patterns but elements like Ca and Mg remain more or less constant because they are controlled by the leaching of sulfates and carbonates from sedimentary rocks. The arrangement of ions along the X-axis is arbitrary, but considers association of species such as adsorption and ion exchange by clay minerals (Y, Cs, Rb), halide minerals (K, Na, Cl, Br), sulfate and carbonates (B, SO_4^{2-}, Ca, Sr, Mg, Ba, HCO_3^-) and the highly mobile element U under reducing conditions. Thus, differences of spider patterns visualize the compositional differences of groundwater environments. For illustrative purposes, normalized element relations for dissolved halite, gypsum and calcite are shown in Fig. 3.1.1-2. The Ca_{sw}-$SO_4^{2-}{}_{sw}$ relationship is variable because Ca is contributed by carbonates and sulfates. The patterns which formed by dissolution of pure gypsum/

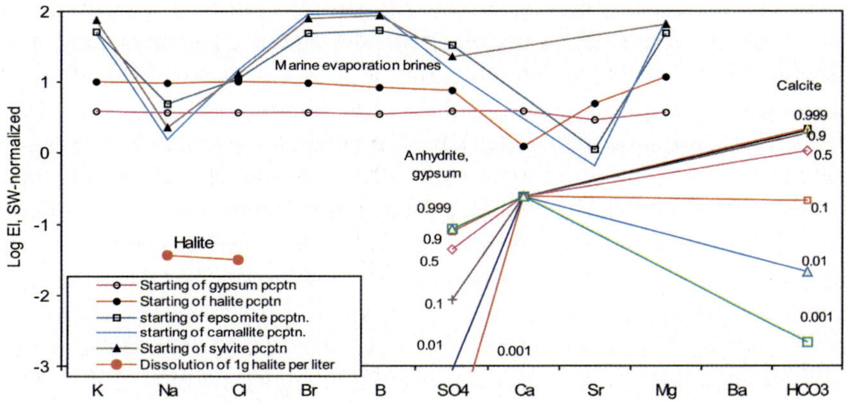

Fig. 3.1.1-2. Spider patterns of evaporation brines of seawater, dissolved halite, calcite/aragonite, and gypsum/anhydride. Averages of elements in seawater were taken from Bruland and Lohan (2004).

anhydrite and calcite/aragonite are indicated by red and blue lines, respectively. If brines only dissolve gypsum, the difference between $\log Ca_{sw}$ and $\log SO_4{}^{2-}{}_{sw}$ should be +0.5 units. If only calcite or aragonite are leached, $\log Ca_{sw}$-$\log SO_4{}^{2-}{}_{sw}$ yields >>0.5 units.

The Na_{sw}-Cl_{sw} trend in seawater is horizontal (Fig. 3.1.1-2). This is also the case in most groundwater, except in the brines associated with the Dead Sea. A slightly inclining trend results from the dissolution of halite, whereas a strongly declining trend proves the presence of mixtures of dissolved $CaCl_2$ and $MgCl_2$, henceforth combined to "(Ca,Mg)Cl_2 brine". Different from K_{sw} and Na_{sw} the enriched normalized heavy alkalis Rb and Cs are leached from clay minerals. Primary brines are therefore marked by spider patterns which show no enrichment of normalized heavy alkalis compared to K_{sw} (Fig. 3.1.1-2).

Ca_{sw} - Sr_{sw}: Limestone is primarily of biogenic origin and biogenic carbonates are often aragonitic or high-magnesian calcitic (Veizer 1969; 1978). Both phases are metastable in contact with meteoric water. Sulfates are always of inorganic origin. Aragonite and anhydrite are richer in Sr than calcite and gypsum because the co-precipitation factors of the latter are <1, whereas for the former they are >1. Calcite and gypsum, which precipitate inorganically from seawater, show co-precipitation factors of about 0.14 (Kinsman and Holland 1969), and 0.3 (Usdowski 1973), respectively. The low factors show that Sr is rejected from being incorporated. In contrast, aragonite shows Ca/Sr ratios of seawater, but releases Sr when converted to calcite during diagenesis (Usdowski 1973). In this process, Sr is either leached by percolating freshwater or precipitated as

strontianite ($SrCO_3$). Thus, geologically young (Eocene) limestone and chalk show horizontal Ca_{sw}-Sr_{sw} trends in the spider diagram, whereas the Mesozoic limestone is expected to show $Ca_{sw} > Sr_{sw}$. Dissolution of anhydrite may yield $Sr_{sw}>Ca_{sw}$. Thus, the Ca_{sw}-Sr_{sw} trend is an important criterion for characterizing waters (and thereby different groundwater aquifers) affected by late diagenetic limestone on the one hand and early diagenetic limestone, chalk and/or gypsum/anhydride on the other.

Ba_{sw} is high, where $SO_4^{2-}{}_{sw}$ is low and vice versa which indicates that Ba is largely controlled by barite dissolution/precipitation. Y (also representing rare earth elements, REE) is only insignificantly leachable from clay minerals, gypsum and carbonates. Like Cs and Rb, Y abundance is controlled by adsorption onto mineral surfaces (Johannesson et al. 1999, 2000; Möller et al. 2003) and co-precipitation with alteration minerals. Thus, mineral surfaces which are in contact with water are greatly enriched in Y (and REE) (Möller 2002). The high scatter of U (Figs. 3.1.1-3–9) results from leaching of various U-bearing minerals which occur in the sediments and particularly in phosphorite-rich Upper Cretaceous limestone. This layer also seems to be the source of the enhanced U contents in groundwater originating from silicified limestone.

Classification of groundwater according to spider patterns

Table 3.1.1-2 presents the main features of spider patterns and ionic ratios. The lithostratigraphic information is taken from Table 3.1.1-1.

In the study area the groundwaters are assembled according to similar shapes of spider patterns in Figs. 3.1.1-3–9. Accordingly, seven types of patterns emerge in the study area. Judging by Na_{sw} and Cl_{sw} most waters are strongly enriched in the seawater-normalized heavy alkalis B, Br, Ba and Y, with the exception of waters from the wells of Haon and Elat 103 and, to a slighter extend, from the Gofra spring (Fig. 3.1.1-9).

Type A groundwaters are characterized by $Ca_{sw} \approx Sr_{sw}$, $B_{sw} > K_{sw} > Na_{sw} \approx Cl_{sw} \approx Br_{sw}$ (Fig. 3.1.1-3) and averaged molar ratios of dissolved $Cl^-:SO_4^{2-}:HCO_3^-$ and Na/Cl are 2.7:0.8:1 and 0.97 in the eastern part of the Rift and 5:4:1 and 0.76 in the western part. Gypsum could have been leached from cements or from lenses of evaporites. In the western part of the Rift *type A* waters occur in the Arava Valley, originating from aquiferous beds in the Lower Cretaceous Kurnub Group, the Neogene Hazeva Formation and the Quaternary alluvial fill. The chemical resemblance between waters derived from different aquifers in the Arava Valley has already been pointed out by Yechieli et al. (1992). In the eastern part of the Rift, the groundwater occurs north of the Araba Valley between Zerqa Ma'in and the Golan Heights. Lithologically, the groundwater mostly

Table 3.1.1-2. General features and averages of molar ratios for each type of spider patterns

	Dead Sea	*Seawater*	*Type A*		*Type B1*		*Type B2*		*Type B3*	*Type B4*	*Type C*	*Type D*
			Western Rift	Eastern Rift	Western Rift	Eastern Rift	Western Rift	Eastern Rift	Eastern Rift	Western Rift	Western Rift	Western Rift
Seawater normalized relationship			$Ca \approx Sr$ $Ca < HCO_3$ $SO_4 > Cl \approx Br$ $B > Br$	$Ca \approx Sr$ $Ca < HCO_3$ $SO_4 > Cl \approx Br$ $B >> Br$	$Ca > Sr$ $Ca << HCO_3$ $Br > SO_4 > Cl$ $B > Br$	$Ca > Sr$ $Ca << HCO_3$ $SO_4 >> Br \approx Cl$ $B > Br$	$Ca \approx Sr$ $Ca < HCO_3$ $Br > Cl > SO_4$ $B > Br$	$Ca \approx Sr$ $Ca < HCO_3$ $Br > SO_4 > Cl$ $B > Br$	$Ca \approx Sr$ $Ca < HCO_3$ $SO_4 \approx Br > Cl$ $B \geq Br$	$Ca \approx Sr$ $Ca > HCO_3$ $Br > Cl \geq SO_4$ $Br > B$	variable	$Ca < Sr$ $Ca << HCO_3$ $SO4 > Cl$ $B > Br$
$Cl:SO_4:HCO_3$		218:11:1	5:2.2:1	2.7:0.4:1	1.8:0.15:1	0.3:0.07:1	10:0.3:1	2:0.3:1	1.1:0.3:1	1200:10:1 (DS) 19:0.2:1 (LT)	1:0.2:1	50:3:1
Na/Cl	0.23	0.86	0.75	0.97	0.77	1	0.7	0.7	0.71	0.2-0.8	1.7	0.6
1000Br/Cl	11	1.5	1.6	1.8	6.2	1.5	4.5	5.8	5	1.3-11	2.2	5.3
1000B/Cl	0.4	0.76	2.5	4.4	2.6	5.5	0.8	1.8	2.5	0.35-1.1	5	1
Ca/Mg	0.24	0.2	1.2	2.5	1.2	1.3	2.4	2.3	2.2	0.1-3	0.6	0.7
Ca/SO_4	127	0.36	0.9	1.7	7.5	5.3	4.9	2.8	2.2	0.5-127	1.1	2-40
$1000B/SO_4$	740	15	6	26	30	10	25	14	10	9-750	40	8-260
Geographical occurrance; localities			Arava Valley	Araba Valley,	Area of Lake Tiberias and Dead Sea; Lower Jordan Valley	Lower Jordan Valley	Area of Lake Tiberias	En Reah; En Makla	Himma; Wadi Araba 1; Wadi Yutum well	Area of Lake Tiberias (LT) and Dead Sea (DS)	Bet Shean area	Haon, Gofra, Elat 103
Aquifer lithology			sandstone; graben fill	sandstones	limestone	sandstone; limestones	limestone; chalk	limestone; chalk	limestone	thermally altered and unaltered limestone	basalt	limestone
Formation			L. Cretaceous; Quaternary	Triassic to L. Cretaceous	U. Cretaceous	L.& U. Cretaceous; Quaternary	Eocene	L. & U. Cretaceous	L. & U. Cretaceous	U. Cretaceous	Neogene	Quaternary

originates there from Triassic, Jurassic and Lower Cretaceous sandstone and other clastics (Table 3.1.1-1).

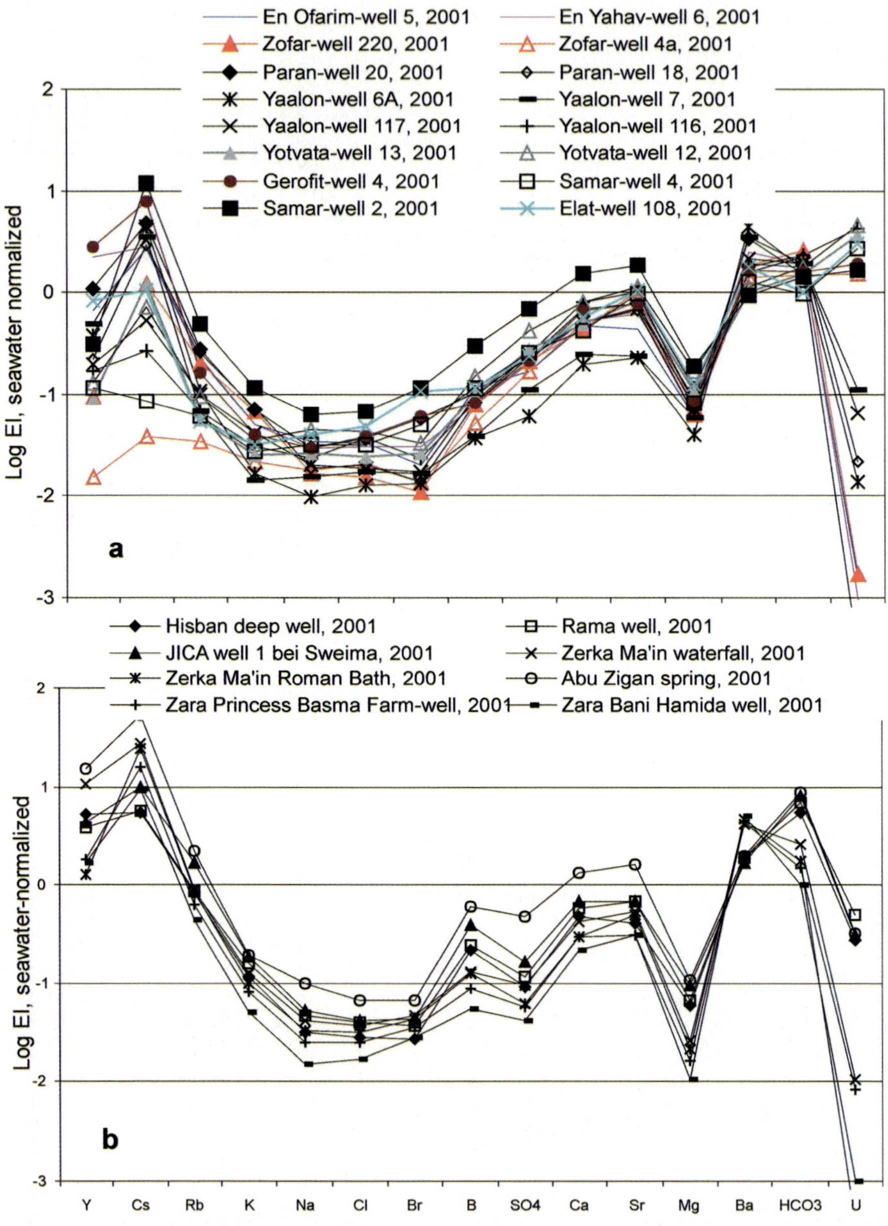

Fig. 3.1.1-3. Spider patterns of *type A* groundwaters. (a) and (b) compile patterns from the western and eastern part of the Rift system, respectively.

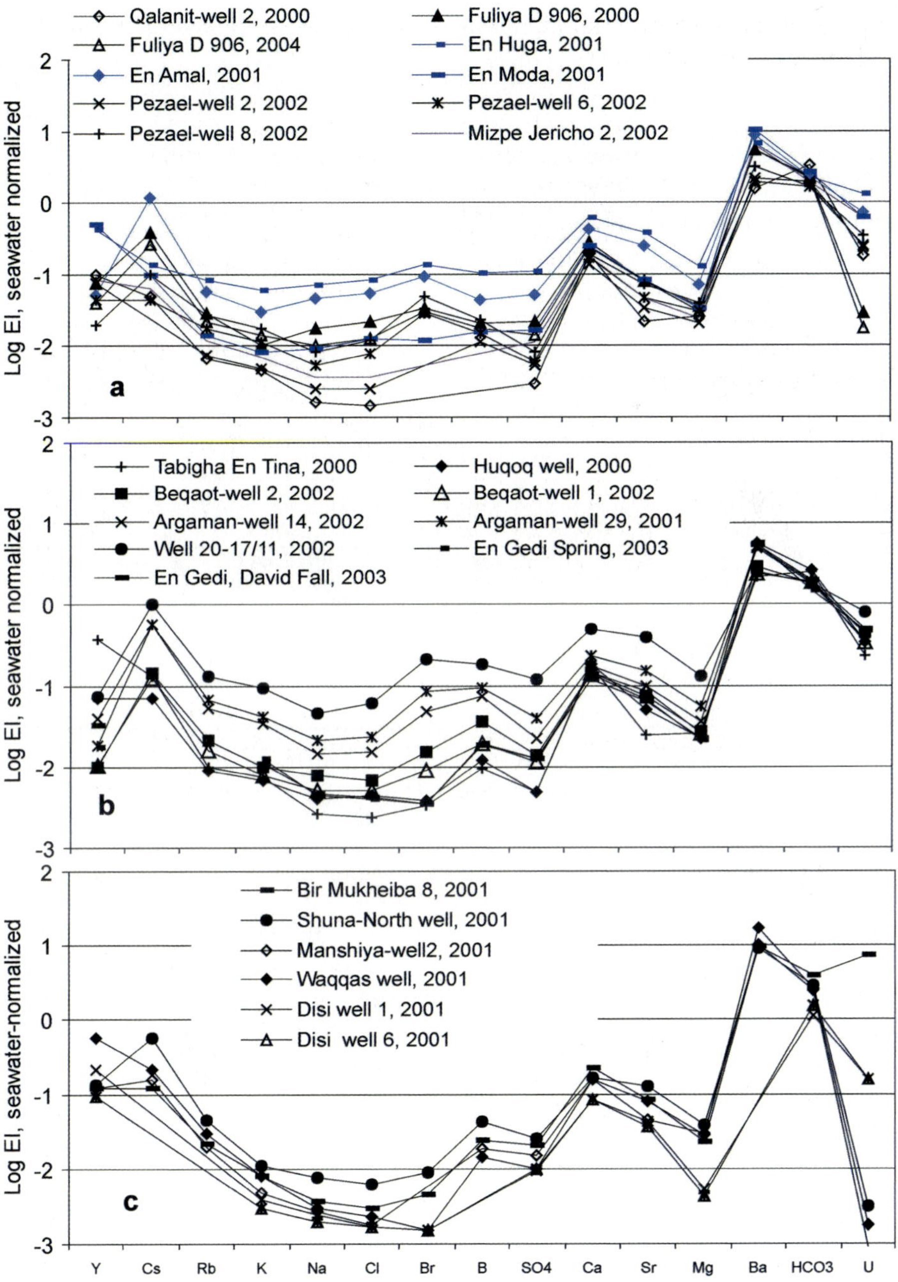

Fig. 3.1.1-4. Spider patterns of *type B1* groundwater. (a), (b), and (c) compile patterns from the western and eastern part of the Rift system, respectively.

Type B groundwaters all originate from limestone aquifers and are subdivided into 4 sub-types:

Type B1 comprises groundwaters with $HCO_3^-{}_{sw} > Ca_{sw} > Sr_{sw} > SO_4^{2-}{}_{sw}$, $Br_{sw} > Cl_{sw}$ (Fig. 3.1.1-4) and molar ratios of $Cl^-:SO_4^{2-}:HCO_3^-$ and Na/Cl are 0.3:0.2:1 and 1.1 in the eastern part of the Rift, and 1.9:0.3:1 and 0.76 in the western part. Saline waters from both sides significantly differ in their relative amounts of dissolved halite.

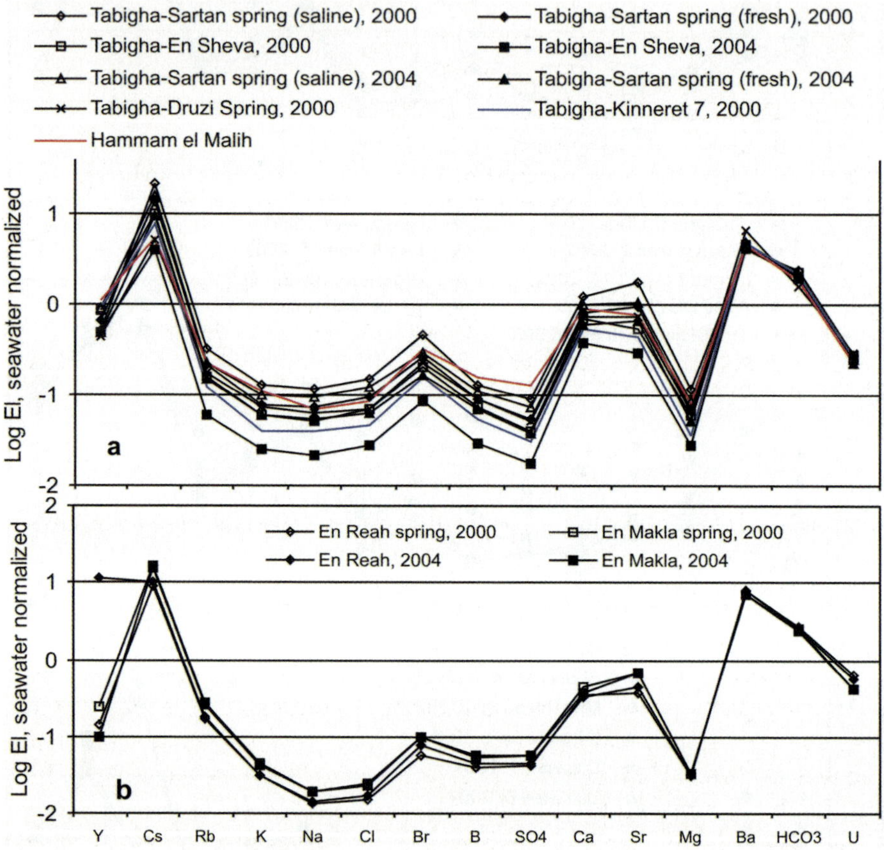

Fig. 3.1.1-5. Spider patterns of *type B2* groundwaters. (a) and (b) compile patterns of the western and eastern side of the Rift, respectively.

On the western side the groundwater mainly originates from the Judea Group aquifer in the surroundings of Lake Tiberias, the lower Jordan Valley, and along the Dead Sea. In the eastern part, this *type* also occurs in Upper Cretaceous limestone and in Lower Cretaceous and Palaeozoic sandstone (Table 3.1.1-1).

Type B2 comprises patterns with $Ca_{sw.} \approx Sr_{sw}$, $Ca_{sw} >> SO_4^{2-}{}_{sw}$, $B_{sw} > SO_4^{2-}{}_{sw}$ and $Cl_{sw} < Br_{sw}$ (Fig. 3.1.1-5). The molar ratios of $Cl^-:SO_4^{2-}:HCO_3^-$

and Na/Cl are 5:0.4:1 and 0.7 in the eastern part of the Rift, and 11:0.8:1 and 0.7 in the western part. This indicates the enhanced influence of the dissolution of halite in the western parts of the Rift. This group includes groundwater from Eocene chalks with high original Sr concentrations that characterize biogenic carbonates. Although the spider patterns are quite similar on both sides of the Rift, major differences exist particularly concerning Ca/Mg and Ca/SO_4^{2-} ratios (Table 3.1.1-2).

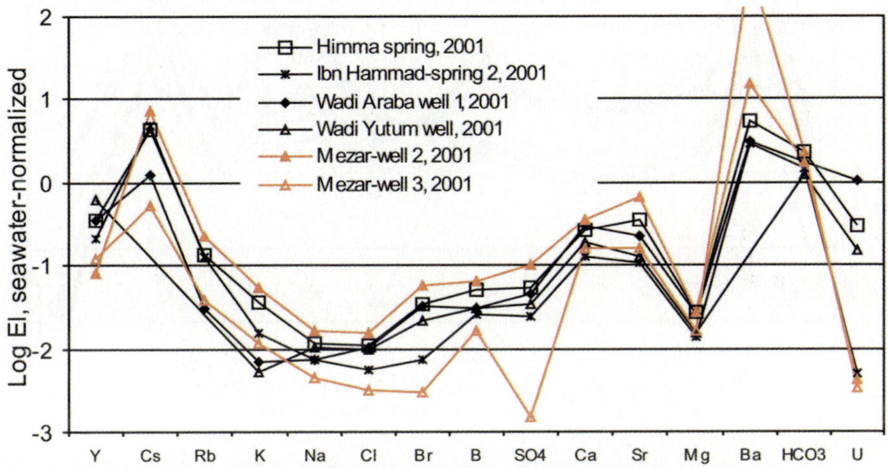

Fig. 3.1.1-6. Spider patterns of *type B3* groundwaters from the eastern part of the Rift

Type B3 comprises patterns with $Ca_{sw} \approx Sr_{sw}$, $Ca_{sw} > SO_4^{2-}{}_{sw}$, $B_{sw} \approx SO_4^{2-}{}_{sw}$, and $Cl_{sw} < Br_{sw}$ (Fig. 3.1.1-6) in the eastern part of the Rift. The waters are characterized by molar ratios of $Cl^-:SO_4^{2-}:HCO_3^-$ and Na/Cl of 1.2:0.6:1 and 0.8, respectively. These ratios point to a dominant presence of halite, followed by dissolved sulfates (Ca/SO_4^{2-} molar ratio ~1.4) and carbonates. The waters differ from *type B2* both by enhanced B contents and Br/Cl ratios, and by diminished B/Cl ratios (Table 3.1.1-2).

Type B4 groundwaters are characterized by variable salinities (TDS) and $Cl_{sw} > SO_4^{2-}{}_{sw}$, $Ca_{sw} \approx Sr_{sw}$, $Br_{sw} > B_{sw}$, and $Br_{sw} \approx Ca_{sw}$ (Fig. 3.1.1-7). The average ratios of $Cl^-:SO_4^{2-}:HCO_3^-$ are about 1220:21:1, but Na/Cl molar ratios of 0.5 prove the presence of other halides besides dissolved halite. Apart from dissolved halides, the most important component is dissolved gypsum. The *type B4* waters do not represent a homogeneous group. The thermomineral (Mazor and Mero (1969) or thermosaline (Rosenthal 1997) waters from the surroundings of Lake Tiberias are distinctly different from the Dead Sea brines and mixtures of recent Dead Sea brines with freshwater.

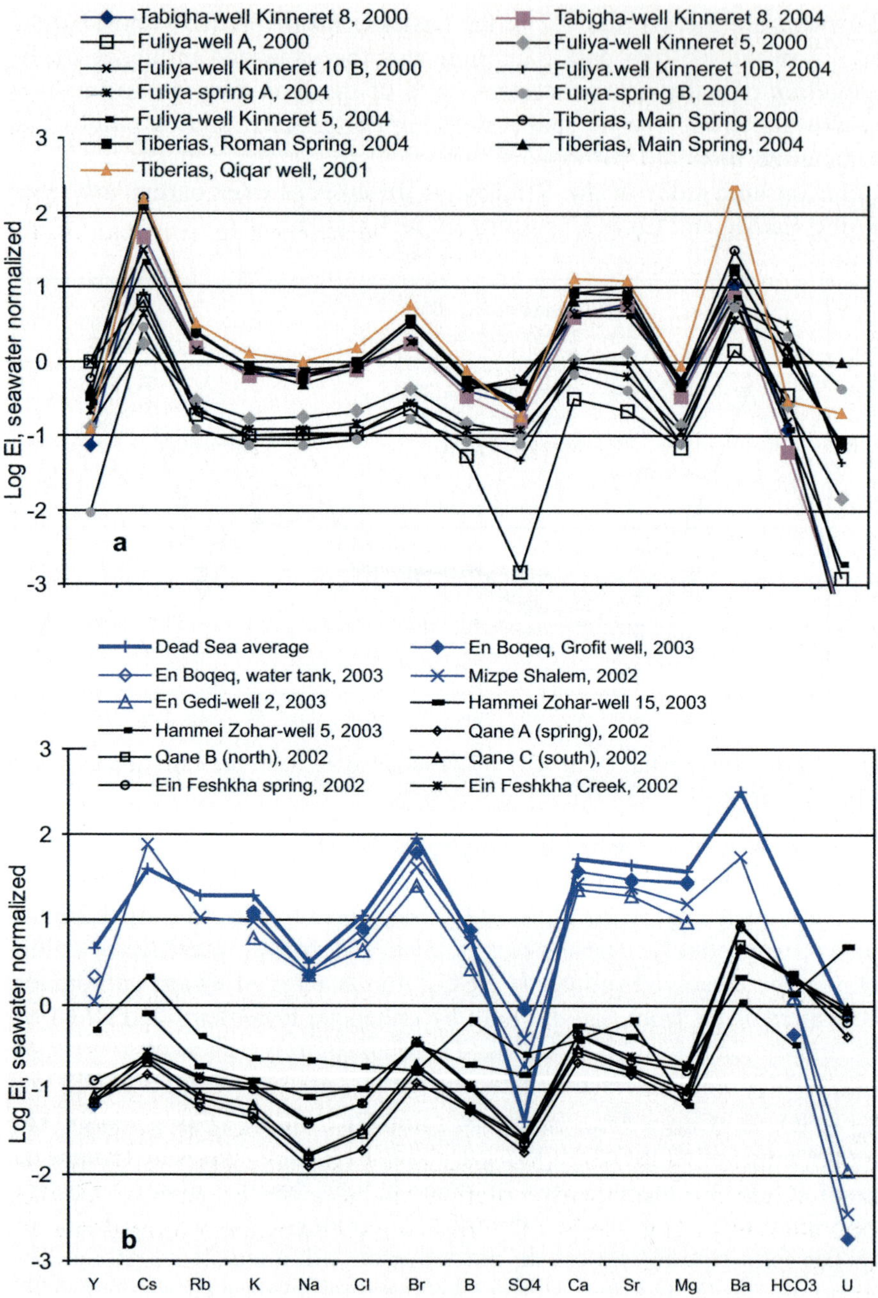

Fig. 3.1.1-7. Spider patterns of *type B4* saline groundwaters from the surroundings of Lake Tiberias (a) and the Dead Sea (b). Dead Sea water is presented as bold line.

The water from the well Zofar 20, which is located in the Arava Valley, also fits into the suite of saline waters in Fig. 3.1.1-7a, which comprises the thermosaline waters of western and north-western shores of Lake Tiberias and the Arava Valley, and thereby indicates the presence of this *type* of saline water throughout the Rift.

Type C manifests highly variable spider patterns of groundwaters (Fig. 3.1.1-8) flowing in basalts of the northern Jordanian Highland and locally in Jurassic sandstones (Möller et al. 2007).

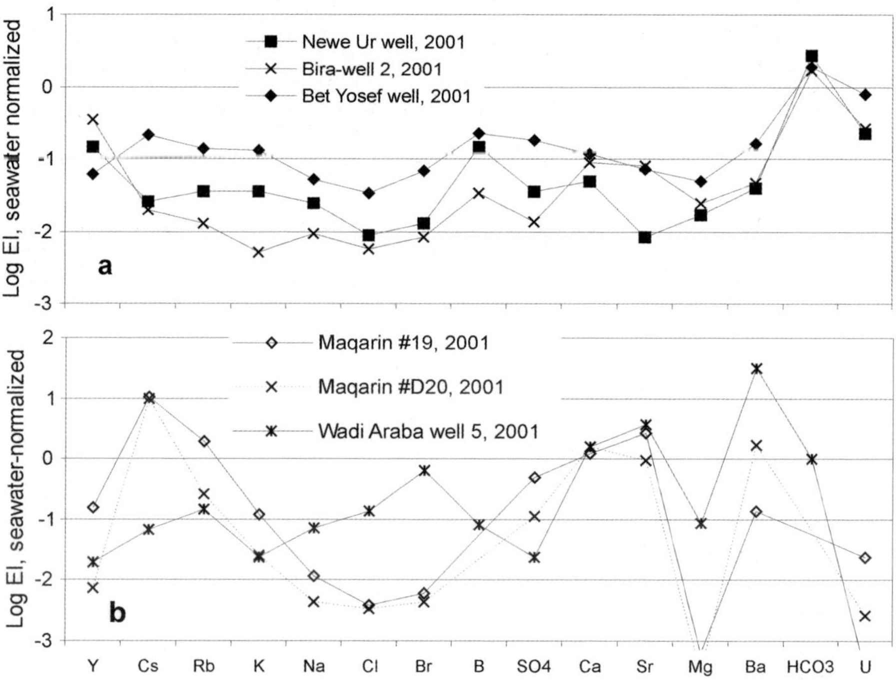

Fig. 3.1.1-8. Spider patterns of *type C* groundwaters from basaltic aquifers. (a) and (b) refer to western and eastern part of the Rift, resepctively.

Type D waters show trends which resemble those of evaporation brines (Fig. 3.1.1-9). Such saline waters occur at Haon and to a lesser extend at the Gofra spring, both along the eastern shore of Lake Tiberias. They represent diluted remnants of evaporation brines of the Sdom Sea. Another evaporation brine is given by the Elat 103 well, which was formed in a local evaporative playa.

With $Cs_{sw} \approx Rb_{sw} \approx K_{sw} \approx Mg_{sw}$ and $Sr_{sw} > Ca_{sw}$ these waters significantly differ from all the others. Most significantly, these waters are not enriched in heavy alkalis such as K_{sw}, Rb_{sw}, and Cs_{sw}. B_{sw} and Mg_{sw} are comparable with the normalized heavy alkalis.

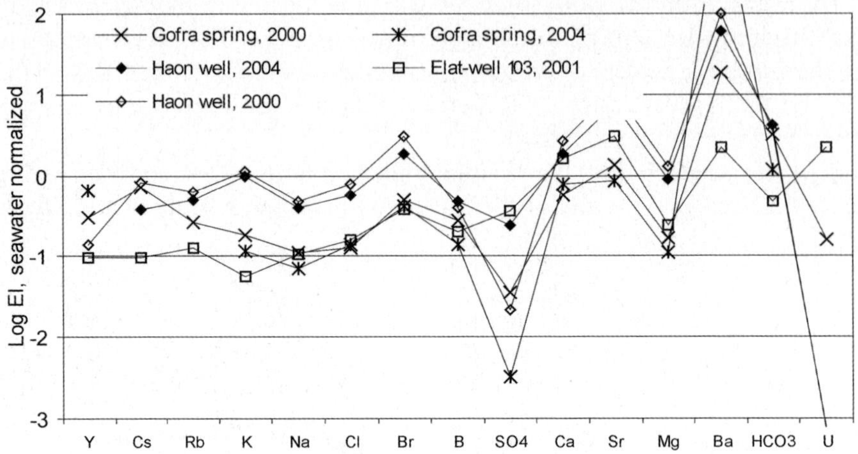

Fig. 3.1.1-9. Spider patterns of *type D*. These patterns resemble those of evaporation brines. Haon and Elat 103 are located on the eastern side of Lake Tiberias and near Elat in the Arava Valley, respectively. The brines definitely derived from different sources.

The composition of these waters is independent of their source rocks (Table 3.1.1-2). Elat 103 has a higher Ca/Mg molar ratio than Haon, which probably indicates advanced dolomitization. In both cases, Ca_{sw} nearly equals $HCO_3^-{}_{sw}$. In evaporated brines, low Na_{sw} and Cl_{sw} would be expected (Fig. 3.1.1-2). The rather high values in Fig. 3.1.1-9 point to the ablation of halite by a diluted evaporation brine. Although located near Haon, the Gofra spring water (Fig. 3.1.1-9) shows less contribution of the evaporation brine than the Haon water.

Element cross plots based on groupings by spider patterns

The trend of *(Na+K) vs. Cl* (Fig. 3.1.1-10) largely corresponds to halite dissolution in the eastern part of the Rift. Only *type B1* waters seem to follow a slightly different slope than the halite dissolution line. In the western part, Cl^- increases with increasing salinity and always exceeds the sum of Na+K equivalent concentration.

For *type B4* water the Cl_{excess} (Eq. 3.1.1-1), which defines the amount of Cl^- according to the presence of $(Ca,Mg)Cl_2$, is very high. Only *B1 type* water wizth low Cl^- concentrations plots near the halite dissolution line, whereas *types A* and *B2* deviate considerably.

$$Cl_{excess} = Cl\text{-}Na\text{-}K \qquad (3.1.1\text{-}1)$$

Plate 3.1.1-1. One of the Fuliya springs (right pool) during the draught late in the year 2000. The spring is visible only under low stands of Lake Tiberias (left to the stone barrier). (Photo: P. Möller, 2000)

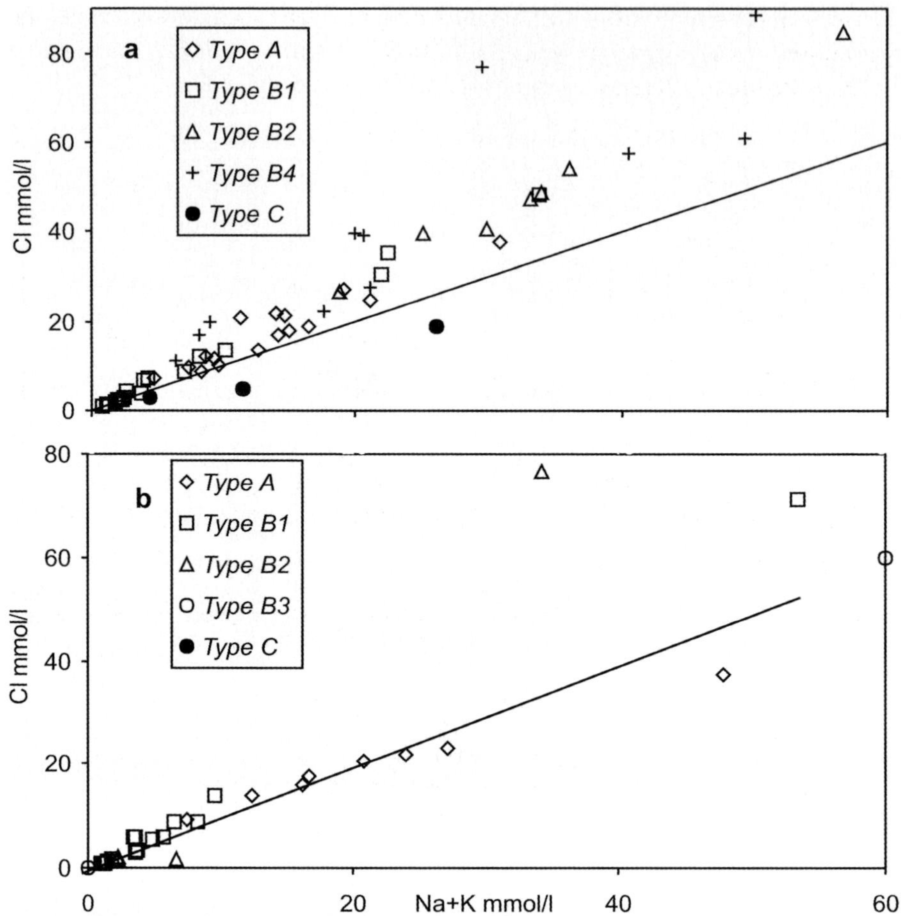

Fig. 3.1.1-10. Cross plots of Na+K vs. Cl. The Na/Cl molar ratio of halite is shown by the diagonal line. (a) and (b) compile samples from the western and eastern part of the Rift, respectively. On the western side, Cl contents are always higher than Na+K contents. The solid line represents halite dissolution.

Br vs. Cl (Fig. 3.1.1-11) displays various linear correlations with 1000Br/Cl between 0.7 and 11.5. In the western part of the Rift, *types A*, *B2* and parts of *B4* show a linear correlation, i.e., more or less constant Br/Cl ratios. At high concentrations, *type B4* yields a non-linear trend, because Br increased in the (now diluted) brines due to halite precipitation. In the eastern part, linear correlations are found for all *types* except *type B3*. Linearity is maintained even at high concentrations.

The low Br/Cl ratios can be explained by dissolution of evaporites, e.g., as in case of Zemah 1 brine with 1000Br/Cl of about 0.25. Residual marine brines vary from 1.5 in seawater to 9.5 in brines precipitating sylvite

(Fontes and Matray 1993). Samples from the Dead Sea and from its surroundings show a narrow spread of 1000Br/Cl ratios of 7 to 11. Waters from volcanic areas show very variable Cl/Br ratios (Fig. 3.1.1-8).

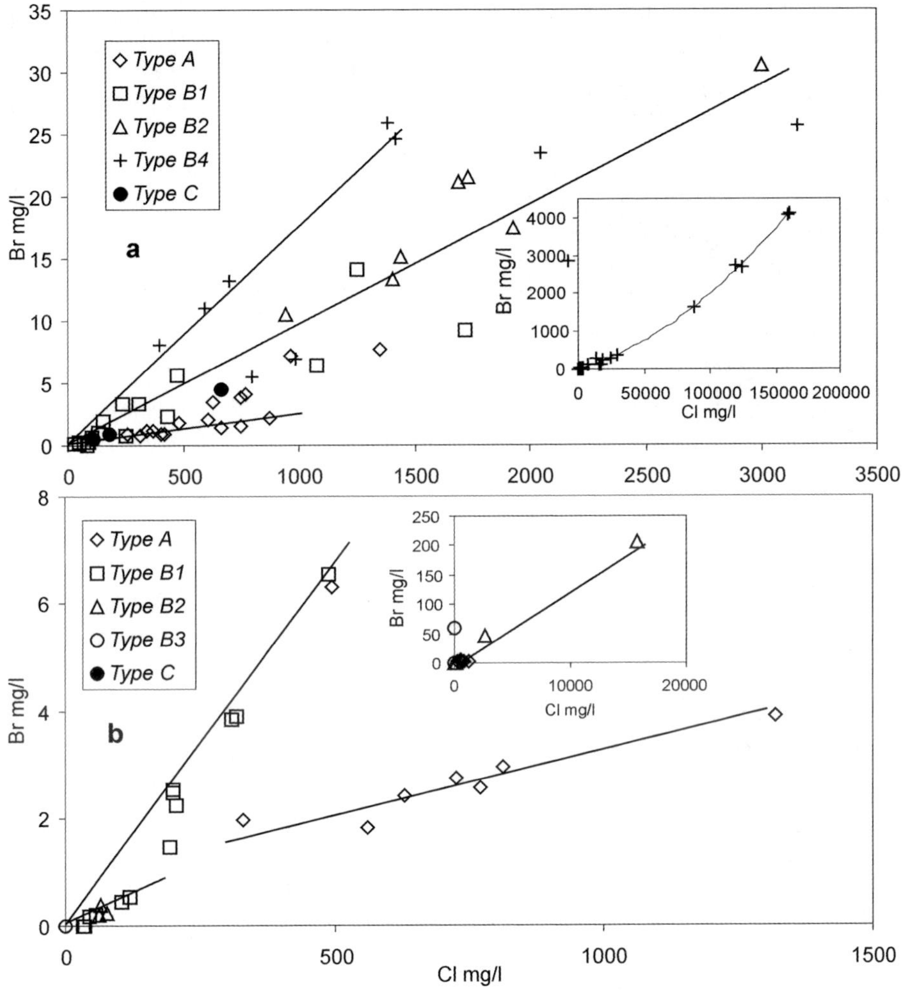

Fig. 3.1.1-11. Cross plots of Br vs. Cl. (a) and (b) compile samples from the western and eastern part of the Rift, respectively. Several correlations are indicated.

The high Br/Cl ratios in water deriving from limestone aquifers are a regional phenomenon. It seems that the enhanced Br contents indicate either residual saline formation waters that are leached from aquifer rocks or from Br-rich bituminous layers and/or from the leaching of phosphorites at the Cretaceous-Tertiary boundary by freshwater (Möller et al. 2003). In many cases, high Br is associated with high U contents, which may preferably be leached from Eocene chalk and limestone. Another source for Br is the regional precipitation which is characterized by 1000Br/Cl molar ratios in the range of 15–190 (Rosenthal 1987; Rimawi and Salameh 1988; Herut 1992). If former Br contents in precipitation are comparable with contemporaneous values, Br must have been enriched in all water bodies which have existed in the Rift since the formation of the Sdom Sea. During this period, huge amounts of meteoric precipitation (particularly during the Pleistocene) mixed with existing lake waters. The high Br/Cl ratio in meteoric water combined with the precipitation of halite continuously increased the Br/Cl ratio in the evaporating water bodies of the lakes with time and thereby of any brine resulting from evaporation.

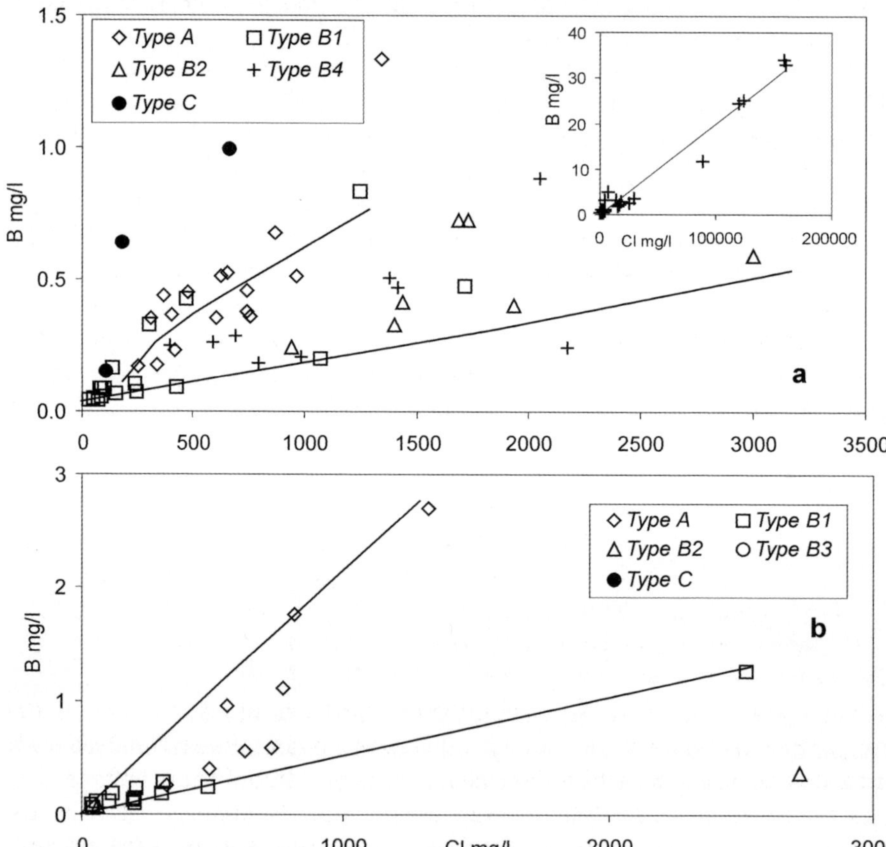

Fig. 3.1.1-12. Cross plots of B vs. Cl; (a) and (b) compile samples from the western and eastern part of the Rift, respectively. Correlations are indicated.

B vs. Cl (Fig. 3.1.1-12) display reasonable correlations for *type B2* waters on both sides of the Rift. Groundwater from both sides by far exceeds the corresponding molar ratio of seawater with 1000B/Cl of 0.76 (Bruland and Lohan 2004). The enhanced ratios in groundwaters either indicate extremely evaporated seawater brines or leachates of evaporites. High B/Cl ratios occur when sabkas repeatedly dry up. Such high-density brines might have replaced less dense brines in the underlying sediments. In the western part, *type B4* shows some correlation at high concentrations. All other *types* scatter widely. Extrapolation of the *type A* waters from the Arava Valley yields a B intercept, which indicates that a low-salinity water is a significant B source. Linear trends are obtained for *types A, B1* and *B2* in the eastern part of the Rift.

The *B vs. Br* cross plot of samples (Fig. 3.1.1-13) from the western part of the Rift shows some correlation for *types B2* and *B4* waters. In the eastern part, linear trends are obtained for *types A, B1* and *B2. Type A* shows a very steep trend with an intercept on the Br axis, whereas on the western side, *type A* waters cluster. Enhanced B and Br contents characterize *types A* and *B4* water in the eastern and western part of the Rift, respectively.

The *B vs.* SO_4^{2-} (Fig. 3.1.1-14) cross plots display individual trends on both sides of the Rift. *Type A* waters from the western and eastern part of the Rift plot below and above the indicated trend line, respectively. The majority of *type B3* and *B4* plot on either side of the trend line. The lowest B/SO_4^{2-} ratios are typical for waters from the Arava Valley, whereas those in the Jordan Valley are enhanced. Thus, B shows a regionally bound distribution irrespective of the lithology of aquifers. This regional grouping indicates that in the cited regions different minerals or processes are in control of B/SO_4 molar ratios.

From SO_4^{2-} and B data of the lower Jordan River water and from data of surface- and groundwater in the lower Jordan Valley between Lake Tiberias and the Dead Sea (Farber et al. 2004), an average molar ratio of $100{\cdot}B/SO_4^{2-}$ of 1.8 is derived from 79 analyses. By way of comparison, gypsum and anhydrite from the German Zechstein contain 20 to 200 µg/g of B, corresponding to $100B/SO_4^{2-}$ molar ratios of 1.4 to 14, respectively (Braitsch 1971). The strong coherence of B and SO_4^{2-} (Fig. 3.1.1-14c) proves a common provenance which coincides with the occurrence of Lisan sediments consisting mainly of aragonite and gypsum. Thus, all groundwater in contact with the Lisan sediments should show enhanced B concentrations compared to seawater. The high B contents of Tiberias ($100B/SO_4^{2-}$ = 1.7) could result from significant eolian input of fine grained Lisan sediments further south. Only *type A* water from the Arava Valley is not affected ($100B/SO_4^{2-}$ = 0.6). Only groundwater mixed with either residual brines, as for instance in the Dead Sea with $100B/SO_4^{2-}$ of 74, or leachates of Lisan or similar sediments (because gypsum contains borate) necessarily show enhanced B/SO_4^{2-} ratios.

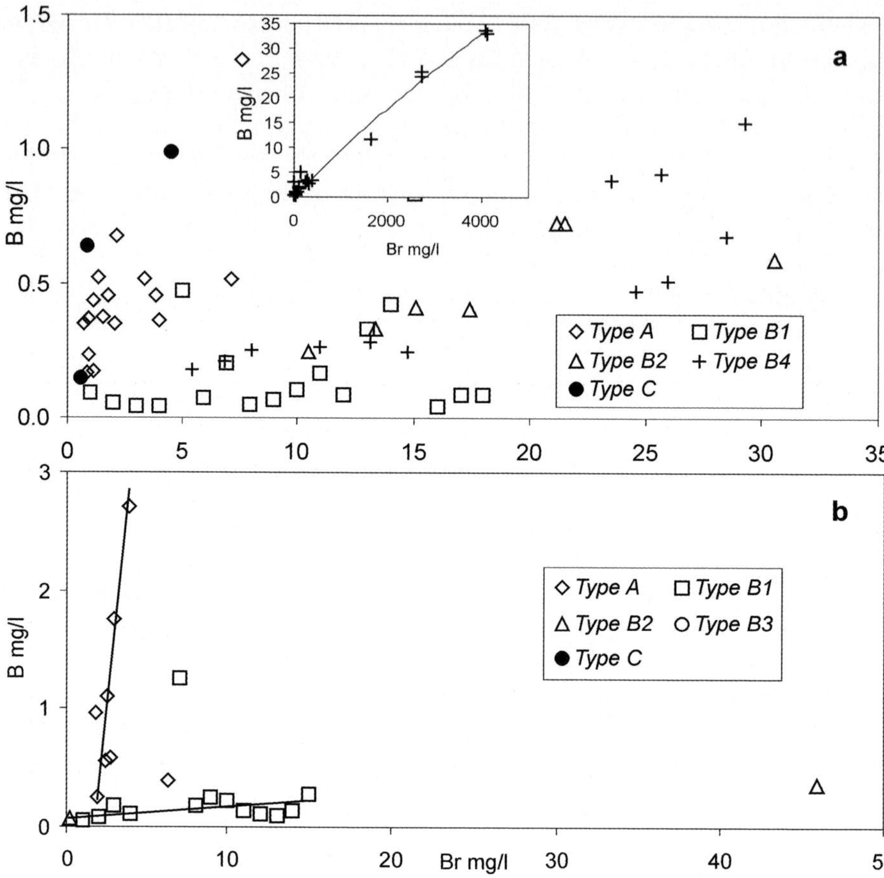

Fig. 3.1.1-13. Cross plots of B vs. Br; (a) and (b) compile samples from the western and eastern part of the Rift, respectively.

Summing up, it seems that spider patterns visualize chemical differences in groundwater from essentially similar aquifer rocks such as young, old, and altered limestones. These differences are due to mineralogical differences of the aquifer rocks, their thermal history, and the mixing with saline waters. As shown in Table 3.1.1-2, the grouping according to spider *types* is not always identical with regional distribution of water. Significantly different trends in various cross plots from the eastern and western part of the Rift prove that mineralization of water on both sides is established by variable contributions of processes which occur on both sides. The variable degrees of salinization result from the relative portion of involved saline waters from different sources. The trends indicate the presence of distinct saline end-members in the Rift facilitating the salinization of groundwater. This aspect will be analyzed in greater detail below.

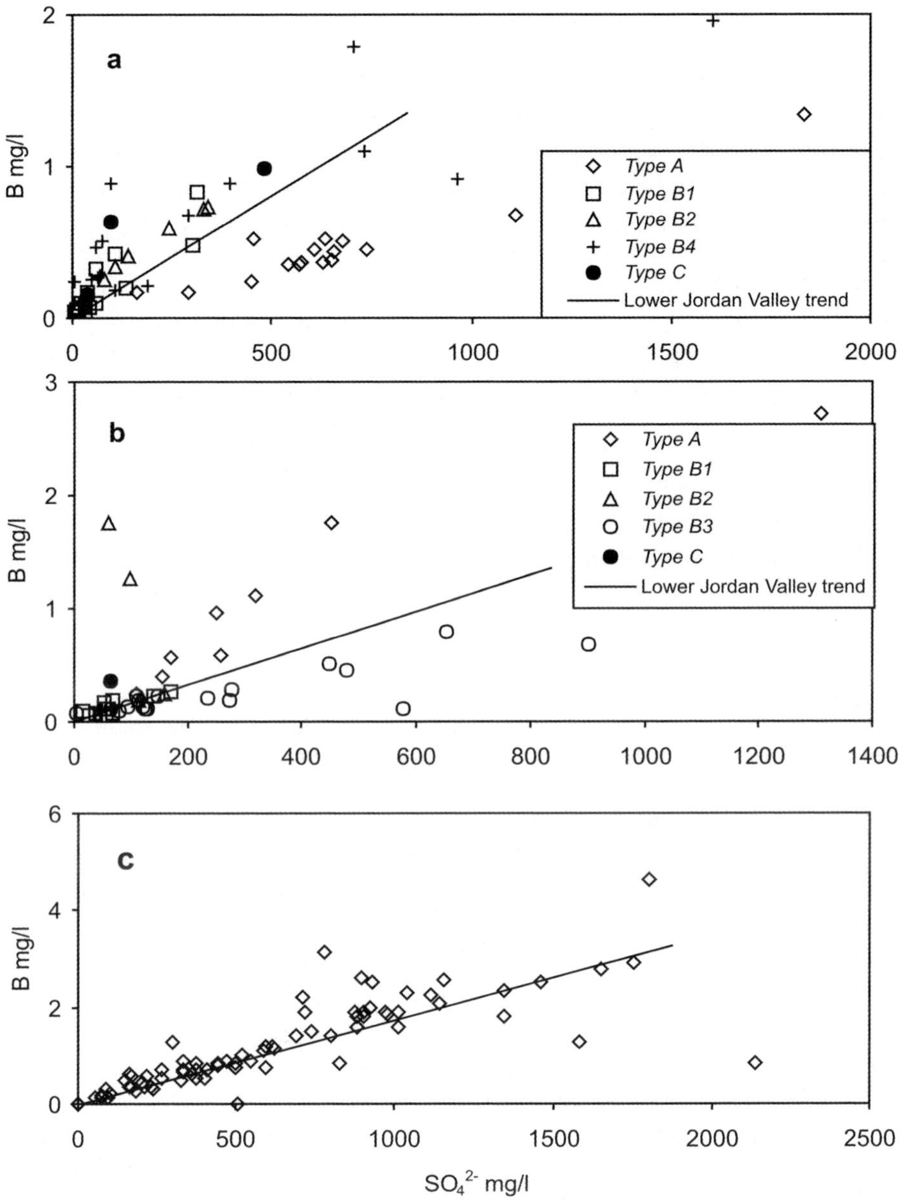

Fig. 3.1.1-14. Cross plots of B vs. SO_4^{2-} in groundwater. The shown line indicates the average value from (c). (a) and (b) compile samples from the western and eastern part of the Rift, respectively; (c): surface and groundwaters from the lower Jordan Valley; data are taken from Farber et al. (2004).

3.1.1.3 Stable isotopes of sulfate

General overview

Dissolved sulfate can be of atmospheric, pedospheric and/or lithospheric origin. Industrial waste products can constitute an additional source. Lithospheric sulfur sources interact with groundwater *in situ*, whereas all other sulfate sources are introduced to aquifers by recharge water. Atmospheric sulfate derived from marine sources shows $\delta^{34}S$ values between +15 and +21‰ (V-CDT). Other sources of atmospheric sulfur are usually depleted in $\delta^{34}S(SO_4^{2-})$ with respect to marine sulfur. $\delta^{34}S(SO_4^{2-})$ between 15 and 20‰ is in agreement with marine sulfur during the Cretaceous (Holser 1977). Lithospheric sulfur compounds are leached from igneous, metamorphic and sedimentary rocks. Igneous and especially sedimentary rocks are the most important sources of sulfates in groundwater. The most soluble and readily available sedimentary sulfur compounds are evaporitic sulfates such as gypsum and anhydrite. Throughout the history of the Earth, $\delta^{34}S$ of sulfate in seawater has changed (Holser 1977). Sulfates are not only restricted to evaporites because limestones often contain significant amounts of sulfate (Staudt et al. 1994) which can be released during water-rock interaction or weathering processes. Basaltic rocks show $\delta^{34}S$ values between 0 and +5‰. Lower values are mostly associated with the oxidation of sedimentary sulfides which had formed earlier as the result of sulfate reduction to H_2S by organic compounds.

The exchange of oxygen between sulfate and H_2O is extremely slow at low temperatures and at normal pH (Chiba and Sakai 1985) but is increasing with temperatures (>100°C) and/or low pH (Mitzutani and Rafter 1969). Sulfate concentrations in aquifers are effectively diminished by dissimilatory bacterial sulfate reduction which causes a preferential enrichment in $\delta^{34}S$ and $\delta^{18}O$ in the residual sulfate.

Evidence from the study area

On the western side of the Rift, a Ca/SO_4^{2-} molar ratio of about 0.9 and a low $\delta^{34}S(SO_4^{2-})$ ranging from 0 to 12‰ dominate, which indicates contributions by oxidation of igneous sulfides. The samples comprise waters from Lake Tiberias, the lower Jordan Valley and from the Elat 103 well in the Arava Valley. En Moda, Qalanit 2, and the Tiberias Main Spring are in the same range of $\delta^{34}S(SO_4^{2-})$ and show enhanced Ca/SO_4^{2-} ratios (Tables 3.1.1-1 and 3.1.1-2). Ca/SO_4^{2-} molar ratios reveal that *type A* water clusters at values ≈1, and that *types B1, B2* and *B4* show ratios between 5 and 120 in the western Rift. All waters from the eastern Rift show ratios between 3 and 5 (Table 3.1.1-2) essentially indicating less amounts of dissolved anhydride.

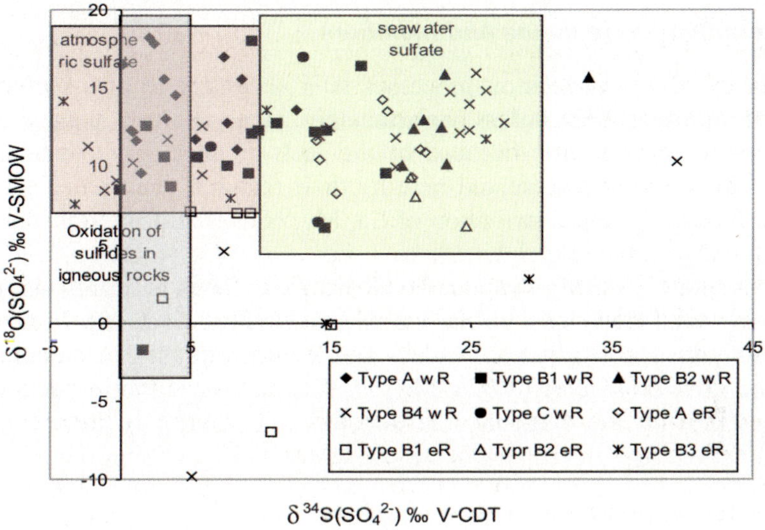

Fig. 3.1.1-15. Cross plots of $\delta^{18}O(SO_4^{2-})$ vs. $\delta^{34}S(SO_4^{2-})$ of groundwaters characterized by their spider patterns. The distribution of sample points elucidate that most of the sulfate is of evaporative and marine origin which is also the source of the atmospheric deposition. Additionally, sulfides in basalts are weathered and form sulfide in the range of 0–5‰. wR , eR = western and eastern Rift.

On the Jordanian side, water from the wells Wadi Araba 1 and Manshiya 2 shows the presence of sulfate from igneous sources. The lowest and the second highest $\delta^{34}S(SO_4^{2-})$ values are shown by waters from the wells Wadi Araba 1 and 5, respectively, which indicate extremes in mixing of sulfate from oxidation of sulfide and dissolution of gypsum. $\delta^{34}S(SO_4^{2-})$ <5‰ are typical for waters related to Tertiary-Quaternary volcanism. These are waters from the wells of Mukheiba, Manshiya, and Wadi Araba 1, which convincingly indicate hydrothermal alteration of limestone by gabbroic intrusions and/or basaltic effusives.

The cross plots of $\delta^{18}O(SO_4^{2-})$ vs. $\delta^{34}S(SO_4^{2-})$ (Fig. 3.1.1-15) show distinct clusters of the thermal waters of *type A* in the western and eastern part of the Rift. An equilibrium distribution of oxygen in sulfate and H_2O has almost been reached. All samples, with only few exceptions, show $\delta^{18}O(SO_4^{2-})$ values between +5 and +19‰. Particularly in the Arava Valley and the lower Jordan Valley $\delta^{34}S(SO_4^{2-})$ strongly indicate weathering of hidden mafic rocks (gabbros) which contributed sulfate by oxidation of sulfides in the past. These sulfates are nowadays leached. Sulfate values in the Rift lie mostly above 100 mg/l, which indicates a rock source rather than an atmospheric origin of the sulfate.

3.1.1.4 Identification of saline end-members

In order to evaluate salinization processes it is necessary to also include the brines from deep bore holes. Unfortunately, these analyses cannot be presented as spider patterns because of the lack of necessary elements. Therefore, all waters are presented here by their regional occurrence. The discussion focuses on the distribution of Ca, Mg, Na, Cl and Br concentrations and the Cl_{excess} (see Eq. 3.1.1-1).

The cross plot Ca vs. Mg reveals two clear mixing lines of highly saline end-members with freshwater in the Ca>Mg field (Fig. 3.1.1-16). In contrast, brines with Ca>Mg scatter widely and do not represent a common mixing line. The cross plot of (Ca+Mg) vs. Cl_{excess} shows that especially deep brines plot on the diagonal line denoting a $(Ca,Mg)Cl_2$ brine (Fig. 3.1.1-17).

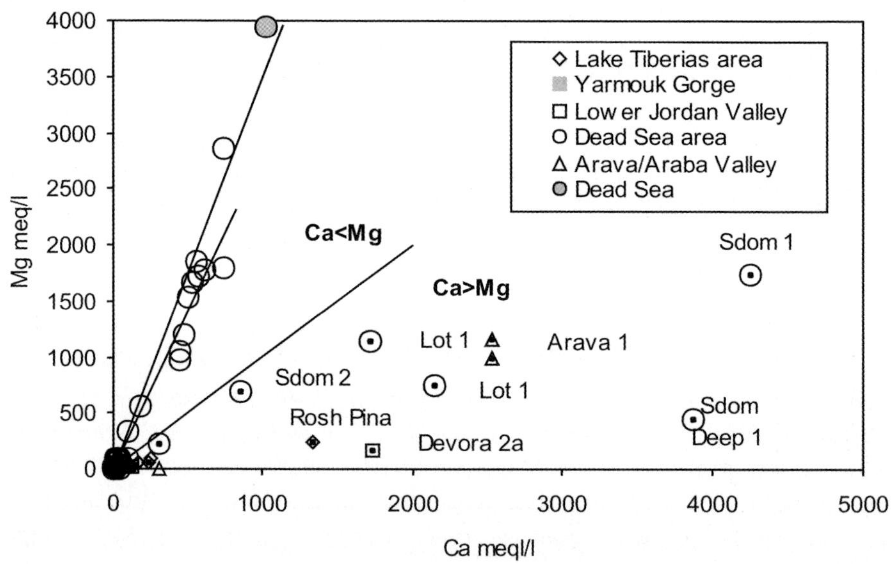

Fig. 3.1.1-16. Cross plots of Mg vs. Ca showing a wide scatter for Ca>Mg brines, which are deep brines. Sym,bols with centered dots indicate deep brines.

Fig.3.1.1- 18 elucidates that mainly the high-salinity waters represent the $(Ca,Mg)Cl_2$ brine, whereas the low-salinity waters are deficient in Cl^- content, which is balanced by sulfate and bicarbonate. Mg-rich but Ca-poor brines are strictly correlated with Cl_{excess} (Fig. 3.1.1-19).

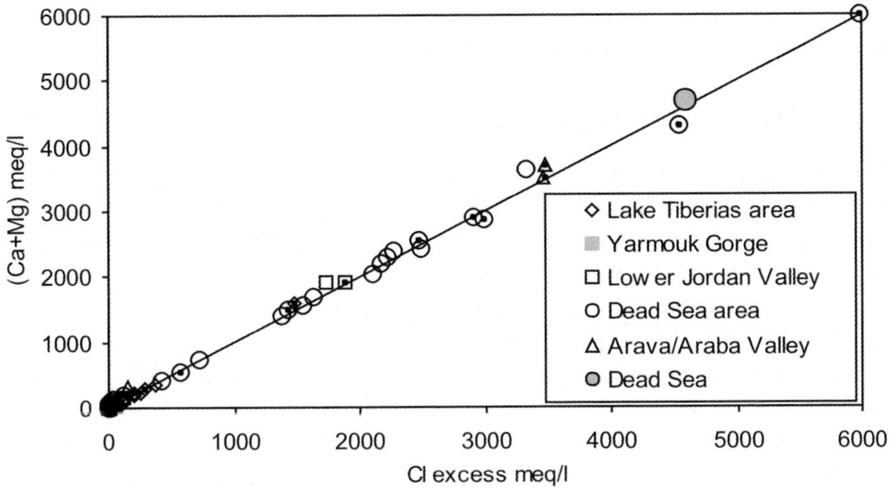

Fig. 3.1.1-17. Cross plots of 2(Ca+Mg) vs. Cl_{excess} Symbols with a centered dot indicate deep brines.

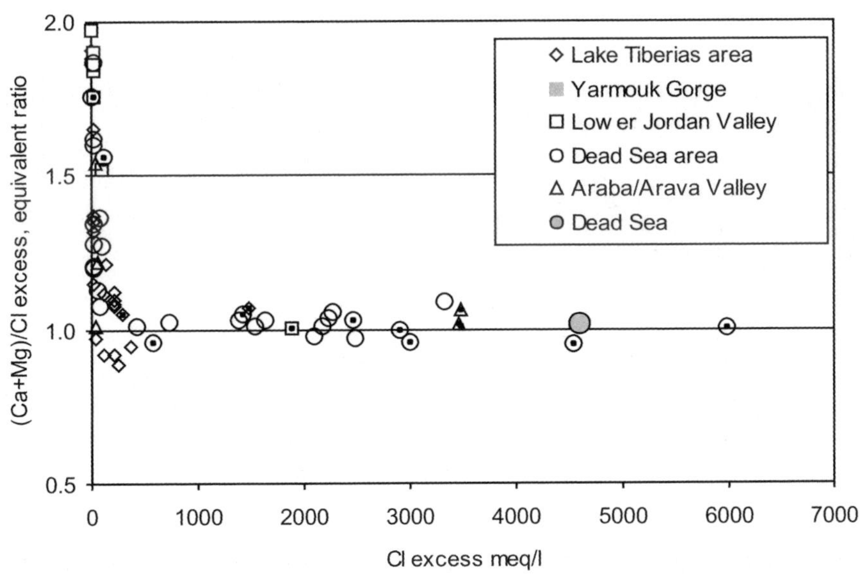

Fig. 3.1.1-18. Cl_{excess} vs. (Ca+Mg)/Cl; symbol with a centered dot indicate deep brines. The high-salinity waters yield nearly stoichiometric $(Ca,Mg)Cl_2$ brine.

Figs. 3.1.1-16 and 3.1.1-19 reveal several mother brines. The Dead Sea brine with the highest Mg/Ca ratio is the youngest brine of the area. The wide scatter of low Mg/Ca ratios may reflect different stages of development of brines by Mg-Ca exchange during dolomitization. In a broad sense, decreasing Mg/Ca ratios qualitatively reflect increasing "age".

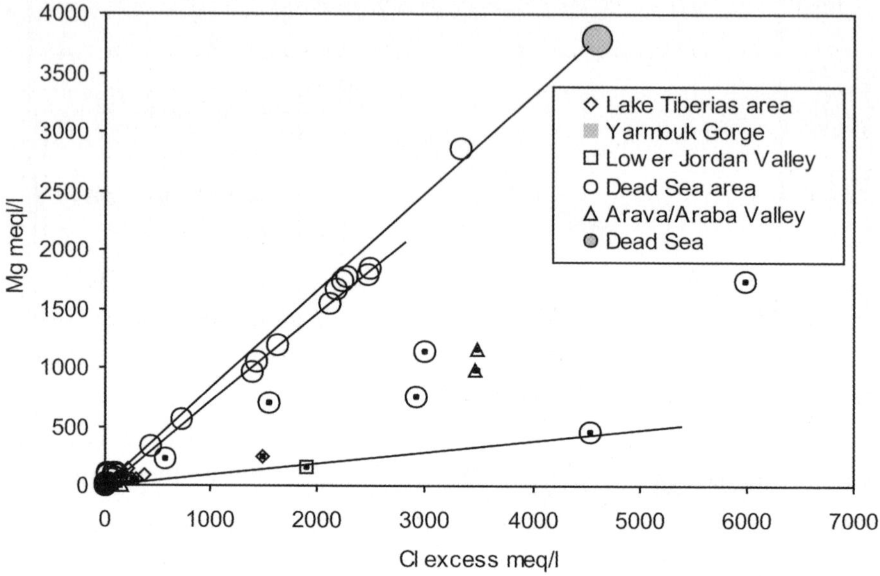

Fig. 3.1.1-19. Cross plots of Mg vs. Cl_{excess}. Symbols with a centered dot indicate deep brines.

The (Ca+Mg)/Cl_{excess} vs. Na/Cl ratios plot shows two aspects (Fig. 3.1.1-20): (i) The solid line represents the ideal mixing of the pure end-members, i.e., the NaCl- and (Ca,Mg)Cl_2 brines. (ii) All data plotting away from the line, indicate contribution of other dissolved components such as carbonates and sulfates by which the (Ca+Mg)/Cl_{excess} equivalent ratio is increased. The Na-rich Zemah brine plots at molar ratios of Na/Cl=1 and 2(Ca+Mg)/Cl_{excess} of about 0 represents an ablation brine of halite (Zemah 1). Fresh and saline waters with Na/Cl ratios >1 contain contributions of waters from basaltic areas.

Mg/Ca vs. Na/Cl

The cross plot of Mg/Ca vs. Na/Cl molar ratios shows three trends (Fig. 3.1.1-21). Modern seawater plots well outside the area covered by all water and brine analyses in the study area. The Na/Cl ratio of seawater is marked by the broken line. Left to this line, the trend of Mg/Ca ratios increases with decreasing Na/Cl ratios, which represents the evolution of brines produced by evaporation. This trend represents the development of the brines of the Lisan period with samples from the Dead Sea which plot at the outmost point. Although it could be interpreted as such, the trend line does not necessarily represent a mixing line of the Dead Sea brine with waters of a low Mg/Ca and Na/Cl ratios such as seawater.

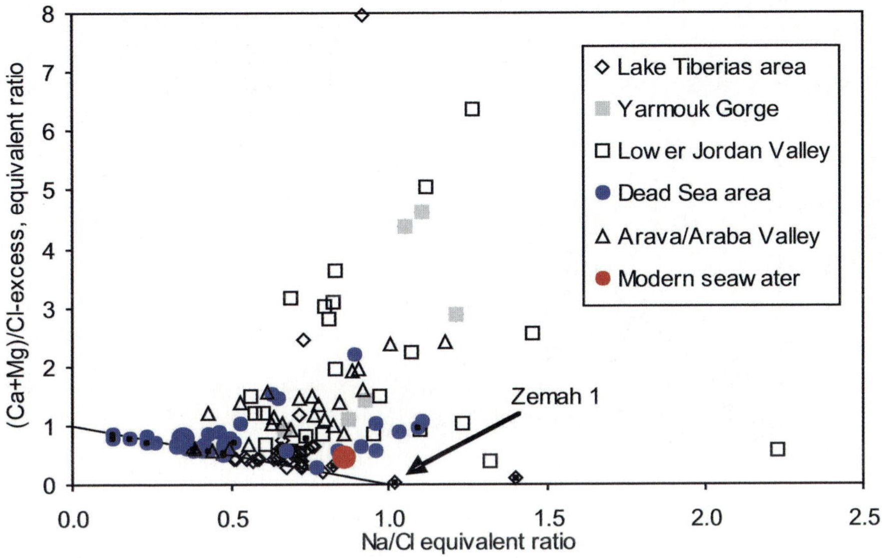

Fig. 3.1.1-20. Na/Cl vs. (Ca+Mg)/Cl_{excess}. Symbols with a centered dot indicate these samples as being deep brines.

The second trend left to the broken line in Fig. 3.1.1-21 is directed towards both low Mg/Ca and Na/Cl ratios and is much more diffuse, meeting many of the deep-seated brines towards the lower end. The decrease of Mg/Ca ratios can be explained by considerable alteration reactions with host rocks. Mg-Ca exchange processes such as dolomitization (Table 3.1.1-3) would shift any saline water towards low Mg/Ca ratios, although concomitant precipitation of calcite and of gypsum buffers this process to a certain degree. The amount of dolomite formed depends on the porosity of the limestone and the Ca and Mg concentration of the primary brine. Assuming a porosity of 30% and an initial Mg content of 2 mol/kg, only 2% of the limestone is dolomitized, which is not detectable by standard XRD. Thus, the limestone is not significantly altered under the assumed static conditions. Due to active Plio-Miocene volcanism south of Lake Tiberias (Rosenthal et al. 2006), Na is consumed by reactions of the anorthite component of plagioclase and pyroxenes, which leads to the formation of albite (spilitization; Table 3.1.1-3). "Brucite" is precipitated as part of chlorite (Table 3.1.1-3), thereby vertically lowering the Mg/Ca ratios. Oxidation of sulfides supports the precipitation of gypsum, thereby increasing the

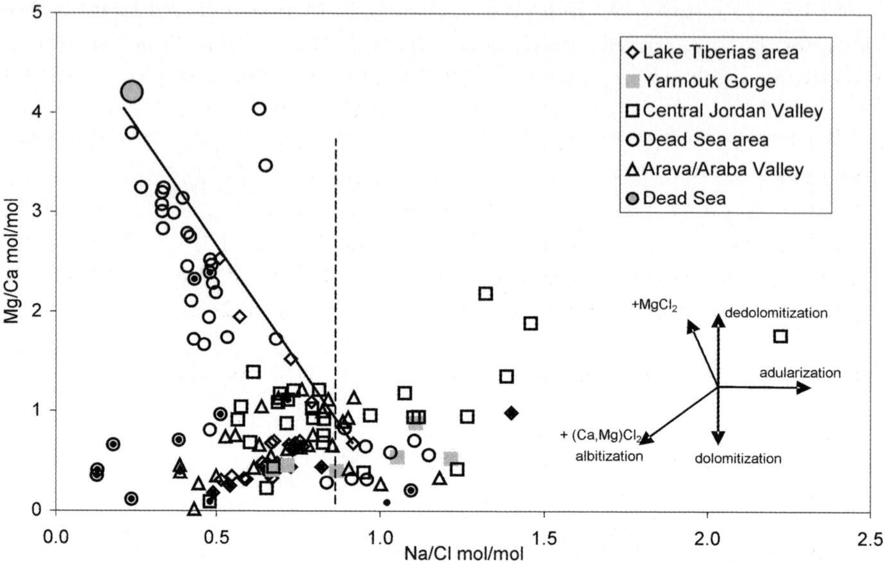

Fig. 3.1.1-21. Mg/Ca vs. Na/Cl. Symbols with a centered dot indicate deep brines.

Mg/Ca ratio. Adularization (Table 3.1.1-3) lowers the Na/Cl ratio. All these reactions take place under subsurface conditions. The trends of Mg/Ca brines in Fig.3.1.1-21 cannot be explained (but may be varied) by the dissolution of carnallite ($KMgCl_3.6H_2O$), bischofite ($MgCl_2.6H_2O$) and/or kainite ($MgSO_4 \cdot KCl \cdot 3H_2O$), because both salts would result in an upward sub-vertical shift of data points. The dissolution of kieserite ($MgSO_4 \cdot H_2O$) also results in an upward vertical shift. Waters with Na/Cl>1 largely originate from interaction with basalts.

All saline waters to the right of the Na/Cl seawater ratio are either affected by ablation of halite (up to Na/Cl~1), or by mixing with Na/Cl>1 solutions as obtained from weathering of basalts, or adularization of albite The last two processes only occur in the presence of mafic matter (basalts, gabbros, mafic tuffs).

Figs. 3.1.1-16–18 and 3.1.1-20 suggest that the $(Ca,Mg)Cl_2$ brines have developed continuously from evaporating lagoonal lakes which also are the ultimate sinks of drainage water from the surrounding areas. This process yielded the Mg>Ca type of primary brines, whereas the Ca>Mg type resulted from brine-rock interaction. This explains why Cl_{excess} is much better correlated with Mg (close to the primary brine) than with Ca (Fig. 3.1.1-19) (tertiary brines). The regression line in Fig. 3.1.1-17 exactly yields the composition of a $(Ca,Mg)Cl_2$ brine, which also indirectly proves the Mg-Ca exchange by alteration reactions, which does not have an influence on the sample distribution along this line.

A mixing of the two types of brines may also occur. The higher the TDS, the heavier is the brine. Such heavy brines replaced lesser saline ones in the subsurface. Initially, all brines are dominated by Mg. Whether they become Ca-dominated depends on the dolomitization of the limestone and/or the albitization of basaltic rocks. If albitization played a considerable role, the linearity in Fig. 3.1.1-17 would not be disturbed in Ca-dominated brines, because by this process Ca is released and the decrease of Na results in an equivalent increase in the Cl_{excess}. However, the samples, which underwent albitization, would plot below the line in Fig. 3.1.1-21, because both Na/Cl and Mg/Ca molar ratios decrease.

Table 3.1.1-3. Typical alteration reactions of groundwater with sediments and rocks.

Alteration reactions
Dolomitization of limestones
$2\ CaCO_3 + Mg^{+2} \Leftrightarrow CaMg(CO_3)_2 + Ca^{2+}$
calcite — dolomite
Albitization of mafic rocks
$CaAl_2Si_2O_8 + 2\ Na^+ + 4\ H_4SiO_4 \Leftrightarrow 2\ NaAlSi_3O_8 + Ca^{2+} + 8\ H_2O$
anorthite — albite
Chloritization of Neogene gabbroic intrusions
$2\ K(Mg)_3[AlSi_3O_{10}(OH)_2] + Mg_2Si_2O_6 + 2\ Mg^{2+} + 2\ CaAl_2Si_2O_8 + 6\ H_2O + 2\ Na^+ \Leftrightarrow$
biotite — pyroxene — anorthite
$2\ (Mg_2Al)[AlSi_3O_{10}(OH)_2]*Mg_3(OH)_6 + 2\ NaAlSi_3O_8 + 2\ Ca^{2+} + 2\ K^+$
clinochlore — albite
or
$CaAl_2Si_2O_8 + 2\ Mg_2Si_2O_6 + 10\ H_2O + Mg^{2+} \Leftrightarrow$
anorthite — pyroxene
$(Mg_2Al)[AlSi_3O_{10}(OH)_2]xMg_3(OH)_6 + Ca^{2+} + 3H_4SiO_4$
clinochlore
Adularization of albite
$NaAlSi_3O_8 + K^+ \Leftrightarrow NaAlSi_3O_8 + Na^+$
albite — adularia
Sulfate reduction by organic matter
$SO_4^{2-} + CH_4 \Leftrightarrow CO_3^{2-} + H_2S + H_2O$

Br/Cl vs. Na/Cl

The diffuse trend of an increasing Br/Cl ratio with decreasing Na/Cl ratios (Fig. 3.1.1-22) contrasts the sharp Dead Sea Rift trend line defined by Klein-BenDavid et al. (2004). Evaporation beyond the beginning of halite precipitation decreases Na/Cl but increases Br/Cl. Evaporation of seawater, however, does not yield a straight line because of precipitation of diverse salts (Fontes and Matray 1993). This seawater evaporation trend borders the data cloud towards the left with only a few samples plotting further left. Few of the deep brines plot along the seawater evaporation curve and only Sdom 1 plots below. The scatter of samples may be due to mineral-brine interaction, mixing of brines with freshwater, and less to analytical uncertainties, particularly of low Br.

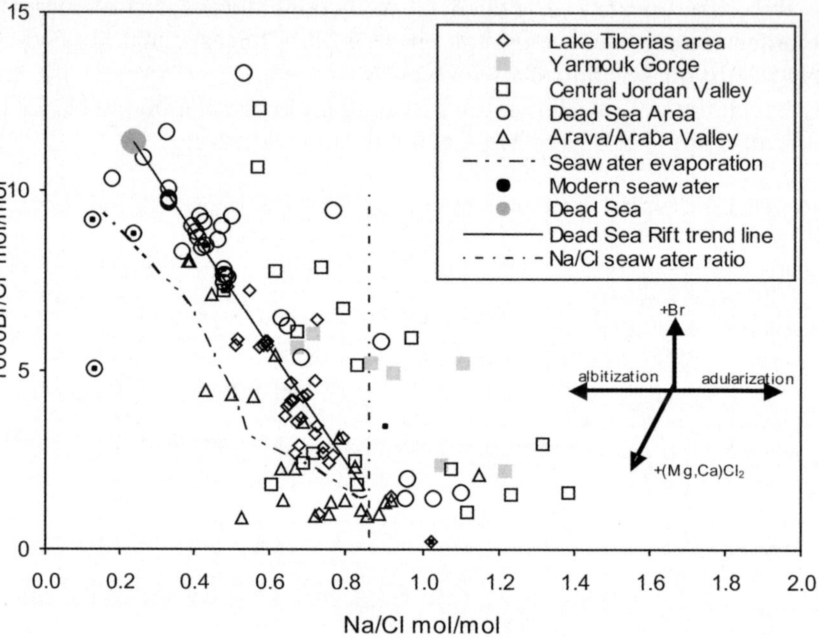

Fig. 3.1.1-22. Br/Cl vs. Na/Cl. Symbols with a centred dot indicate deep brines.

The large scatter of samples suggests that various processes are involved in generating these saline groundwaters and brines. The Dead Sea and Sdom brines could be residual primary brines. They are characterized by high Br/Cl ratios and cannot result from the dissolution of evaporites. The Zemah brine with a very low Br/Cl ratio and Na/Cl about 1 clearly represents dissolution of halite (secondary brine). Data points to the right of Na/Cl≈1 can only be explained by an increase of Na which is not associated with Cl. This could result from dissolution of minerals such as $NaHCO_3$ and/or Na_2SO_4, the latter being formed by oxidative weathering of basaltic rocks. For example, the samples from Haon, En Makla, En Reah, Gofra, Tiberias Hot Springs, and many others (Fig. 3.1.1-23) plot in a horizontal array. Less pronounced in its horizontal distribution is the water from Lake Tiberias which was sampled at various places and depths over a period of three years (Chap. 3.2.1). All these sources largely gain their water either from recharge over basaltic areas or by subsurface interaction with gabbros at enhanced temperature.

A shift along a diagonal line in Fig. 3.1.1-22 could be due to the taking up of a hypothetical $(Ca,Mg)Cl_2$ component. Na exchanged for Ca, albitization horizontally shifts the Na/Cl ratio towards the left. The dissolution of $MgCl_2$-salts such as bischofite ($MgCl_2.6H_2O$) and carnallite ($KMgCl_3.6H_2O$) decreases Na/Cl and Br/Cl ratios, but the shift direction depends on the content of dissolved halite.

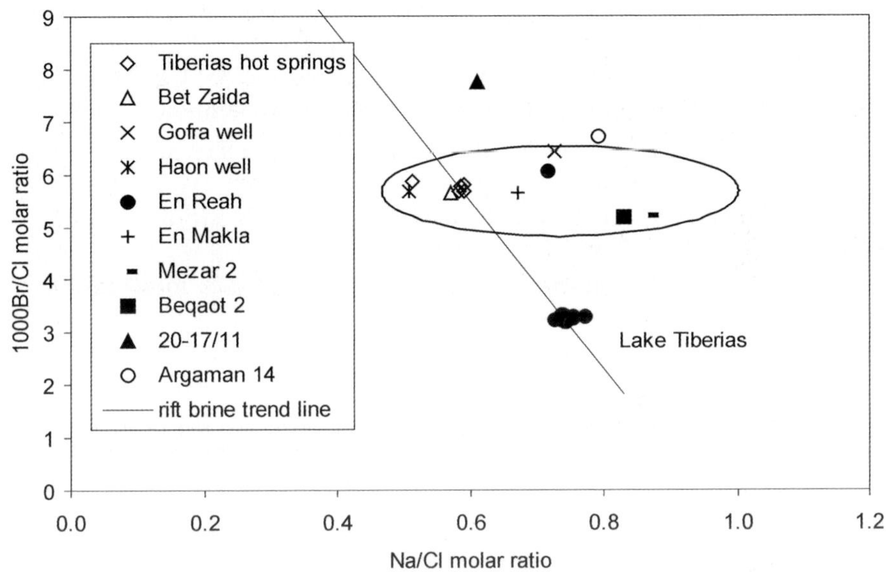

Fig. 3.1.1-23. Br/Cl vs. Na/Cl; selected examples of groundwaters affected by basaltic matter.

Although it could be interpreted as such, the given trend line does not necessarily represent a mixing line of Dead Sea water with altered seawater. A similar trend is obtained when precipitation and evaporation take place in a closed inland basin over a long time.

3.1.1.5 Salinization of groundwater

All evaporation brines are dominated by Mg. As long as these brines are in contact with the atmosphere, the Ca content is controlled by CO_2. Being saturated with aragonite, gypsum and halite, the Dead Sea brine is a typical representative of such processes. Therefore, any brine with a higher Ca+Mg content than the Dead Sea cannot have derived from it. Such saline fluids contain large contributions of ancient brines. In varying proportions, the hot springs and the hot well waters along the Dead Sea represent mixtures of primary brines such as pre-Dead Sea and Dead Sea brines with tertiary $(CaMg)Cl_2$ brines (probably of the Sdom Sea period) or secondary ablation brines (halite and gypsum). Ca/Mg ratios vary locally as a result of dolomitization and of gypsum dissolution. In the Arava Valley, salinization is controlled by a common brine with a (Ca+Mg)/Na ratio of about 1, with the exception of the local occurrences Elat 103 and Zofar 20 which show ratios of about 0.6. A ratio of about 1 seems to be typical for saline waters from the Kurnub and Judea Group. The groundwater of the Jordan Valley is controlled by a brine with (Ca+Mg)/Na $\approx$ 0.2 which turns out to be the lowest ratio in the Rift. In the Lake Tiberias region, Ca > Mg brines with (Ca+Mg)/Na ranging from 0.3 to 0.5 are most common. However, Mg>Ca brines with (Ca+Mg)/Na $\approx$ 0.25 represent a diluted residual evaporation brine, largely of marine origin, which is supposed to be a remnant of the Sdom Sea period but with enhanced Na content due to the ablation of halite.

As pointed out, the increase of the Ca/Mg ratio either indicates an increase in age of the brine (and then all high Ca/Mg brines would be older than those derived from the Sdom Sea period) or these brines have met with more favorable conditions for dolomitization such as higher temperatures. It is suggested that the presence of Ca > Mg waters in sources located along the western shores of Lake Tiberias result from a thermal overprinting of the limestone by the Mio-Pliocene volcanic activity which probably enforced dolomitization.

In summary, several generations of brine can be traced by chemical and isotopic means:

- Primary evaporation brines of the Dead Sea type: Mg > Ca
- Primary evaporation brines of the Sdom Sea period (Haon type): Mg > Ca
- Secondary ablation brines high in Na+K: Zemah type, and
- Tertiary brines with Ca > Mg either of the pre-Sdom- or Sdom- or even Lisan period

In variable proportions, these brines have to be considered the sources of the salinization of groundwater in the Rift.

References

Bentor YK (1969) On the evolution of subsurface brines in Israel. Chem Geol 4:83–110

Braitsch O (1971) Salt deposits. Their origin and composition. Springer Berlin Heidelberg New York, pp 1–297.

Bruland KW, Lohan MC (2004) Controls of trace metals in seawater. In: Holland H.D. and Turekian K.K. (eds) Treatise on geochemistry, Vol 6, Elsevier Amsterdam, pp 23–47

Chiba H, Sakai H (1985) Oxygen isotope exchange rate between dissolved sulfate and water at hydrothermal temperatures. Geochim Cosmochim Acta 49:993–1000

Farber E, Vengosh A ,Gavrieli I, Marie A, Bullen TD, Mayer B, Holtzman R, Segal M, Shavit U (2004) The origin and mechanism of salinization of the Lower Jordan River Geochim Cosmochim Acta 60:1989–2006

Fleischer E, Goldberg M, Gat JR, Magaritz M (1977) Isotopic composition of formation waters from deep drillings in southern Israel. Geochim Cosmochim Acta 41:511–525

Fleischer L, Varshavsky A (2002) A lithostratigraphic data base of oil and gas wells drilled in Israel. Earth Science Research Administration, Oil and Gas Unit Rep.OG/9/02, Jerusalem, pp 1–280

Fontes JCh, Matray JM (1993) Geochemistry and origin of formation brines from the Paris basin, France. 1. Brines associated with Triassic salts. Chem Geol 109:149–175

Greitzer Y (1980) Hydrodynamic investigation of geological formations in Israel to the purpose of oil exploration – stage 1, Vols 1–2. Rep. to the Israel National oil Co., Ltd., pp 1-210

Herut B (1992) The chemical composition and sources of dissolved salts in rainwater in Israel. PhD Thesis, Hebrew University of Jerusalem (in Hebrew, English abstract)

Holser WT (1977) Catastrophic chemical events in the history of oceans. Nature 267:403–408

Johannesson KH, Farnham IM, Guo C, Stetzenbach KJ (1999) Rare earth element fractionation and concentration variation along a ground water flow path within shallow, bassin fill aquifer, southern Nevada, USA. Geochim Cosmochim Acta 63:2697–2708

Johannesson KH, Zhou X, Guo C, Stetzenbach KJ, Hodge VF (2000) Origin of rare earth element signatures in groundwaters of circumneutral pH from southern Nevada and eastern Califiornia, USA. Chem Geol 164:239–257

Kinsman DJJ, Holland HD (1969) The co-precipitation of cations with $CaCO_3$–IV. The co-precipitation of Sr^{2+} with aragonite between 16° and 96°C. Geochim Cosmochim Acta 33:1–17

Klein-BenDavid O, Sass E, Katz A (2004) The evolution of marine evaporitic brines in inland basins: The Jordan Dead Sea Rift Valley. Geochim Cosmomochim Acta 68:1763–1775

Mazor E, Mero F (1969) Geochemical tracing of mineral and freshwater sources in the Sea of Galilee Basin, Israel. J Hydrol 7:276–317

Menashe S (1991) The deep aquifers in the northern part of Israel. MSc Thesis, The Hebrew University, Center for Soil and Water Sciences, Rehovot, Israel.

Mizutani Y, Rafter TA (1969) Oxygen isotopic composition of sulfates, Part 3: Oxygen isotopic fractionation in the bisulfate ion-water system. NZJ Sci 12: 54–59

Möller P (2002) The distribution of rare earth elements and yttrium in water-rock interactions: field observations and experiments. In I. Stober and K. Bucher (eds) Water-rock interaction, Kluwer Acad. Press, Rotterdam, pp 97–123

Möller P, Dulski P, Salameh E, Geyer S (2006a) Characterisation of the sources of thermal spring- and well waters in Jordan by rare earth element and yttrium distribution and stable isotopes of H_2O. Acta Hydrochim Hydrobiol 34:101–116

Möller P, Geyer S, Salameh E, Dulski P (2006c) Sources of mineralisation and salinization of thermal groundwaters of Jordan. Acta Hydrochim Hydrobiol 34:86–100

Möller P, Rosenthal E, Geyer S (2006b) Chemical evolution of saline water in the Jordan-Dead Sea Transform and in adjoining areas. Intern J Geol Sci 96:541–566

Möller P, Rosenthal E, Geyer S, Guttman Y, Dulski P, Rybakov M, Flexer A (2007) Hydrochemical processes in the Lower Jordan Valley and in the Dead Sea area. Chem Geol 239:27–49

Möller P, Rosenthal E, Dulski P, Geyer S, Guttman Y (2003) Rare earths and yttrium hydrostratigraphy along the Lake Kinneret- Dead Sea-Arava transform fault, Israel. Appl Geochem 18:1613–1628

Rimawi O, Salameh, E (1988) Hydrochemistry and groundwater system of the Zerka Ma'in thermal field, Jordan. J Hydrol 98:147–163

Rosenthal E (1987) Chemical composition of rain and groundwater in replenishment areas of the Bet Shean-Harod multiple aquifer system. Israel J Hydrol 89:329–352

Rosenthal E (1997) Thermosaline water of Ca-chloride composition: diagnostics and brine evolution. Environ Geol 32:245–249

Rosenthal E, Flexer A, Möller P (2006) The paleoenvironment and the evolution of the Jordan-Dead Sea transform and in the adjoining areas. Int J Earth Sci 95:725–740

Rosenthal E, Jones BF, Weinberger G (1998) The chemical evolution of Kurnub Group paleowater in the Sinai-Negev province - a mass balance approach. Appl Geochem 27:1–17

Salameh E (2001) Sources of water salinities in the Jordan Valley area/Jordan. Acta Hydrochim Hydrobiol 29:329-362

Starinsky A (1974) Relationship between Ca-chloride brines and sedimentary rocks in Isreal. PhD thesis, Hebrew Univ. Jerusalem

Staudt W, Reeder RJ, Schoonen MAA (1994) Surface structural control on compositional zoning of SO_4^{2-} and SeO_4^{2-} in synthetic calcite single crystals, Geochim Cosmochim Acta 58:2087–2098

Usdowski E (1973) Das geochemische Verhalten des Strontiums bei der Genese und Diagenese von Ca-Karbonat- und Ca-Sulfat-Mineralen. Contrib Mineral Petrol 38:177–195

Veizer J (1969) Reactions controlling strontium abundance in natural waters. In Wedepohl KH (ed.) Handbook of Geochemistry II/4, pp 38-H-1–38-H-13.

Veizer J (1978) Simulation of limestone diagenesis: a model based on strontium depletion: Discussion. Can J Earth Sci 15:1683–1685

Vengosh A, Rosenthal E (1994) Saline groundwater in Israel: Its bearing on the water crisis in the country. J Hydrol 155:103–124

Yechieli Y, Starinsky A, Rosenthal E (1992) Evolution of brackish groundwater in a typical arid region: Northern Arava Rift Valley, southern Israel. Appl Geochem 7:361–374

3.1.2 Characterization of recharge areas by rare earth elements and stable isotopes of H_2O

P. Möller, E. Rosenthal, P. Dulski, S. Geyer

3.1.2.1 Introduction

Wide use is made of the distribution of rare earth elements (REE) and yttrium (Y) (henceforth combined to REY) in order to trace geochemical processes. This method is fairly wide-spread in petrology, ore deposit research, and chemical oceanography, but is relatively new in groundwater studies (Smedly 1991; Banks et al. 1999; Johannesson et al. 1999; 2000; Möller 2000; 2002; Möller et al. 2003a). Conventionally, the REY abundance is presented as distribution patterns normalized to C1-chondrite (Anders and Grevesse 1989). Here, yttrium is included although it is not an REE. Due to the same ionic charge and size in igneous rocks Y behaves exactly like Ho (Bau and Dulski 1995). In aqueous systems, however, Y and Ho are fractionated and Y behaves "anomalously" due to small differences in the solubility products of the hydroxides (Diakonov et al. 1998). The abundance of REY and their distribution patterns are typical for each type of rock and sediment and the anomalous behavior of Ce, Eu, Gd, and Y characterizes processes by which REY are fractionated. In water-rock interaction the most soluble REY-bearing major and accessory minerals determine the REY patterns of the groundwater. Therefore, studying REY distribution in groundwater yields insight into processes by which the REY abundance is controlled (Möller et al. 2003a; 2006).

Stable isotopes of H_2O are useful tools in identifying water bodies which originate from precipitation at different temperatures (latitude effect; Craig 1961; Dansgard 1964), different altitudes (altitude effect: Poage and Chamberlain 2001), and different distances from oceans (continental effect: Dansgaard 1964). Variation also occurs due to seasonal changes of temperature, evaporation rate, source areas along air-mass trajectories and amount of rainfall (amount effect: Gat 1996). All these effects together determine where a water sample plots on the meteoric water line. Thus, modern meteoric and fossil waters (paleo-recharge) and recharge at different altitudes can be distinguished.

The relationship between δ^2H and $\delta^{18}O$ in world-wide precipitation yields the global meteoric water line (GMWL; Craig 1961; Dansgard 1964) which reflects the latitude effect. The local or eastern Mediterranean water line (EMWL; Gat and Carmi 1970), which is considerably different from the Pleistocene or paleo-meteoric water line (palMWL; Gat 1984; Issar et al. 1984; Abd El Samie and Sadek 2001) is much more interesting.

Several processes cause water to deviate from the local meteoric water line. Thus, evaporation, which is a non-equilibrium process, enriches the heavy isotopes in the residual water body. This is the result of humidity, temperature, salt concentration, and of many other factors (Gat and Gonfiantini 1981; Gat 1996). An exchange of oxygen between H_2O and rock minerals only occurs in geologic times but is enhanced at high temperatures. Thus, $\delta^{18}O$ increases in hydrothermal systems. Contrarily, $\delta^{2}H$ does not change, since the hydrogen content in solid rocks is low and only presented by –OH-group-bearing minerals such as micas, chlorites and amphiboles. $\delta^{18}O$ in water decreases in common diagenetic reactions such as silicate hydrolysis, which produces hydrated minerals, whereas $\delta^{2}H$ decreases only insignificantly. These types of water samples plot left to the local meteoric water line. Methane-producing bacterial decomposition of organic matter enhances $\delta^{2}H$ in water. Again, these water samples plot left to the local meteoric water line.

Radioactive tritium is a measure for the residence time of water in the aquifer. Tritium is produced in the upper atmosphere and enters the aquifers by precipitation. The decay half time for tritium is 12.3 years so that measurable tritium contents are only expected in water which is younger than 40 years. The precise age of water can only be derived, if the water in question is not mixed with old water and if determining factors such as recharge, the water's time-dependent concentration of tritium and the predominant flow in the aquifer regime are known.

3.1.2.2 Grouping groundwater by REY patterns

For this study, published (Möller et al. 2003a, 2006) and unpublished REY analyses and isotope analyses of $\delta^{2}H$, $\delta^{18}O$, and ^{3}H were compiled (Table 3.1.2-1). Fig. 3.1.2-1 shows the distribution of sampling points in the Rift. The necessary chemical pre-concentration of REY is described for less and very saline water by Möller et al. (2003b, 2005), respectively. Irrespective of the geographical position of sampling points, REY patterns in groundwater are grouped according to their characteristic trends. The groups of REY patterns typify processes associated with water-rock interaction in different lithological environments which are characterized by the dominance of a typical suite of leachable REY-bearing minerals. For comparison, some REY patterns of the most common rocks in the study area are shown in Fig. 3.1.2-2. From leaching experiments (Möller 2002; Möller et al. 2003a) it is known that due to the dissolution of calcite only

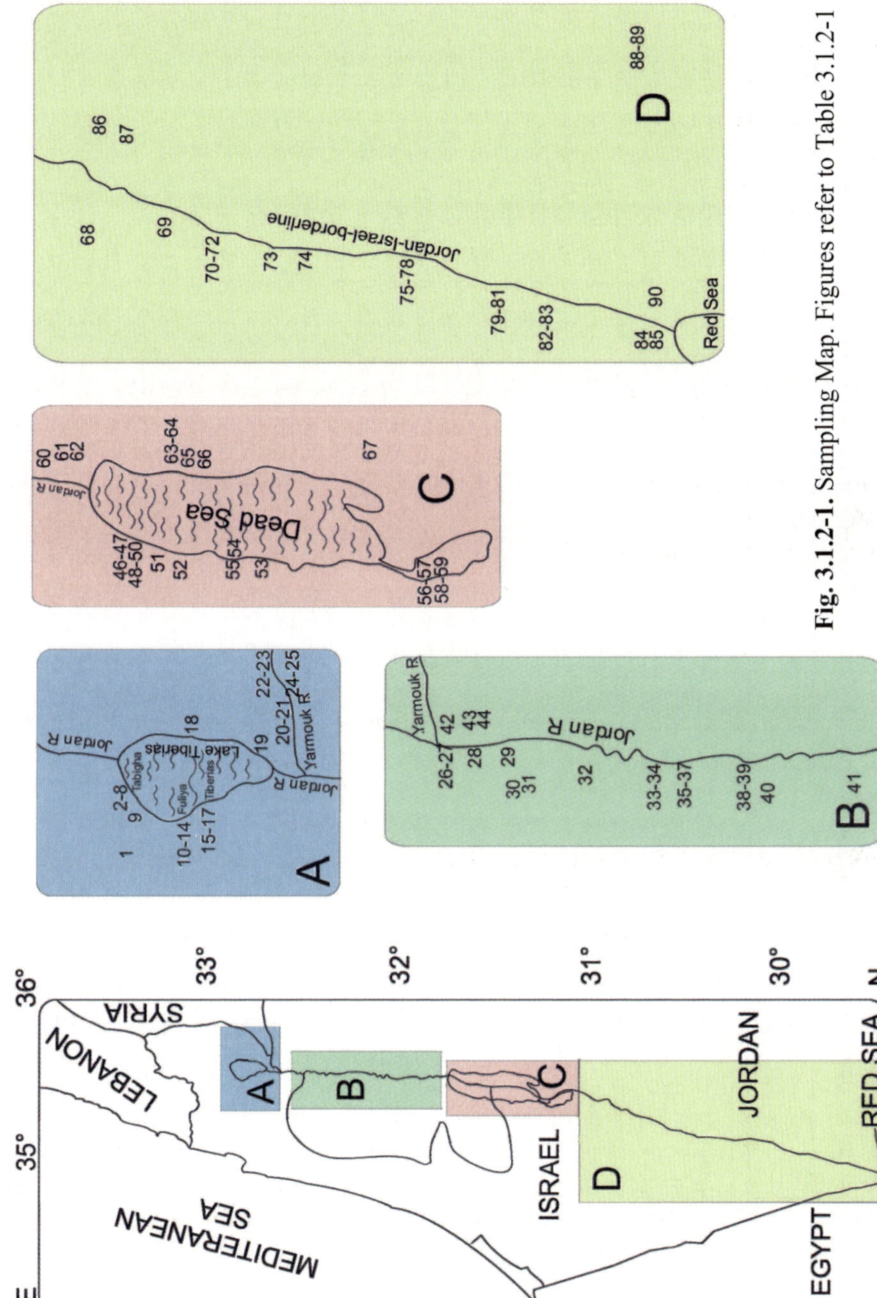

Fig. 3.1.2-1. Sampling Map. Figures refer to Table 3.1.2-1

Table 3.1.2-1. Compilation of chemical analyses of rare earth elements and isotopes of H_2O. Missing values are below the limit of detection or could not be determined due to high interferences by molecular ions in ICP-MS analysis.

Sample	Station	Temp.	Y	La	Ce	Pr	Nd	Sm	Eu	Gd	Tb	Dy	Ho	Er	Yb	Lu	^{18}O	^{2}H	^{3}H
ID		°C							pmol/L								‰	‰	T.U.
1	Well Qalanit 2	25.5	19.8		13.1	1.5	10.0	1.43	0.34	1.26	0.20	1.24	0.24	0.80	0.66	0.09	-5.22	-24.5	0.17
2	Tabigha Druzi Spring	18.3	86.3	14.7	37.1	4.5	24.1	4.65	1.46	5.83	0.84	5.15	1.14	3.62	3.41	0.57	-3.08	-16.8	2.9
3	Tabigha En Sheva	25.2	168	46.2	22.4	9.8	45.7	7.60	2.34	9.90	1.37	8.98	1.96	5.96	4.60	0.69	-5.89	-28.1	2.1
3	Tabigha En Sheva	24.8	126	28.3	31.2	6.5	28.1	5.26	1.68	7.06	0.99	6.58	1.52	4.63	3.83	0.60	-5.95	-27.0	3.2
4	Tabigha Sartan spring (saline)	28.8	104	32.1	41.3	8.3	38.4	6.76	1.95	8.08	1.13	6.68	1.43	4.02	3.06	0.45	-5.36	-26.8	0.8
5	Tabigha Sartan spring (fresh)	27.8	132	33.3	11.6	7.2	35.0	5.92	1.84	7.63	1.11	7.14	1.60	4.73	3.66	0.57	-5.54	-26.2	1.9
6	Tabigha Kinneret 7	24.3	184	47.3	17.3	10.0	45.1	7.86	2.37	10.3	1.43	9.36	2.12	6.41	4.96	0.74	-5.84	-27.9	2.6
6	Tabigha Kinneret 7	24.0	133	26.8	46.3	6.2	28.2	5.42	1.69	7.66	1.10	7.01	1.66	5.07	4.26	0.70	-6.06	-27.1	4.2
7	Tabigha Kinneret 8	27.8	14.2		13.6		9.0	2.30	0.83	2.84	0.34	1.65	0.27	0.61	0.37	0.05	-2.26	-11.1	0.44
8	Tabigha Sapir site, En Tina	19	74.1	14.7	48.4	4.5	22.9	4.31	1.22	4.94	0.80	4.95	1.04	3.22	2.82	0.43			
9	Huqoq well	18.6	14.1		11.2	1.4	8.4	1.39	0.34	1.23	0.19	1.08	0.22	0.65	0.65	0.11	-4.83	-22.0	2.3
10	Fuliya D 906	25	15.4		13.5		9.4	1.57	0.43	1.61	0.21	1.27	0.24	0.66	0.50	0.071	-4.87	-22.6	
10	Fuliya D 906	25	5.8	14.3	21.9	3.2	6.6	0.04	0.06	0.34	0.06	0.30	0.06	0.19	0.17	0.027	-4.7	-19.8	2.1
11	Fuliya A, well	28					5.0	0.55	0.09	0.49	0.05	0.36	0.06	0.15	0.10		-4.02	-11.1	0.53
11	Fuliya A well	24.4	1.8		1.6	0.2				0.12		0.11			0.10		-4.51	-19.4	1.1
12	Fuliya B well	30	7.3	2.3	3.3	0.4	1.3	0.12	0.07	0.35		0.30	0.08	0.28	0.24	0.03	-4.76	-21.3	1
13	Fuliya Kinneret 5	25	25.3		20.6	2.1	11.4	1.43	0.40	1.69	0.25	1.63	0.35	1.15	0.99	0.15	-4.12	-21.1	0.31
13	Fuliya Kinneret 5	27.9	8.1		1.2	0.2	1.3	0.38	0.14	0.50	0.12	0.57	0.12	0.41	0.40	0.072	-5.24	-24.7	2.1
14	Fuliya Kinneret 10 B	27					4.9	2.07		0.33	0.04	0.26	0.04	0.13	0.10	0.025	-1.4	-5.10	0.09
14	Fuliya Kinneret 10 B	36.3	53.4	12.3	31.2	4.3	20.0	5.14	1.36	6.04	0.89	5.12	1.07	3.10	2.78	0.46	-1.7	-5.70	0.7
15	Tiberias, well Qikar	34	56.5		17.3	1.2	10.6	6.55	1.99	6.87	0.84	4.31	0.79	1.92	1.14	0.16	1.8	12.9	2.8
15	Tiberias, well Qikar	40.8	24.0		1.3	0.1		2.48	0.41	1.64	0.22	1.18	0.26	0.68	0.53	0.08	2.6	26.6	2.8
15	Tiberias, well Qikar	37.2	11.9	9.9	13.0	1.2	4.0	0.96	0.18	0.76		0.71		0.42	0.28		-0.34	4.90	2.1
16	Tiberias Roman Spring	57.5	87.3	1.9	8.1	1.4	7.8	2.81	0.89	4.55	0.58	3.68	0.84	2.36	1.62	0.25	-3.26	-13.1	0.7
17	Tiberias Main Spring	59	119		17.8	2.6	14.8	3.76	1.03	4.94	0.74	4.77	1.11	3.25	2.42	0.34	-3.21	-17.7	0.17
17	Tiberias Main Spring	46.7	102	11.7	34.4	4.2	16.3	2.85	0.83	4.36	0.64	4.06	0.95	2.88	2.19	0.34	-2.55	-10.0	

18	Gofra spring	25.5	61.8	18.3	34.3	3.9	19.6	3.63	0.86	3.74	0.56	3.69	0.81	2.48	2.13	0.36	-4.33	-21.0	0.14
19	Haon Beach well	34.8	68.5	29.7	63.2	6.8	27.4	5.20	1.55	5.85	0.88	5.01	1.06	3.21	3.27	0.56	-1.74	-8.0	3.1
20	En Reah	38	27.6	BW	19.7	1.5	9.5	1.35	0.38	1.50	0.24	1.50	0.31	1.17	1.25	0.22	-5.84	-29.9	0.74
20	En Reah	41.5	19.7	4.8	7.7	1.0	3.6	0.62	0.21	0.84	0.15	1.04	0.22	0.73	0.92	0.17	-5.99	-29.9	1.2
21	En Makla	47.3	50.1	12.5	25.6	1.9	11.2	1.81	0.49	2.16	0.33	2.26	0.55	1.91	2.03	0.35	-6	-31.6	0.35
21	En Makla	49.6	47.0	10.3	14.7	1.8	7.4	1.28	0.38	1.83	0.31	1.99	0.48	1.84	1.92	0.33	-6.15	-31.3	0.9
22	Mezar 2	60	15.7	4.4	10.3	1.3	4.8	1.03	0.23	1.05	0.14	0.88	0.17	0.49	0.40	0.06	-6.84	-33.0	
23	Mezar 3	41.8	23.9	10.0	18.1	2.5	10.1	2.24	0.56	2.25	0.31	1.81	0.35	1.02	0.86	0.14	-5.9	-29.4	0.4
24	Himma	41.5	51.8	4.3	23.4	3.5	13.1	2.98	0.78	3.38	0.52	3.33	0.73	2.21	2.02	0.34	-6.08	-32.0	0.5
25	Bir Mukheiba 8	29.1	17.7	2.3	4.3	0.6	2.3	0.48	0.17	0.54	0.09	0.70	0.17	0.72	1.01	0.20	-5.65	-27.4	0.5
26	Bira 2	25.2	70.2	17.4	13.5	4.9	21.2	4.53	1.30	5.52	0.84	5.24	1.11	3.30	2.68	0.38	-3.86	-10.4	0.4
27	Newe Ur	29.2	28.6	4.9	8.4	1.3	5.4	1.24	0.44	1.33	0.23	1.53	0.34	1.15	1.05	0.18	-3.58	-13.5	1.3
28	Bet Yosef	31.1	12.1		1.2	0.3	1.0	0.38	0.10	0.51	0.08	0.57	0.14	0.50	0.60	0.10	-3.24	-10.7	0.4
29	En Huga	25	85.5	9.5	70.3	4.4	18.4	3.99	1.14	4.78	0.74	5.07	1.15	3.77	3.62	0.58	-3.04	-11.5	1.6
30	En Amal	28.2	10.4	2.0	6.6	0.7	3.0	0.71	0.18	0.80	0.11	0.68	0.14	0.46	0.38	0.05	-4.9	-19.4	
31	En Moda	26.3	99.0	17.0	14.7	4.3	19.1	3.96	1.15	5.18	0.79	5.65	1.26	3.90	3.24	0.49	-5.21	-21.0	0.4
32	Hammam el Malih 1	35.1	267	30.7	26.1	11.8	51.8	11.0	3.18	15.6	2.30	15.5	3.47	10.0	7.90	1.21	-4.92	-22.0	
32	Hammam el Malih 2	34.5	113	13.8	31.0	5.52	23.7	4.53	1.36	6.30	0.97	6.37	1.44	4.08	3.14	0.48	-4.88	-22.0	
33	Beqaot 2	31.1	1.97	1.07	1.52	0.19											-5.05	-20.7	0.2
34	Beqaot 1	28.9	2.12	1.27	1.43	0.21						0.15					-5.15	-21.7	0.3
35	Well 20-17/11	29.2	14.5	2.6	3.3	0.7	2.5	0.67	0.15	0.72	0.10	0.69	0.15	0.46	0.43	0.07	-4.83	-23.0	0.4
36	Argaman 14	30	7.79	2.2	3.7	0.6	1.7	0.48		0.44		0.44	0.08	0.27	0.27		-4.85	-20.7	0.2
37	Argaman 29	29.7	3.75	0.9	1.5	0.2						0.18		0.18	0.17		-4.83	-21.2	0.3
38	Pezael 2	25.3	15.5	5.1	7.6	1.2	4.7	1.06	0.22	0.91	0.13	1.06	0.22	0.70	0.60	0.07	-5.49	-24.2	0.9
39	Pezael 6	26.1	8.88	2.2	5.4	0.6	2.1	0.55		0.62		0.62	0.13	0.39	0.34		-5.63	-23.4	1.5
40	Pezael 8	28.5	3.95	1.4	2.6	0.3	0.9			0.30		0.30		0.16	0.15		-5.55	-23.1	0.1
41	Mizpe Jericho 2	24.3	16.8	3.3	4.3	0.6	2.2	0.60	0.15	0.72	0.10	0.83	0.22	0.75	0.81	0.14	-5.57	-24.3	1.3
42	Shuna-North 1010 m	54.8	19.1	4.0	10.4	1.5	8.6	3.36	0.83	4.12	0.47	2.24	0.38	0.86	0.45	0.06	-5.7	-27.5	
43	Manshiya 2	48.8	18.0	6.3	10.7	1.3	4.9	1.48	0.32	1.62	0.21	1.22	0.24	0.64	0.47	0.06	-5.68	-26.8	
44	Waqqas	49.4	83.2	32.7	51.1	6.4	27.3	5.80	1.28	7.12	0.95	5.91	1.14	3.13	1.99	0.26	-5.75	-27.8	

45	Abu Zigan	34.8	2242	26.3	125	26.2	150	85.7	19.7	116	16.7	106	23.1	66.5	47.9	7.12	-6.32	-39.6	
46	Ein Feshkha creek	28.2	14.6	1.3	3.4	0.4	1.4	0.41		0.52	0.07	0.60	0.15	0.59	0.67	0.12	-5.25	-21.6	1
47	Ein Feshkha spring	26.6	24.8	3.5	4.5	0.7	2.4	0.58	0.14	0.73	0.11	0.91	0.24	0.83	0.76	0.12	-5.39	-22.6	1.3
48	Qane A (spring)	25.8	14.5	2.1	3.1	0.4	1.4	0.40		0.54	0.08	0.59	0.16	0.56	0.58	0.10	-5.64	-22.9	0.6
49	Qane B (north)	26.9	13.7	1.7	2.9	0.4	1.3	0.43		0.55		0.64	0.15	0.56	0.58	0.08	-5.5	-23.9	0.9
50	Qane C (south)	26.5	17.0	2.2	2.7	0.4	1.8	0.57	0.14	0.58	0.10	0.69	0.18	0.64	0.65	0.11	-5.47	-22.9	0.8
51	Nahal Qidron		46.2	31.5	47.8	6.64	22.8	3.91	0.84	5.14	0.45	2.65	0.52	1.65	1.63	0.26	-4.56	-18.4	
52	Mizpe Shalem	43.6	221		1.6	0.3	1.3	2.41	1.24	9.09	1.06	6.30	1.35	3.72	2.16	0.33	0.96	-3.4	0.5
53	En Gedi Spa, well 2	39.4	100	2.04	10.0	1.86	10.1	6.40	1.55	7.35	0.92	5.29	1.08	2.86	2.07	0.32	-2.12	-13.7	
54	Dead Sea, 1-45 m depth	25.3	887	1371	1623	80.4	326	45.0	10.5	53.6	5.12	29.0	6.76	19.9	13.8	2.51	4.33	23.3	3.8
55	En Gedi spring	27.9	6.45	2.55	4.73	0.56	1.93	0.48	0.10	0.50	0.07	0.47	0.09	0.30	0.28	0.05	-5.46	-23.0	
55	En Gedi waterfall	21	5.10	2.06	3.97	0.44	1.58	0.37	0.08	0.42	0.05	0.35	0.07	0.22	0.20	< BG	-5.27	-21.1	
56	En Boqeq, Gofrit well	27.3	419	53.5	168	14.8	64.2	10.7	2.75	17.4	1.92	11.6	2.91	7.97	3.93	0.67	0.21	0.4	
57	En Boqeq, PB-26	31.1	14.5	1.47	5.62	0.73	2.86	0.77	0.19	0.87	0.14	0.94	0.21	0.54	0.42	0.05	-0.08	-26.7	
58	Hammei Zorah 5	27.1			1.27	0.13		0.09				0.06					-6.34	-41.9	
59	Hammei Zorah 15	34	13.1	1.73	4.51	0.59	2.25	0.50	0.11	0.58	0.09	0.64	0.14	0.47	0.39	0.06	-5.68	-28.3	
60	Rama well	33.5	566	11.6	77.7	11.6	56.2	17.6	3.17	27.5	4.42	29.9	6.92	21.3	17.8	2.72	-5.33	-35.5	
61	Hisban art. deep well	32.0	756	56.2	234	31.9	142	32.3	7.97	41.8	6.84	46.0	10.1	30.4	24.5	3.57	-5.67	-36.3	
62	JICA well 1	33.7	641	6.7	47.9	9.6	53.5	18.8	5.05	30.4	4.98	33.9	7.68	22.8	17.0	2.53	-5.58	-42.2	
63	Zerka Ma'in Waterfall	57.0	1538	11.2	82.3	19.7	137	52.2	16.2	107	14.6	87.5	18.0	46.6	27.6	4.08	-4.11	-31.8	
64	Zerka Ma'in Roman Bath	61.7	189	8.6	20.9	3.2	17.8	10.7	4.29	28.2	3.11	14.1	2.24	4.67	2.10	0.27	-4.21	-34.5	
65	Zara Princess Basma Farm	51.7	266	76.4	129	16.2	68.1	16.0	4.36	22.1	2.96	16.8	3.32	8.50	6.00	0.82	-4.56	-37.1	
66	Zara Bani Hamida	54.6	245	10.0	27.1	5.5	45.3	18.5	5.28	33.4	3.75	18.7	3.30	7.32	2.81	0.38	-5.79	-44.4	
67	Ibn Hammat 2	38.8	31.0	2.2	7.0	1.4	8.5	3.34	0.80	4.04	0.55	3.06	0.56	1.38	0.87	0.12	-7.17	-41.8	
68	En Ofarim 5	25.0	134	1.9	14.0	3.0	20.1	7.19	2.42	11.7	1.48	7.70	1.44	3.24	1.56	0.21	-7.0	-40.8	
69	En Jahav 6	44.3	443	5.8	50.9	10.6	68.3	25.8	8.18	44.4	5.17	24.8	4.12	8.85	3.90	0.51	-7.21	-46.4	
70	Zofar 220	39.4	19.2	5.0	20.7	2.7	11.4	2.50	0.64	2.25	0.33	1.80	0.34	0.86	0.67	0.08	-6.79	-42.1	
71	Zofar 20	35.3	12.2		0.9	0.2	1.1	0.97	0.22	0.98	0.12	0.68	0.11	0.34	0.23	0.03	-8.08	-55.3	
72	Zofar 4a	26.5	3.0		1.2	0.2		0.17	0.04	0.20		0.15	0.03	0.12	0.13	0.02	-4.46	-23.2	1.8
73	Paran 20	59.2	220		8.4	2.6	25.4	20.5	6.97	38.1	4.12	17.9	2.65	5.03	2.12	0.27	-7.98	-53.0	

74	Paran 18	36.3	43.6	4.6	16.7	2.2	8.3	1.73	0.45	2.12	0.30	2.18	0.48	1.62	1.40	0.20	-7.49	-49.8	
75	Yaalon 116	32.6	33.2	2.2	5.3	1.4	5.7	1.24	0.32	1.54	0.23	1.55	0.33	1.01	0.87	0.12	-5.92	-34.5	
76	Yaalon 6A	42.2	73.9	2.6	7.5	1.3	7.5	3.05	0.92	5.68	0.85	5.11	0.93	2.28	1.34	0.19	-5.87	-28.2	
77	Yaalon 7	41.6	98.9		6.3	1.1	8.2	3.54	1.13	6.72	0.90	5.51	1.05	2.50	1.39	0.19	-5.86	-29.3	
78	Yaalon 117	39.6	39.5	4.3	19.0	2.6	10.6	2.51	0.67	2.75	0.42	2.71	0.54	1.62	1.38	0.19	-6.93	-39.8	
79	Yotvata 13	29.1	18.1	1.1	1.1	0.5	2.1	0.45	0.13	0.74	0.11	0.84	0.21	0.72	0.78	0.12	-6.49	-40.7	
80	Yotvata 12	29	25.6		0.8	0.6	2.8	0.78	0.21	1.15	0.17	1.19	0.29	0.89	0.73	0.12	-6.4	-38.2	
81	Gerofit 4	33.3	564		11.9	8.5	81.4	38.4	11.8	68.7	7.67	37.1	6.51	13.8	5.64	0.77	-7.58	-46.9	
82	Samar 4	31.8	22.9	2.0	20.1	1.0	3.8	0.91	0.22	1.12	0.16	1.18	0.27	0.82	0.66	0.09	-6.82	-38.9	
83	Samar 2	28.3	62.7		1.1	0.6	3.8	1.91	0.53	2.78	0.39	2.79	0.64	1.98	1.37	0.22	-6.67	-46.2	0.4
84	Elat 103	32	19.9	1.7	7.6	0.9	4.2	1.11	0.23	1.20	0.16	0.99	0.22	0.68	0.61	0.11	-7.19	-45.6	
85	Elat 108	31.8	169	34.6	106	14.4	56.3	9.34	2.33	10.4	1.36	8.66	1.89	5.57	3.87	0.60	-6.81	-42.1	
86	Wadi Araba 1	29.8	52.4	2.8	12.0	2.3	15.4	5.71	1.93	6.10	0.74	3.78	0.67	1.71	1.18	0.18	-6.1	-29.4	
86	Wadi Araba 1	29.9	20.1	1.4	3.8	0.4	2.3	0.55	0.19	0.99	0.13	0.77	0.19	0.52	0.41	0.07	-6.16	-33.0	
87	Wadi Araba 5	27.3	2.8		1.3	0.3	1.5	0.32	0.11	0.38	0.05	0.31	0.06	0.14	0.11	0.01	-6.97	-35.2	
88	Well Disi 6	27.9	13.5		2.4	0.3	1.4	0.59	0.12	0.55	0.10	0.68	0.16	0.49	0.39	0.06	-6.57	-40.2	
89	Well Disi 1	28.5	30.6	13.2	22.8	3.8	15.6	3.26	0.68	2.93	0.42	2.34	0.46	1.24	0.83	0.13	-6.54	-40.2	0.4
90	Wadi Yutum	29.7	94.1	16.0	3.2	2.6	11.7	2.38	0.66	3.34	0.46	3.38	0.86	2.85	2.56	0.43	-5.79	-30.5	0.5

REY patterns of water from limestone resemble those of their host rocks. Water from all other types of host rocks shown in Fig. 3.1.2-2 is mainly characterized by leachable accessory minerals.

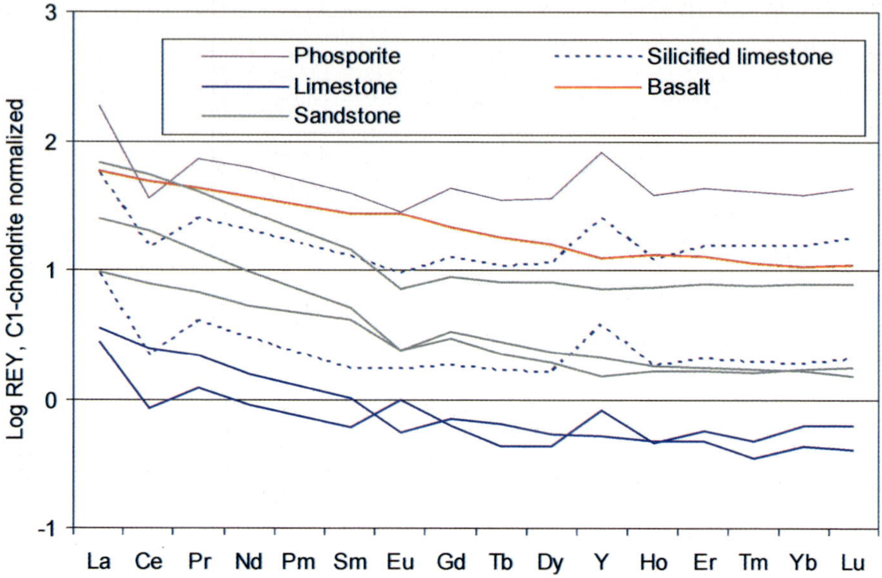

Fig. 3.1.2-2. Rare earths and yttrium distribution patterns of aquifer rocks of the study area

Type a REY patterns are convex with reference to the X-axis, i.e., the intermediate REE are preferentially leached (Fig. 3.1.2-3). These patterns are unlike those of the host rocks, i.e., Paleozoic to Mesozoic sandstone (Fig. 3.1.2-2). Experimental leaching of sandstone at pH 3 proves that the middle REE are more easily leached than the end members (Möller et al. 2003a). This is typical for the leaching of phosphates (Irber 1996; Hannigan and Sholkovitz 2001) and gypsum/anhydrite (Kagi et al. 1993; Toulkeridis et al. 1998). Groundwater with a conspicuous maximum of the intermediate REE with negative Eu anomalies can neither be derived from limestone nor basalt. The deficiency of Eu is inherited but the excess of Y, if present, is acquired during groundwater migration (Möller et al. 2003b; 2004). The high solubility of gypsum yields this type of REY distribution pattern in correspondence with the lithological origin of the groundwater in Lower Cretaceous to Triassic sandstone, which contains either gypsum lenses or gypsiferous cements.

Type b groundwater indicates interaction with limestone and is subdivided into four sub-types. Two sub-types (Figs. 3.1.2-4 and 3.1.2-5) strongly resemble those of limestone (Fig. 3.1.2-2). They show REY patterns which decrease from La to Lu with variable negative Eu, positive Y, and slightly positive Gd anomalies. The next two sub-types (Figs. 3.1.2-6 and 3.1.2-7) indicates interaction with basalt/limestone reaction zones such as silicified and mottled limestone.

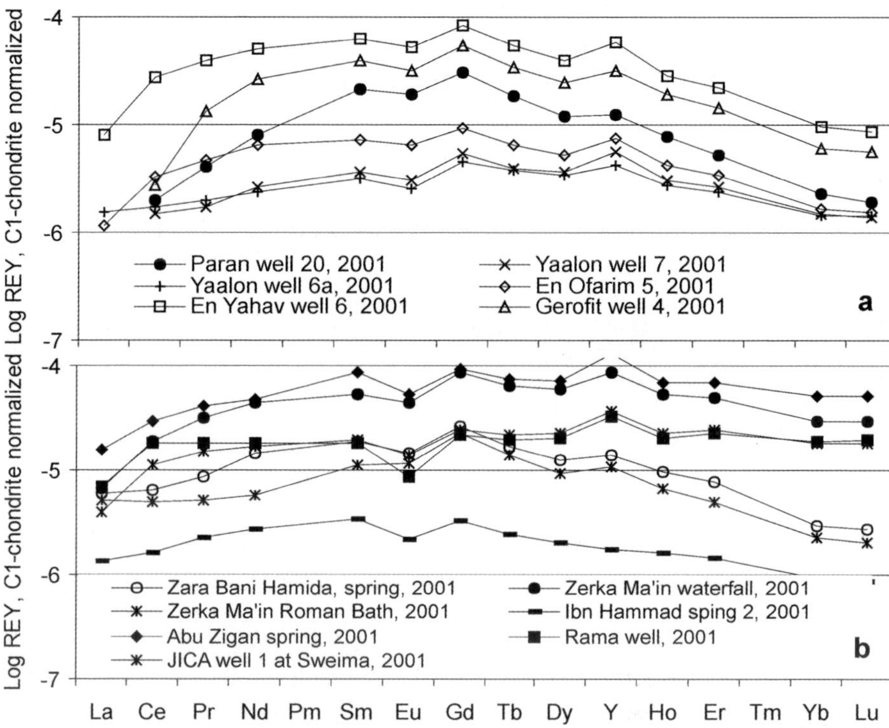

Fig. 3.1.2-3. Rare earths and yttrium distribution patterns of *type a* groundwater (mainly from sandstone). Patter of water samples from the western and eastern part of the Rift are shown in (a) and (b), respectively.

Type b1 trends of groundwater (Fig. 3.1.2-4) correspond to those of the Judea limestone (Fig. 3.1.2-2) and of their leachates (Möller et al. 2003a). Y, Ce, and Gd anomalies are present due to the dissolution of calcite. These anomalies only disappear in marly limestone due to ion exchange with the high abundance of REY in clay minerals. This characterizes the recrystallized Upper Cretaceous limestone with its abundance of marly layers. In the eastern part of the Rift some of the REY patterns of the groundwaters (Fig. 3.1.2-4c) show slightly different trends, which suggests leaching of mildly hydrothermally altered limestone, by which REY deriving from various sources are co-precipitated by accessory minerals and recrystallized calcite. The general trends of the patterns in Figs. 3.1.2-4b and

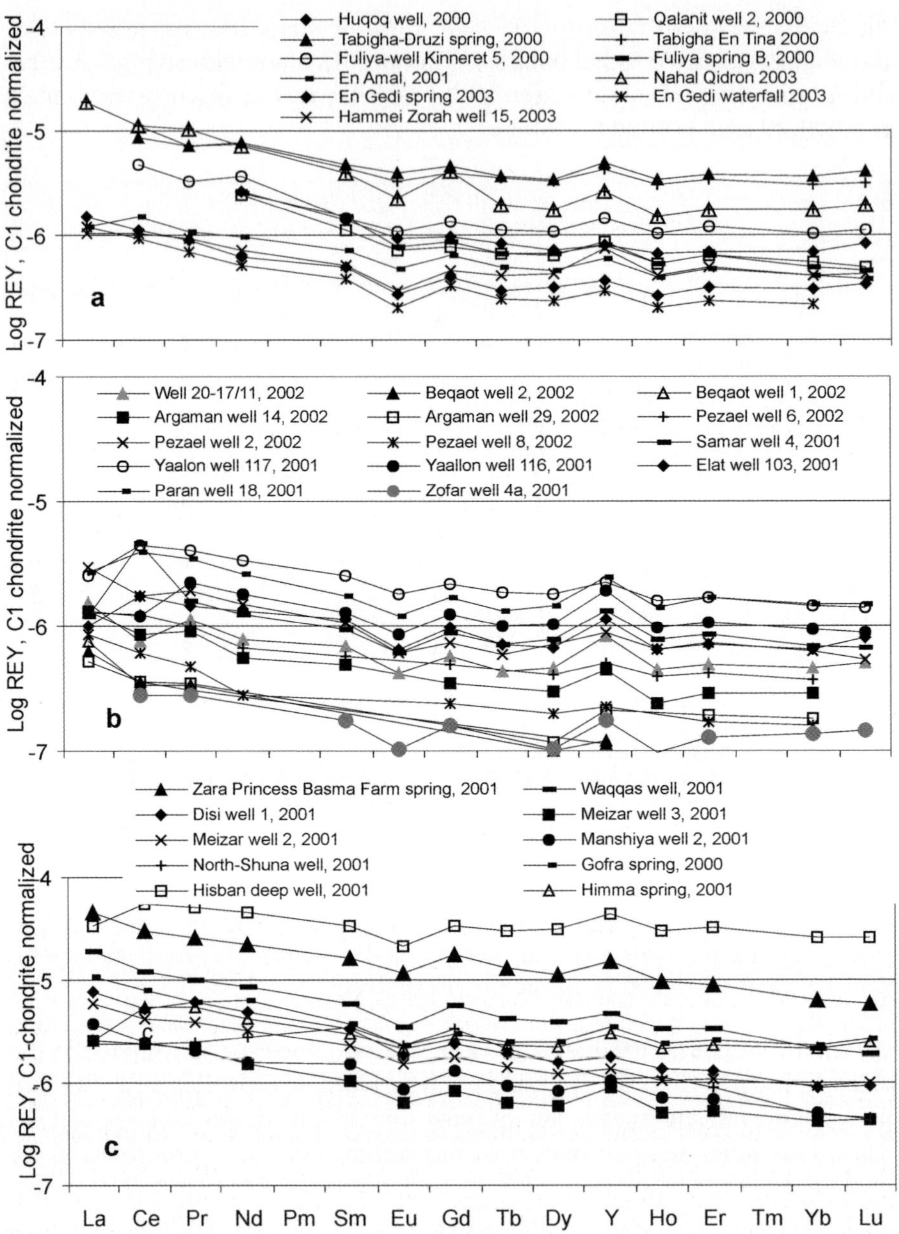

Fig. 3.1.2-4. Rare earths and yttrium distribution patterns of *type b1* groundwater from Upper Cretaceous limestone. (a), (b) and (c) compile samples from both sides of the Rift.

3.1.2-4c are sub-parallel in each group, but the patterns in Fig. 3.1.2-4c show enhanced REY abundance. This could be the result of fluid circulation during hydrothermal alteration of limestone induced due to contact with basalts.

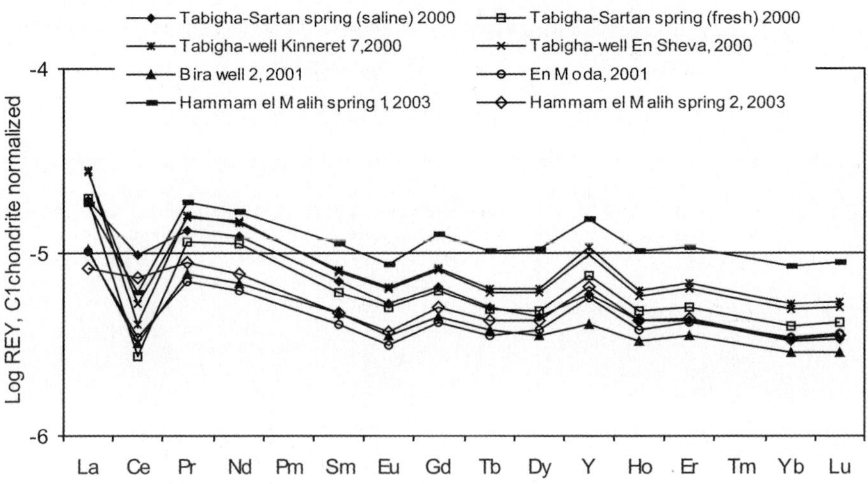

Fig. 3.1.2-5. REY distribution patterns of groundwater from Eocene limestone and chalk in the western part of the Rift.

Type b2 groundwater is characterized by pronounced negative Ce and positive Gd and Y anomalies (Fig. 3.1.2-5). In the Lake Tiberias area, these characteristics are encountered in the groundwater samples of the Tabigha cluster, with the exception of the very saline water from well Kinneret 8. They also occur in the waters from well Bira 2 and En Moda spring, both located in the lower Jordan Valley. There are no equivalents on the eastern side of the Rift. All this groundwater manifests features such as one would expect from water which circulates through karstic limestone (Möller 2002). The negative Ce anomalies are either inherited from the Eocene limestone or acquired by precipitation of FeOOH (Bau 1999; Kawabe et al. 1999a, b). Y, Ce, and Gd anomalies occur in groundwater as the result of dissolution of calcite by meteoric water. Only in cases where the dissolved limestone is rich in clay minerals, these anomalies may disappear because of the high REY abundance in clay minerals without anomalous Ce, Eu and Y contents. Although there is an abundance of marly layers in the Avedat Group aquifer, Ce anomalies are still present, probably because of FeOOH precipitation.

Type b3 REY patterns of water in Jordan (Fig.3.1.2-6b) are variable but similar to patterns of water from the studied reaction zone of basalt with

Plate 3.1.2-1 Preconcentration of rare earth elements in the field. Here the necessary equipment is shown in a hotel room. 5L of sampled waters are passed through an ion exchange cartridge of about 1 ml volume using a multi-channel peristaltic pump (in the background). The cartridges are visible between the two rows of bottles on the table. Analytical details are given by Möller et al. (2003a). (Photo: S. Geyer)

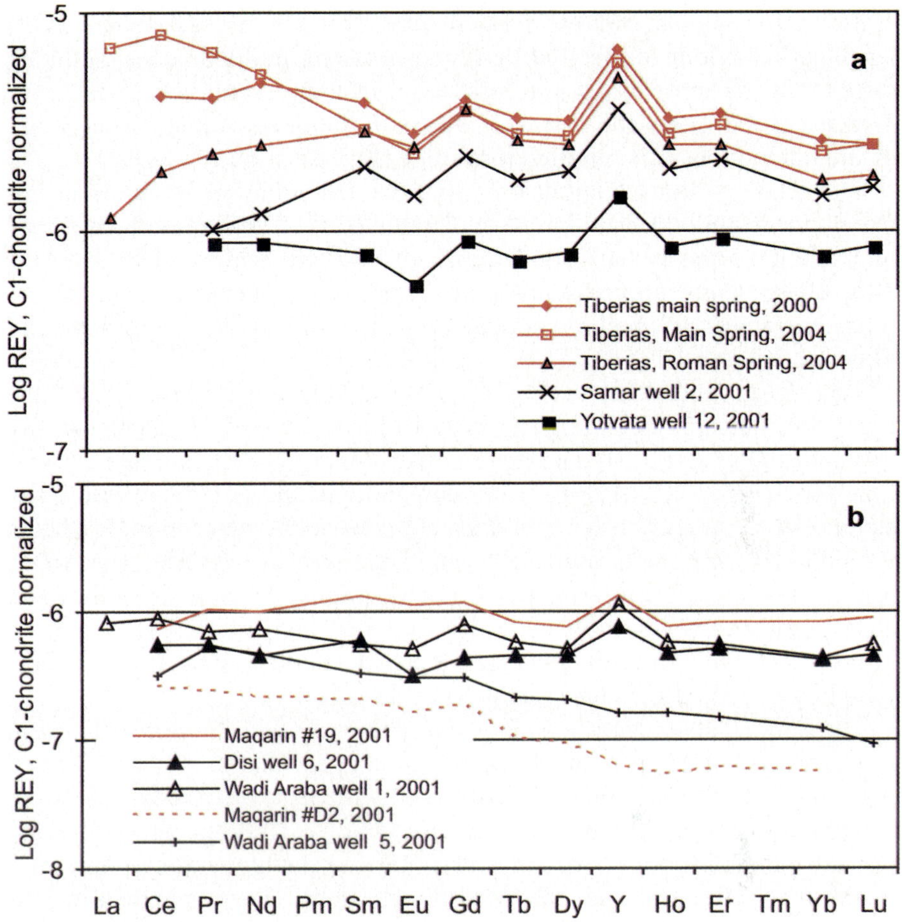

Fig. 3.1.2-6. Rare earths and yttrium distribution patterns of *type b3* patterns of groundwater from silicified limestone. The patterns represent water samples from the western (a) and eastern part (b) of the Rift.

limestones at Maqarin (Koury et al. 1998). The Maqarin samples are emphazised by red lines. Most water samples show rather flat patterns. All patterns elucidate that REY are leached from an inhomogeneous group of secondary accessory minerals in silicified limestone. The negative Ce and Eu anomalies and positive Y anomalies are inherited from the limestone. The local presence of fluorapatite indicates that the silicified limestone was penetrated by fluorine-rich hydrothermal fluids, which may also have been in contact with the phosphorites near the Cretaceous-Tertiary boundary. Due to hydrothermal fluids resulting from the interaction of limestone with overlying or intruded basalt, REY are redistributed by circulating hydrothermal fluids and by mineral alteration. As a result, the natural leaching behaviour of the silicified limestone and mottled zones distinctly differ from those of basalt and limestone. In the western part of the Rift, Tiberias Hot Springs and two well waters in the Arava Valley strongly indicate interaction with silicified limestone (Fig. 3.1.2-6a).

The REY patterns compiled in Fig. 3.1.2-6 allow differentiation between water derived from basalt and water derived from basalt-limestone interaction zones, because both types of aquifers widely differ in their mineralogy. Conventionally, most of these types of water are ascribed to limestone aquifers, which do not distinguish them from the water shown in Figs. 3.1.2-3 and 3.1.2-4.

Type b4 groundwater, so far only known from the western side of the Rift, shows steeply declining trends from La to Lu, with and without any anomalies (Fig. 3.1.2-7). The general trend is typical for interaction with limestone (Fig. 3.1.2-2). Enhanced temperatures and late diagenetic limestone lead to increasing REY abundance patterns in water. Some patterns resemble the trend and anomalies which are typical for Dead Sea water. Common to all of these waters is the high salinity. From the presence and absence of anomalies it may be deduced that at least two sub-types of brines exist which are diluted to generate the saline waters from the surroundings of either the Dead Sea or Lake Tiberias.

The spread of patterns may be caused by variable dilution with fresh groundwater. Patterns without anomalies suggest that they were derived from limestone which recrystallized at enhanced temperatures that were generated by the proximity of mafic in- and/or extrusions. The originally present anomalies were levelled off by REY exchange with clay minerals in the marly limestone at enhanced temperatures. Some patterns are incomplete because of their high Ca contents interfering with light REE in the analytical pre-concentration process (Shabani et al. 1992; Möller et al. 2005).

Type c: The common REY patterns (Fig. 3.1.2-8) are typical for waters which experienced (Fe,Mn)OOH precipitation, e.g., in the course of weathering of the basalt (Paces et al. 2001), basaltic debris in sediments or during dissolution of Fe-bearing limestone (Möller et al. 2003b). They show a minimal abundance of normalized intermediate REE with variable anomalies of Ce, Gd, and Y. If the water is rich in oxygen, a negative Ce anomaly is acquired due to precipitation of (Fe,Mn)OOH. (Fe,Mn)OOH precipitation is precluded and Ce anomalies are not developed where the oxidation of sulfides in basalt yields acidic solutions. The acid is only neutralized in the presence of carbonates. The REY patterns of these types of waters can be explained in two ways: (i) (Fe,Mn)OOH precipitation under oxidizing conditions (Bau 1999) or (ii) leaching of rocks after significant loss of phosphates (Irber 1996; Hannigan and Sholkovitz 2001) and sulfates (Kagi et al. 1993; Toulkeridis et al. 1998). The presence of Gd and Y anomalies supports explanation (i).

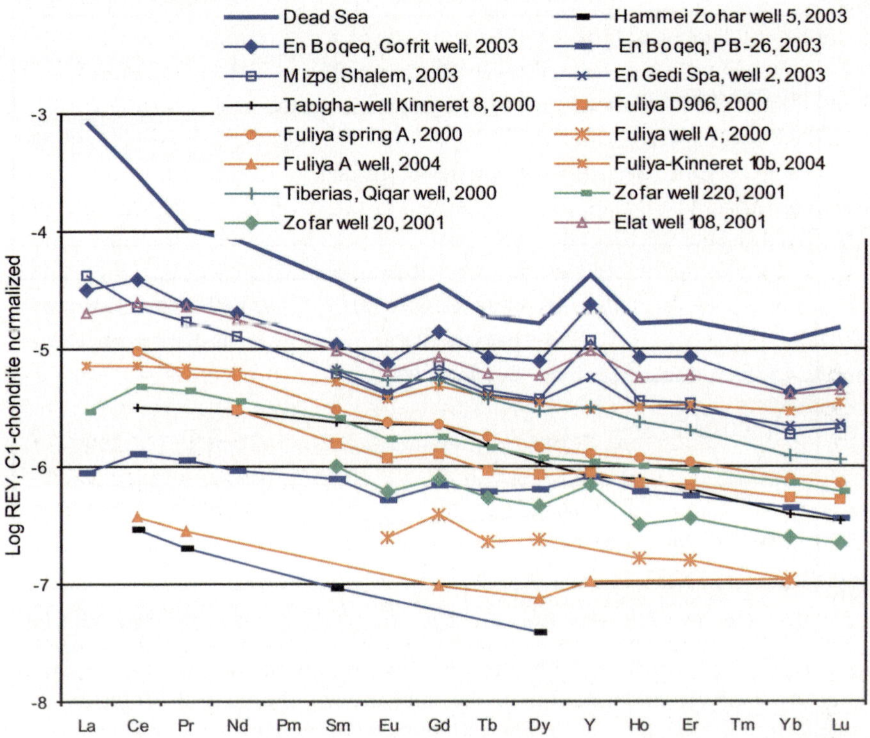

Fig. 3.1.2-7. Rare earths and yttrium distribution patterns of *type b4* of thermosaline groundwater from limestone aquifers in the western part of the Rift.

The negative Ce and positive Gd anomalies prove FeOOH precipitation in the Wadi Yutum water with its high redox potential (Eh=218 mV relative to standard hydrogen electrode), in the eastern part of the Rift. Basaltic dikes criss-cross the Precambrian basement rocks outcropping on both sides of the wadi. At Bir Mukheiba 8, basalt extends to the east. On the western side, waters from the springs at Qane and Ein Feshkha at the north-western Dead Sea are strongly influenced by FeOOH precipitation, whereas waters from Newe Ur and En Huga in the Bet Shean region represent a mixture of *type b1* and *type c*.

Table 3.1.2-2 summarizes the typical features of REY patterns, the lithology, the meteoric water interacted with, and the final occurrence of the groundwater.

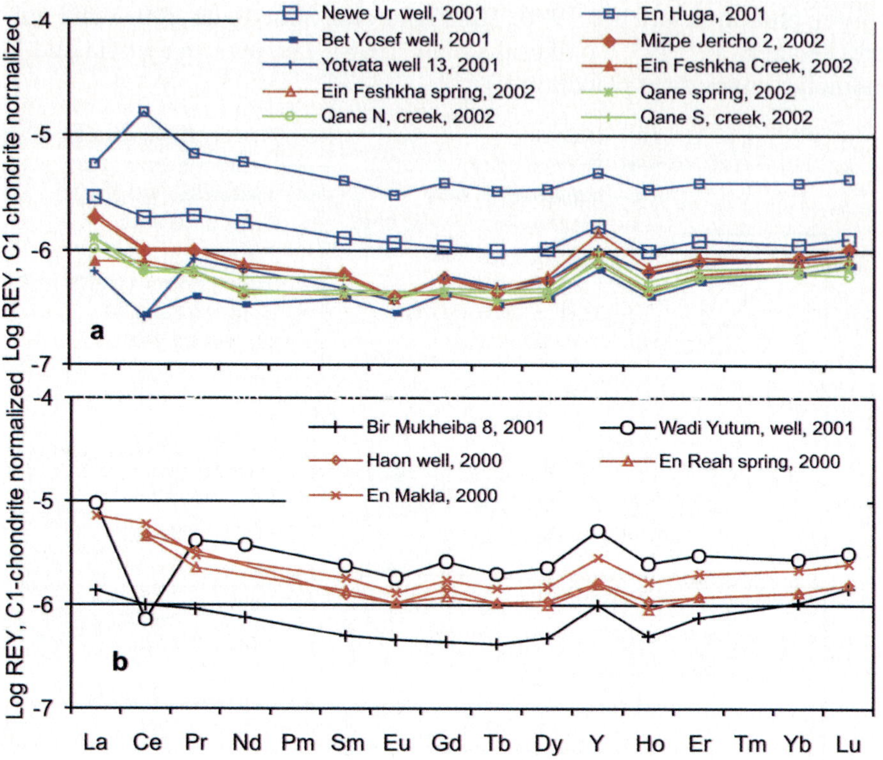

Fig. 3.1.2-8. Rare earths and yttrium distribution patterns of *type c* groundwater, altered by FeOOH precipitation in the subsurface. The pattern represent water samples from the western (a) and eastern part (b) of the Rift.

3.1.2.3 Anomalies of rare earth elements

Bau (1999) and Kawabe et al. (1999a, b) showed in experiments that Ce anomalies develop by induced oxidation of Ce(III) to Ce(IV) during the aging of FeOOH precipitates. Therefore, all oxygen-rich surface waters and many karst waters show variable but negative Ce anomalies (*type b2*).

Most negative Eu anomalies in waters are inherited from sediments. For instance, seawater shows a negative Eu anomaly and so do all marine precipitates. Groundwaters from limestone aquifers or marine evaporites such as gypsum or anhydrite are inevitably marked by a small negative Eu anomaly.

Geogene Gd anomalies result from the enhanced stability of inner-sphere aqua-complex Gd which is less reactive than those of the near-neighbors (Chopin et al. 2002). In systems with increasing salinity, the concentration of Gd does not increase but the abundance of the other REE

Table 3.1.2-2. The typical features of REY patterns, the lithology with which the meteoric water interacted, and final occurrence of groundwater.

	Type a		*Type b1*		*Type b2*	*Type b3*		*Type b4*	*Type c*	
Shape of REY patterns	Convex; negative Eu anomalies		Declining		Declining; negative Ce anomalies	Flat		Declining	Concave shaped	
Feature	Dissolution of anhydrite, cement in sandstones or gypsum lenses in limestone		Dissolution of late diagenetic limestones		Dissolution of young limestones and chalks	Dissolution of late diagenetic limestones		Dissolution of re-crystallized limestones	FeOOH precipitation during weathering Fe bearing minerals in limestones and basalts	
	West	East	West	East	West	West	East	West	West	East
Occurrence	Arava Valley	Lower Jordan Valley, Dead Sea area	Lake Tiberias, Lower Jordan Valley, Dead Sea area, Arava Valley	Golan Heights, Yarmouk Gorge, Lake Tiberias, Lower Jordan Valley, Disi area	Lake Tiberias	Lake Tiberias, Arava Valley	Wadi Araba wells, Disi area	Declining, some with strong Gd anomalies	Lake Tiberias Lower Jordan Valley, Dead Sea area, Arava Valley	Yarmouk Gorge,
Ground waters occur in format-ions:	Lower and Upper Creta-ceous	Triassic to Quater-nary	Lower Cretaceous to Quaternary	Lower Cretaceous to Quaternary.	Eocene	Upper Cretaceous	Palaeozoic to Jurassic	Upper Cretaceous to Quaternary	Upper Cretaceous to Quaternary	Upper Cretaceous to Quaternary

decreases by co-precipitation. Highest geogenic Gd anomalies are observed in the Dead Sea. Small geogenic Gd anomalies develop along with FeOOH precipitation as displayed in Figs. 3.1.2-5 and 3.1.2-8. The Fe(II) originates from basalt (Paces et al. 2001) or limestone (Möller 2002). Today, strong anthropogenic anomalies occur beside geogene Gd anomalies (Bau and Dulski 1996; Fuganti et al. 1996; Nozika et al. 2000; Möller et al. 2003a; Hennebrüder et al. 2004; Verplanck et al. 2005). Anthropogenic anomalies are significant in the Jordan River but cannot account for the high anomaly in the Dead Sea (Möller et al. 2007).

Y-Ho fractionation develops during alteration processes. The small dissimilarity of sorption onto mineral surfaces controlled by surface complexation (Bau and Dulski 1995; Diakonov et al. 1998) enhances in migrating fluids.

3.1.2.4 Behavior of REY in water-rock interaction

REY abundance in water, unless not acidic, is in the nmol/L to sub-pmol/L range, which is extremely low compared with REY abundance in typical Ca minerals of interest such as gypsum or calcite. Thus, if the REY bearing mineral dissolves, more than 99% are immediately adsorbed onto mineral surfaces which is the primary removal mechanism (Möller 2002). This immediately puts REY in the water into static equilibrium with its surroundings. As long as the groundwater is confined to a specific aquifer, changes of REY patterns are negligible. Changes in the REY pattern are to be expected, if the water subsequently moves into different rock environments. However, after a long time of water-rock interaction, a static equilibrium is established all along the flow-paths of water and all surfaces of minerals are adjusted to the initial amount of REY in the recharge water. In well-established hydrological systems, changes in REY do no longer occur or are at least insignificant. Once the water has attained its REY abundance pattern, it will be difficult to change it. Along the flow path, the wall-rock minerals behave similarly to an ion exchanger with the REY load on the walls of minerals being very high as compared to the amount of REY in the passing water. From a chemist's perspective, the system behaves like an ion exchange column after the breakthrough of REY ions. Whatever REY are released at the top of the local lithological column finally appears in the effluent. This way, the grouping of REY patterns facilitates a more precise definition of the origin of groundwater and, thereby, its flow-path. Characterizing groundwater according to its aquifer lithology often does not allow a clear distinction of its source, because the rocks of the replenishment area and the producing strata are not necessarily the same. Thus, the REY patterns have a memory for the initial water-rock interactions in the recharge area which defines the primary REY characteristics of the water. Inconsistencies between the lithology of the water-producing strata and the interpretation of REY distribution patterns with respect to the lithology of the replenishment area reveal interaquifer flow. Details will be discussed in Chap. 3.1.3. Interaquifer flow is a common phenomenon on both sides of the Rift. The dominant flow direction is towards the lower horizons of the stratigraphic column.

It appears that only small amounts of brines (excepting those from wells at the Dead Sea) are involved in cases where salinization occurs due to mixing with brines. Insignificant variability in the REY patterns is expected where these brines are equilibrated with rocks of similar mineralogical composition. Therefore, the REY patterns reveal the type of rocks (sandstone, limestone or basalt) which predominantly make up the main water replenishment areas of the groundwater that flows into the Rift. Moreover, due to circulating hydrothermal fluids, it is possible to identify zones of interaction between basalt and limestone which host numerous new soluble REY-bearing minerals.

Where gypsum/anhydrite is absent, REY patterns reveal the recharge area of groundwater. Only when water from a basaltic or limestone area contacts gypsiferous layers, the much higher solubility of gypsum overprints the original REY patterns with its own. Therefore, patterns of groundwater from sandstone aquifers, which are characterized as *type a* REY patterns, do not unequivocally represent the signature of the recharge area but also the interaction of any kind of replenishment water with anhydrite along the flow path.

The behaviour of REY is different from that of all major elements that control the salinity of groundwater (Capt. 3.1.1). First, REY are largely released from accessory minerals in the recharge area and second, saturation of groundwater with respect to REY is quickly achieved. In contrast, the amount of major elements increases only slowly, so that it takes much longer until saturation of the groundwater is achieved. Therefore, REY patterns indicate the recharge area, whereas the major element chemistry characterizes the aquifer rocks that host the groundwater for a long period of time.

3.1.2.5 Stable isotopes of H_2O

Based on REY grouping, δ^2H vs. $\delta^{18}O$ cross plots yield insight into the relationships between groundwater bodies (Fig. 3.1.2-9). It also reveals two significant trends. Most of the waters plot between the eastern Mediterranean water line, EMWL (Gat and Carmi 1970) and the paleo-eastern Mediterranean water line, palEMWL (Gat 1984; Issar et al. 1984). In the western part of the Rift, most samples from the sandstone and from the unaltered limestone plot on a straight line crossing both the EMWL and palEMWL. The thermosaline waters (*type b4*), some waters of *type b2* in the western and eastern part of the Rift plot along evaporation and/or mixing line. The local meteoric water line for the area of Lake Tiberias (LMWL after Bergelson et al. 1999; Bajjali and Jaber 2001) is less steep than the given MWLs. It fits the groundwater lines from the basaltic areas and most of the Judea Group (*type b1*) waters, but not so well most of the well and spring waters from the area of Lake Tiberias.

Type a waters (mostly from the Kurnub Group) in the western Arava Valley spread between the palEMWL and the LMWL. Their lowest isotope ratios are characteristic for Pleistocene precipitation (Gat 1984; Abd El Samie and Sadek 2001). Water from the Kurnub Group plots along a line starting from these low values and running through to the clusters of *types b1* and *b2* waters, i.e. waters from the Judea and Avedat Groups. Obviously, *type a* waters represent a mixture of Pleistocene waters and recent waters from the Judea Group. The slope of the trend line of *type b4* is much less steep than any of the indicated MWLs, which indicates that this trend characterizes either an evaporation line or a mixing line with an unspecified saline water which is isotopically similar to the Dead Sea brine. The thermosaline *type b4* groundwaters of the western part of the Rift are mixtures of brines with regional freshwater. Around Lake Tiberias, groundwater is salinized by three types of brines: residual primary evaporation brines (present in diluted form in Haon well water), secondary ablation brines of the Zemah type, and tertiary altered evaporation brines (presented in diluted form by Tiberias Hot Spring water). For a detailed discussion refer to Chap. 3.1.1. The brines mix in varying proportions with recent freshwater from either the Avedat or the Judea Group. In the Arava Valley, the saline groundwaters are mixtures of Pleistocene water and freshwater from the Judea Group. Locally, brines interplay as, e. g., in Zofar 20 and Elat 103 (Chap. 3.1.1).

Although identified as *type B4*, the groundwater from the Zofar 20 and Elat 103 wells (both located in the Arava Valley) plots among *type a* waters, which shows that they are mixtures of Pleistocene water with another saline end member. The chemical composition of water from Zofar 20 corresponds to that of the Tiberias Hot Springs. Water from Elat 103 resembles that of Haon, both represent diluted primary evaporation brines (Chap. 3.1.1).

On the other hand, the waters from the Druzy and Gofra springs (*type b1*) and the well of Haon (*type c*) cluster with thermosaline waters (*type b4*). The Mezar 2 groundwater from the Golan Heights, identified as *type b1* water, plots on the thermosaline trend line. All these "exotic" samples prove that the isotopes are less controlled by water-rock interaction than by climatic and geomorphological conditions at the time of replenishment.

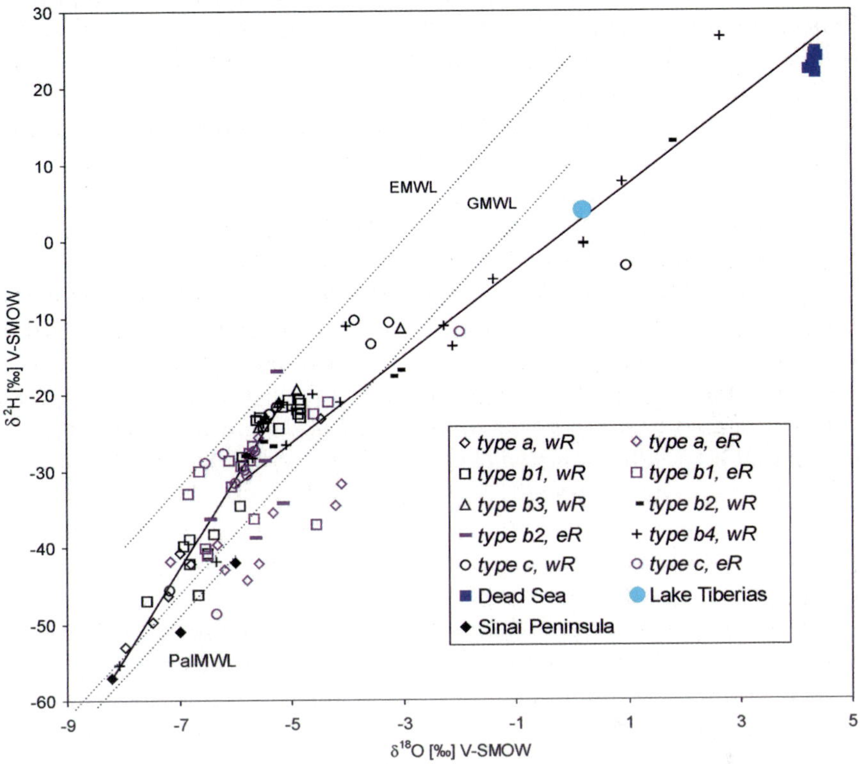

Fig. 3.1.2-9. Cross plot of $\delta^{18}O$ vs. $\delta^{2}H$. Characterization according to rare earth element patterns. eR and wR denote western and eastern Rift. Data for groundwater from the Sinai Peninsular are taken from Abd El Samie and Sadek (2001)

The average water from the Avedat Group is isotopically lighter than that from the Judea Group aquifer. Since most of the Judea and Avedat groundwaters are the result of ongoing replenishment, the enriched isotopic composition of the Judea Group waters may be explained by the "altitude effect". With an average height difference of about 300 m between the Avedat and the Judea replenishment areas, the altitude effect shifts $\delta^{18}O$ of the Avedat area towards smaller values by about 1‰.

In the eastern part of the Rift, nearly all data plot below the EMWL, which corresponds to results reported by Salameh and Gedeon (1999). Three trends can be recognized: *type b2* and *b3* waters largely plot along a mixing line of recent precipitation and Pleistocene water (see lower bold line in Fig.3.1.2-9). Waters from Mesozoic sandstone (*type a*) are among the isotopically lightest in δ^2H but due to hydrothermal water-rock interaction, $\delta^{18}O$ is shifted towards enhanced values (Taylor 1976). Waters from Zerka Ma'in are the most hydrothermally altered. Among the sandstone-derived waters, only the very saline Wadi Araba 5 plots near the EMWL. *Type b4* groundwater splits into two different trends: they either plot along the EMWL or at much enhanced $\delta^{18}O$ values.

3.1.2.6 Conclusions

The amount of REY in groundwater depends on soluble REY bearing compounds in the replenishment area, where the meteoric water first reacts with minerals. Due to the very low occurrence in water, the amount of REY/Ca in groundwater is much smaller than in rocks. Most of the released REY are immediately adsorbed or surface-complexed. With time, all surfaces along the flow-paths of water show compositions of REY gained from rocks with highest solubility of REY minerals. This process can be compared with an ion exchange column. When breakthrough of all REY occurs, the quantity and composition is the same as in the water of which the chemical composition is established in the replenishment area. REY have a strong remembrance with respect to the source of water. Thus, recharges over young (here: Eocene) and old (here: Cretaceous) limestone, basalt-limestone interaction zones, sandstone and basalt can be distinguished in the study area. Knowledge of the true recharge area is important for understanding the hydraulic system. The lithology of the recharge area is not necessarily the same as the strata from which the water is finally drawn.

Interaquifer flow is derived from the inconsistencies between the lithology of the water-producing strata and the interpretation of REY distribution patterns with respect to the lithology of the replenishment area.

Stable isotopes record the climatic and morphological condition during precipitation. Mixing of water bodies is only traceable, if they originate from climatically or geomorphologically different areas. In contrast, REY only respond to different mineralogical compositions of soils, sediments and rocks of the recharge area. REY patterns are only altered by highly soluble REY bearing minerals such as gypsum which can occur as lenses or cements in sandstone.

References

Abd El Samie SG, Sadek MA (2001) Groundwater recharge and flow in the lower Cretaceous Nubian sandstone aquifer in the Sinai Peninsula, using isotopic techniques and hydrochemistry. Hydrogeol J 9:378–389

Anders E, Grevesse N (1989) Abundance of elements: Meteoric and solar. Geochim Cosmochim Acta 53:197–214
Bajjali W, Abu Jaber N (2001) Climatological signals of the paleogroundwater in Jordan. J Hydrol 243:133–147
Banks D, Hall G, Reimann C, Siewers U (1999) Distribution of rare earth elements in crystalline bedrock groundwaters: Oslo and Bergen regions, Norway. Appl Geochem 14:27–39
Bau M (1999) Scavenging of dissolved yttrium and rare earths by precipitating iron oxyhydroxides: Experimental evidence for Ce oxidation, Y-Ho fractionation, and lanthanide tetrad effect. Geochim Cosmochim Acta 63:67–77
Bau M, Dulski P (1995) Comparative study of yttrium and rare earth element behaviour in fluorine-rich hydrothermal fluids. Contrib Mineral Petrol 119:213–223
Bau M, Dulski P (1996) Anthropogenic origin of positive gadolinium anomalies in river waters. Earth Planet Sci Lett 143: 245–255
Bergelson G, Nativ R, Bein A (1999) Salinization and dilution history of groundwater discharging into the Sea of Galilee, the Dead Sea Transform, Israel. Appl Geochem 14:91–118
Chopin F, Berger G, Bauer A, Castet S, Loubet M (2002) Sorption of lanthanides on smectite and kaolinite. Chem Geol 182:57–68
Craig H (1961) Isotopic variations in meteoric waters. Sci 133:1702–1703
Dansgaard W (1964) Stable isotopes in precipitation. Tellus 16:436–468
Diakonov II, Ragnarsdottir KV, Tagirov BR (1998) Standard thermodynamic properties and heat capacity equations of rare earth hydroxides: I. Ce(III)-, Pr-, Sm-, Eu(III)-, Gd-, Tb-, Ho-, Er-, Tm-, Yb-, and Y-hydroxides. Comparison of thermodynamical and solubility data. Chem Geol 151:327–347
Fuganti A, Möller P, Morteani G, Dulski P (1996) Gadolinio ed altre terre rare usabili come traccianti per stabilire l'eta il movimento ed i rischi delle acque sotterranee: esempio dell'area di Trento. Geol Tecn Ambien 4:13–18
Gat JR (1984) The stable isotope composition of Dead Sea waters. Earth Planet Sci. Lett. 71:361–376
Gat JR (1996) Oxygen and hydrogen isotopes in the hydrologic cycle. Annual Review of Earth and Planetary Sci. 24:225-262
Gat JR, Carmi I (1970) Evolution of the isotopic composition of atmospheric waters in the Mediterranean Sea area. J Geophys Res 75:3039–3048
Gat JR, Gonfiantini R (1981) Stable Isotope Hydrology. Deuterium and Oxygen-18 in the Water Cycle, Technical Reports Series No 210, IAEA Vienna, pp 1–337
Hannigan RE, Sholkovitz ER (2001) The development of middle rare earth elements in freshwaters: weathering of phosphate minerals. Chem Geol 175:495–508
Hennebrüder K, Wennrich R, Mattusch J, Stärk HJ, Engewald W (2004) Determination of gadolinium in river water by SPE preconcentration and ICP-MS. Talanta 63:309–316

Irber W (1996) Laugungsexperimente an peraluminischen Graniten als Sonde für Alterationsprozesse im finalen Stadium der Granitkristallisation mit Anwendung auf das Rb-Sr-Isotopensystem. PhD thesis, Free Univ. Berlin, pp 1–319

Issar A, Nativ R, Karnieli A, Gat JR (1984) Isotope evidence of the origin of groundwater in arid zones. Isotope Hydrology 1983. Int Symp Isotope Hydrology in Water Resources Development, Vienna, 1983, pp 85–104

Johannesson KH, Farnham IM, Guo C, Stetzenbach KJ (1999) Rare earth element fractionation and concentration variation along a ground water flow path within shallow, bassin fill aquifer, southern Nevada, USA. Geochim Cosmochim Acta 63:2697–2708

Johannesson KH, Zhou X, Guo C, Stetzenbach KJ, Hodge VF (2000) Origin of rare earth element signatures in groundwaters of circumneutral pH from southern Nevada and eastern Califiornia, USA. Chem Geol 164:239–257

Kagi H, Dohmoto Y, Takano S, Masuda A (1993) Tetrad effect in lanthanides partitioning between calcium sulfate crystal and its saturated solution. Chem Geol 107:71–82

Kawabe I, Ohta A, Miura N (1999a) Distribution coefficients of REE between Fe oxyhydroxide precipitates and NaCl solutions affected by REE-carbonate complexation. Geochem J 33:181–197

Kawabe I, Ohta, A, Ishii S, Tokumura M, Miyauchi K (1999b) REE partitioning between Fe-Mn oxyhydroxide precipitates and weakly acid NaCl solution: Convex tetrad effect and fractionation of Y and Sc from heavy lanthanides. Geochem J 33:167–179

Khoury HN, Salameh E, Mazurek M, Alexander WR (1998) Geology and hydrogeology of the Maqarin area. In:. Smellie JAT (ed) Maqarin natural analogue study Swedish Nuclear Fuel and Waste Management Co. Stockholm. ISSN 0284-375, Techn. Rep. TR-98-04

Möller P (2000) Rare earth elements and yttrium as geochemical indicators of the source of mineral and thermal waters. In: Stober I. and Bucher K. (eds.) Hydrology of crystalline rocks. Kluwer Acad. Press, Rotterdam, pp 227–246

Möller P (2002) The distribution of rare earth elements and yttrium in water-rock interactions: field observations and experiments. In: Stober I and Bucher K (eds) Water-rock interaction, Kluwer Acad. Press, pp 97–123

Möller P, Dulski P, Salameh E, Geyer S (2006) Characterisation of the sources of thermal spring- and well waters in Jordan by rare earth element and yttrium distribution and stable isotopes of H_2O. Acta Hydrochim Hydrobiol 34:101–116

Möller P, Dulski P, Savacin Y, Conrad M (2004) Rare earth elements, yttrium and Pb isotope ratios in thermal spring and well waters of East Anatolia, Turkey: a hydrochemical study of their origin. Chem Geol 206:97–118

Möller P, Morteani G, Dulski P (2003b) Anomalous Gadolinium, cerium, and yttrium contents in the Adige and Isarco river waters and in the water of their tributaries (Provinces Trento and Bolzano/Bozen, NE Italy). Acta Hydrochim Hydrobiol 31:225–239

Möller P, Rosenthal E, Dulski P, Geyer S, Guttman J (2003a) Rare earths and yttrium hydrostratigraphy along the Lake Kinneret – Dead Sea – Arava transform fault, Israel. Appl Geochem 18:1613–1628

Möller P, Rosenthal E, Geyer S, Guttman J, Dulski P, Rybakov M, Flexer A (2007) Hydrochemical processes in the Lower Jordan Valley and in the Dead Sea area. Chem Geol 239:27–49

Möller P, Woith H, Dulski P, Lüders V, Erzinger J, Kämpf H, Pekdeger A, Hansen B, Lodemann M, Banks D (2005) Main and trace elements in KTB-VB fluid: composition and hints to its origin. Geofluids 5:28–41

Nozika Y, Lerche D, Alibo DS, Tsutsumi M (2000) Dissolved indium and rare earth elements in three Japanese rivers and Tokyo Bay : Evidence for anthropogenic Gd and In. Geochim Cosmochim Acta 64:3975–3982

Paces T, Möller P, Fuganti A, Morteani G, Pecek J (2001) Sparkling mineral water at western rim of the Doupowske hory Mountains (Czech Republic): genesis by water rock interaction and deep-seated CO_2. Bull Czech Geol Surv 76:189–202

Poage MA, Chamberlain CP (2001) Empirical relationships between elevation and the stable isotope composition of precipitation and surface waters; Considerations for studies of paleoelevation change. Am J Sci 301:1–15

Salameh E, Gedeon R (1999) Renewability study of Disi – Wadi Yutum aquifers' water using isotopes and hydrological analyses. Hydrogeol Umwelt, 18:1–16

Shabani MB, Akagi T, Masuda A (1992) Preconcentration of trace rare-earth elements in seawatee by complexation with bis(2-ethylhexyl) hydrogen phosphate and 2-ethyl dihexyl dihydrogen phosphate adsorbed on a C18 cartridge and determination by inductively coupled plasma mass spectrometry. Anal Chem 63:751–767

Smedley P (1991) The geochemistry of rare earth elements in groundwater from the Carnmenellis area, southwest England. Geochim Cosmochim Acta 55:2767–2779

Taylor CB (1976) Isotope hydrology laboratory technical procedure. Note No.19, International Atomic Energy Agency, Vienna.

Toulkeridis T, Podwojewski P, Clauer N (1998) Tracing of the source of gypsum in New Caledonian soils by REE contents and Sr-Sr isotopic compositions. Chem Geol 145:61–71

Verplanck PL, Taylor HE, Nordstrom DK, Barber LB (2005) Aqueous stability of Gadolinium in surface waters receiving sewage treatment plant effluent, Boulder Creek, Colorado. Environ Sci Techn 39:6923–6929

3.1.3 Regional hydrochemical and hydrogeological aspects of groundwater in the Jordan-Dead Sea Rift system

P. Möller, E. Rosenthal, S. Geyer

3.1.3.1 Introduction

Based on the results on gross chemical compositions as displayed by spider patterns (Chap. 3.1.1), on rare earth element distribution patterns and on stable isotopes of water (Chap. 3.1.2), a synthesis is attempted for the regions of the surroundings of Lake Tiberias, the lower Jordan Valley, the surroundings of the Dead Sea, and the Arava Valley. In the subsurface of all these regions Ca-Mg-Cl brines predominate. Only along the south-eastern and eastern shores of Lake Tiberias and locally near Elat, Mg-rich brines occur (Chap. 3.1.2). Mg>Ca is typical for freshwaters deriving from basaltic aquifers in the Bet Shean area, northern lower Jordan Valley, and Golan (Sandler 1981). With groundwater being the most important socio-economic factor, deterioration of freshwater by inflow of saline water-bodies presents a true economic danger for any community in the Rift. In order to minimize salinization the local sources of the saline brines have to be known.

The present discussion is based on the chemical and isotopic results presented in previous Chaps. 1.1 and 3.1.2, in which the chemical analyses and their hydrochemical interpretations are given. There, the various types of waters were subdivided according to major and minor elements into *types A* to *D* (Chap. 3.1.1), and were classed into *types a* to *c* by the distribution of rare earth elements and yttrium (henceforth combined to REY) (Chap. 3.1.2). In brief, the major and minor elements characterize the aquifer in which the gross chemical composition is attained, whereas the REY reflect largely the recharge area (Table 3.1.3-1).

3.1.3.2 The region of Lake Tiberias

Lake Tiberias is located in one of the morphotectonic depressions formed as a pull-apart basin in the Rift. The Lake is 20 km long, 12 km at its widest, and has a maximum depth of 46 m. Its surface is at present below -209 m msl. This low elevation makes it a regional drainage base for precipitation which falls on the surrounding hills and percolates further through the stepwise down-faulted blocks, flowing finally through the subsurface to the Lake. At present it is the main freshwater storage body in the Rift and

Table 3.1.3-1. Characterization of groundwaters by spider- and rare earth element patterns being affiliated to given lithology

Type	Lithology in control of composition of groundwater acc. to spider patterns	*Type*	Lithology of recharge area acc. to REY patterns
A	Gypsum bearing strata; mostly sandstones	*a*	Gypsum bearing strata; mostly sandstones
B1	Cretaceous limestone	*b1*	Cretaceous limestone
B2	Eocene limestones; chalks	*b2*	Eocene limestones, chalks
B3	Thermally altered limestone	*b3*	Limestone - basalt contact zones
B4	Cretaceous limestone hosting thermosaline waters	*b4*	Cretaceous limestones hosting thermosaline waters
C	Basalts	*c*	Basalts
D	Limestones hosting evaporation brines		

the main component of Israel's water supply system. Its quality is constantly endangered by saline springs discharging along its shore and on the bottom of the Lake (Chap. 3.2.3).

Around the Lake and close to the water line, four large clusters of springs are discerned (Fig. 3.1.3-1). Three such clusters are located along the western shore, positioned on- or close to the western faults of the Rift and mostly at intersections with diagonal, NW-SE directed ones.

The Tiberias Hot Springs (Hammei Tveria) stand out by their high Cl^- content of up to 18,000 mg/l and temperatures of up to 60°C. These springs emerge along a zone of step-faults which cut through Upper Cretaceous limestone and dolomite beds. Similar thermosaline waters of *type B4* emerge to the east of the Hot Springs from the bottom of the Lake as submarine springs. However, REY patterns of *type b2* indicate recharge over slightly altered limestone. The water from well Qiqar *(type B4)* presents the least diluted saline component. All saline waters of the Tiberias cluster plot on the nearly vertical trend in the insert of Fig.3.1.3-2, which distinguishes these saline waters from those plotting on the diagonal trend of the Dead Sea brine type. From Fig. 3.1.3-2 it is evident that the thermosaline waters from Tiberias show an intercept on the ordinate. This intercept quantity of Ca is independent of any Ca-Mg exchange process and is explicable by albitization of basalts that occur south of Tiberias (Rosenthal et al. 2006). As shown by Fig. 3.1.3-3 the saline source of the hot springs is similar to that of the eastern cluster.

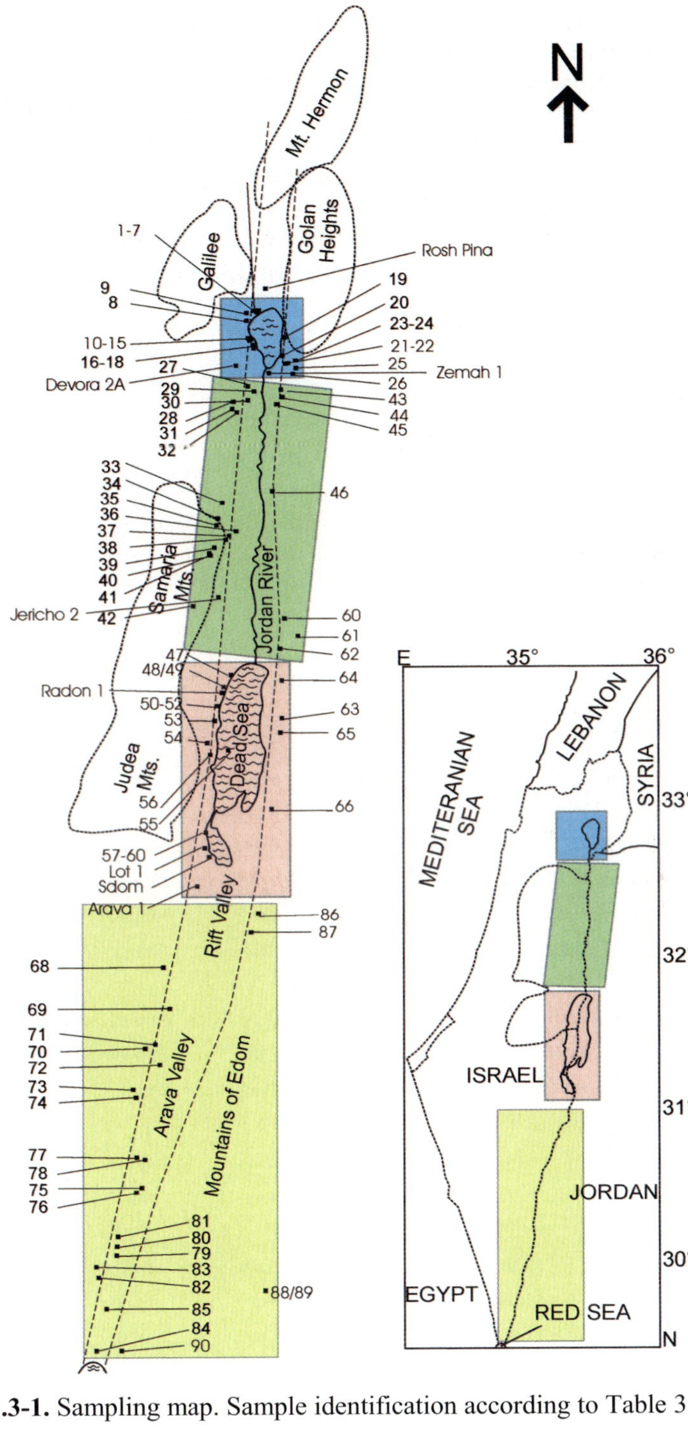

Fig. 3.1.3-1. Sampling map. Sample identification according to Table 3.1.3-2

Table 3.1.3-2. Sample origin, lithology of aquifers, hosting formation and the interpretation of spider- (Chap. 3.1.1) and rare earth patterns (Chap. 3.1.2) of groundwater. Differences in interpretation in REY and spider pattern elucidate flow from recharge areas to main aquifers. Differences between lithology indicated by spider pattern and the lithology of groundwater extraction indicate inter-aquifer flow. Lithological background of spider and rare earth patterns are shown in Table 3.1.3-1. S=spring; W=well

Sample ID	Location	Sample origin	Hosting Formations	Lithology	Type of spider pattern	Type of REY pattern
LAKE TIBERIAS AREA						
Tabigha Group						
1	Tabigha-Druzi Spring	S	U. Cret.	limest.	*B2*	*b1*
2	Tabigha-En Sheva	S	Tertiary	chalk	*B2*	*b2*
3	Tabigha-Sartan spring (saline)	S	U. Cret.	limest.	*B2*	*b2*
4	Tabigha-Sartan spring (fresh)	S	Tertiary	chalk	*B2*	*b2*
5	Tabigha-well Kinneret 7	W	U. Cret.	limest.	*B2*	*b2*
6	Tabigha- well Kinneret 8	W	U. Cret.	limest.	*B4*	*b4*
7	Tabigha, Sapir Site, En Tina	S	U. Cret.	limest.	*B1*	*b1*
8	Huqoq well	W	U. Cret.	limest.	*B1*	*b1*
9	Qalanit 2	W	U. Cret.	limest.	*B1*	*b1*
Fuliya Group						
10	Fuliya-spring A	S	U. Cret.	limest.	*B4*	*b4*
11	Fuliya- well D 906	W	U. Cret.	limest.	*B1*	*b4*
12	Fuliya-well A	W	U. Cret.	limest.	*B4*	*b4*
13	Fuliya-well B	S	U. Cret.	limest.	*B4*	*b4*
14	Fuliya- well Kinneret 5	W	U. Cret.	limest.	*B4*	*b1*
15	Fuliya-well Kinneret 10 B	W	U. Cret.	limest.	*B4*	*b4*
Tiberias Hot Springs						
16	Tiberias, Qiqar well	W	U. Cret.	limest.	*B4*	*b4*
17	Tiberias-Main Spring	S	U. Cret.	limest.	*B4*	*b3*
18	Tiberias -Roman Spring	S	U. Cret.	limest.	*B4*	*b3*
Eastern Shore of Lake Tiberias						
19	Gofra spring	S	Quat.	clastics	*D*	*b1*
20	Haon beach, well	W	Quat.	clastics	*D*	*c*
YARMOUK GORGE						
21	Yarmouk Gorge-En Reah	S	Tertiary	chalk	*B2*	*c*
22	Yarmouk Gorge-En Makla	S	Tertiary	chalk	*B2*	*c*
23	Yarmouk Gorge-well Mezar 2	W	U. Cret.	limest.	*B3*	*b1*
24	Yarmouk Gorge- well Mezar 3	W	U. Cret.	limest.	*B3*	*b1*
25	Yarmouk Gorge-Bir Mukheiba 8	W	U. Cret.	limest.	*B1*	*c*

26	Himma spring	S	U. Cret.	limest.	*B3*	*b1*
LOWER JORDAN VALLEY						
Western side						
27	Bira 2	W	Neogene	basalts	*C*	*b2*
28	Newe Ur	W	Quaternary	basalts	*C*	*c*
29	Bet Yosef	W	Neogene	basalts	*C*	*c*
30	En Huga	S	Tertiary	chalk	*B1*	*c*
31	En Amal	S	U. Cret.	limest.	*B1*	*b1*
32	En Moda	S	U. Cret.	limest.	*B1*	*b2*
33	Hammam el Malih 1	S	Jurassic	sandst.	*B2*	*b2*
34	Beqaot 2	W	U. Cret.	limest.	*B1*	*b1*
35	Beqaot 1	W	U. Cret.	limest.	*B1*	*b1*
36	Well 20 17/11	W	Cret.-Tert.	limest.	*B1*	*b1*
37	Argaman 14	W	Cret.-Tert.	limest.	*B1*	*b1*
38	Argaman 29	W	Cret.-Tert.	limest.	*B1*	*b1*
39	Pezael 2	W	U. Cret.	limest.	*B1*	*b1*
40	Pezael 6	W	U. Cret.	limest.	*B1*	*b1*
41	Pezael 8	W	U. Cret.	limest.	*B1*	*b1*
42	Mizpe Jericho 2	W	U. Cret.	limest.	*B1*	*c*
Eastern side						
43	Shuna-North deep well	W	U. Cret.	limest.	*B1*	*b1*
44	Manshiya 2	W	L. Cret.	sandst.	*B1*	*b1*
45	Waqqas	W	U. Cret.	limest.	*B1*	*b1*
46	Abu Zigan	S	L. Cret.	sandst.	*A*	*a*
DEAD SEA AREA						
Western side						
47	Qalia beach, seepage	S	Quaternary	clastics	*B4*	*b4*
48	Ein Feshkha creek	S	Quaternary	clastics	*B4*	*c*
49	Ein Feshkha spring	S	Quaternary	clastics	*B4*	*c*
50	Qane A (spring)	S	Quaternary	clastics	*B4*	*c*
51	Qane B (north)	S	Quaternary	clastics	*B4*	*c*
52	Qane C (south)	S	Quaternary	clastics	*B4*	*c*
53	Mizpe Shalem-Mineral beach	W	Quaternary	clastics	*B4*	*b4*
54	En Gedi spring	S	U. Cret.	limest.	*B1*	*b1*
55	Dead Sea, 31° 28.664' N; 35° 24.980' E; Teufe: -60 m	L			*B4*	*b4*
56	En Gedi Spa	W	Quaternary	clastics	*B4*	*b4*
57	En Boqeq, Nirvana Hotel, Gofrit	W	Quaternary	clastics	*B4*	*b4*
58	En Boqeq, PB 26	W	Quaternary	clastics	*B4*	*b4*
59	Hammei Zohar 15	W	Quaternary	clastics	*B4*	*b4*

60	Hammai Zohar 5	W				
Eastern side						
61	Rama	W	Pal/L. Cret.	sandst.	*A*	*a*
62	Hisban artesian deep well	W	L. Cret.	sandst.	*A*	*b1*
63	JICA well 1	W	Pal/L. Cret.	sandst.	*A*	*a*
64	Zerka Ma'in Waterfall	S	Pal/L. Cret.	sandst.	*A*	*a*
65	Zara Princess Basma Farm	S	L. Cret.	sandst.	*A*	*b1*
66	Zara Bani		L. Cret.	sandst.	*A*	*a*
67	Ibn Hammat 2	S	L. Cret.	sandst.	*B3*	*a*
ARAVA/ARABA VALLEY						
Western side						
68	En Ofarim 5	W	L. Cret.	sandst.	*A*	*a*
69	Ein Jahav 6	W	L. Cret.	sandst.	*A*	*a*
70	Zofar 220	W	U. Cret.	sandst.	*A*	*b4*
71	Zofar 20	W	L. Cret.	sandst.	*B4*	*b4*
72	Zofar 4a	W	U. Cret.	limest.	*A*	*b1*
73	Paran 20	W	L. Cret.	sandst.	*A*	*a*
74	Paran 18	W	L. Cret.	sandst.	*A*	*b1*
75	Yaalon 116	W	U. Cret.	limest.	*A*	*b1*
76	Yaalon 6A	W	L. Cret.	sandst.	*A*	*a*
77	Yaalon 7	W	L. Cret.	sandst.	*A*	*a*
78	Yaalon 117	W	Quaternary	clastics	*A*	*b1*
79	Yotvata 13	W	U. Cret.	limest.	*A*	*c*
80	Yotvata 12	W	U. Cret.	limest.	*A*	*b3*
81	Gerofit 4	W	U. Cret.	limest.	*A*	*a*
82	Samar 4	W	U. Cret.	limest.	*A*	*b1*
83	Samar 2	W	U. Cret.	limest.	*A*	*b3*
84	Elat 103	W	Quaternary	clastics	*D*	*b1*
85	Elat 108	W	Quaternary	clastics	*A*	*b1*
Eastern side						
86	Wadi Araba 1	W	Jurassic	sandst.	*B3*	*b3*
87	Wadi Araba 5	W	Jurassic	sandst.	*C*	*b3*
SOUTHERN JORDAN						
88	Disi 1	W	Paleozoic	sandst.	*B1*	*b1*
89	Disi 6	W	Paleozoic	sandst.	*B1*	*b3*
90	Wadi Yutum	S	Quaternary	clastics	*B3*	*c*

The springs contain considerable contribution of strongly Ca-enriched brine with very high $\delta^{18}O$ and $\delta^{2}H$ values (Fig. 3.1.3-4). This Ca-Mg-Cl saline water is highly concentrated by evaporation. Mixing of all these brines result in a Ca+Mg concentration which is balanced by Cl_{excess} as defined in Chap. 3.1.1 (see also Fig. 3.1.3-5).

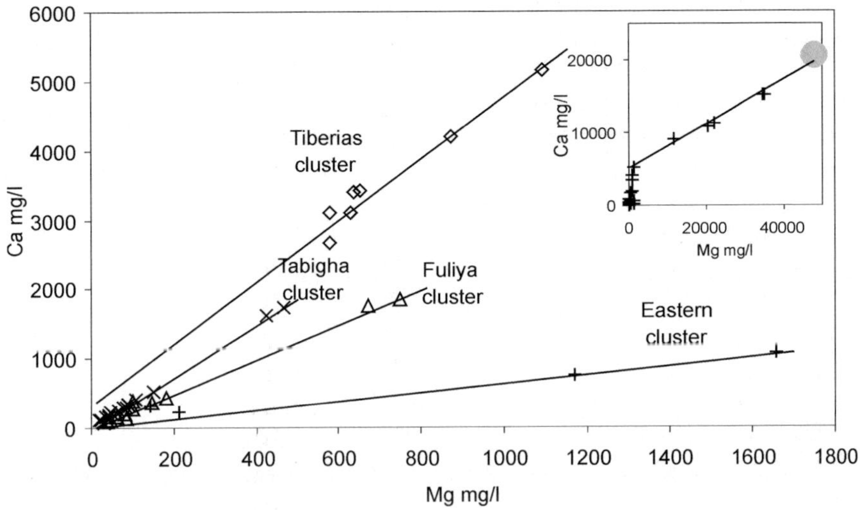

Fig. 3.1.3-2. Ca vs. Mg cross plots show distinct trends for each cluster in the surroundings of Lake Tiberias. Each cluster has a well defined Ca/Mg ratio. The Y-axis intercepts are different.

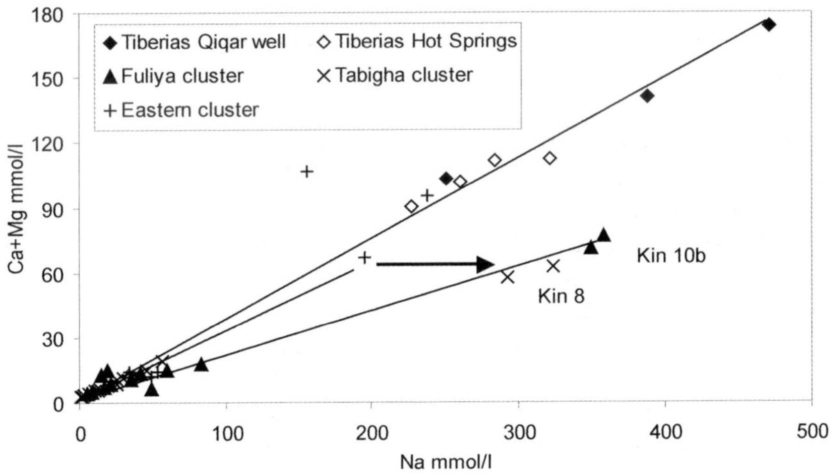

Fig. 3.1.3-3. Cross plots of Ca+Mg vs. Na of the various spring clusters and their wells in the surroundings of Lake Tiberias. Kinneret 8 is enriched in Na and does not fit the trend of all the other groundwaters from the Tabigha area. The less saline groundwaters from Fuliya area plot together with the Tabigha waters, whereas the more saline ones build a new mixing line with Kinneret 10b onto which also plots Kinneret 8.

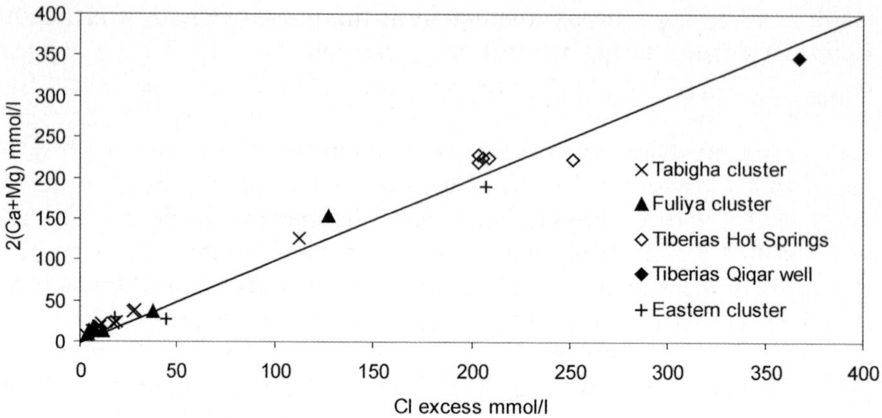

Fig. 3.1.3-4. Cross plots of Ca+Mg vs. Cl_{excess}

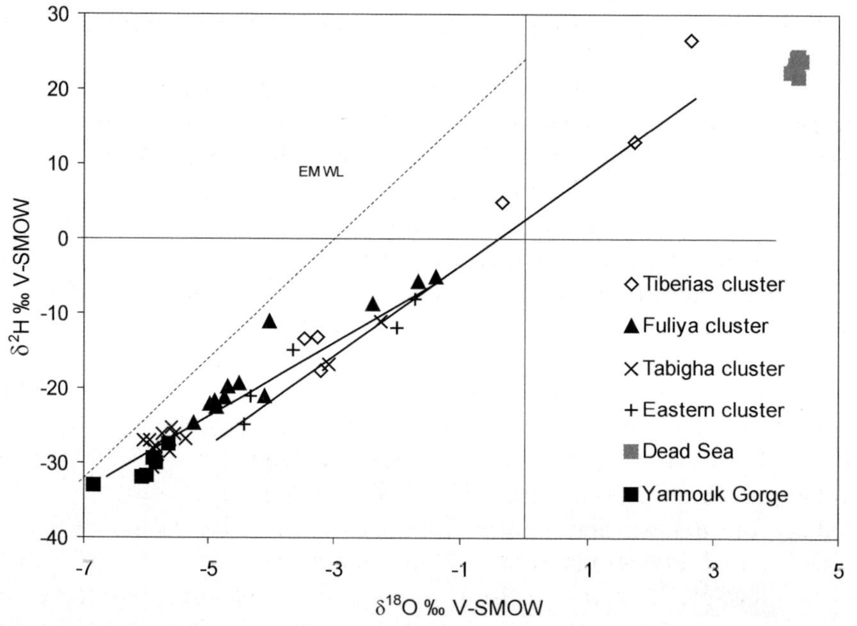

Fig. 3.1.3-5. The $\delta^{18}O$ vs. δ^2H isotope plots yield a common mixing line that fits all clusters of springs and wells in the surrounding of Lake Tiberias.

Further northwards, near *Fuliya* (Fig. 3.1.3-1) springs emerge from a tilted fault-block built of Upper Cretaceous carbonates. The salinities of the emergent springs are lower (17 g/l Cl^-) and so are their temperatures (28°C). Submarine seepages also emerge around the eastern margins of this fault-block. The gross chemical composition indicates thermosaline waters of *REY-type b4* originating from limestones (*type B4)* (excepting Fuliya-D906 positioned between *types B1* and *B4*). Fig. 3.1.3-3 suggests that the Fuliya cluster has two saline sources. Fig. 3.1.3-3 shows that Kinneret 10b is the most saline water of this cluster but is only the source of salinity of few saline waters of the Fuliya cluster. The low-salinity waters are mixtures between a different Ca-Cl brine and recent rainwater precipitating over Upper Cretaceous limestones. Irrespective of mixing of freshwater with Ca rich saline water, it is also mixed with a Na rich brine (probably of the Zemah type). The mixed brine is also diluted by freshwater which causes in turn the high variety of saline waters at Fuliya. The low-salinity water of Fuliya cluster corresponds in (Ca+Mg)/Na with those of the Tabigha cluster but Fig. 3.1.3-2 shows that Ca/Mg ratios are different. Measurable tritium indicates young ages of water (Table 3.1.2-1).

Springs producing the largest quantities of saline water emerge along the north-western shore of the Lake near *Tabigha.* There, the Nur and Sheva springs flow out from fault-blocks built of Upper Cretaceous and Tertiary limestone beds.

The gross chemical composition indicates that the limestone is slightly hydrothermally altered (*type B2).* The REY-*type b3* pattern indicates recharge over Tertiary (Eocene Avedat Group) limestone. Further to the west, the wells Qalanit 2 and Huqoq are replenished from Upper Cretaceous Judea Group limestone. These waters show low chlorinities (< 100 mg/l), whereas other wells and spring range up to 16 g/l. Kinneret 8 is the least diluted saline water in this cluster. The two Sartan springs differ only in mixing ratios with a common saline brine of type Kinneret 8. The different Mg-Ca trends of the Tabigha and Fuliya cluster are explicable by different degrees of dolomitization. Kinneret 8 is affected by Na rich but low Ca brine like many waters of the Fuliya group (Fig. 3.1.3-3). The fact that Ca/Mg ratios of the Tabigha cluster are uniform and also include Kinneret 8 elucidates that this ratio is locally controlled (Fig. 3.1.3-2).

Along the eastern and southern shores of the Lake, saline waters in boreholes and in springs reach Cl^- concentrations up to 20 g/l. Their chemical composition differs from that of waters encountered along the western shore of the Lake mainly by Mg/Ca ratios >1. These saline waters emerge mostly from Neogene clastic beds. The REY patterns indicate that the recharge of the Gofra spring (*type b1*) occurs over Cretaceous limestone but emerges mostly from Neogene clastic beds but the nearby Haon water mainly originates from the basalts of the Golan Heights (*type c*). The chemical composition of the Haon water is the one that shows best the composition of a residual brine from the Sdom Sea. All other saline waters may carry a fraction of this original brine but with varying degrees of Mg-Ca exchange by dolomitization. In this view, the most saline groundwater of the Tabigha-, Fuliya-, and Tiberias clusters could be regional derivatives of a brine, best presented by Haon water.

Each cluster represents mixtures of the locally altered saline waters and meteoric water from Upper Cretaceous Judea- and Tertiary Avedat Groups. Locally, high-Na brine enters the system by which the waters of the wells Kinneret 8 and 10b gain some Na. Thus, along the western shore of Lake Tiberias, at least two brines from different sources interact with freshwater and by alteration (mostly by dolomitization) yield the high variety of saline waters in the area. With the exception of the brine from borehole Zemah 1 characterized by molar ratios (Ca+Mg)/Na~0.01and Na/Cl~1, all other brines in the Rift show Na/Cl<<1 (Goldschmidt et al. 1967). Their evolution is based on the evaporation of mixtures of seawater, drainage water and of meteoric precipitation. The chemical development of the Mg-dominated brine of the Haon type is not the same as would be expected for evaporating seawater but fits the trend of Rift brines as outlined by Klein-BenDavid et al. (2004). It is generated by the drying-up of the Sdom Sea. Drainage water plays an important role in explaining the enrichment in Br. The very low Mg- and high Ca- content and the fact that Ca+Mg equals Cl excess (Fig. 3.1.3-4) strongly indicate Mg-Ca exchange by local dolomitization resulting in tertiary brines. The tertiary brines are not explicable as ablation brine from the Zemah evaporite body as suggested by Simon and Mero (1992). Such a process would result in low Ca concentrations because of dissolution of Mg sulfates. Any reduction of sulfate by organic matter would enhance bicarbonte and finally cause precipitation of calcite. The Haon brine resembles much more a residual marine evaporation brine than any other saline water of the area. The Haon water contains a fraction of the Zemah brine because Na_{sw} and Cl_{sw} are rather high. This fraction may be the result of leaching the salt plug of Zemah by freshwater percolating into the Rift and rising as the water becomes heated (Flexer et al. 2000). The deep-seated Zemah brine is not a residual liquor but an ablation brine. Furthermore, the reported Zemah brine is of the Ca>Mg typc. Flexer et al. (2000) suggested that Ca-chloride thermosaline waters could be dilution products of deep-seated Ca-chloride brine. They also considered that the "western" Ca-chloride brine could have geochemically evolved from the "eastern" Mg-rich brine by water/rock interaction.

There is no need to assume constant penetration of seawater into the Rift in order to explain salinization of groundwater as suggested by Mazor and Mero (1969b) and Kafri and Arad (1980). Brine evolution started neither with a diluted Ca-brine (Bergelson et al (1999) nor from Ca-rich brines which migrated northward within the Rift from the Dead Sea area (Starinsky, 1974). Using the (Ca+Mg)/Na index the Dead Sea water (1.7) compared with 0.2 to 0.4 for Lake Tiberias waters, the Zemah brine with 0.01

and the Rosh Pina brine with 0.55, exclude any contribution of the Dead Sea. Based on similarly high $^{11}B/^{10}B$ ratio ($\delta^{11}B$ = 44‰) in the water of Tiberias Hot springs and in the Dead Sea (56 ‰), Vengosh et al. (1991) suggested mixing with Dead Sea water. The semblance of B-isotope ratios, however, is not a convincing argument for the presence of Dead Sea brine in the water of Tiberias Hot Springs because all evaporating saline water finally yields high $^{11}B/^{10}B$ ratios.

Klein-BenDavid et al. (2004) suggested that the brines were formed by mixing of highly evaporated marine brines with less evaporated ones in a dual-mode evaporation basin. This scenario overlooks the influence of drainage water and rainwater during the long time necessary for the evolution of the inland lake and cannot explain the observed different enrichments of Br and B shown in the spider patterns (see Fig. 3.1.1-3–9).

In the nearby gorge of the Yarmouk River, 7 km east of Lake Tiberias, emerge the springs of En Reah and En Makla (Hammat Gader) which differ by their temperatures (25–50°C) and by Cl^- contents of 82–527 mg/l. All springs flow out of chalks of the Eocene Avedat Group, which outcrop along the gorge and on the northern slopes of the Ajlun Mountains. As a rule, in those springs salinity increases in direct relation to temperature. South of the river on the Jordanian side, the groundwater flows westwards in the limestone aquifers of the Upper Cretaceous Belqa Group. The marly aquicludal layers overlying this aquifer create confining conditions with piezometric heads tens of meters above ground-level. These waters are exploited by the Mukheiba and Wadi Arab wells and discharge artesian water. The wells at Mukheiba are extracting water of *type B1* from Upper Cretaceous limestone, whereas REY indicate a basaltic recharge area which is extending to the east of the locality. At Himma, the water of *type B3* is extracted from altered Upper Cretaceous limestone but the recharge area is characterized by *REY- type b1* patterns.

The spider patterns of water from the wells Mezar 2 and 3 are of *type B3*. REY patterns of *type b2* indicate replenishment over limestones. The low $\delta^{18}O$ and $\delta^{2}H$ isotopic ratios in the waters of Hammat Gader springs and of the Mezar wells (Fig. 3.1.3-5) fit an earlier proposed recharge at high altitudes, probably on the Golan Heights (Arad and Bein 1986) and in the Mt. Hermon region (Möller et al. 2003). The chemical composition of most saline groundwaters encountered is represented by Na>Ca>Mg, low Na/Cl molar ratios of 0.75, high 1000Br/Cl molar ratios of 2.5, and by high $\delta^{11}B$ values of 36‰ (Bergelson et al. 1999). Rare gases such as Xe, Ar, Kr and Ne which are dissolved in the thermal and saline waters have an isotopic composition indicating equilibrium with the atmosphere (Mazor 1972). In that sense, it resembles the chemistry of saline springs emerging along the western shores of Lake Tiberias. This chemical composition is indicative for mixtures of meteoric water with Ca-Cl brines (Starinsky et al. 1979). Enhanced temperatures and pressures, correlated dissolution of minerals, and water salinization were recorded on the Jordanian side in the confined parts of the aquifer. Dissolution of U-bearing minerals and of U-decay products, result in high concentrations of radon gas in the discharged water (Salameh 1996). Few samples from both sides of the Yarmouk Gorge show similar trends in cross plots of Ca vs. Mg and Ca+Mg vs. Na (Fig. 3.1.3-6). Mukheiba plots aside in Fig. 3.1.3-6b which could be due to recharge over basalts, whereas the other groundwaters - according to REY patterns - are recharged over limestone areas.

The lower Jordan Valley

The lower Jordan Valley includes two rhombic pull-apart basins - the Bet Shean Valley and the Pezael depression. The Dead Sea area, which is of similar morphotectonic origin, will be dealt with separately. The lower Jordan Valley as a whole is the regional recipient and mixing zone for groundwater originating in adjacent calcareous and basaltic aquifers.

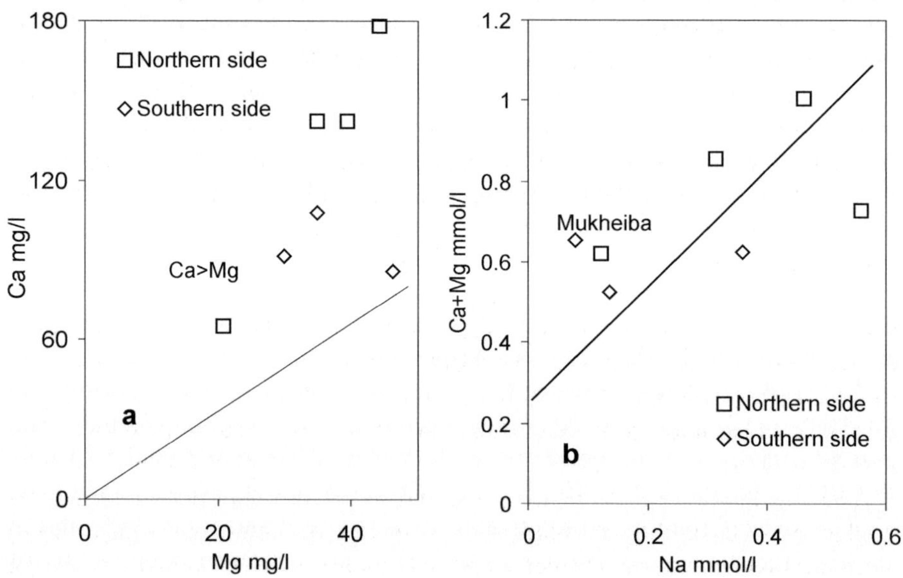

Fig. 3.1.3-6. Cross plots of Ca vs. Mg and Ca+Mg vs. Na of well water samples from the Yarmouk Gorge.

The eastern and western margins of the valley are outlined by the longitudinal faults of the Rift and by step faults. In the valley, the fault-blocks are buried under clastic beds, the thickness of which increases progressively towards the middle parts of the Valley reaching thousands of meters. The thick clastic beds are locally intruded by igneous bodies.

On both sides of the Valley, rock-formations are exposed which include Triassic gypsum beds, Jurassic sandstones, thick sequences of Cretaceous-Tertiary limestones and of basalts of Neogene to Pleistocene age. Depending on physical properties of these formations, they act as regional aquifers. In the centre of the Rift Valley, the upper part of the clastic sequence is formed by 100–150 m thick beds of the Lisan Formation characterised by low transmissivities. Locally, Neogene salt structures are buried beneath the clastic fill. The large Zemah structure is buried at a depth of over 1600 m. Two other salt-structures were identified: the Zaharat el-Qurein structure located south of the Bet Shean area (Belitzky and Mimran 1996) and another one in the subsurface between Jericho and Pezael (Shulman et al. 2003; Chap. 3.2.1) are identified.

Because of structural conditions specific to the western part of the Valley, the Judea-, Avedat- and two basaltic groundwater bodies drain to the Valley through interconnected outlets which are common to these aquifers (Rosenthal 1987, 1988). Upon reaching the interconnecting zones created by the marginal faults of the Rift, fresh groundwater deriving from different aquifers may intermix. East of the Jordan River, groundwater flows from the foothills westwards towards the River through alluvial fans and other young clastic sediments interfingering with the beds of the Lisan Formation, which in these areas contain gypsum and halite. Recharge takes place through lateral flow from aquifers extending east of the foothills. Deep groundwater in the Lower Cretaceous aquifer is saline and under artesian conditions. It seeps upwards through the young surficial sediments forming saline springs such as in Wadi Mallaha. The salinity of these springs reaches to a few tens to thousand mg/l (Salameh 1996). Locally, alluvial fans contain water of excellent quality. However, the extension of these fans towards the River contains increasingly brackish and salty waters which originate from irrigation return flow and mixing with salty formation waters.

West of the Jordan River most groundwaters are chemically characterized as *types B1* and *B2* typically hosted in Upper Cretaceous aquifer limestone and REY patterns of *type b1;* occasionally *type b3* indicate recharge over Upper Cretaceous and Eocene limestones. At En Huga the recharge occurs over a basaltic area. The groundwaters from the Bira, Newe Ur and Bet Yosef wells (Neogene and Quaternary aquifers) are chemically controlled by leaching of basalts but Bira 2 is recharged over Eocene limestone (*type b3*). This water flows laterally into the basaltic aquifer from which it is exploited (Möller et al. 2003). In the eastern Jordan Valley the groundwaters are drawn from Lower Cretaceous sandstones and Upper

Cretaceous limestones. The deep well of North-Shuna produces water from Upper Cretaceous limestone aquifer which corresponds with types *B1* and *b1*. At Manshiya, the chemistry of major ions and the REY, characterize the water as originating from Upper Cretaceous limestones although it is drawn from Lower Cretaceous sandstones. Here young water descends into deeper rock-horizons. At Abu Zigan, all chemical indicators correspond with the occurrence in Lower Cretaceous sandstone.

In the $\delta^{18}O$ vs. $\delta^{2}H$ cross plots (Fig. 3.1.3-7) the data from the western side of the Jordan River split into two groups. The upper group denotes waters from basaltic aquifers (Bet Shean), whereas the lower one comprises waters from various lithologies such as Jurassic limestone (Hamman el Malih), Judea Group limestones (Beqaot, Pezael, Jericho), and Eocene Avedat limestone Group (Argaman). The isotopic difference is caused mainly by an altitude effect of the recharge area. Both groups also differ in Na and Br vs. $\delta^{18}O$ (Fig. 3.1.3-8). Ca+Mg vs. Na+K (Fig. 3.1.3-9) and Ca vs. Mg (Fig. 3.1.3-10) highly scatter. The high scatter of samples from basaltic aquifers is caused by weathering processes of mafic rocks. Hammam el Malih and well 20-17/11 are not shown because they plot outside the diagram. The waters from Beqaot, Argaman, and Pezael are slightly infested by Ca-Cl-Br brines. Comparing the lithological sources of the Jordanian waters with those of the REY interpretation, it follows that groundwater from Zara Princess Basma Farm (Lower Cretaceous) is recharged by water from Upper Cretaceous limestones (Fig. 3.1.2-4). In Fig. 3.1.3-7, δ2H and δ18O show lowest isotope ratios. In Fig. 3.1.3-8, δ18O

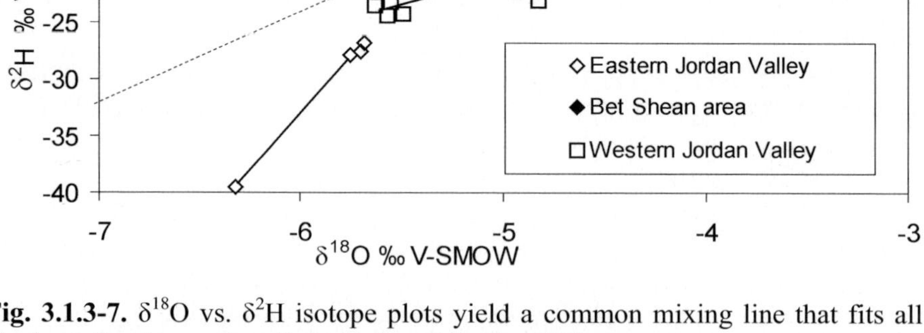

Fig. 3.1.3-7. $\delta^{18}O$ vs. $\delta^{2}H$ isotope plots yield a common mixing line that fits all clusters of springs and wells in the lower Jordan Valley

shows a negative trend with Na and Br indicating dilution of saline groundwater by freshwater. The correlations of Ca+Mg vs. Na+K (Fig. 3.1.3-9) are similar with those of the western waters. Ca vs. Mg (Fig. 3.1.3-10) show different ratios for high concentration in both parts of the Rift.

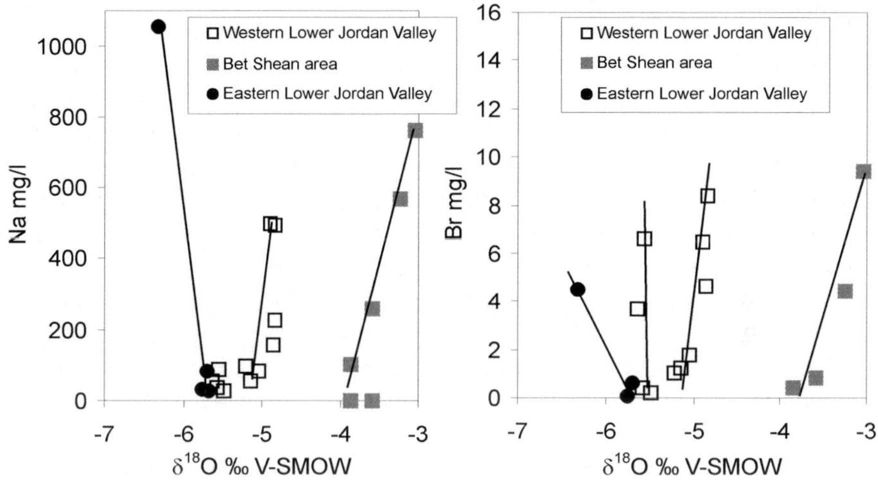

Fig. 3.1.3-8. The Na and Br vs. $\delta^{18}O$ cross plots reveal two distinctly different trends for groundwater from basaltic (Bet Shean area) and from limestones aquifers (Western lower Jordan Valley).

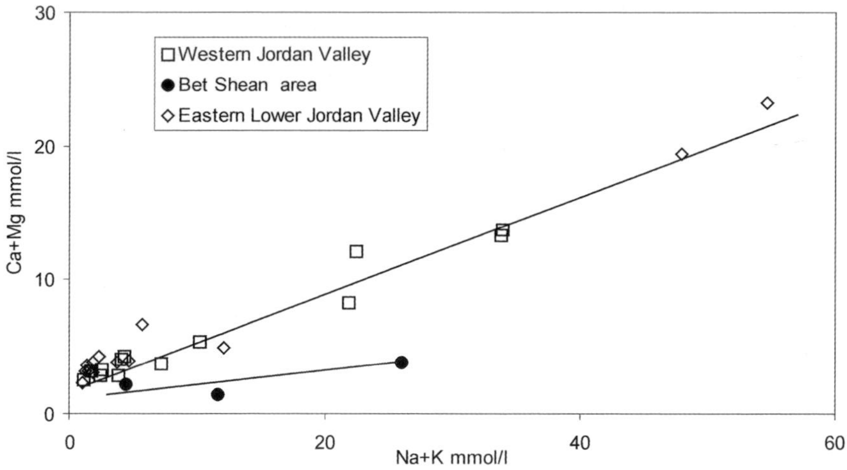

Fig. 3.1.3-9. Cross plots of Ca+Mg vs. Na+K of the various spring clusters and their wells in the surroundings of the lower Jordan Valley. The groundwater from basaltic aquifers scatters widely, whereas those from Upper Cretaceous limestone aquifers are more homogeneous.

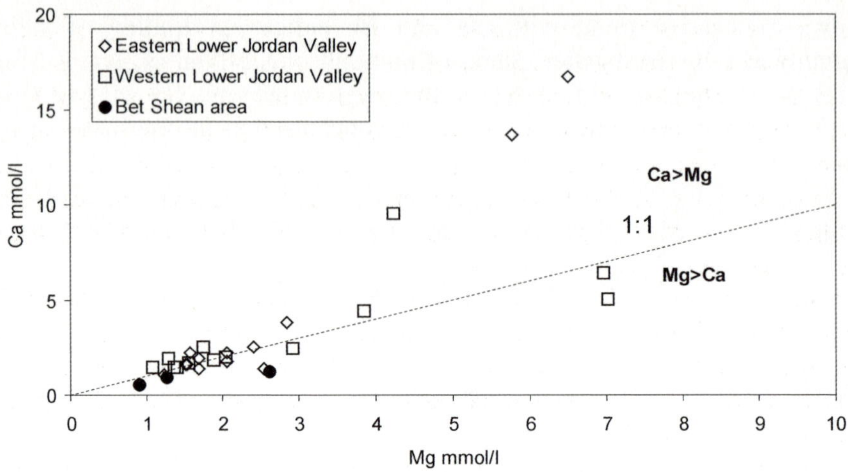

Fig. 3.1.3-10. Ca vs. Mg cross plots show distinct trends for all clusters in the lower Jordan Valley. The groundwater from basaltic aquifers is characterized by Mg>Ca, whereas the limestone waters with higher concentrations of Ca and Mg scatter. The groundwater from the eastern side shows a tendency to Ca>Mg.

The Ca-chloride brines originating in the western lower Jordan Valley from deep-seated and confined reservoirs leak along regional faults and contaminate the relatively fresh groundwater which originates from the two calcareous and two basaltic aquifers bordering the Valley. It was suggested that these brines are residual ones from the Sdom Sea period (Starinsky 1974; Rosenthal 1988), heated up by reaction with basalt bodies (Vengosh and Rosenthal 1994).

Considering the prevalently clastic character of the beds filling the centre of the Valley, hydraulic contact between the areas located east and west of the River Jordan, cannot be excluded *a priori*. However, the chemical composition of water from lithologically comparable aquifers differs on both sides of the Jordan River (Fig. 3.1.1-10). This denies significant exchange of water beneath the Jordan River bed. Enhanced B contents in Jordanian groundwater are controlled by dissolution of gypsum, probably originating from wind-blown sediments possibly from outcrops of the Lisan Formation or from other distant sources much further to the south.

3.1.3.4 The surroundings of the Dead Sea

The Dead Sea is located within a rhomb-shaped pull-apart basin bordered by the various fault systems of the Rift. It extends at the lowermost elevation in the world (-410 m msl) and it is the terminal lake for surface and groundwater flow in the area.

In areas west of the Dead Sea mostly calcareous rock-formations of Upper Cretaceous age outcrop. West of the Rift, exceptional geological features are several breached and eroded anticlinal structures (Makhtesh Hatira and Hazera) in which Jurassic and Lower Cretaceous beds are exposed. The Neogene salt diapir of Mt. Sdom protrudes along the south-western margins of the Sea. Similar buried salt structures are identified further northwards along the western shores of the Sea and also beneath the Lisan Peninsula (Bender 1974).

The central part of the Rift contains thousands of meters of clastic fill covered by the Pleistocene Lisan Formation the thickness of which does not exceed 100 m. The Dead Sea is imbedded into this sequence and its bottom is covered by very thick layers of rock-salt (Neev and Emery 1966).

Three main sources of water originating from the west drain to the Dead Sea:

- Paleowater flowing in the Lower Cretaceous Kurnub Group sandstones. These waters are characterized by chlorinity of the 500–600 mg/l, and become progressively saline along their flow paths towards the Dead Sea reaching Cl^- concentrations as high as 5 g/l. This is due to progressive mixing with brines in the fault zones outlining the Rift.
- Flash floods occurring over the eastern and north-eastern Negev and infiltrating into and recharging the Upper Cretaceous limestone and dolomite of the Judea Group aquifer. The chlorinity varies widely reaching from 200 mg/l to maximal concentrations of 7.5 g/l.
- Groundwater originating from the Arava Graben fill aquifer which consists of coarse gravel deposited as terraces or as alluvial fans spreading out at the outlets of the wadis towards the Dead Sea. The aquiferous horizons in the alluvial fill are mainly recharged by flash floods in the wadis and by springs and seepages emerging along the Dead Sea shore. The Cl^- concentrations of these waters vary between 1.5 to about 26 g/l. The high salinity is related mainly to mixing with remnants of the Dead Sea water when it was at higher stands.

Along the western shore of the Dead Sea, waters of *types B1* and *B4* are drawn from Quaternary sediments of dominantly Upper Cretaceous limestone. REY patterns of *type b4* indicate presence of very saline brines. Only the freshwaters of En Gedi spring and waterfall are from Upper Cretaceous limestone aquifer, a fact confirmed by REY indications as the recharge area. The freshwater at En Gedi and En Boqeq is recharged by current rainfall over the eastern flanks of the Judea Mts. (Arad 1964).

The water of En Gedi Spa 2 and of En Boqeq-Gofrit is high in $\delta^{18}O$ and shows limestone signatures by REY patterns. In contrast, Zohar 5 is extremely low in δ^2H (-42 ‰) indicating contribution of Pleistocene water from Kurnub sandstones. The thermal waters from the western shoreline plot along a mixing line of a very saline component like the Dead Sea brine and waters from the Kurnub- and Judea Group aquifers (Fig. 3.1.3-11 and Fig. 3.1.2-9). This water mix consists of Pleistocene and recent meteoric water flowing in the Judea Group. The cross plot of Na and Br vs. $\delta^{18}O$ show two widely separated clusters and a poor trend line (Fig. 3.1.3-12). The Ca+Mg vs. Na+K cross plots (Fig. 3.1.3-13) show wide ranges of trends. The lowest one is given by the En Gedi Spa trend with $(Ca+Mg)/(Na+K) \approx 0.5$. Only very few samples follow the Dead Sea trend with $(Ca+Mg)/(Na+K) \approx 1.5$, which represents the evaporation of the late Lisan Lake.

Most of the brines plot between the two denoted trends. All samples below the evaporation trend may have gained alkalis by ablation of halite, whereas those above the line acquired their chemistry either by albitization of gabbros or by mixing with an old $(Ca,Mg)Cl_2$ brine. More details may be derived from Ca vs. Mg plot Fig. 3.1.3-14). Most of the deep brines are of the Ca>Mg type, whereas spring and well waters along the western shore of the Dead Sea belong to the Ca<Mg type. With few exceptions, most of the saline waters represent a mixing line of En Boqeq-Gofrit brine with a Ca>Mg water of lower salinity, which itself may be a mixture of a high-salinity Ca>Mg brine with freshwater.

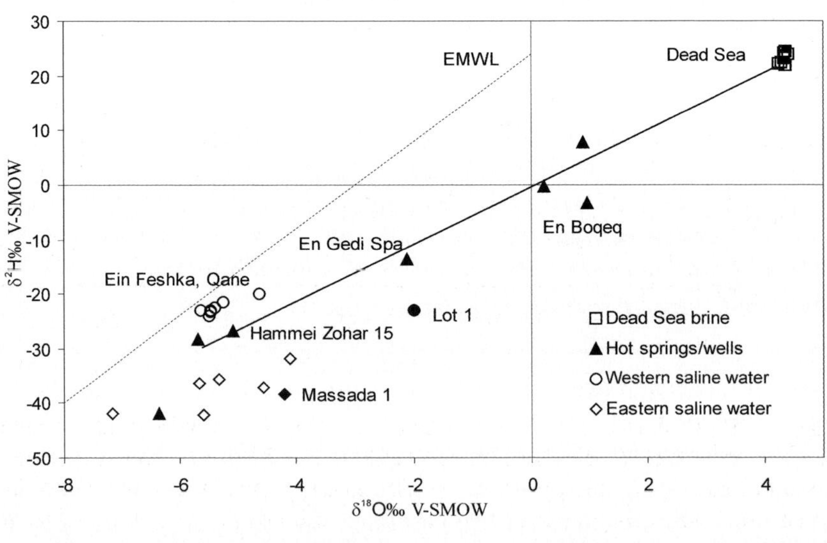

Fig. 3.1.3-11. $\delta^{18}O$ vs. δ^2H isotope plots for spring and well waters of low to high salinity in the surroundings of the Dead Sea.

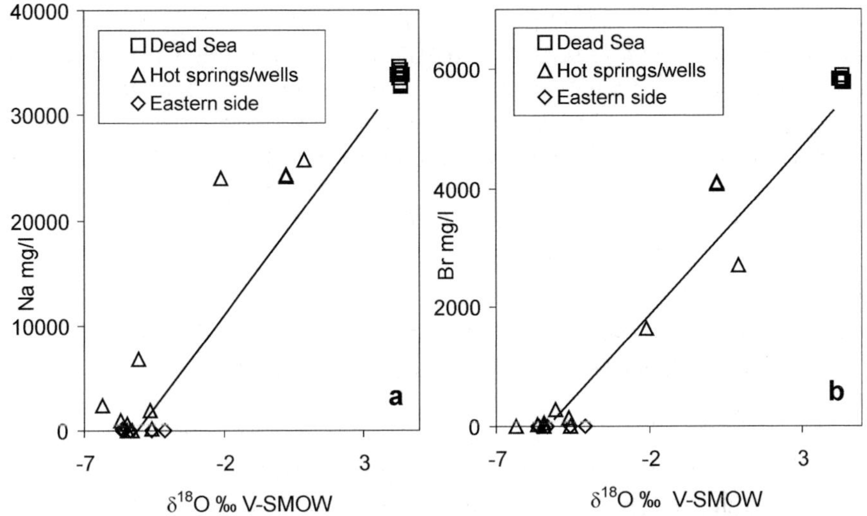

Fig. 3.1.3-12. The Na and Br vs. $\delta^{18}O$ plots show two widely separated clusters and a separate trend line for the hot spring brines from the surroundings of the Dead Sea.

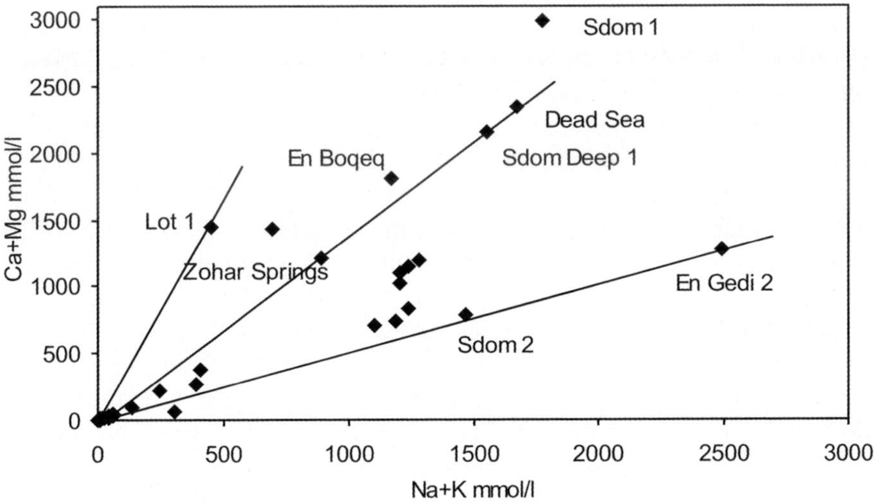

Fig. 3.1.3-13. Cross plots of Ca+Mg vs. Na+K show a wide scatter of (Ca+Mg)/(Na+K) framed by those of En Gedi 2 and Lot 1. Three trends are discernable: The Dead Sea analyses cluster narrowly; the hot brines show highly variable Ca+Mg contents at nearly constant Na content; the low salinity waters show a trend at low (Ca+Mg)/Na ratio.

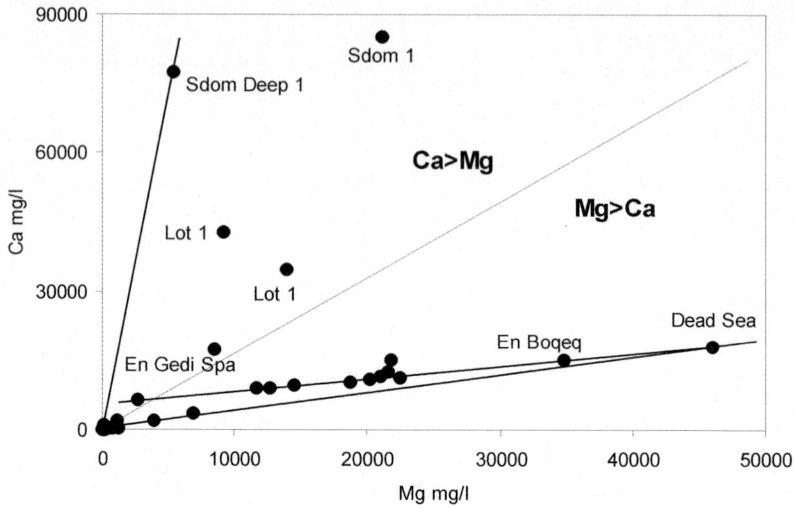

Fig.3.1.3-14. Ca vs. Mg plots shows distinct trends. All brines plot left to the Dead Sea - En Boqeq line. On the other hand En Boqeq, En Shalem, and En Gedi Spar (=En Gedi 2) show increasing Ca/Mg ratios.

In the low sulfate but Ca-rich waters (Fig. 3.1.3-15: Qane, Ein Feshkha, En Gedi 2, Sdom 2) Ca is 40 times higher than sulfate content, which strongly points to contributions of deep-seated Ca-Cl brines. The association of halite and anhydrite may indicate leaching from the sediments of the late Lisan period. In the deep brines, Sdom 1, Sdom Deep 1, and Lot 1, sulfate increases with decreasing Ca (Fig. 3.1.3-15b) and a much higher sulfate concentration is reached than in the Dead Sea brine itself. Thus, Figs. 3.1.3-10–14 reveal that at least three different saline sources define the chemical composition of waters: primary Dead Sea brine, tertiary $(Ca,Mg)Cl_2$ brine, and a secondary ablation brine of halite and anhydrite.

The springs of Ein Feshkha and Qane, emerge at the north-western corner of the Dead Sea, at the fault contact between Upper Cretaceous limestones and the fill of the Rift. Their chlorinity varies from 1.7–40.2 g/l. $\delta^{18}O$ of -23‰ is typical for water from the Judea Group (Fig. 3.1.3-11). They are non-thermal waters and mixing products between freshwater draining from the west and probably water of the Dead Sea at higher stands. Ein Feshkha receives water with the same REY pattern as the well 20-17/11 north-west of Dead Sea (Fig. 3.1.2-8; Chap. 3.3.2). The REY patterns of these waters indicate that they experienced (Fe,Mn)OOH precipitation probably caused by weathering of basaltic material (Chap. 3.1.2).

Plate 3.1.3-1. Sampling of brines seeping from varved sediments exposed after the retreat of the Dead Sea near Qalia recreation centre. (Photo: S. Geyer, 2000)

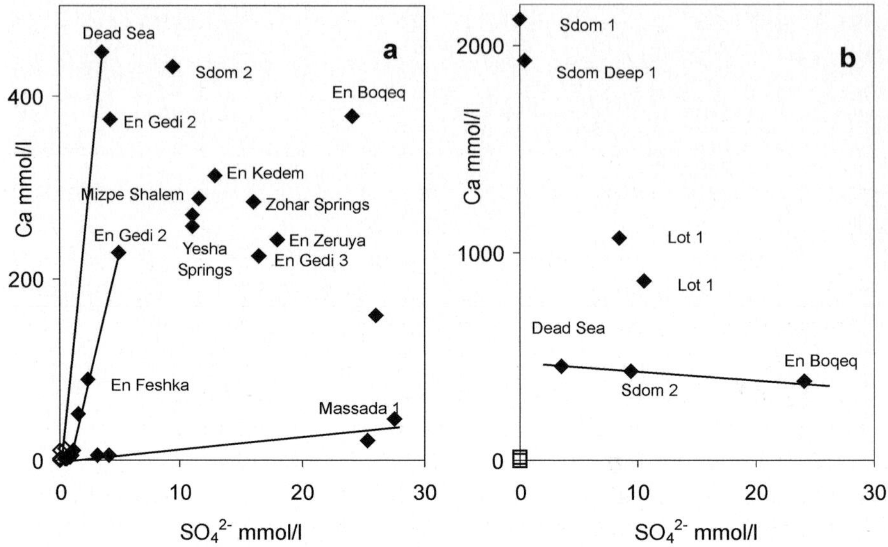

Fig. 3.1.3-15. Ca vs. SO_4^{2-} cross plots showing a wide scatter of trends

Three major clusters of thermosaline springs emerge along the western shores of the Dead Sea: Hammei Zohar in the south, Hammei Yesha south of En Gedi and Hammei Shalem further north. The temperatures of these waters reach 42°C and their Cl^- content ranges from 30–130 g/l. The waters from wells En Boqeq-Gofrit, En Gedi 2, and Mizpe Shalem show subparallel REY patterns with pronounced anomalies of Gd and Y, similar to that of the Dead Sea average (see Fig. 3.1.2-7). They could well represent mixtures of recent Dead Sea water with freshwater from adjoining aquifers in the west and from flash floods in the wadis flowing to the Dead Sea. As to the thermosaline springs along the western shores of the Dead Sea (Hammei Zohar, Hammei Yesha and Hammei Shalem) Bentor (1969), Arad (1964) and Mazor and Mero (1969a) suggested that they are the result of mixing of freshwater derived from adjacent aquifers and upflowing brines.

Rosenthal et al. (1991) found that water from the Kurnub Group flow laterally into the down-faulted blocks built of Judea Group beds. Rosenthal et al. (1998) delineated different flow paths and geochemical processes that could shape the chemical composition of the Kurnub Group water along the flow path (dissolution of soluble salts from the Kurnub Group rocks and chemical reduction of sulfates). Close to the fault zone bordering the Rift, these waters mix with the brines circulating mainly in the fault zone and with remnants of the Dead Sea from the times it was at higher stands (Yechieli et al. 1992).

All Dead Sea groundwaters and Dead Sea water itself plot on the Ca+Mg vs. Cl_{excess} line (see Fig. 3.1.1-17), which confirms a uniform development of $(Ca,Mg)Cl_2$ brines. In the Mg vs. Ca plot (Fig. 3.1.3-4), however, the Dead Sea and related En Boqeq-Gofrit water follow a trend different from that of all the other samples from the surroundings of the Dead Sea. The difference could be the fact that the groundwaters are involved in dolomitization processes and therefore loose Mg but gain Ca. The Dead Sea water represents the result of evaporation of the Lisan Lake - a continental lake fed by drainage water and by rain flushing out residual brines and dissolving evaporites from the Sdom Sea period. Seawater infiltration as suggested by Mazor and Mero (1969b) can neither be proved nor is it necessary to explain the hydrochemical processes in the area.

Starinsky (1974) suggested that the salts dissolved in the Dead Sea were derived from the springs discharging during geological times. Considering the variations of B and of its isotopes, Vengosh et al. (1991) confirmed Starinsky's view and excluded the possibility that the waters of the thermosaline springs were derived from dilution of the Dead Sea. However, the great similarity of both spider- and REY patterns of Mg>Ca waters from En Gedi 2, Mizpe Shalem, En Boqeq-Gofrit with the Dead Sea brine (Fig. 3.1.2-13) strongly suggests that these waters represent mixtures of subrecent Dead Sea brines with other saline waters in various proportions. Any old, deep-seated brine would yield thermosaline springs with Ca>>Mg.

East of the Dead Sea, groundwater occurs in two different aquifer complexes, i.e., the upper limestone aquifer complex and the lower sandstone complex of which the Palaeozoic to Lower Cretaceous sandstones are the main aquifers. Major ion- and REY chemistry show that these waters originate from sandstones, excepting Hisban deep well and Zara Princess Basma spring for which REY of *type b1* indicate replenishment over Upper Cretaceous limestone. At Ibn Hamad 2, REY clearly indicate the sandstone as the source but the spring waters discharge from hydrothermally altered limestone (*type B3*) indicating ascending water or lateral connection between these formations. Hisban deep well water (*type A*, REY-*type b1*) represents water from Upper Cretaceous aquifer infiltrating Triassic/Jurassic sandstones, where the gross chemical composition is attained.

The upper limestone aquifer receives replenishment by infiltrating precipitation. Large volumes of groundwater discharge through springs along the upper reaches of the wadis of Zerqa Ma'in, Walla, Mujib, Karak, Shaqiq, and Ibn Hamad. The lower aquifer (sandstone complex) does not receive any appreciable amounts of direct recharge by precipitation. Its water originates from different sources such as leakage from the upper aquifers and subsurface flow from the Disi area. Most of the waters are regarded as thermomineral (Salameh and Udluft 1999).

3.1.3.5 The Arava Valley

The whole Arava Valley is a graben-like structure. It is bordered to the east by the Mountains of Edom (Fig. 3.1.3-1) built of Precambrian igneous rocks overlain by sandstones and to the west by calcareous beds of the Judea Group overlying the sandstones of the Kurnub Group. The Arava Valley constitutes a drainage base for the large wadis running from eastern Sinai, the Negev and the Mountains of Edom. On both sides of the Valley a dense swarm of faults created fault blocks buried under thousands of meters thick sequences of clastics related to the Neogene Hazeva Formation overlain by Quaternary to Recent alluvium. In the central parts of the Arava, groundwater is extracted from aquiferous horizons in the Neogene Hazeva Formation and in the young alluvial fill. These waters are recharged mainly by flash floods in the wadis draining the mountainous areas east and west of the Rift and also by groundwater originating in the Kurnub Group leaking into the buried fault block. The chlorinity of these waters varies and ranges from 200–4,700 mg/l. The main source of salinization is the upflow of a regional brine (represented by wildcat borehole Arava 1 in Fig. 3.1.3-16) along the various systems of faults (Yechieli et al. 1992). All Jordanian groundwaters tend to fit the trendline defined by Arava 1 and Zofar 20. Zofar 20 resembles the omnipresent Rift brines of the Tiberias Hot Spring type (see Fig. 3.1.1-7a), whereas Elat 103 is a local occurrence caused by evaporation under playa conditions, which were similar to those that created the brines controlling the salinity of Haon well (Fig. 3.1.3-9).

In the western Arava Valley, water of *type A* from En Ofarim, En Yahav 6, Paran 20, Yaalon 6A and 7 is recovered from the Lower Cretaceous aquifer and REY distribution patterns indicate recharge over sandstone areas (Table 3.1.3-2; REY-*type a*). The groundwater from well Paran 18 is recharged over Upper Cretaceous limestones but drawn from the Lower Cretaceous sandstones. The water from Zofar 20, although flowing in the Lower Cretaceous sandstones and close to a fault, is macrochemically controlled by hydrothermally altered limestone characterized by *type B4* spider pattern (see Fig. 3.1.1-7a) and *type b4* REY pattern (see Fig. 3.1.2-7). Many wells extracting water from the Upper Cretaceous aquifer, actually yield water that passed Lower Cretaceous sandstones with anhydrite cement or lenses (Table 3.1.3-2; *type A*: Zofar 4a, Yaalon 116, Yotvata 12, and Samar 2 and 4) but their recharge is over Upper Cretaceous limestone (REY-*types b1* and *b3*). Groundwater drawn from Quaternary sediments either originate from sandstones (Elat 108, Yaalon 117) or from altered limestones (*type B4*). However, REY-*type b1* indicates recharge over Upper Cretaceous limestone. These waters are recharged by floods flowing through wadis embedded in calcareous formations of the Judea Group. On the other hand, as a result of step-faulting, hydraulic contact is locally established between aquiferous beds of the Kurnub and the Judea Groups facilitating flow of the confined Kurnub paleo-water into the karstic formations of the Judea Group and vice versa (Chap. 2.2).

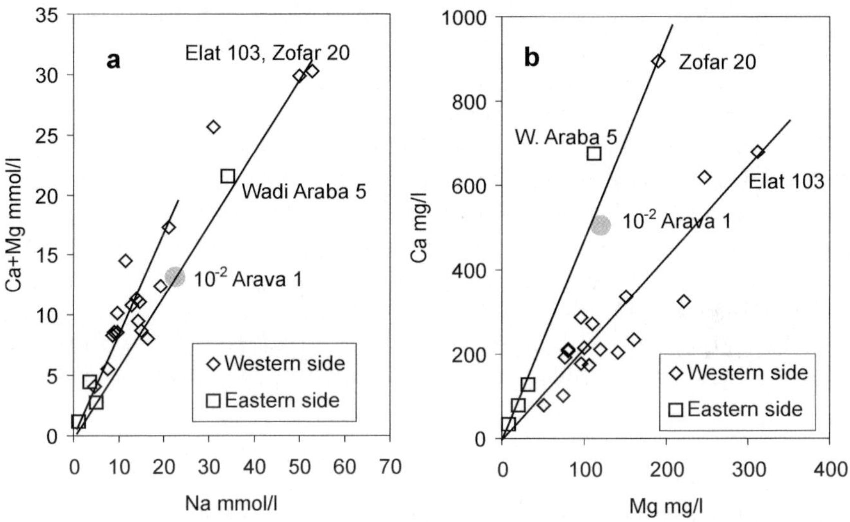

Fig. 3.1.3-16. Cross plots of Ca+Mg vs. Na and Ca vs. Mg of groundwater from the Arava/Araba Valley. In both figures two different trends occur. Note that Arava, Elat 103 and Zofar 20 are located in the western part of the Valley, whereas borehole Araba 5 is on the eastern part.

In the southern Arava, groundwater from wells Samar 2 and Yotvata 12 (*type A* indicating Kurnub sandstone origin) derives from the alluvial fill of the Rift which is chiefly composed of limestone debris and sand (Table 3.1.3-2). However, the REY patterns of these waters are chemically similar to waters of the main spring of Tiberias Hot Springs (see Fig. 3.1.1-6a) and water from a basalt-limestone contact zone Maqarin in Jordan (see Fig. 3.1.2-6b). In view of this chemical similarity it occurs that the above mentioned Arava waters are probably recharged by flood events occurring in the Wadis Daraba and Turhan in Jordan, the drainage network of which is prevalently within volcanic rocks and their interaction zones with limestone and in limestone-rich alluvium (Rosenthal et al. 1998).

In southern Jordan adjacent to the Arava Valley the Disi wells yield water which is pumped from Palaeozoic sandstone. However, spider (see Fig. 3.1.2-4) and REY patterns (see Figs. 3.1.2-4c and 3.1.2-6b) indicate that these waters are recharged over Upper Cretaceous limestones (*type B1*). This finding contradicts the assumption that it is ancient water as suggested by Salameh and Gedeon (1999). Considering the position of these waters in the δ^2H vs. $\delta^{18}O$ plots, these waters may contain only small amounts of Pleistocene water (Fig. 3.1.3-17). Changes in chemical composition from one well to another are probably due to the variability of local recharge water which could be either floodwater over areas covered by granites/sand or by direct recharge of precipitation over areas covered by sandstones or by infiltration of floodwater accumulating in playas and exposed to evaporation before infiltration.

In a well in Wadi Yutum, water is extracted from the gravel fill. REY indicate dominant interaction with basalts which are present as abundant dikes penetrating the Precambrian basement rocks outcropping on both sides of the wadi. The concave-shaped REY-*type c* patterns show the high influence of mafic material. Thus, one could consider the basement rocks as the recharge area but actually the water is chemically controlled by the limestone debris of the wadi fill (Table 3.1.3-2).

Analysed waters from Paleozoic aquifers in Jordan do not show REY patterns typical for a sandstone aquifer containing anhydrite cement. They are similar to those described as being typical for waters of REY-*types b1* and *b3* from Upper Cretaceous limestone and its silicified equivalent. For instance, the drilling of Wadi Araba 5 was stopped when entering a basaltic sill in sandstone of the Palaeozoic Ishrin Group. This water with *type C* spider- and REY-*type b3* patterns is controlled in its gross chemistry by basalts, but the recharge occurred over Upper Cretaceous limestone.

In the Jordanian part of the study area, i.e., in the northern Wadi Araba, the subsurface of the Valley is also built of thousands of meters thick sequence of alluvial sediments. At great depth, the water is saline due to the effects of the Dead Sea interface and to the relics of its predecessors. The groundwater in the area occurs in alluvial deposits, talus and in alluvial fans the thickness of which attains 250 m. Groundwater flows from the mountains in the east, west- and north-westwards to the Dead Sea. Generally, the salinity of water increases in the direction of flow. In the confined parts of the aquifer, higher temperatures and pressures enhance dissolution of minerals and water salinization. Dissolution of U minerals and of its decay products results in high concentrations of radon gas in the discharged water (Salameh 1996). In the southern Wadi Araba, the subsurface is similarly built of very thick (thousands of meters) sequence of alluvial sediments. The fresh and the brackish groundwater are found in the uppermost portions of the aquifer. Here, groundwater flow is directed southwards to the Red Sea. Recharge is from precipitation falling on the surrounding mountains in the east there it infiltrates the barren rocks and flows laterally into the old fluviatile alluvial deposits covering the floor of the Valley. Some recharge takes place along the wadi courses of the side wadis and of

Wadi Araba itself. Salinity increases in the direction of groundwater flow, i.e., from north to south.

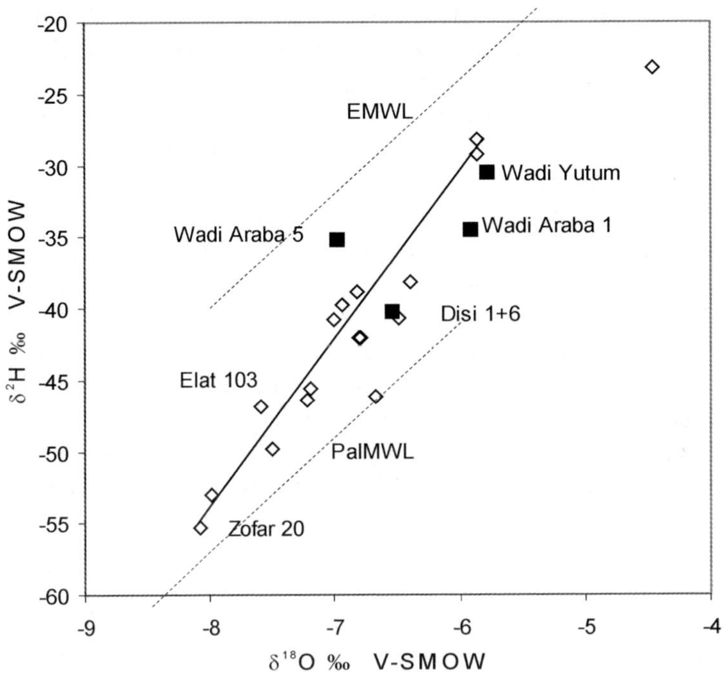

Fig. 3.1.3-17. $\delta^{18}O$ vs. $\delta^{2}H$ isotope plot yields a common mixing line for all groundwaters from the Arava/Araba Valley and south-eastern Jordan. Samples from the eastern part of the Valley are given by solid symbols. Compared with waters from the western part, they contain much less contribution of Pleistocene water. EMWL=eastern meteoric water line; PalMWL=Paleozoic (Pleistocene) meteoric water line.

3.1.3.6 Origin of saline water bodies in the Rift

The chemical evolution of the brines generated during the Sdom Sea period, comprises evaporation of inland transgressing seawater (which was already loaded with dissolved Messinian salts) and of drainage water. This created the Sdom Sea brine with Mg>Ca composition (Chap. 2.3).

As shown in Fig.3.1.3-3, the saline waters from sources along the eastern side of Lake Tiberias (Haon, Gofra) plot together with the water of the Tiberias Hot Spring. This points to a common source. The waters from sources located along the eastern and the western shores of the Lake correspond in the Br/Cl, B/Cl, and Na/Cl ratios (Table 3.1.3.-3). They differ, however, in Mg/Ca and (Ca+Mg)/(Na+K) ratios because of alteration processes such as albitization which could have taken place in the deep structural low of the eastern Lower Galilee. Water resembling that of Tiberias Hot Spring was also identified in well Zofar 20 in the Arava Valley (Chap. 3.1.2). This implies that the Tiberias type of saline water is omnipresent throughout the Rift occurring in the area of the former Sdom Sea (Fig. 2.3-1). Considering the geological evolution of the area as described in 2.1, these saline waters originate from the same mother brine which was formed during the latest stage of evaporation of the Neogene Sdom Sea. As part of this evaporative process, salt bodies formed in the Rift (Zemah, Zahrat el Qurein, Pezael, Lisan Peninsula, Mt. Sdom). Further subsurface ablation of salt bodies such as the Zemah one produced a secondary brine characterized by Na/Cl ≈1 and by low Br/Cl ratios (Table 3.1.3-3).

Table 3.1.3-3. Comparison of ionic ratios of high-salinity groundwater (standard) and regional deep brines (*italics*)

Period	Na/Cl molar ratio	Mg/Ca molar ratio	(Ca+Mg)/(Na+K) equivalent ratio	1000Br/Cl molar ratio	1000B/Cl molar ratio
late Lisan Lake-Dead Sea period					
Dead Sea	0.3	3.8	1.5	26.5	
Radon 1	*0.38*	*3.15*	*0.85*	*9.34*	
Dead Sea Hot Springs	0.4	2.7	0.6	9	
Sdom Sea period					
Representatives of primary brines					
Haon	0.51	2.5	0.4	5.7	0.36
Elat 103	0.56	0.8	0.58	3.6	0.95
Representative of secondary brine					
Zemah 1	*1*	*0.09*	*0.01*	*0.2*	
Representatives of tertiary brines					
Tiberias-Main Spring	0.51	0.32	2.8	5.7	0.5
Zofar 20	0.5	0.4	1.7	5.8	0.38
Pre-Sdom Sea periods, probably Triassic-Jurassic					
Devora 2A	*0.48*	*0.09*	*0.54*	*7.2*	
Rosh Pina	*0.53*	*0.2*	*0.45*	*7.4*	
Sdom 2	*0.3*	*0.79*	*0.9*	*7.5*	
Arava 1	*0.85*	*0.39*	*1*	*8.1*	
Sdom Deep 1	*0.25*	*0.12*	*1.4*	*8.8*	
Lot 1	*0.16*	*0.5*	*2.6*	*9.7*	
Arava Valley	0.77		0.85	1.4	

Considering the overall presence of saline waters of the Tiberias type throughout the Rift, the question should be raised whether even older brines than those related to the Sdom period take part in the salinization of groundwater. Such deep brines are higher in Br/Cl ratio and are pressurized (partial pressure release test in Devora 2A and Rosh Pina 1 up to 480 bar). Any dilution of these brines by less saline waters would have enabled ablation of halite resulting in lower Br/Cl ratios. However, such a process could hardly explain the uniformity and omnipresence of the Tiberias type of brine throughout the Rift. Nevertheless, it cannot be excluded that deep-seated brines from the Triassic/Jurassic strata may locally ascend along deep-reaching faults and penetrate into active aquifers causing their salinization. These very ancient, prevalently Ca-Cl type of brines, are not of uniform composition and may reflect different genetic processes in the past.

The influence of Dead Sea water is best discussed by its unique Gd anomaly. Any brine of this type under subsurface conditions looses the Gd anomaly by ion exchange with REY-bearing minerals. Thus, the thermomineral waters of the springs located along the western shores of the Dead Sea (En Boqeq-Gofrit and Hammei Shalem) with their strong Gd anomalies, indicate considerable contribution of Dead Sea water flushed out from its previous higher stands and mixing with high-temperature saline water from deep horizons. Going further back into the development of the previous Dead Sea, the saline water from well Radon 1 located in the Jericho area (Guttman 2000), seems to be of a similar composition as the thermomineral springs at the Dead Sea (Table 3.1.3-3). The contribution of the brine encountered in the deep borehole Arava 1 to the salinization of groundwater bodies in the Arava, does not seem to be significant. Comparison of Br/Cl ratios of groundwater and Arava 1 brine, reveals that the contribution of the latter is negligible in the former (Table 3.1.3-3). As previously indicated by Yechieli et al. (1992) and by Rosenthal et al. (1998), salinization of Arava groundwaters is mainly due to water-rock interaction.

By comparing Br/Cl ratios of water from borehole Radon 1 with those of the Dead Sea thermomineral springs, it occurs that the salinization of groundwater in the Jericho area is mainly by flushing out of the remnants of the late Lisan Lake and of the previous Dead Sea.

Summing up, one may derive that deep-seated brines of the Devora type do not take part in the groundwater salinization process in the study area. Similarly, the brine of Arava 1 is not reflected in the chemical composition of the waters in the Arava Valley. Groundwater salinization is manly influenced by the omnipresent Rift brines as encountered at Haon, Tiberias Hot Springs and Zofar 20. They are also influenced by flushing out of the remnants of late Lisan Lake and of the Dead Sea from the times when it was at higher stands

3.1.3.7 Conclusion

By considering chemical and isotope data, it occurs that the saline water bodies contributing to saline groundwater in the Rift are as following:

- Recent Dead Sea water which is a primary evaporation brine with Mg>Ca,
- Neogene Sdom Sea primary evaporation brine as represented by the Ha-on saline water characterized by Mg>Ca,
- Neogene to Recent secondary ablation brine of the Zemah type with Na/Cl≈1,
- Neogene tertiary brines with Ca>Mg encountered in Tiberias Hot Spring and in well Zofar 20 and present throughout most of the study area, and
- Deep seated brines of the $(Ca,Mg)Cl_2$ type.

References

Arad A (1964) The geology and hydrogeology of the Lower Cretaceous of the Northern Negev and Judean Mts. PhD Thesis, Hebrew University of Jerusalem, in Hebrew, English abstract, pp1–154

Arad A, Bein A (1986) Saline versus freshwater contribution to the thermal water of the Northern Jordan Rift Valley, Israel. J Hydrol 83:49–66

Belitzky S, Mimran Y (1996) Active salt diapirism at the Zahrat el Qurein Dome, lower Jordan Valley. Isr J Earth Sci 45:11–18

Bender F (1974) Geology of Jordan. Bebrüder Borntraeger, Berlin, pp 1–196

Bentor YK (1969) On the evolution of subsurface Brines in Israel. Chem Geol 4:83–110

Bergelson G, Nativ R, Bein A (1999) Salinization and dilution history of groundwater discharging into the Sea of Galilee, the Dead Sea Transform, Israel. Appl Geochem 14:91–118

Flexer A, Yellin-Dror A, Kronfeld J, Rosenthal E, Ben-Avraham Z, Artzstein P, Davidson L (2000) Neogene salt-body as the primary source of salinity of Lake Kinneret. Arch Hydrobiol Spec Issues Limnol 85:69–85

Goldschmidt M, Arad A, Neev D (1967) The mechanism of saline springs in the Sea of Galilee depression. Hydrol Serv Hydrological Paper 11, Geol Surv Bull pp 1–45

Guttman, J (2000) Hydrogeology of the Eastern Aquifer in the Judea Hills and Jordan Valley. Final Report 1997-2000-Multi Lateral Project, Mekorot Report, pp 1–468

Kafri U, Arad A (1980) Current subsurface intrusion of Mediterranean Sea water, possible ource of ground water salinity in the Rift Valley system, Israel. J Hydrol 44:267–287

Klein BenDavid O, Sass E, Katz A (2004) The evolution of marine evaporitic brines in inland basins: The Jordan Dead Sea Rift Valley. Geochim Cosmochim Acta 68:1763–1775

Mazor E (1972) Paleotemperatures and other hydrological parameters deduced from noble gases dissolved in groundwaters: Jordan Rift Valley, Israel. Geochim Cosmochim Acta 36: 1321–1336
Mazor E, Mero F (1969a) The origin of the Kinneret-Noit water association in the Kinneret-Dead Sea Valley, Israel. J Hydrol 7:318–333
Mazor E, Mero F (1969b) Geochemical tracing of mineral and freshwater sources in the Sea of Galilee Basin, Israel. J Hydrol 7:276–317
Möller P, Rosenthal E, Dulski P, Geyer S, Guttman Y (2003) Rare earths and yttrium hydrostratigraphy along the Lake Kinneret – Dead Sea – Arava transform fault, Israel. Appl Geochem 18:1613–1628
Neev D, Emery KO (1966) The Dead Sea. Sci J 2(12):50–55
Rosenthal E (1987) Chemical composition of rainfall and groundwater in recharge areas of the Bet Shean-Harod multiple aquifer system, Israel. J Hydrol 89:329–352
Rosenthal E (1988) Ca-chloride brines at common outlets of the Bet Shean multiple aquifer system, Israel. J Hydrol 97:107–128
Rosenthal E, Flexer A, Möller P (2006) The paleoenvironment and the evolution of the Jordan-Dead Sea transform and in the adjoining areas. Int J Earth Sci 95:725–740
Rosenthal E, Jones BF, Weinberger G (1998) The chemical evolution of Kurnub Group paleowater in the Sinai-Negev province - a mass balance approach. Appl Geochem 27:1–17
Rosenthal E, Weinberger G, Berkovicz B. Flexer A, Kronfeld J (1991) The Nubian sandstone aquifer in the western Negev, Israel: delineation of the hydrgeological model underconditions of scarce data. J Hydrol 132:107–135
Simon E, Mero F (1992) The salinization mechanism of Lake Kinneret. J Hydrol 138:327-343
Salameh E (1996) Water quality degradation in Jordan. Friedrich Ebert Stiftung, pp 1–178
Salameh E, Gedeon R (1999) Renewability study of Disi – Wadi Yutum aquifers' water using isotopes and hydrological analyses. Hydrogeol Umwelt 18, pp 1–16
Salameh E, Udluft P (1999) Towards a water strategy for Jordan. Geologie und Umwelt 18, pp 1–19
Sandler A(1981) Geochemistry of groundwater from basaltic aquifers in the Lower Galilee and the Golan. Isr Geol Surv Rep Hydro 2/81, pp 1–91
Shulman H, Flexer A, Guttman Y, Yellin-Dror A, Rosenthal E, Anker Y (2003) The hide-and-seek game between geological factors and groundwater flow in the Jericho-Fazael Region. Proc. Ann Meeting Isr Geolog Soc, Ein Boqeq, 2003, p 118
Starinsky A (1974) Relationship between Ca-chloride brines and sedimentary rocks in Isreal. PhD thesis, Hebrew Univ. Jerusalem
Starinsky A, Katz, A, Levitte, D (1979). Temperature, composition-depth relationships in Rift Valley hot springs, Hammat Gader, Northern Israel. Chem Geol 27:233–244

Vengosh A, Rosenthal E (1994) Saline groundwater in Israel: its bearing on the water crisis in the country. J Hydrol 156:389–430
Vengosh A, Starinsky A, Kolodny Y, Chivas AR (1991) Boron Isotope geochemistry as a tracer for the evolution of brines and associated hot springs from the Dead Sea area, Israel. Geochim Cosmochim Acta 55:1689–1695
Yechieli Y, Starinsky A, Rosenthal E, (1992) Evolution of brackish groundwater in a typical arid region: Northern Arava Rift Valley, southern Israel. Appl Geochem 7:361–374

3.2 Specific regional hydrological aspects

3.2.1. Geophysical investigations of the hydrogeological basins in the central Jordan Valley

Y. Anker, H. Shulman, J. Guttman, A. Yellin-Dror, A. Flexer

3.2.1.1 Introduction

In the present study the natural water resources along the central Jordan Valley and the future development of these resources (Fig. 3.2.1-1) were investigated. The hydrogeological modeling of the area was examined against the background of the geological history of the Dead Sea Rift Valley. The reconstruction of major geological events in the history of the basin enables the identification of tectonic and sedimentary processes which defined the regional hydrogeological system. Thus, events such as compression and decompression and the resulting folding and faulting which occurred during the uplift of the eastern and western borders of the Dead Sea basin, seawater invasion, fast subsidence of the basin floor and the sedimentary events which followed, had to be considered both in time and space.

The characterization of the hydrogeological system was done by combining multidisciplinary geoscience methods such as classic field geology, geochemical and geophysical field methods, and climatological analysis of precipitation systems. The interpolation of various models yielded several data sets. In this particular chapter only the geophysical results are presented.

The Dead Sea Rift is a unique region in which powerful tectonic processes created a deep (about -400 m msl) and narrow valley, bordered on its eastern and western margins by high mountainous ridges (up to about +1000 m msl). An active transform fault, which runs along its central part, shifts the eastern (Arabian) plate about 105 km. northwards relative to the western one (African).

This structural complexity bears naturally upon the hydrogeological and hydrochemical system. Thus, the western aquiferous complex includes the Upper Cretaceous calcareous karstic formations and younger rock-units and recent alluvial sequences. The eastern water-bearing strata are of Jurassic age and younger, ending with the same alluvial complex. On both sides of the Rift, the regional aquifer occurs in the Upper Cretaceous karstic limestone and dolomite formations.

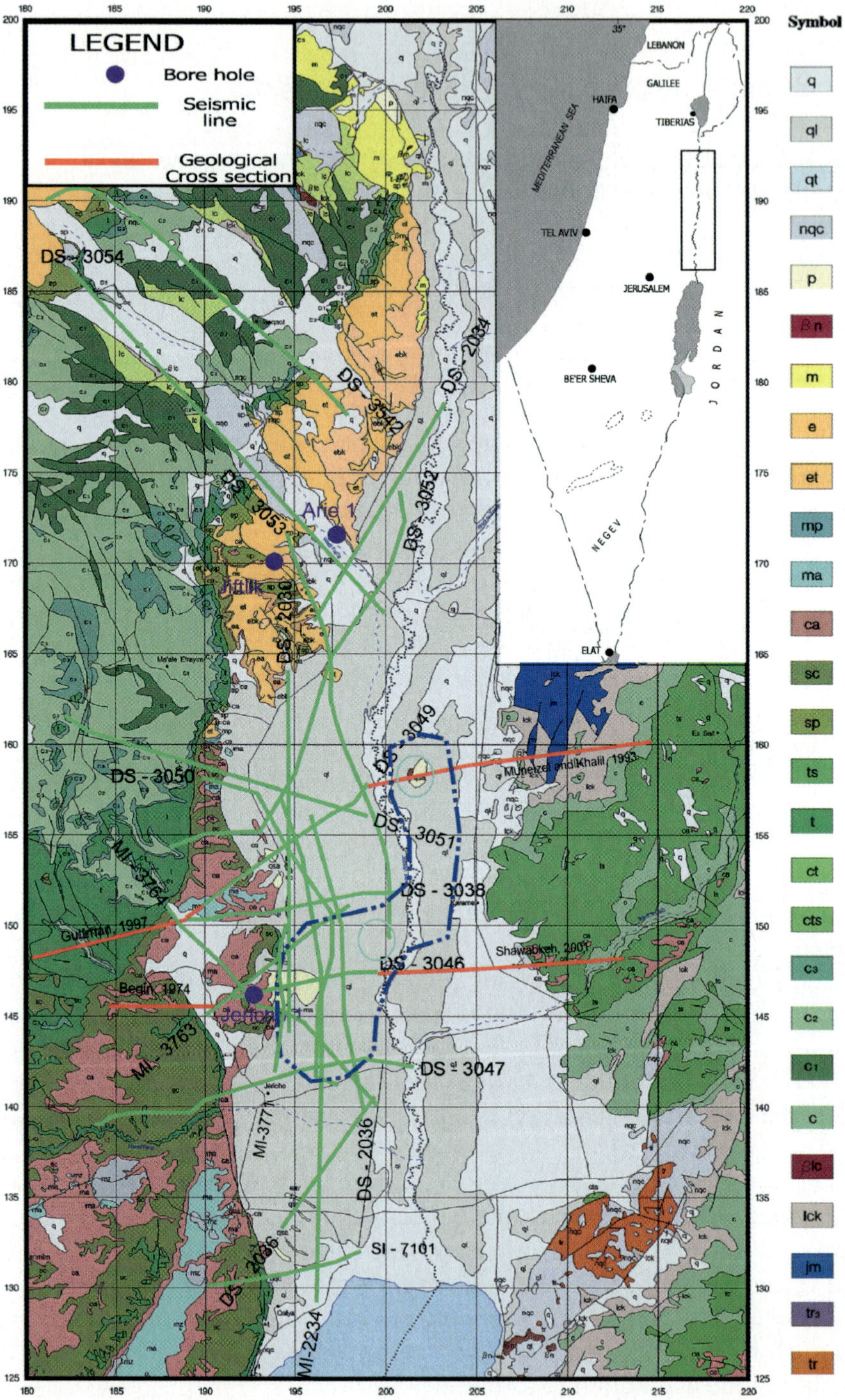
LEGEND
Bore hole
Seismic line
Geological Cross section
DS - 3054
DS - 3540
DS - 2034
DS - 3053
DS - 3052
Arie 1
Jiftlik
DS - 2030
DS - 3049
DS - 3050
Muneizel and Khalil 1993
DS - 3051
MI - 3764
DS - 3038
Gutman 1997
Shawabkeh 2001
DS - 3046
Begin 1974
Jericho 1
MI - 3763
DS - 3047
MI-3771
DS - 2036
SI - 7101
DS - 2036
MI-2234
LEBANON
GALILEE
HAIFA
TIBERIAS
MEDITERRANEAN SEA
TEL AVIV
JERUSALEM
BE'ER SHEVA
JORDAN
NEGEV
ELAT
Symbol
q
ql
qt
nqc
p
βn
m
e
et
mp
ma
ca
sc
sp
ts
t
ct
cts
C3
C2
C1
c
βlc
lck
jm
tr3
tr

Fig. 3.2.1-1. Geological map (Sneh et al. 1998) with general location map in the insert and location of seismic sections (green lines), significant boreholes (blue points), geological cross-sections (red lines) that were combined with seismic interpretation and also location of an inferred evaporite body (dotted blue line), light blue circles mark the location of known saline ponds.
Description of symbols: **q** - Alluvium; **ql** - Lisan Fm.; **qt** - travertine; **nqc** - conglomerate units (undiff.); **p** - Bira, Gesher and Pleshet Fms.; **βn** - volcanic rocks (undiff.); **m -** Hordos Fm. and Umm Sabune Conglomerate; **e** - Eocene (Jordan); **et** - Timrat, Meroz and Yizre'el Fms.; **mp** - Maastrichtian - Paleocene (Jordan); **ma** - Ghareb Fm.; **ca** - Mishash Fm., (Jordan); **sc** - Menuha Fm.; **sp** - Mt. Scopus Group; **ts** - Turonian–Santonian (Jordan); **t** - Bina, Derorim, Shivta and Nezer Fms.; **ct** - Cenomanian-Turonian; **cts** - Cenomanian-Turonian-Senonian (Jordan); **c3** - Weradim and Tamar Fm.; **c2** - Bet Meir, Moza, Amminadav, Kefar Shaul, En Yorqe'am, Zafit and Avnon fms.; **c1** - Givat Ye'arim, Soreq, Kesalon and Hevyon F,ms.; **c** - Albian–Cenomanian (Jordan), **βλc -** Tayasir Volcanics; **lck** - Kurnub Group; **jm** - Middle Jurassic (Jordan); **tr3** - Upper Triassic (Jordan); **tr** - Permian–Triassic (Jordan)

The previously mentioned clastic aquifer complex which makes up the sedimentary sequence, accumulated in the valley since the last marine transgression (Pliocene). It contains a thick sequence of salt-rich sediments and evaporite layers, which salinize the groundwater draining to the Rift from the adjacent aquifers.

Since there are no reliable sub-surface data regarding the Central Rift, geophysical research based on interpretation of seismic profiles, was carried out. Geophysical data such as seismic reflection and gravity surveys were used to reveal and delineate major subsurface structural elements that control the geohydrological setting, and enable its reconstruction and evaluation (Fig. 3.2.1-1). Compilation of this data yielded a sufficient framework for the delineation of a conceptual hydrogeological model for the region. In addition, several important implications regarding the structure and tectonic setting of the subsurface were revealed (Fig. 3.2.1-2).

3.2.1.2 Geophysical methods

Subsurface modeling is based on geophysical research of the area that was carried out during three different periods. The first one during the mid 1970's, was the initial phase of the Dead Sea - Jordan Valley project (DS-2#). The survey was done with a 12–48 Hz sweep-frequencies-Vibroseis (Four Vibrators) source, a geophone-array spread-length of about 5 km, with 48 channels and a group interval of 100 m. The second phase was done during 1980s with 8–62 Hz Vibroseis sources, spread length of 4455 m used 96 channels and a group interval of 45 m. In two lines the group interval was 35 m, hence the spread length was 3,465 m (Kashai and Croker

1987; Rotstein et al. 1991). The last set was done during the project of the Ministry of Energy and Infrastructure (MI) in 1991 and 1994 with 8–48 Hz Vibroseis sources; a geophone-array spread-length of about 6 km, with 120 channels and a group interval of 50 m. In one line (MI-2234/1994) an array of 240 channels was used (Shulman et al. 2003; Shamir et al. 2006). It is important to note that the DS surveys were done for hydrocarbon prospection, while the MI was done for structural geological research. Therefore, data acquisition parameters in both sets were aimed at deeper sections of the strata rather than for relatively shallow groundwater prospection. The lines were processed by a standard processing sequence, and when necessary, by migration and crooked-line processing. Generally, data can be regarded as fair to good but it deteriorates slightly northward.

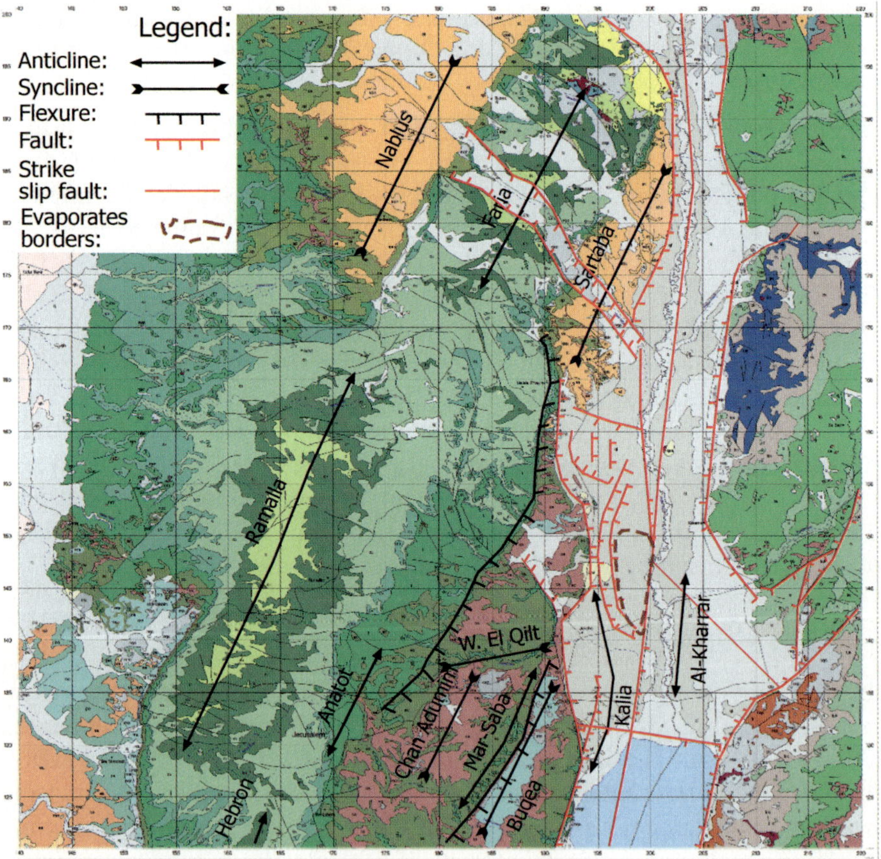

Fig. 3.2.1-2. Tectonic map showing structures, folding and faulting features. The structural elements within the Valley were deciphered from surface and subsurface results of the seismic interpretation (geologic legend see Fig. 3.2.1-1).

By correlating with the stratigraphy of Jericho 1, (the only deep borehole in the region which penetrated the Cretaceous, was bottomed in the Jurassic Kidod Formation and was later deepened up to the Mid-Jurassic Zohar Formation), three major horizons were identified. These horizons were: top Eocene, Middle Cretaceous (Judea Group), and LC-3 (near top Lower Cretaceous electric log marker). Other reflectors, such as top Jurassic and reflectors within the Neogene sequence, were not consistent in all sections, and therefore were not regarded as major reflectors (Table 3.2.1-1).

Table 3.2.1-1. Reflector legend, indicating the color of the markers, which delineate the top of the layers.

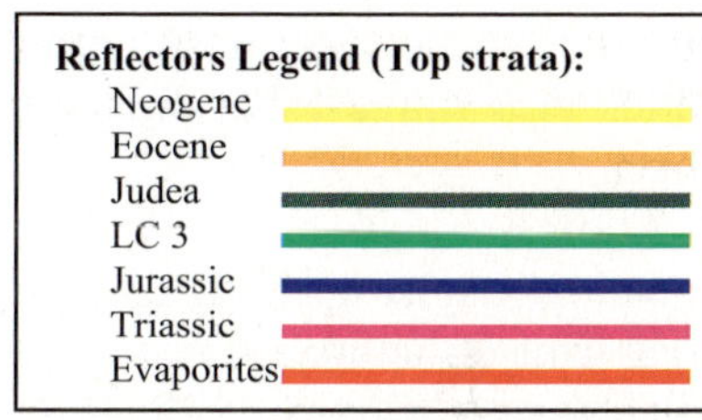

3.2.1.3 Seismic interpretation results

Out of about 20 seismic profiles which covered the study area (Fig. 3.2.1-1), the study was focused on several specific seismic key-profiles for interpretation and geological modeling. As part of the geological modeling procedure, the interpreted time sections were converted to depth, and several geological cross sections were produced. Five reflectors were picked as key horizons: top Neogene, which indicates the transition from the Tiberias Group to the Dead Sea Group, top Eocene, which underlies it, top Judea Group, LC3 (which indicates a stratigraphic boundary within the Lower Cretaceous) and finally, top Jurassic. Additionally, in some profiles the top Triassic reflector was also outlined. It is important to mention that no structural maps were prepared. Only one first order anchor (Jericho 1 deep oil well) points to the existence of the reflectors. All reflectors were correlated with the geological strata defined in the Jericho1 borehole. The current interpretation is tied to and partially overlaps laterally a recent survey by Lazar (2004), which investigated the Dead Sea basins south of the study area.

Shaliv (1983) implied that the study area can be divided into three basins: the Dead Sea, the intermediate Fari'a, and the Argaman basins. Another basin (Mallaha) could be added at the northern edge of the Dead Sea basin. This assumption is supported by evidence from the Bouguer gravity anomaly map (Fig. 3.2.1-3; Ten Brink et al. 2001). This sequence of the basins is located along the Rift.

In this (southern) region that reaches to the contemporary shore of the Dead Sea, Lazar (2004) defined two sub-basins (the Southern and the

Northern Province) both filled with sediments accumulated during the time when the Dead Sea and its progenitor lakes covered that area. Seismic line Si-7102, which reveals a monocline (Qalia) with pronounced eastwards dipping, is correlated to strata at the same level in line DS-3046 of the Northern Province (Fig. 3.2.1-4).

The comparison of seismic profile SI-7102 with seismic profile DS-3046 implies that the Upper Cretaceous reflector along the Qalia monocline axis is slightly shallower in the northern section (DS-3046), which indicates that the Qalia monocline axis dips slightly southward. It appears that the Cretaceous strata within the Valley are at about the same depth, which may suggest that the inclination of the layers becomes steeper to the north.

Another small but deep low can be noticed in the Bouguer gravity anomaly map (Fig. 3.2.1-3) along latitude 3º00' in the middle of the study area. This structure is the steep Fari'a graben, which is a part of the Fari'a basin system (Shaliv 1972; 1983). Seismic sections DS-3052 and DS-2034 show a possible eastward continuation in the sub-surface of this structure. Interpretation of seismic sections (DS-2034, Fig. 3.2.1-5) indicates a possible juxtaposition of the Eocene and Upper Cretaceous strata, and also indicates that the Fari'a graben, as it extends in the NW-SE direction, exhibits a rather deep structural low that is covered with young clastic sediments (DS-3052, Fig. 3.2.1-6).

Seismic lines DS-3053 and DS-3034 (Fig. 3.2.1-1) are located along the southern and northern margins of the Fari'a structure, in relatively perpendicular positions to the DS-3052 (Fig. 3.2.1-6) and DS-2034 (Fig. 3.2.1-5) seismic lines. Their interpretation suggests that, within the graben, the inclination of the layers towards the Rift is not accompanied by significant normal faulting. Nonetheless, projection of the flower structure (occurring in the Fari'a northern faulting system) on the N-S and E-W directed seismic sections (Fig. 3.2.1-5) indicates strike-slip movement along several faults within this structure. The fault-displacement of the layers in the graben, and the size of the graben itself, seems to be larger in profile DS-3052 (Fig. 3.2.1-6) than in profile DS-2034. This may hint at an N-S tensional regime inducing the reinforcement of the Fari'a graben structure in the east, which is exhibited by greater normal displacement along the graben. In addition, possible strike-slip movement was found along the northern border fault system of the Fari'a graben, which could be related to the north-western branching of the Jericho Fault (Fig. 3.2.1-2).

By considering the field relations between the fault-systems around the Fari'a graben, it seems that the N-S tensional stage is younger than the E-W compressional stage that created the sets of monoclines along the Rift. The structural saddle as seen in the Bouguer gravity anomaly map (Fig. 3.2.1-3)

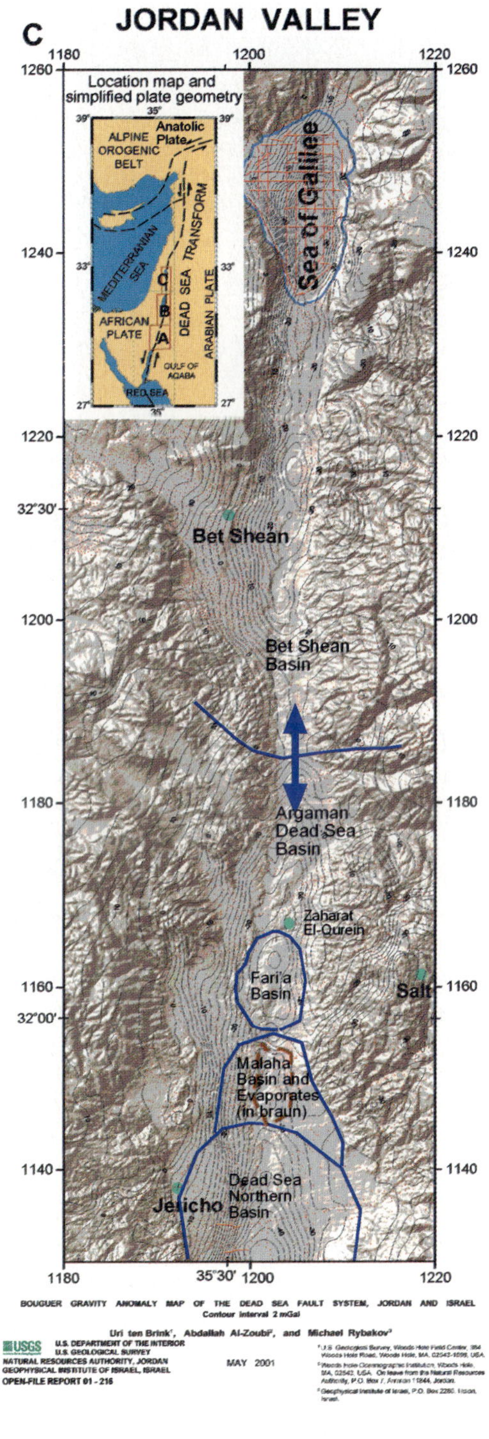

Fig. 3.2.1-3. The distribution of hydrogeological basins along the study area, on the background of Bouguer gravity anomaly map after Ten Brink et al. (2001). The blue line with perpendicular double head arrow indicate the supposed north south surface and subsurface water divide line.

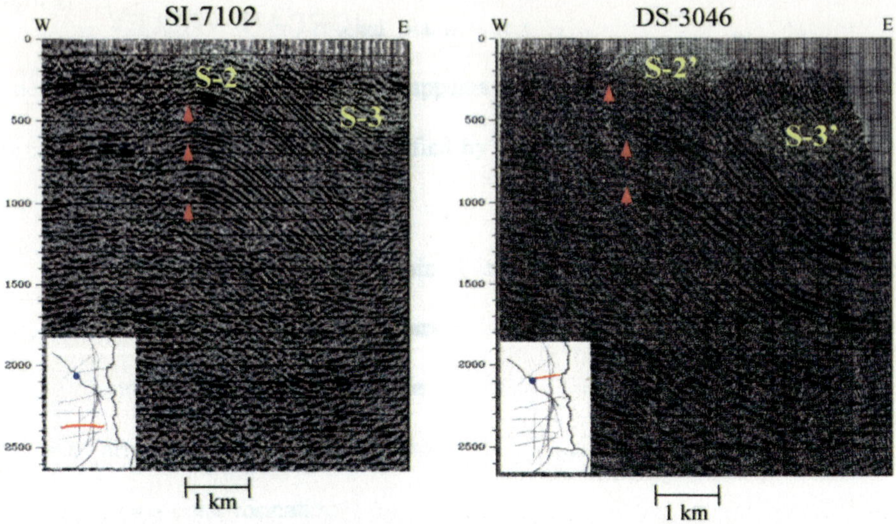

Fig. 3.2.1-4. Comparison of the eastern part of seismic profile SI-7102, situated in the southern province of the Northern Dead Sea basin, with seismic profile DS-3046, located at the Northern Province. S-2 and S-3 indicate correlative sections of the strata. The red arrows point to three strong reflectors inclining eastwards in both profiles. In the location map: JR= Jordan River, J1= Jericho 1 borehole, red line= location of the profile (after Lazar 2004).

outlines the Argaman basin to the north. Furthermore, it may also act as the subsurface water divide line between the Bet She'an and Dead Sea depressions (Ten Brink et al. 2001). While every structural sub-basin has its own typical secondary faulting system, several major faulting systems prevail along the entire region. The most predominant faulting system is that of the border faults of the Rift (Fig. 3.2.1-2). The projection of the Cretaceous reflector east of the Jordan River, seems to be at about the same depth as its projection in the west. Nevertheless, the eastern margin is lacking step-block structures. Therefore, the inclination of the Mesozoic strata towards the Rift occurs across a narrow faulting zone implying that the offset of strata is steeper than it is in the west.

The main segment of the Dead Sea transform faulting system (Kashai and Croker 1987; Rotstein et al. 1991; Lazar 2004) was not definitely encountered in any of the interpreted seismic sections. Nonetheless, it probably emerges north of the (northern) Dead Sea province on the eastern side of the Jordan River Gorge and runs northwards (Salameh et al. 2003).

The last major faulting element in the study area is the Jericho fault (Begin 1975; Reches and Hoexter 1981; Gardosh et al. 1990), which can also be discerned in aerial photos as a linear feature (Anker 2007). The Jericho fault starts where the rhombic feature of the contemporary Dead

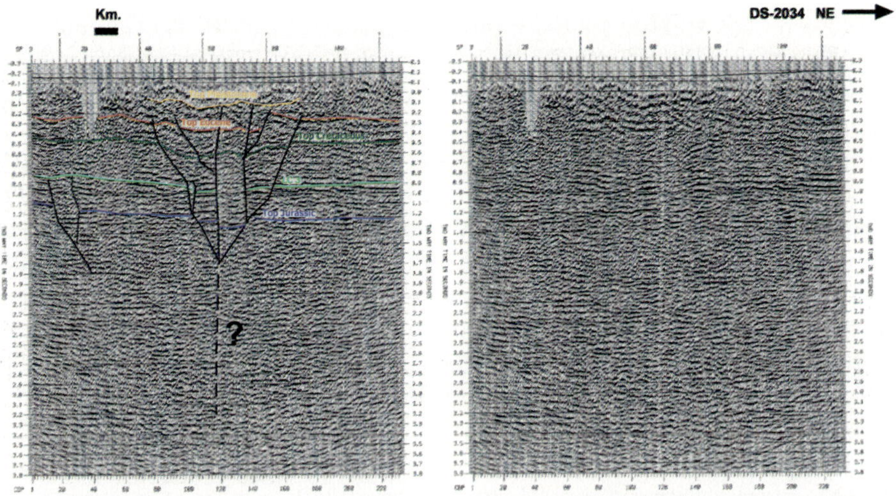

Fig. 3.2.1-5. Geological interpretation of seismic profile DS-2034 showing an aproximate N-S cross section of the Fari'a graben. The flower structure is correlated with the northern boundary of the graben.

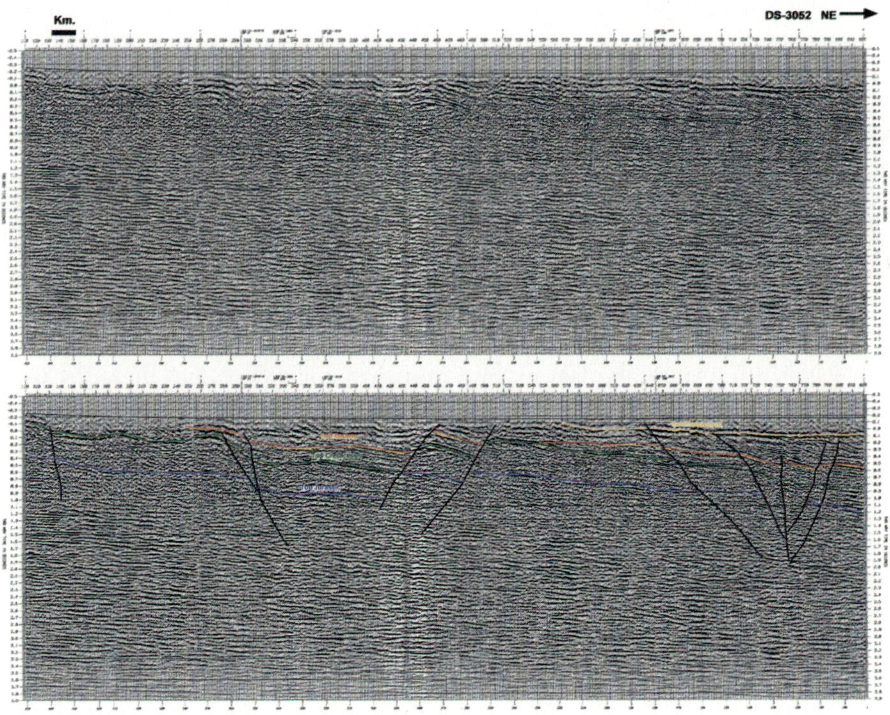

Fig. 3.2.1-6. Geological interpretation of seismic profile DS-3052 showing subsurface eastward continuation of the Fari'a Graben

Sea basin ends and the southern province starts (Lazar 2004). The fault extends northwards, sub-parallel to the Jordan River. The appearances of a flower structure in the DS-3047 seismic profile (Fig. 3.2.1-7) could indicate some lateral movement along the Jericho fault.

Interesting findings emerged from the interpretation of five seismic lines, in which a subsurface lenticular body was observed (Shulman et al. 2003). The overall configuration of the body might imply that this lenticular body is composed of evaporites that are interbedded with clays, marls and aragonite (Figs. 3.2.1-8 and 3.2.1-9). Further support for this assumption was extrapolated from soils, soil crusts and seasonal lakes which were investigated during the current work and by Lazar (2004), who made a methodological study of trenches. Lazar described the sediments of the trenches as "salt saturated". A similar scenario was observed during the current study in the bed of the Mallaha seasonal lake, which was also found to be "salt saturated".

The interpretation of several profiles presented, was done following reprocessing by Shamir et al. (2006). Additional verification for the location of the Jericho fault was done by re-interpretation of the seismic profiles DS-3047 and DS-3046. Nevertheless, their interpretation of seismic profiles DS-3046 and MI-2234 has not discerned the reported evaporite body, but rather indicated the existence of a much larger evaporite sequence in deeper parts of the strata section. If these evaporite domes indeed exist, then they are of an older age, and of an earlier stage than the previously reported evaporite body. Other important observations revealed in both studies are small intermediate basins found in most profiles at various depths. While tracing the Jericho fault, Al Zoubi et al. (2006) discovered indications of small intermediate basins on the eastern part of the Jordan Valley (Fig. 3.2.1-1). They also noted that the trace of the Jericho fault does not coincide with that of the transform, and that the exact trace of the latter was not revealed.

The proposed mechanism for the precipitation of evaporites in the Jordan Rift Valley since the last marine transgression (Pliocene) includes several stages. The hyper-saline water body extending into the evolving valley either partially evaporated or infiltrated into the subsurface precipitating the evaporite domes that were reported in several papers (Zak 1967; Shamir et al. 2006). The next stages of development were as follows: As the freshwater Lisan and Samra Lakes evaporated, the salinity of the solution increased until evaporites precipitated. These sediments accumulated mainly in local structural lows as the valley floor subsided. During a later stage, these local evporitic basins were covered by fluvial clastic sedimentation. During this same fluviatile scenario, evaporites which were precipitated in these basins, were flushed and dissolved and were further reprecipitated in newly evolving local basins. In a way, this process of fluviatile remobilization of salts was on-going ever since the Pliocene. The high concentrations of halite, gypsum and other evaporites found in soil samples from all around the study area indicate that this process has been continuous throughout the subsequent history of the Rift. As previously mentioned, the entire clastic sequence of the Jordan Valley is saturated with salts of the evaporated saline Lisan Lake and of other water bodies which existed in the valley after (and probably before) these Lakes. This observation

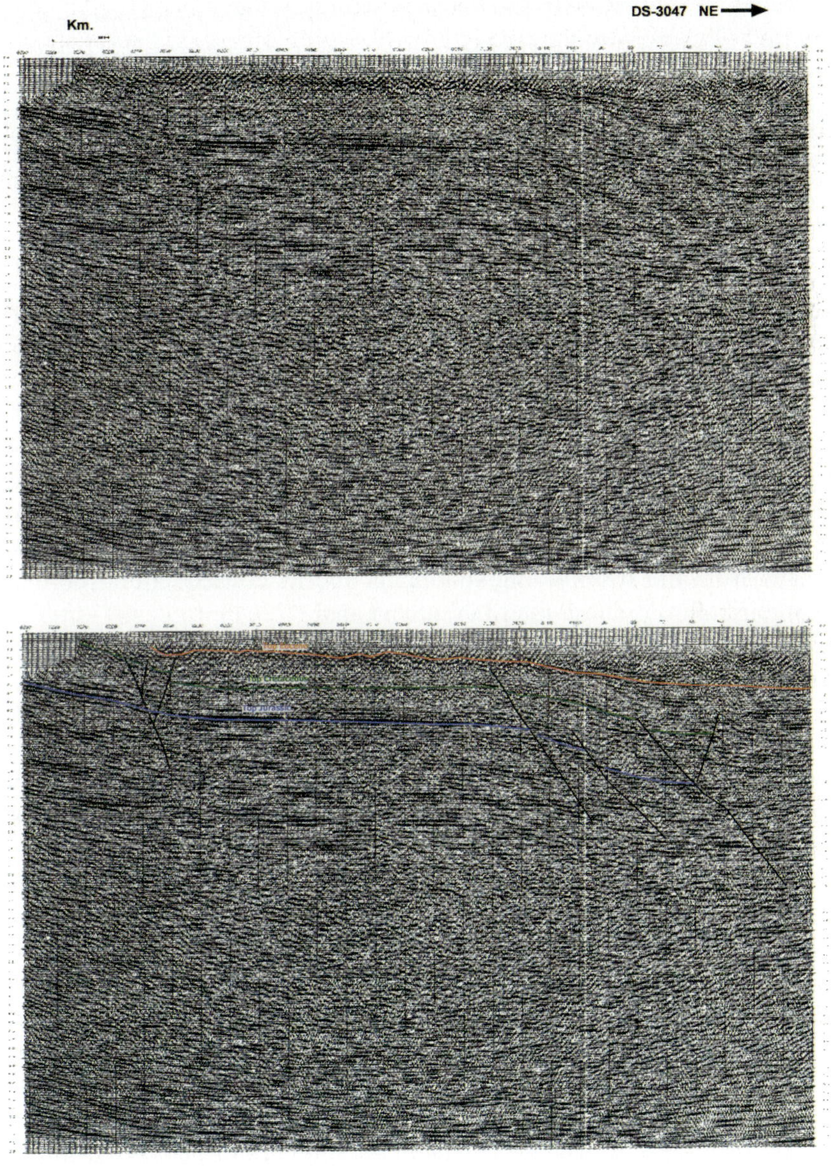

Fig. 3.2.1-7. Geological interpretation of seismic profile DS-3047.

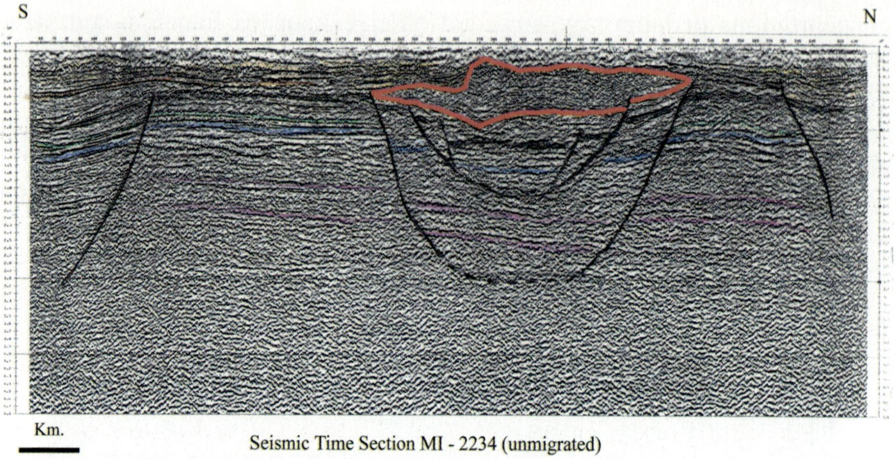

Fig. 3.2.1-8. Geological interpretation of seismic line MI-2234, indicating a possible lenticular shaped body consisting of evaporated mineral accumulation (Red polygon), as well as halokinetic phenomena such as reflector irregularity and transparency of strata in addition to upwelling along its margins.

indicates that an evaporite sequence such as the one found in the Mallaha basin, was not created by a direct mechanism of evaporation. Small-scale local basins, such as that of Mallaha and those observed in the seismic profiles, might have produced such evaporite lenses by runoff, flushing salts from the surrounding sediments and its consequent accumulation and precipitation in these intermediate basins.

Finally, due to the fast subsidence of the Jordan Valley floor, the ancient basins were buried and preserved as evaporite lenses. Moreover, new intermediate basins are created by the ever-changing drainage system. Indications for this sequence of events were observed in seismic profiles. A good example of these buried basins is the Mallaha-Uja evaporite body (Fig. 3.2.1-8). The seismic study revealed that halokinetic phenomena such as upwelling of strata, were encountered at the margins (Figs. 3.2.1-8 and 3.2.1-9) due to overload of the denser clastic sediments above the evaporite body. The available seismic grid is sufficient to delineate the boundaries of the lenticular body (with the exception of its eastern side, which was not seismically surveyed. This local Mallaha basin is delineated by faults, which seem to be crescentic (Fig. 3.2.1-8).

The Mallaha area acts as an intermediate basin between the Fari'a and the Dead Sea basins. Seismic section DS-2030 (Fig. 3.2.1-1) outlines at around longitude 3°28' the western border of the evaporite body. However, since its main axis is NW-SE oriented, its southern extremity is shifted by up to two kilometers. The northern border is indicated by section DS-3038 (Fig. 3.2.1-1) at latitude 31°58'. The Mallaha local seasonal lake (Fig. 3.2.1-2) is probably created due to dissolution and compaction of the near-surface evaporite body. The southern extent of the body is indicated by its absence from section DS-3047 (Fig. 3.2.1-7) at latitude 3°50'. In the south, the western extent of the lenticular body can be indirectly derived by the drop in groundwater salinity as observed in the Jericho Mus'a Allami well field (Anker 2007), where the location of wells crosses longitude 35°30'. Moreover, the geochemical type of groundwater changes from Na-Mg-Cl (which would be due to dissolution of the evaporite body) to Ca-HCO_3^- dominated groundwater which most probably originates from the Judea Group aquifer. Using interval velocities that are typical for evaporites, the thickness of the body is calculated to be of a few hundred meters and may reach to 1000 m at maximum thickness (Fig.s 3.2.1-8 and 3.2.1-9).

From correlating this finding with the works of Lazar (2004) and Ten Brink et al. (2001), it can be implied that the evaporite body is located at the northern point of the Dead Sea Northern Province which can be identified on the Bouguer gravity anomaly map as a local structural low (Fig. 3.2.1-3). About 10 km northwards - at Zahrat el Qurein - Belitzky and Mimran (1996) interpreted a gravimetric anomaly as being the result of halokinetic movement in an evaporite body. As implied from the geological interpretation (Fig. 3.2.1-10) of the area, branching of the Mallaha evaporite body to the nearby Zahrat el Qurein system cannot be ruled-out. Further more, Salameh (2004) suggested that a similar scenario exists along a structural low located east to the Jordan River parallel to the evaporite body in a basin of the same name (Mallaha). This region was used for the construction of the Karama Dam and reservoir, and suffers probably due to the alleged evaporite region from extreme salinization of the water in the reservoir and from structure problems, and as indicated by Salameh (2004), might be connected to the Zahrat el Qurein complex.

West of the Jordan River, the evaporite body is delimited by a local hydrological divide line from the intermediate Fari'a basin (Shaliv 1972 and Ten Brink et al. 2001). As indicated by former studies (Begin 1975; Garfunkel and Ben-Avraham 1996; Salameh et al. 2003) the Jericho fault could be the boundary between these two evaporite provinces (Fig. 3.2.1-2). Though estimation of the horizontal movement on this segment is not yet possible, this setting seems feasible.

3.2.1.4 Discussion

The subsurface structure (Fig.s 3.2.1-9 and 3.2.1-10) is a major factor in the reconstruction and evaluation of the geohydrological setting. Mimran

(1969); Shaliv (1972); Benjamini (1973); and Begin (1975) studied the mountainous margins of the area in the west; Muneizel and Khalil (1993), and Shawabakeh (2001) studied those in the east. The geological structure and stratigraphic column of the area - as derived from these studies - were presented in chapters 2.1 and Fig. 3.2.1-1. Shaliv (1972) and Guttman (1997) studied the regional hydrological pattern of the area and both indicated that there is a distinct southern groundwater flow component. A significant eastern component was deciphered in several locations. In the Fari'a graben area, for instance, the eastern component seems to be controlled by the local structure of the graben (Shaliv 1972). A general picture of the entire region suggests that the south-eastern groundwater flow component is consistent along the margins, and then turns southward where the Dead Sea basin is predominant. Along the eastern margins, the hydrological setting is somewhat different. The flow pattern there seems to be mainly directed westward, other than in places where it is shifted due to local structural elements. Along the western flank of the Rift the descent to the Valley is by several normal step-blocks (Figs. 3.2.1-7 and 3.2.1-8). Contrarily, the slope of the eastern flank is steeper and it occurs along several segments of a major normal fault (Salameh et al. 2003; Al-Zoubi et al. 2006).
Several east-west directed fault systems penetrate the Rift, causing discontinuance along the eastern border fault system and as such, act as conduits for groundwater flow (Fig. 3.2.1-2). Two such distinctive systems are the Wadi Shueib (Shawabakeh 2001) and Wadi Zerka (Muneizel and Khalil 1993) systems. In the same way, the western step-block margins of the Rift are interrupted by the structural basins found in their center (Fig. 3.2.1-10). The faulting systems of the basins in the Dead Sea, Faria, and Argman regions differ from each other. The main runoff and groundwater drainage base in the region is the Dead Sea. The Mallaha basin is located in a subsurface structural low and acts as an intermediate basin, draining to the Dead Sea after passing over a certain threshold there, the water level is higher than the lithological boundary. Within the regional hydrological setting, the Argaman-Fari'a groundwater basin is unique because of it's large storage capacity. This structural basin acts as a semi-terminal drainage basin and is neither part of the larger basin of Bet She'an basin to the north nor part of the Dead Sea basin to the south.

Fig. 3.2.1-9 (left). Geological interpretation of seismic profile DS-3046 combined with geological cross-sections by Begin (1975) to the west and by Shawabakeh (2001) to the east (Location and sources in Fig. 3.21-1).

Fig. 3.2.1-10 (right). Geological cross section across the Fazael and Zahrat el Qurein regions, after Muneizel and Khalil (1993) and Guttman (1997) (Location and sources in Fig. 3.2.1-1).

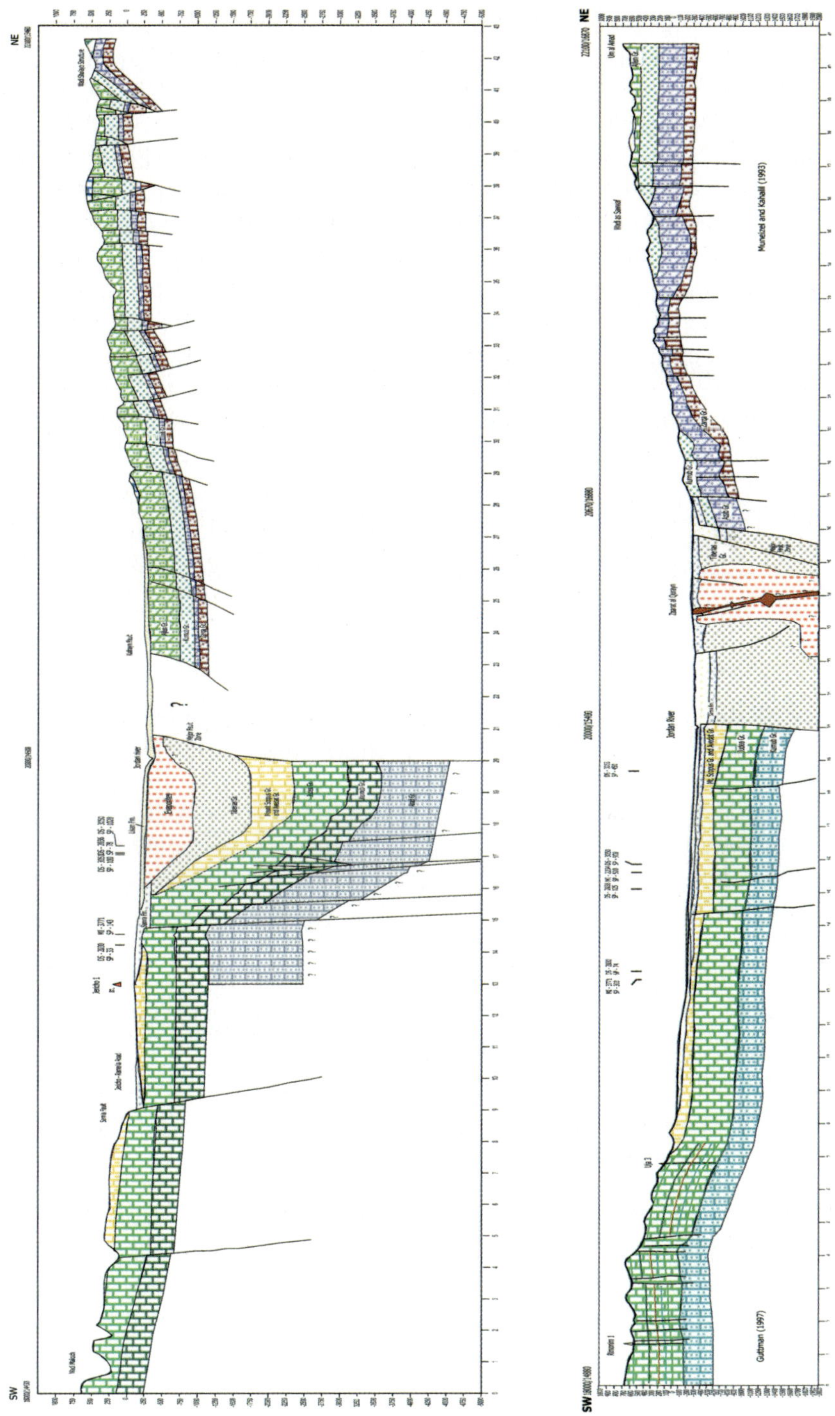
SW
NE
SW
NE
Muneizel and Kahalil (1993)
Guttman (1997)

From the flow lines in the geohydrological map of Shaliv (1983), the hydrological impact is easily discernable. North of the Fari'a structure groundwater flow is diverted because of the dip of the strata. In the area, the Eocene Avedat sequence (Fig. 3.2.1-10) overlies directly the Upper Cretaceous aquifer. This particular situation allows extraction of groundwater (originating in the Upper Cretaceous aquifer) by the Argaman wells which are drilled into Eocene Avedat layers. The area is located along the hydrological divide line, between the Fari'a basin and the Argaman basin located to the north of it.

3.2.1.5 Conclusions

By analyzing the results from this part of the study, several important implications were recognized:

- Several newly found faulting systems were discovered in the subsurface of the Rift; the most intriguing is that of the Fari'a graben. The graben was found to be continuous to the east, cutting through a buried north–south extending anticlinical structure. In addition, the northern margin of the graben shows indications of strike-slip movement. The groundwater divide extends north of the Fari'a graben into the Argaman region. Hence, south of it, groundwater drains via several intermediate basins to the Dead Sea basin, while north of it drainage is to the Bet She'an and Kinnarot basins.
- The eastern margins of the Rift are outlined by a narrow fault-zone, there the alluvial sediments on-lap the marine sequence which makes up the stratigraphic column of the area. In the west, the approach to the Valley is by a set of step-blocks that extend eastwards in the subsurface. This setting facilitates fresh groundwater, which flows (in the Upper Cretaceous Judea Group aquifer) from the replenishment areas in the west, to the Valley and could be exploited there from aquiferous strata which are not as deep as previously suspected.
- The sub-surface topography indicates the presence of intermediate hydrological basins located along the study area. In one of these basins, a lenticular evaporite body was found which is up to one kilometer thick, up to 10 km long and about 4 km wide. This natural reservoir of the Wadi Mallaha marsh was probably created in a structural low that evolved due to halokinetic processes and to leaching of an evaporite body, which was identified by seismic methods. The marsh is replenished during the rainy season and is completely evaporated at the beginning of the hot season. Moreover, overflow of the upper water layer of the marsh to the Jordan River during (rare) rainy winters makes this sys-

tem even more complex. The occurrence of the subsurface evaporite complex bears also on the geochemical characteristics of the water in the marsh and on its morphological setting.

- The geophysical modeling in the regions of Uj'a and Argaman revealed two saddle-like structures built of aquiferous layers at shallow depths beneath the young sedimentary cover. Geochemical modeling indicated that the exploitation of these aquifers is more than feasible and is actually necessary in order to prevent fresh groundwater of the Cretaceous (Naaran) and Eocene (Argaman) aquifers from becoming saline, as they flow eastwards and replenish the Neogene and Quaternary aquifers.

References

Al-Zoubi AS, Heinrichs T, Sauter M, Qabbani I (2006) Geological structure of the eastern side of the lower Jordan Valley/Dead Sea rift: Reflection seismic evidence. Mar Petrol Geol 23:473–484

Anker Y (2007) Hydrogeochemistry of the Central Jordan Valley. Ph.D. thesis, Tel Aviv University, pp 1–130

Begin ZB (1975) The geology of the Jerico sheet (Geological map series 1:50,000). Bull Geol Surv Isr 67:1–35

Belitzky S, Mimran Y (1996) Active salt diapirism at the Zahrat El-Qurein dome, Lower Jordan Valley (Jordan). Isr J Earth Sci 45:11–18

Benjamini C (1973) The stratigraphy and structural geology of the Sartaba area, Samaria. M.Sc. thesis, Hebrew University, Jerusalem, (in Hebrew unpubl.)

Gardosh M, Reches Z, Garfunkel Z (1990) Holocene tectonic deformation along the western margins of the Dead Sea. Tectonophysics 180:123–137

Garfunkel Z, Ben-Avraham Z (1996) The structure of the Dead Sea basin. Tectonophysics, 266:155–176

Guttman J (1997) Saliniation along the western border of the Jordan Valley and the Dead Sea. Proc. of the 13th GIF Meeting on the Dead Sea Rift as a unique global site, p 46

Kashai EL,Croker PF (1997) Structural geometry and evolution of the Dead Sea – Jordan rift system as deduced from new subsurface data. Tectonophysics 141:33–60

Lazar M (2004) Tectonic processes along the northern edges of pull-apart basins in the Dead Sea rift: A case study from the northern Dead Sea. PhD. Thesis Tel Aviv University, pp 1–140

Mimran Y (1969) The Geology of Wadi el Malih region. M.Sc. Thesis, Hebrew University, Jerusalem, (in Hebrew unpubl.)

Muneizel S, Khalil B (1993) As Salt Geological map 1:50,000. Hashemite Kingdom of Jordan, Ministry of Energy and Mineral Resources, Natural Resources Authority - Geology Directorate

Reches Z, Hoexter DF (1981) Holocene seismic and tectonic activity in the Dead Sea area. Tectonophysics 80:23–254

Rotstein Y, Bartov Y, Hofstetter A (1991) Active compressional tectonics in the Jericho area, Dead Sea Rift. Tectonophysics 52:65–76

Salameh E (2004) The tragedy of the Karama Dam project/Jordan. Acta hydrochim Hydrobiol 32:249–258

Salameh E, El-Naser H, Al-Zoubi, AS (2003) German-Israeli-Jordanian-Palestinian Joint Research Program for Sustainable Utilization of aquifer systems. Final Report 2001-2003, submitted to the BMBF, pp 1–22

Shaliv G (1972) The hydrogeology of the Wadi Fari'a region. MSc thesis Hebrew University, Jerusalem, THAL Rep. HR/72/097, pp1–200

Shaliv G (1983) The Geology of the eastern Shomron, *in:* IGS, Fieldtrip proceedings, 15p (in Hebrew)

Shamir G, Eyal Y, Bruner I (2006) Localized vs. distributed shear in transform plate boundary zones – The case of the Dead Sea Transform in the Jericho Valley. Geochem Geophys Geosys 6:1–21

Shawabakeh KF (2001) Al Karame Geological map 1:50,000. Hashemite Kingdom of Jordan, Ministry of Energy and Mineral Resources, Natural Resources Authority-Geology Directorate

Shulman H, Flexer A, Guttman Y, Yellin-Dror A, Rosenthal E, Anker Y (2003) The game of hide and seek between the geological factors and the flow patterns of groundwater in the Fazael-Jericho area. Isr Geol Soc Ann Meeting En Bokek (abstract), p 118

Sneh A, Bartov Y, Weissbrod T, Rosensaft, M (1998) Geological Map of Israel, 1:200,000. GSI

Ten Brink U, Al-Zoubi A, Rybakov M (2001) Bouguer gravity anomaly map of the Dead Sea fault system, Jordan and Israel. USGS, NRA, Jordan, GII, Israel, Open file report, pp 1-216

Zak I (1967) The geology of Mount Sedom. PhD thesis, The Hebrew University in Jerusalem, pp 1–193

3.2.2 Hydrogeology and flow pattern along the Fazael–E-Salt cross-section

J. Guttman, Y. Anker, A. Flexer, E. Salameh, H. Shulman

3.2.2.1 Background

This chapter describes the hydrogeological situation in the Jordan Valley and in the two lateral mountainous belts along a virtual cross section going from En Samiya in the west to the E-Salt area in the east (Fig. 3.2.2-1).

The expected outcome of this chapter is to understand the flow mechanism and the relationships between groundwater bodies flowing from the regional aquifers extending east and west of the Rift, towards the Rift Valley. The understanding of the flow pattern from the regional aquifers into the Rift Valley is an important tool for the sustainable exploitation of the groundwater resources for the benefit of the population in the study area.

3.2.2.2 Geo-hydrological outline

The formation of the Dead Sea basin and of the adjacent mountain provinces are consequences of major geological processes which generated several horizontal and vertical fault systems, major fold systems and a series of pull-apart basins (see Chap. 2.1). In the study area there are two basic physiographic provinces: the highlands and the low-lying areas of the Jordan Rift Valley. A close correlation exists between physiographic features, geological structures and climate both on regional and local scales. Surface drainage of rainwater (surface runoff) has cut steep "canyon-like" wadis extending from the mountain region towards the Rift Valley. Wadis generally cut into the rocks in an east-west direction, following the overall drainage pattern.

The stratigraphic column on both sides is asymmetric and consists of carbonates, chert, chalk, gravels, sandstones and evaporites which range in age from Triassic to Holocene (see Fig. 2.1-7 in Chap. 2.1). On the eastern side, the Triassic to Eocene succession is exposed, whereas on the western side the exposed sequence starts with Lower Cenomanian beds. Inside the Jordan Rift Valley the alluvial fill is composed of gravel, sand, clay and evaporites. A detailed description can been found in chapters 2.1 and 2.2.

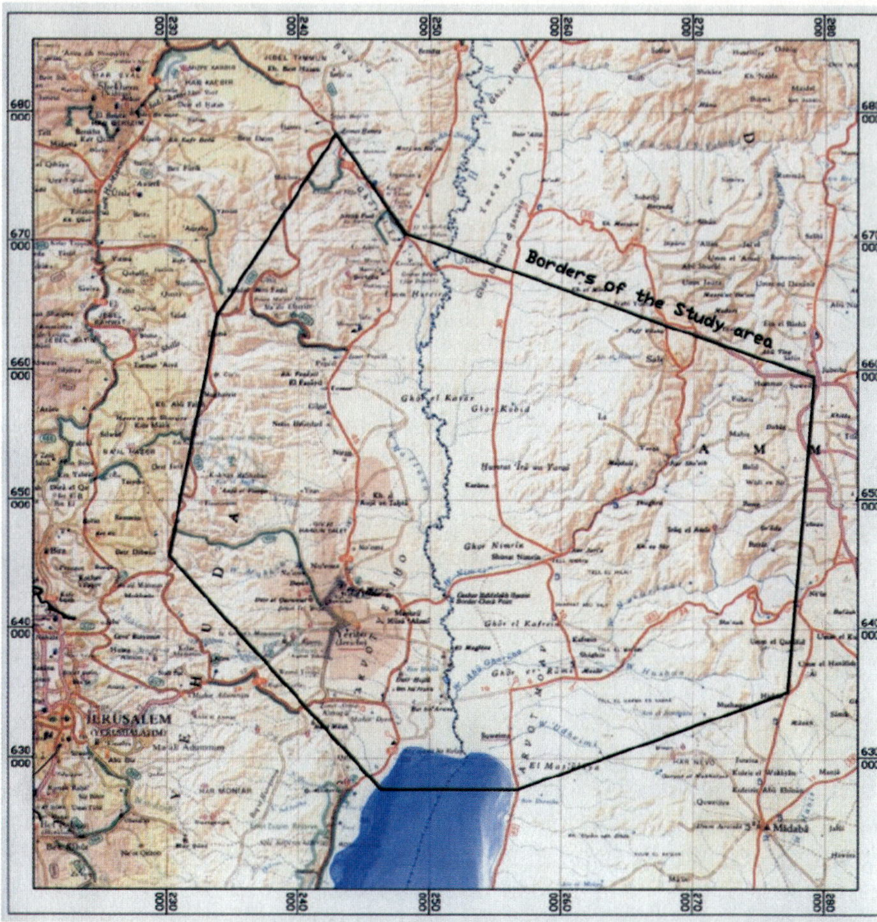

Fig. 3.2.2-1. Location map

The principal fault structure in the area is that of the Dead Sea-Jordan Rift Valley. Along the western side of the Valley, the fault runs along the western shore of the Dead Sea up to Jericho city and its direction is N-S. Around Jericho, the fault direction changes to NW, continuing as the En Samiya fault strip (Begin 1976).

Along the eastern side of the Dead Sea, the major fault runs along the eastern side of the Dead Sea and near the village of Suweimeh, it changes direction northeastwards continuing as the Sweima-Amman-Halabat fault. The fault passes north of the village Al Addasiyeh and goes further to Amman. This fault zone is 1–5 km wide and consists of many structural elements including faults and folds controlled by the same stress system. The downthrown block caused by this fault extends northwestwards with a vertical displacement up to 400 m (Shawabekeh 2001). This fault outlines the southeastern border of the study area.

Following the interpretation of seismic measurements, several buried fault systems were found inside the Rift Valley.

In the mountain provinces along both sides of the Rift, the predominant elements governing the geological structure are anticlines and synclines and associated fault systems crossing the study area. On the western side, the Ramallah anticlinal axis represents the maximal structural elevations in the area. It extends from the Bar Giora area in the Jerusalem area to the Laban-a-Sharqiya area to the north of Ramallah where the axis splits into the Anbata-Um el Fahm anticlinal axis, running N-NW and the Fari'a anticlinal axis, running N-NE. The Fari'a anticline descends towards the Jordan Valley and is buried there. The mountain structure is dissected by fault systems that are perpendicular to the anticlinal axis, creating horsts and grabens such as in the Fari'a River Valley.

The Ramallah anticline is an asymmetric structure, dipping steeply westwards and moderately eastwards. The study area is located on the eastern side of this structure, which plunges in the direction of the Jordan Rift Valley and the Dead Sea, forming a series of undulations and secondary structures. The eastern syncline called the Uja-Sartaba syncline has an important impact on the flow and salinity mechanism along the eastern foothills and will be discussed in the next paragraph.

On the eastern side of the Rift (the Jordanian structural side) the main structures running from Amman–E-Salt up to Deir Alla, represent the maximal elevations in the area. The structure descends in the direction of the Jordan Rift Valley and the Dead Sea in a series of undulations, forming secondary structures. There are few principal faults. The Dead Sea transform fault is the most distinctive fault in the study area. Other faults are the Wadi Shueib structure and Sweima-Amman-Halabat structure. These two structures are clearly visible in the region and also consist of several faulted synclines and anticlines. These structural elements have an important impact on the flow and salinity mechanism on the eastern side of the study area and will be discussed in the next paragraph.

The Jordan-Dead Sea Valley which is the drainage basin of the whole study area, is filled with clastic sediments that have been accumulated in the evolving basin since the process began during the Miocene (Garfunkel and Ben-Avraham 1996). The Dead Sea Group clastic sedimentary succession creates a multi-aquifer system which consists of three major units. Aquifer layers exist in the Samra Formation, in the Lisan Formation and in the Quaternary recent alluvial clastic sediments (see Table 2.1-1). Additional clastic aquifers which are not exposed in the study area exist in the Tiberias Group aquifer. The Samra Formation is divided into two sub-units, both built of conglomerates, sandstone and siltstone. Conglomerates are common in the lower unit, and bigger portions of silt layers together with interfingering with the Lisan Formation characterize the upper unit. The Lisan thin-bedded layers of aragonite, chalk, gypsum and clays act as aquitards separating the Samra permeable layers and the alluvial clastic sediments, imbedded on the Lisan sediments, and in the Lisan layers as lobes of conglomerates (Begin 1974).

Geological interpretation of seismic lines carried out as part of this study, indicates the occurrence of local basins in the Jordan Valley, in which evaporated minerals accumulated. These bodies mainly consist of cycles of evaporites containing aragonite, gypsum, halite and more soluble salts. Although the sets are laminated, they show clear halokinetic features and possible contact with other rock-units and with adjacent aquifers (see Chap. 3.2.1). A preliminary study of such a system in the Eastern Jordan Valley might hint at a possible physical and genetic connection to the evaporites mentioned above. Another important observation from the seismic interpretation indicates that within the Jordan Valley, the old (Triassic to Eocene) formations are situated at depths of a few hundred to a few thousand meters below surface.

3.2.2.3 The conceptual flow regime

The western part of the study area (Fazael-En Samiya) is characterized by intensive fault systems that form grabens, horsts and step structures. This tectonic pattern allows in some places hydraulic connections between sub-aquifers and groundwater can flow from one sub-aquifer to another. In some places, the faulting system upthrows the aquifer base above the regional water table and forms local barriers to groundwater which is flowing downstream towards the Rift Valley.

The Judea Group is the main aquifer in the western side of the study area. The total thickness is from 800 to 850 meters. Considering water level measurements, in most of the study area two sub-aquifers may be identified: an upper and a lower aquifer. The upper aquifer generally includes the Bina, Weradim and Aminadav Formations, and occasionally the aquiferous horizons in the Bet Meir tongue (See Fig. 2.1-7 for the Palestinian correlative stratigraphic units). In the Jericho area (south of the study area), the clay layers in the bottom of the Bina Formation divide the upper aquifer into two sub-aquifers: the uppermost aquifer and the upper aquifer. The Jericho springs that emerge from the Turonian layers are the outlet of the uppermost aquifer and will be discussed in the regional case study dealing with the Jericho region (see also Chap. 3.3.1.4).

The lower aquifer includes the Givat Yearim and Kefira Formations and occasionally the aquiferous horizons in the Soreq Formation (Guttman 2000). According to the stratigraphy of the boreholes, the hydraulic separation between the two aquifers is composed of blueish-greenish clay, marl and chalk and is situated primarily in the mid-lower section of the Bet Meir Formation. The separation is manifested by the difference in water levels in the two sub-aquifers, the level in the upper aquifer being higher than the level in the lower one. Thus, in the En Samiya and Uja area, there is a difference of about 200 m between the En Samiya and En Uja springs, which are the outlets of the upper aquifer and the wells which exploit water from the lower aquifer (Shaliv 1980; Guttman 1980; 2000; Guttman and Zuckerman 1995).

The Uja spring represents the flow system of the upper aquifer in the study area. The discharge of the Uja spring shows a quick response to climate changes (periods of drought and very rainy periods). The strong correlation between rainfall (Jerusalem station) and the fluctuation of spring flow is shown in Fig. 3.2.2-2. The aquifer which feeds the spring is a local, very karstic system that immediately reacts to changes in the monthly and yearly precipitation in this recharge area. The maximum discharge appears a short time after the rainiest month and dies out towards the end of the winter season. Most of the time (except in very rainy years such as 1992) large storage is not required in order to maintain a high and steady flow for a long period. Contrarily, the Uja wells draw water from a deep, separate aquifer. A similar hydrological situation exists in the En Samiya area, water is exploited from the En Samiya spring (upper aquifer) and from the En Samiya wells (lower aquifer).

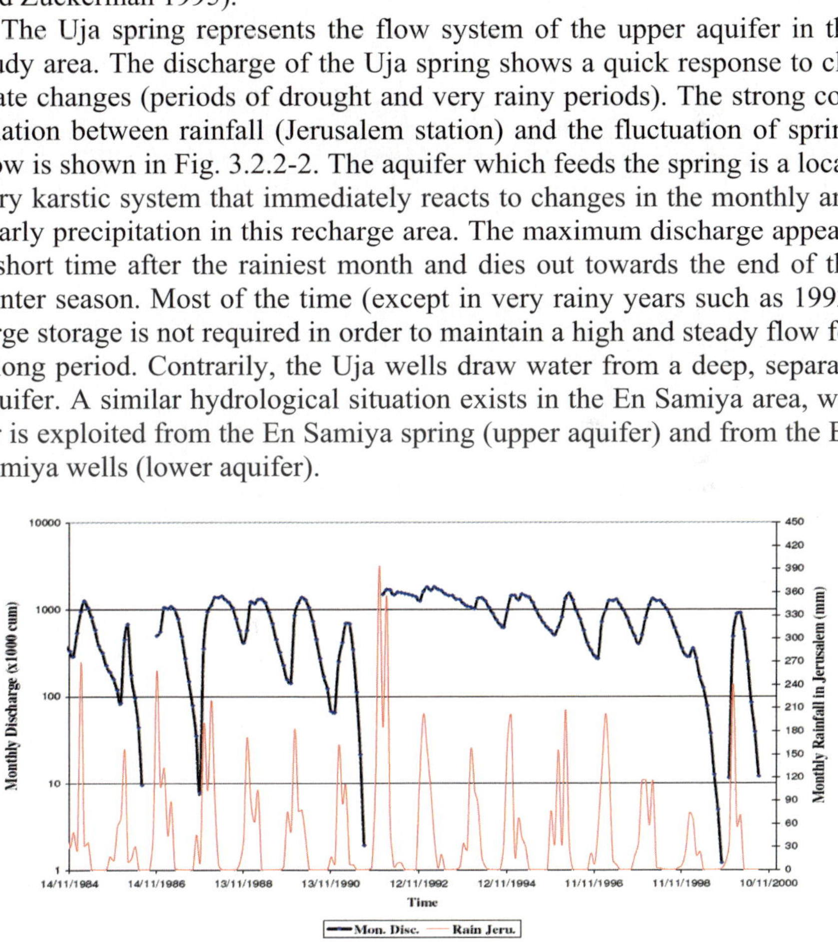

Fig 3.2.2-2. The discharge behavior of Uja Spring as an example of the upper aquifer flow mechanism (cum = m^3)

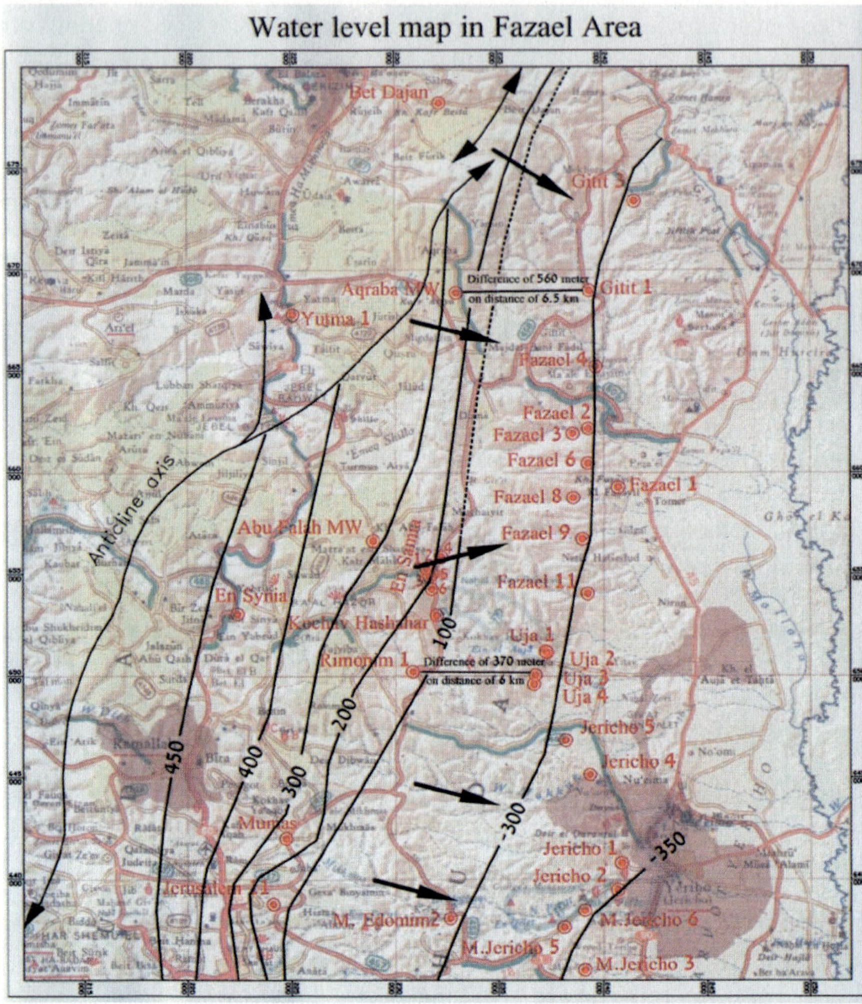

Fig 3.2.2-3. Groundwater contour lines of the lower Cenomanian aquifer in the Fazael area

The lower aquifer is the regional aquifer and most wells are pumping from it. The wells in the study area are located in two clusters. The first cluster is located between Rimonim, En Samiya and Aqrabe with a groundwater elevation between 100–250 m above sea level. The second cluster is located along the foothills (Gitit, Fazael, Uja wells). The water levels in this cluster are between 200 meters below sea level in the Uja wells to 300 meters below sea level in the Fazael and Gitit wells. A difference of 300–500 meters along 5–7 km exists between the two clusters (Fig. 3.2.2-3). This hints at a very steep gradient (about 7% on average), which, under natural regional conditions, cannot be maintained. Therefore, a mechanism is needed to represent the flow patterns in the aquifer. In wells located close to the foothills (Gitit, Fazael, Uja) the water levels raised 2–2.5 years after the very rainy year of 1992 (Fig. 3.2.2-4). The large delay is a result of the long distance between the recharge area and the wells, the gradient of the water level and the permeability of the aquifer. In contrast to the wells, springs like En Uja response much quicker to rainfall, i.e., within few months (Fig. 3.2.2-2).

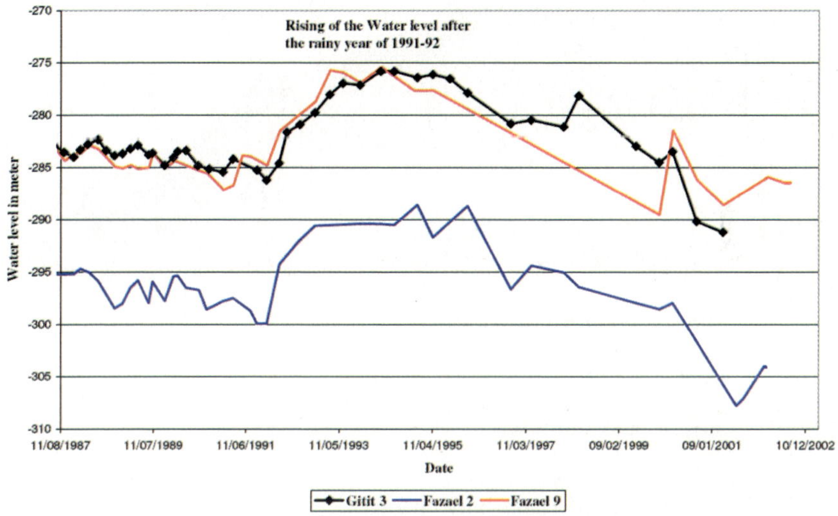

Fig 3.2.2-4. Water levels (msl) in pumping wells in the Fazael-Gitit area

Several geological cross-sections were drawn in order to understand the geological situation in the study area (Fig. 3.2.2-5). This information enables to locate the structural elements that form the underground cascades and the places where the regional water level is above the base of the aquifer. The cascades are zones of low permeability that force groundwater to bypass them and flow along individual paths rather than along a uniform flow front. The main factors that influence the location and the amount of water that flows via these individual paths are geological and structural elements, the regional water level, the saturated thickness of the aquifer, and the hydraulic characteristic of the aquifer.

This flow mechanism can explain the differences in the chemical compositions and the salinity fluctuations between the wells located along the eastern edge of the Fazael-En Samiya area (Fazael, Gitit and Uja wells). In the natural replenishment zones Cl^- concentrations ranged from 30–50 mg/L and remained constant throughout the period of exploitation. In wells located along the eastern foothills (Fazael, Gitit and Uja wells), a wide range of Cl^- concentrations (from about 30–40 mg/L to about 600 mg/L) was observed (Fig. 3.2.2-6).

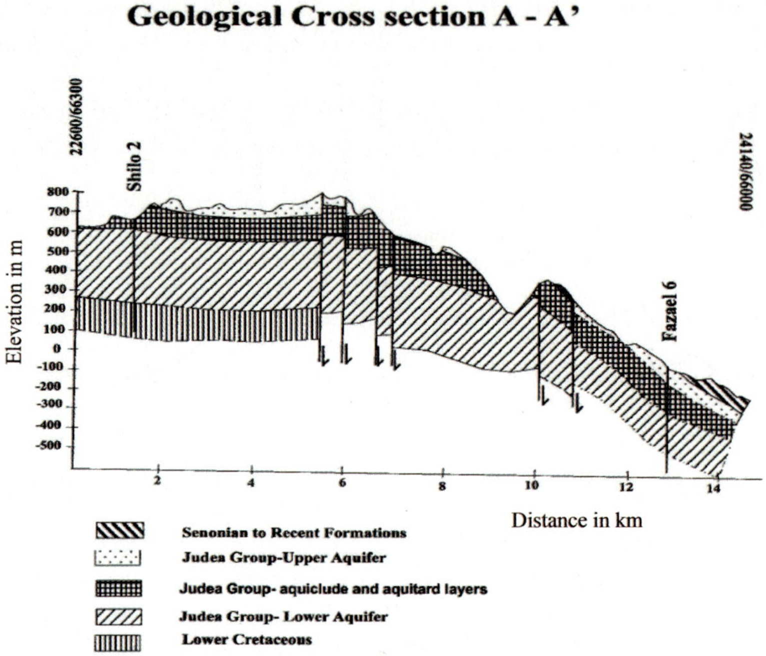

Fig 3.2.2-5. Geological cross section A-A'. Description of the aquifer formations is given in the Geology and Hydrology chapters

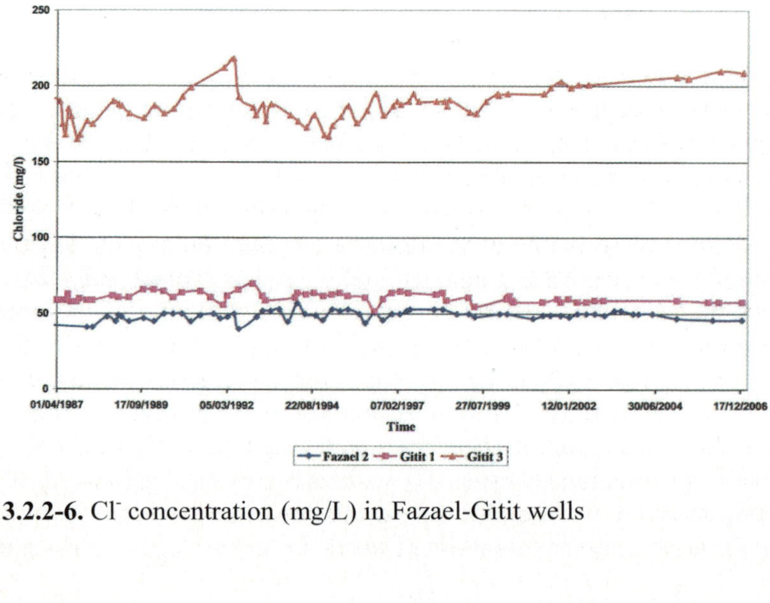

Fig 3.2.2-6. Cl^- concentration (mg/L) in Fazael-Gitit wells

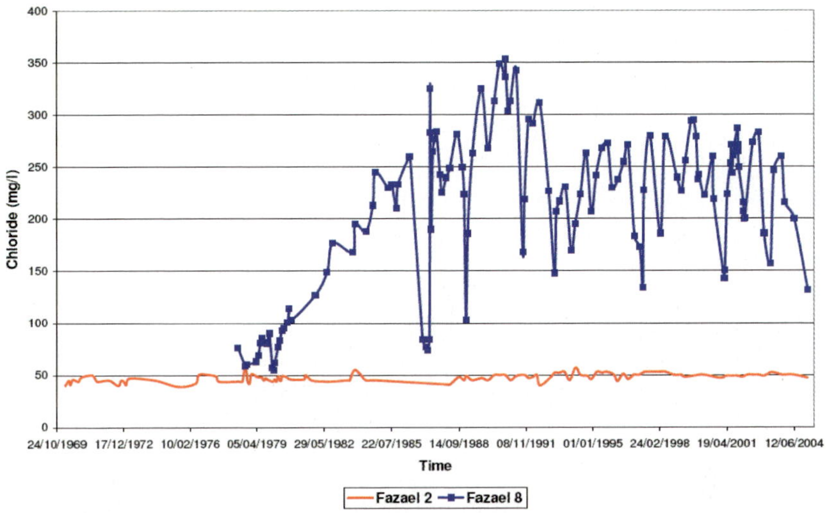

Fig. 3.2.2-7. Two types of Cl^- concentration (mg/L) behavior in Fazael wells

Fig. 3.2.2-7 shows two types of salinity behavior: wells with constant salinity over time and wells with salinity fluctuations of tens to hundreds of mg/L per season. It can be seen that salinity was low when the water production from these wells was first started but rose constantly as time went by. The increase in salinity is accompanied by seasonal salinity fluctuations. Fig. 3.2.2-8 presents a correlation between monthly pumping and salinity fluctuations in Fazael 6. Such behavior means that there is a saline body very close to the radius of influence of the well. There is an increase in salinity during periods of very high monthly pumpage because the saline water flows towards the well. After reducing pumping, the freshwater, which comes from the recharge area, pushes the saline water out of the radius of influence and the salinity in the well is reduced. Nevertheless, the salinity never returns to its initial state.

In order to understand the relationship between the fresh and the saline water, chemical and isotopic data have been collected from various wells in the region (Tables 3.2.2-1 and 3.2.2-2). The use of chemical ratios such as HCO_3^-/Cl (Fig. 3.2.2-9) and stable isotopes helps in understanding the relationships between the water bodies (for more detail on the salinity sources, see Chaps. 2.3 and 3.1). The most logical sources of salts are thermal, pressurized Ca-chloride brines which eventually flow up from depths and penetrate into the aquifers. Starinsky (1974); Rosenthal (1988); Guttman and Rosenthal (1991); Vengosh and Rosenthal (1994); Guttman (1997) and others, studied the geochemical characteristics of the Ca-chloride brines. Although Ca-chloride brines were found in all rock-units

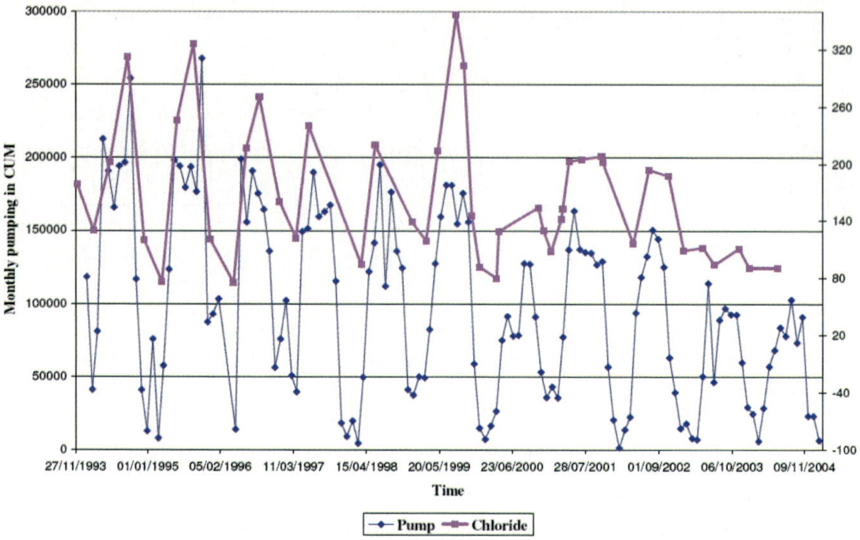

Fig 3.2.2-8. Correlation between pumping and abstraction (CUM = m^3)

along both sides of the Rift (El-Naser 1991), they probably originate from different processes and the exact mechanisms by which they are produced are still under dispute. Nevertheless, Salameh et al. (2003) noted that the major sources for salts in the east are Na-chloride type brines. It seems that the mixing of freshwater with Ca-chloride or other deep-seated brines occurs in the border faults region (mostly near major faults) and deteriorates the quality of the groundwater. Some observations indicate that the deterioration is accelerated by overpumping (Rosenthal 1988; Guttman 1997).

Other brines have a chemical composition which resembles that of the Dead Sea water. These solutions are Mg-rich residues of seawater, which penetrated the Jordan Valley several times since the Oligocene (Starinsky 1974; Bein 1999). Further increase in Mg/Ca ratio is due to seawater evaporation, which precipitates gypsum, thereby enriching the solution with Mg. Additional rise in Mg, K and SO_4^{2-} concentrations occur during the percolation of the resultant solute to the aquifer (e.g., in Fazael 1 and 11). Final mixing of the solute with relatively fresh water occurs as it percolates into the aquifer (Starinsky 1974; Raab 1998; Bein 1999).

The HCO_3^-/Cl ratio as is shown in Fig. 3.2.2-9 was chosen to show the mixing line between the water sources in the case study area. The HCO_3^-/Cl ratios in carbonate aquifers are high in places close to the natural replenishment zones and decrease in the flow direction (Table 3.2.2-2 and Fig.3.2.2-9). The high ratio can point to the proximity of natural replenishment

Table 3.2.2-1. Chemical data from various wells (mg/L)

Location	Na^+	K^+	Ca^{2+}	Mg^{2+}	SO_4^{2-}	Cl^-	HCO_3^-
K. Hasahar	14	1.3	55.1	25.9	12	29	261
Uja 2	21	1.6	58.6	28.4	15	39	282
Fazael 1	880	79	305.1	267	81.9	2536	234
Fazael 2	26	2.3	57.5	27.5	15	50	260
Fazael 6*	44	4	56.7	31.6	17	91	255
Fazael 6**	110	7	98	60	18	330	251
Fazael 8*	32	4	61	32	24	76	266
Fazael 8**	100	8	87	55	24	304	250
Fazael 11	610	32	237.2	208	52.7	1798	282
Gitit 1	28	1.6	51	25.5	16	59	218
Jericho 1	167	15	111	76.3	41	485	328
P-Meholah 2	260		126	56	92	577	199
Sa'id-Abu Mansour A-6	37		35	26	12.7	71	215
Anton Nazzal 10	100		43.9	81.3	46	212	306

Main Source: Guttman 2000 and TAU Geohydrological Information Center. During low abstraction*; During high abstraction**

Table 3.2.2-2. Chemical ratio (meq/L) and isotopic data

Location	Na/Cl	HCO_3^-/Cl	Mg/Ca	$\delta^{18}O$ ‰
K. Hasahar	0.74	5.23	0.78	-5.8
Uja 2	0.83	4.2	0.8	-5.6
Fazael 1	0.54	0.05	1.45	
Fazael 2	0.8	3.02	0.79	-5.4
Fazael 6*	0.75	1.63	0.92	-5.7
Fazael 6**	0.51	0.44	1.01	
Fazael 8*	0.65	2.03	0.87	
Fazael 8**	0.51	0.48	1.08	-5.58
Fazael 11	0.52	0.09	1.45	
Gitit 1	0.73	2.15	0.82	
Jericho 1	0.56	0.47	1.05	
P_Maholah2	0.74	0.14	1.56	
Sa'id-Abu Mansour A-6	0.8	1.76	1.23	
Anton Nazzal 2 10	0.67	0.99	0.31	

zones that receive "clean" meteoric water. In areas in which salinization takes place, the ratios decrease as a result of the increase in the chloride values. It can be seen that the decrease in these ratios becomes asymptotic as salinity increases. According to Fig. 3.2.2-9, two end-members may be discerned: natural replenishment water and saline water. A glance at the graph in Fig. 3.2.2-9 reveals that the change in this trend occurs at Cl^- concentration of 400–600 mg/L. The less saline wells, plot close to the freshwater end-member zone and vice versa. In certain wells along the border between the foothill and the Rift Valley, e.g., Uja, Fazael 2, 3 and 4, and Gitit 1, the values of the chemical ratios are as high as those in wells located in the natural replenishment zone (Jerusalem, Kochav Hachar, En Samiya wells). This means that these wells apparently receive a contribution of fresh groundwater from adjacent areas (possibly from the upper horizons) and are also isolated from sources of salinization that cause low ratios in the neighboring wells in which there is an increase in salinity (more detail on the salinity sources: see Chaps. 2.3 and 3.1).

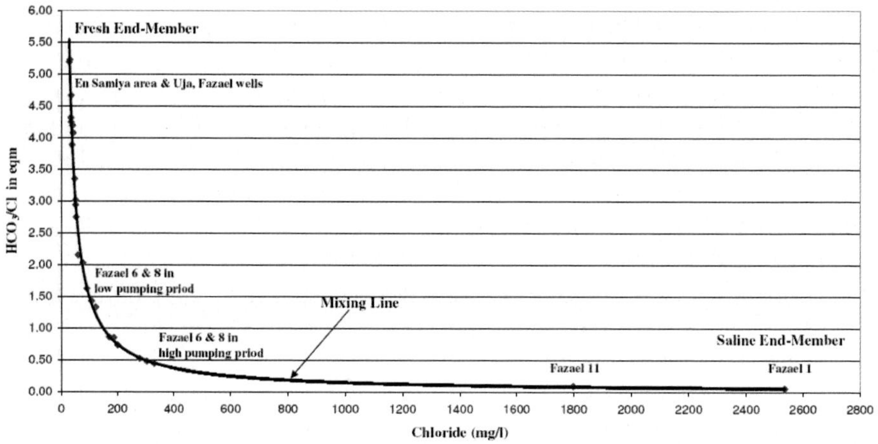

Fig 3.2.2-9. HCO_3^-/Cl^- molar ratio showing the end-members in the Fazael Area

The values of $\delta^{18}O$ in the regional aquifer indicate that the water in the aquifer was derived from rainfall on the aquifer outcrops in the recharge areas. More detailed information on stable isotopes are given in Chap. 3.1.2. The waters of En Feshkha and Uja springs are more enriched in $\delta^{18}O$ than other samples. This is due to the fact that the samples were taken from the flow in the open channels that might be influenced by evaporation. The isotopic values are quite similar to the isotopic values from the recharge area measured by Kroitoru (1987) and Schilman et al. (2004). This means that most waters abstracted by the wells along the eastern edge of the basin originate from the recharge area. Because the percentage of the saline water is very small, it does not affect the isotopic composition of the water, which remains more or less constant in the basin. According to the chemical ratios and the stable isotopes it can be said that the abstracted water in the Fazael-Gitit wells is a product of mixing of large quantities of freshwater (about 95 %) with small percentages of brine (about 5 %). The wells that pump saline water (such as Fazael 6, 8, 9 and Gitit 3) are located in the "mixing" zones between the fresh and saline water bodies (see also Chap. 3.1.3).

The border between the mountain province and the Rift Valley is the Uja-Sartaba flexure. Understanding the flow mechanism of groundwater after it crosses this flexure is an interesting issue. It is particularly interesting to evaluate the natural outlet of the aquifer. According to the geological situation (see Fig. 3.2.2-5 and Chap. 2.1) the aquifer inside the Rift Valley is at a depth of few hundred meters below surface and is covered by young formations of Mount Scopus, Avedat, Tiberias and the Dead Sea Groups that contain many aquiclude layers. This geological situation prevents any hydraulic connection between the regional aquifer (Judea Group aquifer) and the Quaternary aquifer or the Jordan River. Therefore, the Quaternary aquifer and/or the Jordan River are not the outlets of the regional aquifer. However, it is not excluded (though not yet proven) that part of the groundwater could mix with brines which are located inside the Rift, while most of the water, after crossing the Uja-Sartba flexure, flows eastward and, according to structural elements, splits northward but mainly southward towards the Dead Sea, parallel to the flexure and the Jericho fault.

On the eastern side of the study area the outcrops of Triassic and Jurassic Formations in the Deir Alla area create a watershed and divide the regional aquifers into two basins: a northern basin that dips towards the Yarmouk River and a southern basin that dips towards Karama, Kafrein, and the northern Dead Sea. The E-Salt–Kafrein area is a kind of "syncline" where young layers of Upper Cretaceous (Ajlun Group) are exposed (Fig. 3.2.2-10). The southern border of this basin ("syncline") is the Amman-Halabat fault belt which was described previously. The regional aquifers in the E-Salt–Kafrein area are in the Lower Cretaceous and in the Upper Cretaceous (especially in units A4, A7–B2; see Table 3.4.4-1). The uplift of the "Deir Alla" anticline forces the groundwater to flow westward towards the Jordan Valley in a NE–SW direction (Fig. 3.2.2-11). The "syncline" acts as a groundwater drainage system. The outlet of the Ajlun aquifer is the Pleistocene layer in the Jordan Rift Valley.

The sinistral movement of the Rift Valley transform fault in this area, brings the Pleistocene Formations in contact with the Ajlun Group layers (Fig. 3.2.2-12). This creates a direct hydraulic connection so that groundwater can easily flow from the Ajlun aquifer into the Pleistocene layers and reach the Jordan River. It is totally different in the Fazael area, where thick impermeable layers of the Mt. Scopus and Avedat Formations prevent

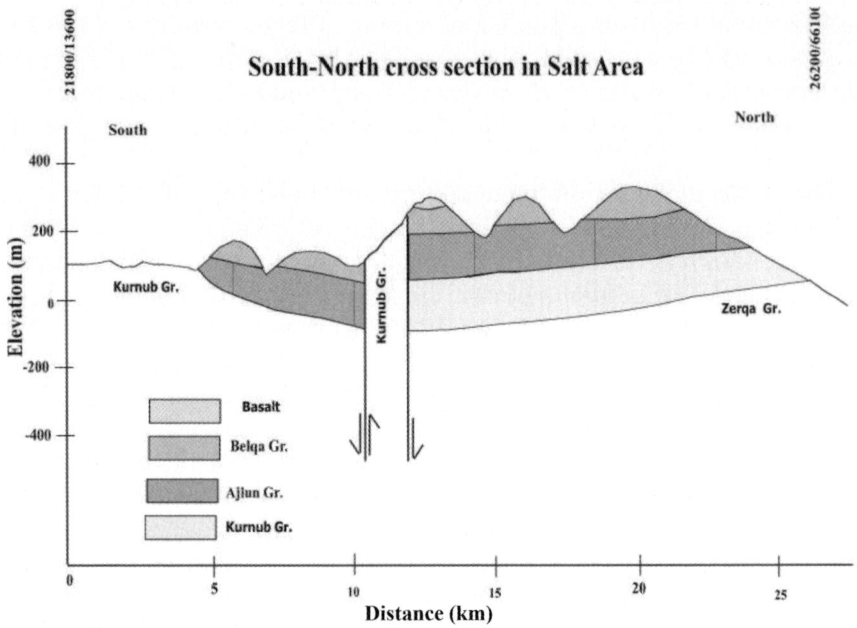

Fig 3.2.2-10. Geological cross-section in the E-Salt-Kafrein area showing the "syncline"

any direct connection between the Judea Group aquifer (equivalent to Ajlun Group aquifer) and the Quaternary layers.

On the western side of the Valley, the recharge into the Quaternary aquifer complex is mainly from leakage of surface runoff and irrigation return flow (Marie and Vengosh 2001), while on the eastern side, there is also contribution of lateral flow from the regional aquifers. The flow pattern in the Quaternary aquifer is dictated by the lithological variability, by the geological structure and by local tectonic factors (Belitzky 1996). The horizontal and vertical lithological changes turn the Quaternary aquifer into a multi-aquifer system. The recharge from surface water to this multi-aquifer system mainly consists of low salinity water. However, most of the groundwater in this system is brackish (see Table 3.2.2-1). Former studies (Rosenthal, 1988; Vengosh and Rosenthal 1994; Guttman 1997) showed that fresh groundwater which flows across the western Dead Sea Rift border faults to the Quaternary aquifer system, becomes saline with relatively high concentrations of chloride and sulfate. Likewise, unique ionic ratios such as Mg/Ca (> 1) and Mg/Cl (of approx. 0.6) distinguish the groundwater of the aquifer complex from the water replenishing it (Rosenthal 1988; Yechieli et al. 1992). Guttman (1997) noticed this phenomenon in the groundwater of the Uja-Fazael region (see Fig. 3.2.2-6 and 3.2.2-7).

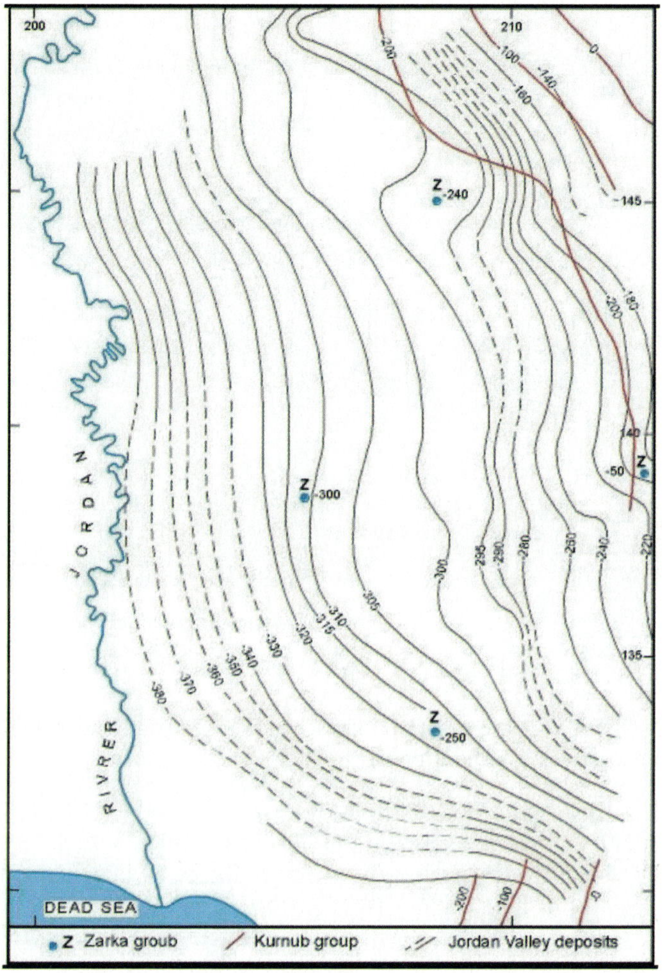

Fig 3.2.2-11. Water level map in the E-Salt–Kafrein area (Salameh et al. 2003)

On the eastern side of the Jordan Valley the replenishment of the multi-aquifer system derives from surface runoff, spring flow and from lateral flow from the regional aquifers, mainly the Ajlun Group limestone aquifers. The groundwater in the Ajlun Group aquifer is considered fresh to brackish with chloride concentrations ranging from about 20 mg Cl^-/l to 140 mg Cl^-/l (Abu-Jaber et al. 1997). Awad et al. (1997) concluded that the fresh groundwater which replenishes the Pleistocene aquifer complex from the east, deteriorates and reaches Cl^- concentrations of up to about 50,000 mg Cl^-/l due to mixing with the Ca-chloride brine that is the saline end-member (see Chap. 2.3).

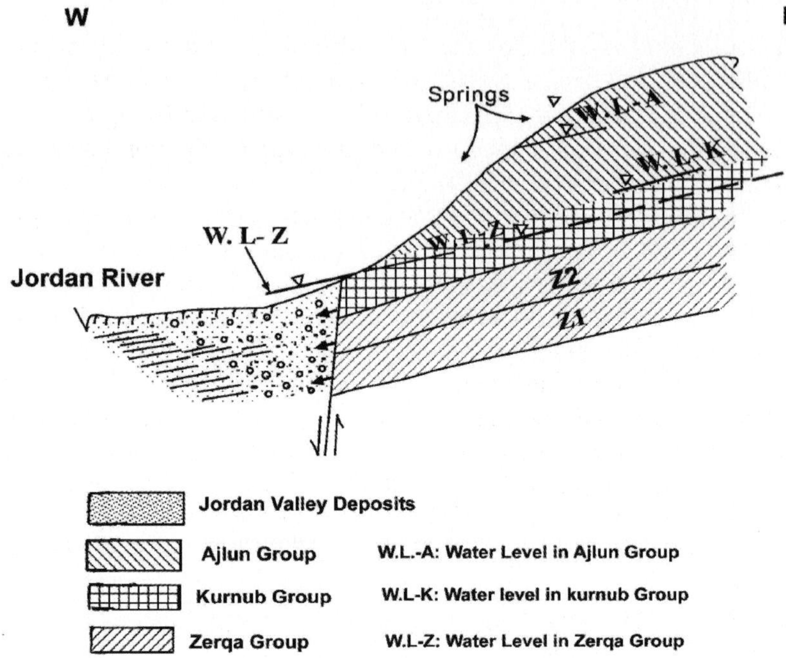

Fig 3.2.2-12. Geological cross-section showing the hydrological connection between regional aquifers and the Pleistocene layers

A certain saline input to the Quaternary aquifer complex probably derives from deep-seated brines via the faulting zones which are located mainly in the west. Another process which is likely to take place was suggested by Begin et al. (1974) and Bein (1999). They assume the mixing of freshwater with residual brines from the ancient Dead Sea and the Lisan Lake in addition to the dissolution of gypsum layers in the Lisan Formation of the Dead Sea Group. Others noted that freshwater infiltrating the aquifer leaches salts from the ground-surface which were probably deposited as soil crusts after runoff evaporation (Yechieli et al. 1992; Anker 2003) and irrigation return flow. Flexer et al. (2000) suggested that a major saline source to the hydrogeological system of Lake Kinnert is a subsurface evaporite body (the Zemah salt beds) that contributes salts in high concentration to percolating groundwater. Considering the fact that ancient evaporating lakes also existed in the study area, there is reason to suspect that salinization due to dissolution of buried evaporite bodies, such as the one which was found east of Fazael, should not be excluded. This "evaporite body" might be genetically related to the Zahrat el Qurein salt body (see Chap. 3.2.1).

3.2.2.4 Conclusion

The understanding of the flow pattern from the regional aquifers into the Rift Valley is an important tool for the sustainable exploitation of the groundwater resources for the benefit of the population living in the study area. Two basic physiographic provinces are typical: the highlands and the low-lying areas of the Jordan Rift Valley. There is a close correlation between physiographic features, geological structures and climate, both on a regional and local scale. Surface drainage of rainwater (surface runoff) has cut steep-sided wadis extending from the mountain region to the Jordan Rift Valley. Wadis generally cut into the rocks in an east-west direction, following the overall drainage pattern.

On the western side, the Judea Group aquifer, with a thickness of about 800–850 m is the main aquifer. According to water level measurements, two sub-aquifers may be identified in most of the study area: the upper aquifer and the lower aquifer. On the eastern side, the E-Salt–Kafrein area is a kind of "syncline" where young layers of Upper Cretaceous (Ajlun Group) are exposed. The regional aquifers are in the Lower Cretaceous and in the Upper Cretaceous (especially A4, A7–B2; see Table 3.4.4-1).

The groundwater flow from the recharge areas to the Jordan Valley is influenced by the tectonic pattern. The tectonic pattern creates zones of low permeability ("cascades") that force the groundwater to bypass them and to flow along individual paths and not as a uniform flow front. The main factors that influence the location and the amount of water that flows via these individual paths are the geological and structural elements, the regional water level, and the saturated thickness of the aquifer and the hydraulic characteristics of the aquifer. The sinistral movement of the Rift Valley transform fault brings the Pleistocene Formations on the eastern side in contact with the Ajlun Group layers. This creates a direct hydraulic connection so that the groundwater can easily flow from the Ajlun aquifer into the Pleistocene layers and reach the Jordan River. On the western side (Fazael area) thick impermeable layers of the Mt. Scopus and Avedat Formations prevent any direct connection between the Judea Group aquifer (equivalent to Ajlun Group aquifer) and the Quaternary layers.

A wide range of Cl^- concentrations of about 30–40 mg/L to about 600 mg/Lwas observed in the Uja-Fazael-Gitit wells. According to the chemical compositions and isotopic values, there are two end-members: the natural replenishment water and the saline water. The abstracted water in the Fazael-Gitit wells is a product of the mixing of large quantities of freshwater (about 95%) with small percentages of brine (about 5%). The wells that pump saline water (such as Fazael 6, 8, 9 and Gitit 3) are located in the "mixing" zones between the fresh and saline water bodies.

The recharge into the Quaternary aquifer on the eastern side of the Jordan River is mainly from leakage of surface runoff and irrigation return flow. On the western side of the Jordan River, there is also contribution of lateral flow from the regional aquifers. The flow pattern in the Quaternary aquifer is controlled by the variability in the lithology, the geological structure and local tectonic factors. The horizontal and vertical lithological changes transform the Quaternary aquifer into a multi-aquifer system. The recharge from surface water to this multi-aquifer system mainly consists of low-salinity water. However, most of the groundwater in this system is brackish. Inside the Jordan Valley there are several sources of salinity such as shallow brines, deep-seated brines and evaporite bodies. In the Jordan Valley the freshwater coming from the mountain provinces mixes with saline water in different ratios and contacts.

References

Abu-Jaber NS, El Aloosy AS, Jawad AA (1997) Determination of aquifer susceptibility to pollution using statistical analysis. Environ Geol 31:94–106

Anker Y (2003) The evolution of the carbonate system in hyper-arid environments (Central Arava area, Israel), and its implication for ^{14}C groundwater dating, GSI report GSI/19/03, pp 1–111

Awad AR, Taqieddin SA, Al-Homoud AS, Barakat EA (1997) Hydrochemical assessment of a deep Jordanian aquifer, Environ Geol 30:198–208

Begin BZ (1974) A. Ehrlich and Y. Nathan, Lake Lisan – the Pleistocene precursor of the Dead Sea, Geol Soc Isr Bull 63:1–38

Begin ZB (1976) Geological Map - Jericho Sheet (1:50,000), published by the Geological Survey (in Hebrew)

Bein A (1999) Salting processes of lake Kinneret, and their implication on the water management policy of the lake, and of the groundwater exploitation in the region. Water and irrigation, 397:3–12

Belitzky S (1996) Tectonic geomorphology of the lower Jordan Valley – an active terrestrial fault. PhD. Thesis, Hebrew University, Jerusalem.

El-Naser II (1991) Groundwater Resources of the Deep Aquifer System in NW – Jordan: Hydrgeological and Hydrogeochmical Quasi 3-Dimensional Modelling. Hydrogeol Umwelt, 3, pp 1–144

Flexer A, Yelin-Dror A, Kronfeld J, Rosenthal E, Ben-Avraham Z, Artsztein PP, Davidson L (2000) A Neogene salt body as a primary source of salinity in Lake Kinneret. Arch Hydrobiol Spec Iss Advanc Limnol 55:1–21

Garfunkel Z, Ben-Avraham Z (1996) The structure of the Dead Sea basin. Tectonophysics 266:155–176

Guttman J (1980) Gitit–Kane region: Calibration of flow model in the lower aquifer of the Judea Group. TAHAL Conculting Engeneers Ltd 01/80/96

Guttman Y (1997) Salinization along the western border of the Jordan Valley and the Dead Sea. Proc. of the 13th GIF meeting on the Dead Sea Rift as a unique global site, pp 1–6

Guttman J, (2000) Hydrogeology of the Eastern Aquifer in the Judea Hills and Jordan Valley- Final Report. Mekorot no. 468, pp 1–87

Guttman J, Rosenthal A (1991) Mizpe Jericho region – Study of the salinity sources and water potential. TAHAL Consulting Engeneers Ltd 01/91/46 (in Hebrew)

Guttman J, Zuckerman H (1995) Flow Model in the Eastern Basin of the Judea and Samaria Hills, TAHAL Consulting Engineers Ltd. 01/95/66 (in Hebrew)

Kroitoru L (1987) Characterization of Flow System in Carbonate Rock by Means of Groundwater Parameters in Central Israel. PhD thesis, Weizmann Institute of Science, Rehovot-Israel. pp 1–109

Marie A, Vengosh A (2001) Sources of salinity in ground water from Jericho area, Jordan Valley. Ground Water 39:240-248

Raab M (1988) The origin of the evaporites in the Jordan – Dead Sea Valley in view of the evolution of brines and evaporates during seawater evaporation. GSI Rep. 1/98, pp 1–142

Rosenthal E (1988) Ca chloride brines at common outlets of the Bet Shean – Harod multiple aquifer system, Israel. J Hydrol 97:89–106

Salameh E, El-Naser H, Al-Zoubi A (2003): German-Israeli-Jordanian- Palestinian Joint Research Program for Sustainable Utilization of Aquifer System 2nd year report 2001-2003 (unpubl)

Shawabakeh KF (2001) Al Karame Geological map 1:50,000. Hashemite Kingdom of Jordan, Ministry of Energy and Mineral Resources, Natural Resources Authority – Geology Directorate.

Starinsky A (1974) Relation between Ca-chloride brines and sedimentary rocks in Israel. PhD. Thesis, Hebrew University, Jerusalem, pp 1-176

Shaliv G (1980) Bet Shean and Eastern Samaria Basins - Updating of Hydrogeological Model, TAHAL Consulting Engineers Ltd. 01/80/51 (in Hebrew), pp 1–122

Schilman B, Teplyakov N, Gavrieli I (2004) Identification of nitrate contamination sources in groundwater using nitrogen and oxygen isotopes in nitrate: Wadi Kelt- Jericho area as a case study. GSI/25/2004, pp 1–10

Vengosh A, Rosenthal E (1994) Saline groundwater in Israel: Its bearing on the water crisis in the country. J Hydrol 156:389–430

Yechieli Y, Starinsky A, Rosenthal E (1992) Evolution of brackish groundwater in a typical arid region: Northern Arava Valley, southern Israel. Appl Geochem 7:361–374

3.2.3 Lake Tiberias and its dynamic hydrochemical environment

C. Siebert, S. Geyer, P. Möller, E. Rosenthal, D. Berger, J. Guttman

3.2.3.1 Introduction

Lake Tiberias (LT) represents the largest (170 km^2) surface freshwater reservoir in the southern Levant, with an average depth of 25 m and an average volume of 4.2×10^9 m^3 (Fig. 3.2.3-1). The National Water Carrier (NWC) withdraws about one third of the water supply of Israel and adjoining territories from the lake. The warm and monomictic lake is mainly fed by the River Jordan (about 66%), the rest derives from creeks, surface runoff, direct precipitation, fresh- and saline groundwater inflow. Originally it was a through-flow lake, today it is completely dammed. Mainly NWC, local consumers and the high evaporation withdraw water from Lake Tiberias.

The high evaporation rates in this semi-arid region enforce salinization of the lake. Its chlorinity (200–340 mg/l) is much higher than that of the inflowing River Jordan (13–30 mg/l). Only 20% of this enormous discrepancy in salinization is caused by evaporation, while the major part is the result of unspecified saltwater inputs through the bottom of the lake and by the springs emerging along the shore.

The Salinity Diversion Channel (SDC) collects most of the waters from saline onshore-springs in Tabigha and Tiberias. As a result of the diversion, the chlorinity of the Lake has decreased from 400 mg/l in 1964 to 240 mg/l today. During the drought period from the mid 1990's to 2002, the level of the Lake dropped down to -215 m m.s.l and its chlorinity increased to 300 mg/l. During the following rainy seasons, the lake was filled up and chlorinity decreased stepwise from about 260 mg/l (2003) to 240 mg/l (2004). In order to understand the mechanisms of saltwater input to LT, it is decisive to know the time-bound behaviour of the discharge of the different spring systems of the Kinneret basin and with respect to hydrochemistry.

The emerging groundwaters differ in their hydrochemistry and in the mixing of ascending brines with modern fresh- and lake water. The amount and element concentrations of saline water depend on many factors such as the elevation of the groundwater table in the aquifers and the aquifer confinement.

The aim of the study is to search for saline inflows to Lake Tiberias by using rare earth elements and Y (REY), water isotopes and major elements. For this purpose, water samples were repeatedly collected from 27 springs and wells around the lake and 127 samples from different locations within the Lake in the years 2000 and 2004 (Fig. 3.2.3-1) (Siebert 2006).

3.2.3.2 The study area and its hydrogeological setting

Lake Tiberias is situated in one of the rhomb-shaped pull-apart basins of the Dead Sea transform, which is part of the 6,000 km long Syrian-East African Rift system. Folding of the Syrian Arc, subsidence of the graben, and lateral movement of the transform and volcanism are responsible for the morphology and the hydrogeology of the Kinneret basin and of its watershed. The intense tectonic history of the area is represented by deep reaching internal, crescentic and border faults, troughs and folds (Michelson et al. 1987; Kashai 1988; Frieslander et al. 1997; Ginzburg and Ben-Avraham 2001; Hurwitz et al. 2002). South of Tiberias, in the south-eastern Lower Galilee, a more than 1000 m deep trough of pre-Miocene age is filled up with basalts, which are postulated to have an important effect on the composition of brines ascending along the western margin of the graben (Siebert 2006; ref. Chapt. 2.3; 3.1.1).
Ben-Avraham et al. (1990; 1996) postulated an E-W fault on the southern end of LT, blocking fluid movements from the Zemah 1 salt body towards the Kinneret basin. The wildcat drilling Zemah 1, situated south of the Lake (Marcus and Slager 1985), hit a large evaporite body with post-halite minerals and associated residual brines (Flexer et al. 2000). These very probably enter the Lake (Hurwitz et al. 2002) through the graben fill consisting of lacustrine and hydraulic conductive sediments and igneous rocks (Marcus and Slager 1985), and also along two longitudinal faults, which penetrate the Lake from the south and connect it with the Zemah-1 structure (Flexer et al. 2000).

With a maximum altitude of -209 m m.s.l. Lake Tiberias represents the local base level of erosion. Surface- and groundwater drain from the Eastern Galilee, the Hula Valley, parts of the Hermon Massif and the Golan Heights toward the Lake (Fig. 3.2.3-1). The River Jordan drains around 1,700 km^2 of the 2,730 km^2 large catchment area. Especially during the rainy wintertime, the fast runoff is strong and flows through wadis or directly towards the Lake.

The primary source of freshwater in the drainage basin is the occasionally high amount of precipitation (400–500 mm/a) on the mountain ranges built of permeable limestones and basalts. The flat morphology on the tops of the highlands enhances recharge of groundwater. During the year,

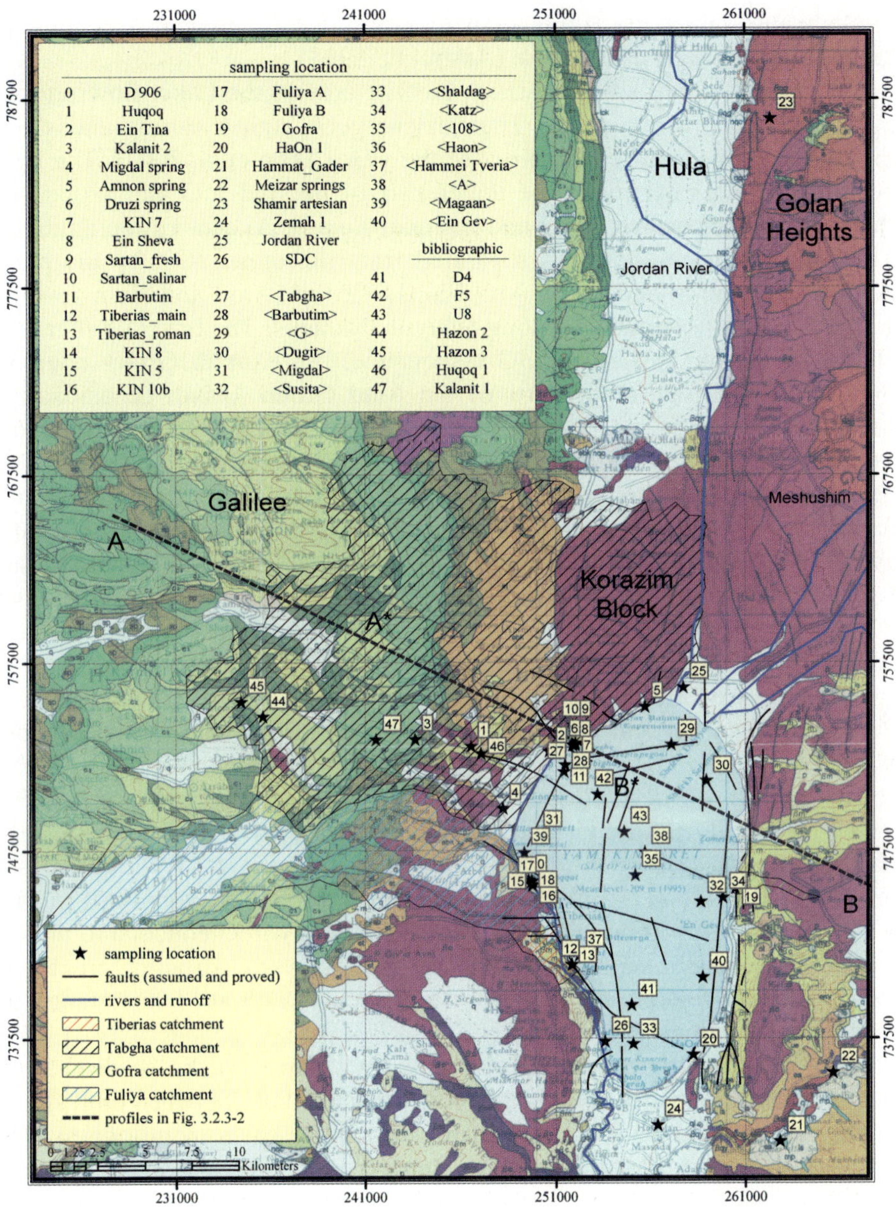

Fig. 3.2.3-1. Geological map of the study area (for detailed legend ref. Fig. 2.1-6). The catchments of spring clusters Tabigha, Fuliya, Tiberias and Gofra are indicated by dashed areas. All sampling locations of this study are given as stars. Coordination system is the New Israel Grid

groundwater tables in aquifers surrounding Lake Tiberias are always higher than the level of the Lake (Assouline 1993).

On- and off-shore springs, which emerge around the Lake, flow out of the upper and lower aquifers. Their discharge is highest during and after the rainy season in the first quarter of the year (Abbo et al. 2003). The upper karst aquifer occurs in the upper part of the Judea Group sequence (henceforth abbreviated JGA) and regional Eocene Avedat Group aquifer. The Avedat Group aquifer is separated from the upper JGA by the predominantly chalky and marly Mt. Scopus Group (Arad and Bein 1986), which is partially permeable along tectonic fractures. The lower aquifer includes the lower part of the JGA sequence, the Kurnub Group aquifer (KGA) and the aquiferous parts of the Arad Group. Brines occasionally occur in pockets of the Eocene Avedat aquifer built of marly limestone. Marly strata shield them against fast flushing out by naturally replenished groundwater. Both (upper and lower) Judea Group aquifers are predominantly separated by the Dir-Hana aquitard. From deeper parts of the crust, ascending (thermo-) saline and pressurised waters intermix in proximal parts of both aquifers with freshwaters from the surrounding mountains. Some of the resulting saline waters emerge as springs inside the Kinneret basin (Fig. 3.2.3-2). The springs are mainly related to faults, which highly increase the secondary porosity of aquifers. Faults, associated with the major faults of the Syrian-African Rift Valley, act as preferred flow paths for waters. In zones, where such tectonic structures are accumulated or intersected, springs occur in clusters. Spatial geological differences, especially in the morpho-tectonic blocks of Tiberias, Tel Rakkat and Tabigha at the base of the western graben shoulder, cause sophisticated hydraulic systems. These differences influence the ascent of brines, the supply of freshwater and their resulting mixing. As a result, the hydrochemical and physicochemical characteristics of spring waters vary strongly in time (seasonal) and region (between spring clusters) (Siebert 2006). Hence it follows that important differences in electrical conductivity (< 2–56 mS/cm) and temperature (19–60°C) occur. The systems of discharge and the hydrochemistry of springs differ fundamentally between both sides of the graben. The Dead Sea transform crosses the basin meridionally and probably separates the hydraulic systems on both sides (Gvirtzman et al. 1997a), even though the graben is permeable for fluids at least in geological times. While gravitation dominates groundwater flow in the Galilee, in the Golan Heights buoyancy flow acts as the prominent force (Gvirtzman et al. 1997a; 1997b). The hydraulic differences on both sides are mainly caused by different lithologies caused by the 105 km N-S transform movement (Michelson et al. 1987) and the lower incidence of tectonic fractures on the eastern side as compared to those on the western side.

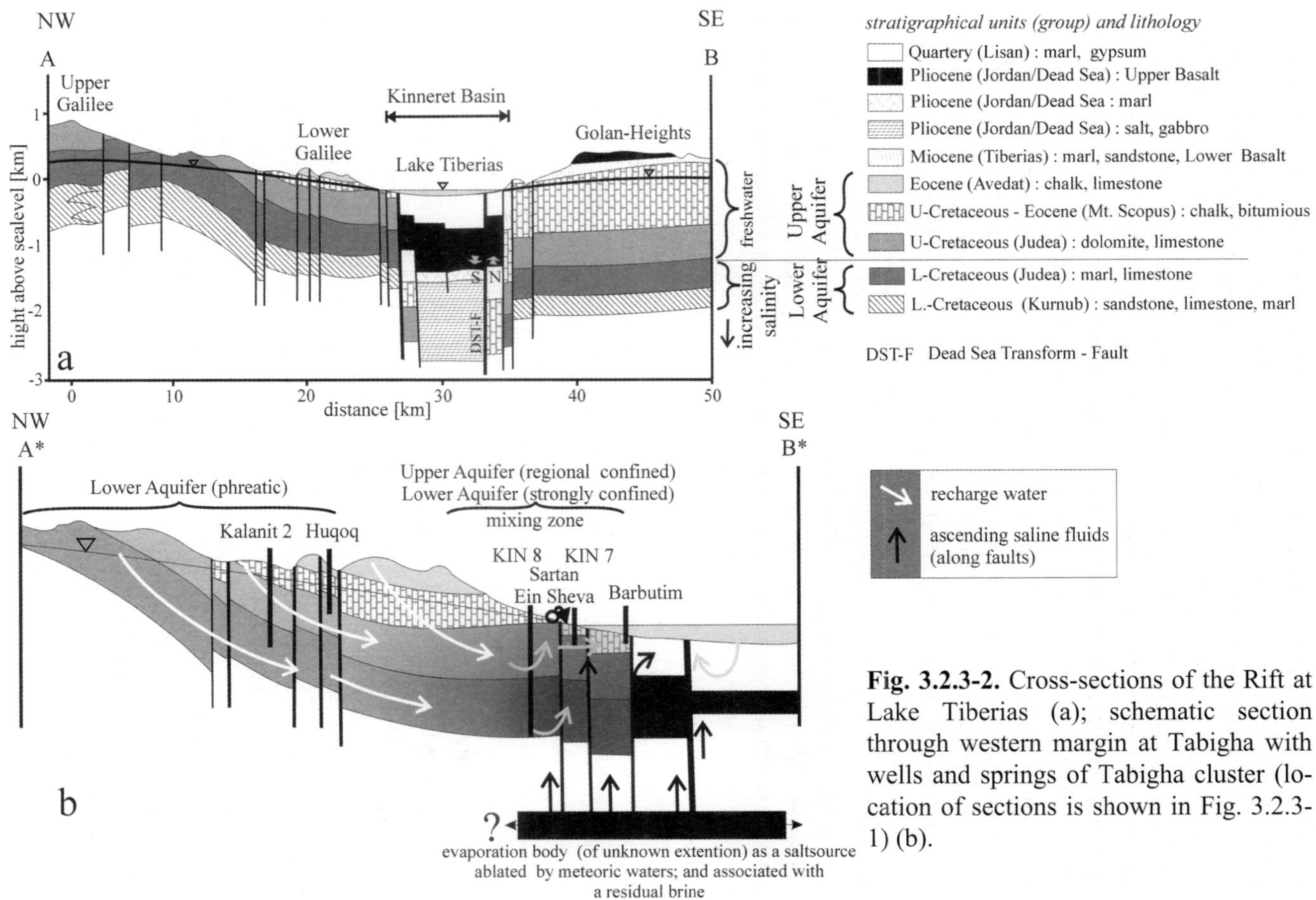

Fig. 3.2.3-2. Cross-sections of the Rift at Lake Tiberias (a); schematic section through western margin at Tabigha with wells and springs of Tabigha cluster (location of sections is shown in Fig. 3.2.3-1) (b).

The phreatic upper aquifer along the western margins contains mainly calcareous freshwater. Close to the shore of the Lake, the aquifer becomes regionally confined. Its water becomes brackish because of uprising saline water from the lower aquifer. Inside the Kinneret basin, the lower aquifer is strongly confined and its water is saline. In distal regions like in western Galilee, the aquifer is phreatic and contains freshwater, but closer to the graben, it becomes more saline than the upper aquifer (Fig. 3.2.3-2a). Pressurised and deep-seated brines penetrate into the confined lower aquifer. The resulting mixed waters rise into the upper aquifer along faults and there mix again with freshwater (Fig. 3.2.3-2b).

Fresh groundwaters of Ca-Mg-HCO_3-type, which are typical for the predominantly calcareous upper aquifer, occur in the lower aquifers only in distal regions of the basin. On the other hand, saline waters inside the Kinneret basin, which are strongly influenced by brines, generally are of two types: (1) Na-Mg-Ca-Cl-waters dominate along the NE, E, S, and SSW Lake-shores and (2) Na-Ca-Mg-Cl-waters emerge along the western and northwestern shores of the Lake. These two types point to different deep-seated brines or to different evolution-paths of a single brine by water-rock interactions. Contrarily, most rift brines (Starinsky 1974) are of alkaline earth type (Ca, Mg). After geological times, the inescapable end stage of the brines is a Ca-Cl type, like the one occurring in the Rosh Pina 1 or Zemah 1 wells (Starinsky 1974; Rosenthal 1988; Marcus and Slager 1985; Simon and Mero 1992; Flexer et al. 2000). The Neogene Kinneret brines are Na-dominant. The outstanding result is a Na-source like Zemah 1 evaporation body. Ablation fluids penetrate the subsurface of Lake Tiberias.

The Mg enrichment in the saline water types occurring along the eastern shore cannot be the result of weathering of basalts and halite ablation does not yield low Cl/Br wt. ratios (<200). Therefore, a residual Br-rich Mg-Cl brine of seawater evaporation must be present (ref. 3.1.1). Ca-dominated, saline waters emerge along the western side of the basin. They are mainly the result of dolomitisation of limestones and intense interaction with basalts and gabbroic bodies located in the pre-Miocene basin south of Tiberias and in the graben.

3.2.3.3 Results and discussion

Unspecified influx of saline groundwater into Lake Tiberias

Nishri et al. (1999) and Rimmer and Gal (2003) reported yearly unspecified inflow of saline groundwater into the Lake of 61×10^6 m³ and 78×10^6 m³, respectively. Out of these, the Barbutim area with about

16×10^6 m³/a (Mercado and Mero 1984) represents about one-fourth of the total input. Estimated discharges of known off-shore sources like Barbutim are indirectly determined and are subjected to seasonal and annual fluctuations like the on-shore springs of the related cluster. Mass- and volume balances of large hydrological systems like Lake Tiberias are subject to high uncertainties of the involved parameters (Assouline 1993; Rimmer and Gal 2003). Mero (1978), Nishri et al. (1999) and Smith et al. (1989) estimated the annual Cl^- supply to the Lake to be about 65, 72 and 90×10^6 kg, respectively. In this study, calculations of mass and volume contribution to the Lake in the hydrological year 2002/03 yield an amount of 84×10^6 m³, 66×10^6 kg Cl^- and 0.5×10^6 kg Br^-. However, the uncertainties for one-year-calculations on a monthly base cover more than two thirds of the yearly balanced amounts. There are two other ways to determine the minimum inflow of unspecified water into the Lake: evaluation of Ce anomalies and estimation of diffusion-advection transport.

Ce balance

The most prominent compositional changes are with Ce at the sediment-water interphase in the area of the hypolimnion. The drafted trends in Figs. 3.2.3-3 and 3.2.3-4 indicate that in autumn the maximum observed Ce-anomaly A_{max} of the lower hypolimnion is about 1.562, whereas during springtime the Ce-anomaly of the exfiltrating pore water is a = 1.2 to 1.25. Assuming that the former value is typical for the system from May to December, the balance Eq.(3.2.3-1) yields the percental rise of volume of the hypolimnion, caused by the amount of exfiltrating pore water ε.

$$A(1+\varepsilon) = a\varepsilon + A_{max} \tag{3.2.3-1}$$

The volume-weighted average of Ce-anomaly in the lower 18 m of the Lake, A=1.554, is estimated in layers using (a) the hypsographic curve (Ben-Avraham et al. 1990) for estimating the surface of each layer of 3 m thickness (Fig. 3.2.3-5) and (b) the Ce-anomalies derived from Fig. 3.2.3-4. In order to increase the correspondence between determined Ce-anomalies and the weighted averages of each layer, the latter is adjusted to the former. By this procedure a volume-weighted minimum of A = 1.552 is derived. The maximum of the Ce-anomaly is found to be 1.562. From Eq.(3.2.3-1) it follows Eq.(3.2.3-2):

$$\varepsilon = (1.562 - 1.552) \Big/ \left(1.552 - \begin{matrix}1.20\\1.25\end{matrix}\right) = \begin{matrix}0.029\\0.033\end{matrix} \tag{3.2.3-2}$$

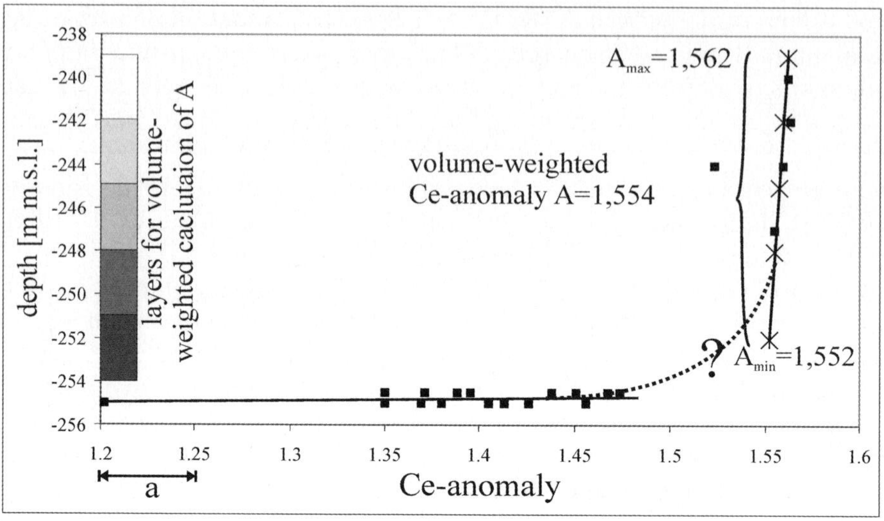

Fig. 3.2.3-3. Principle description of calculation of volume-weighted Ce-anomaly at the sediment-water-interface

The considered hypolimnion of 543×10^6 m^3 (up to a level of -237 m m.s.l.) is enlarged by 3.1% resulting in 16.8×10^6 m^3/a. With a derived area of 65×10^6 m^2 the average linear flow rate is 0.26 m/a. Using this flow rate for the entire area of the Lake with an area of about 170×10^6 m^2, the annual inflow of water from below is about 44×10^6 m^3/a, which corresponds to about 1.4% of the total volume of the Lake (4.2×109 m^3). Using this quantity and multiplying it with an average Cl^- content (350 mg/l) in pore water of the upper 0.3 m of the sediment column (Stiller et al. 1975, Siebert 2006), a minimum of 15.4×10^6 kg of Cl^- is thus added, which corresponds to $100\times(44\times106\times0.35/4.2\times109\times0.26) = 1.4\%$ of the Cl^- content of the Lake and is therefore below the standard analytical limit of detection.

Diffusive and advective flux

The diffusive flux based on concentration gradients in the sediment column, and advective fluxes (Stiller 1994; Dror et al. 1999) in areas, where they were measured, are only estimated. This does not include fault-bound inflow. According to Stiller (1994), analyses of pore waters show the potential penetration of chloride from the deeper parts of the sediment into the Lake by diffusive and advective fluxes F_{AD} after Eq.(3.2.3-3). Here, D_i is the diffusion coefficient of Cl^-, v is the velocity of the medium, C_z the concentration of the element at the depth z. Porosity p of sediment is given by Stiller (1994) as 0.89.

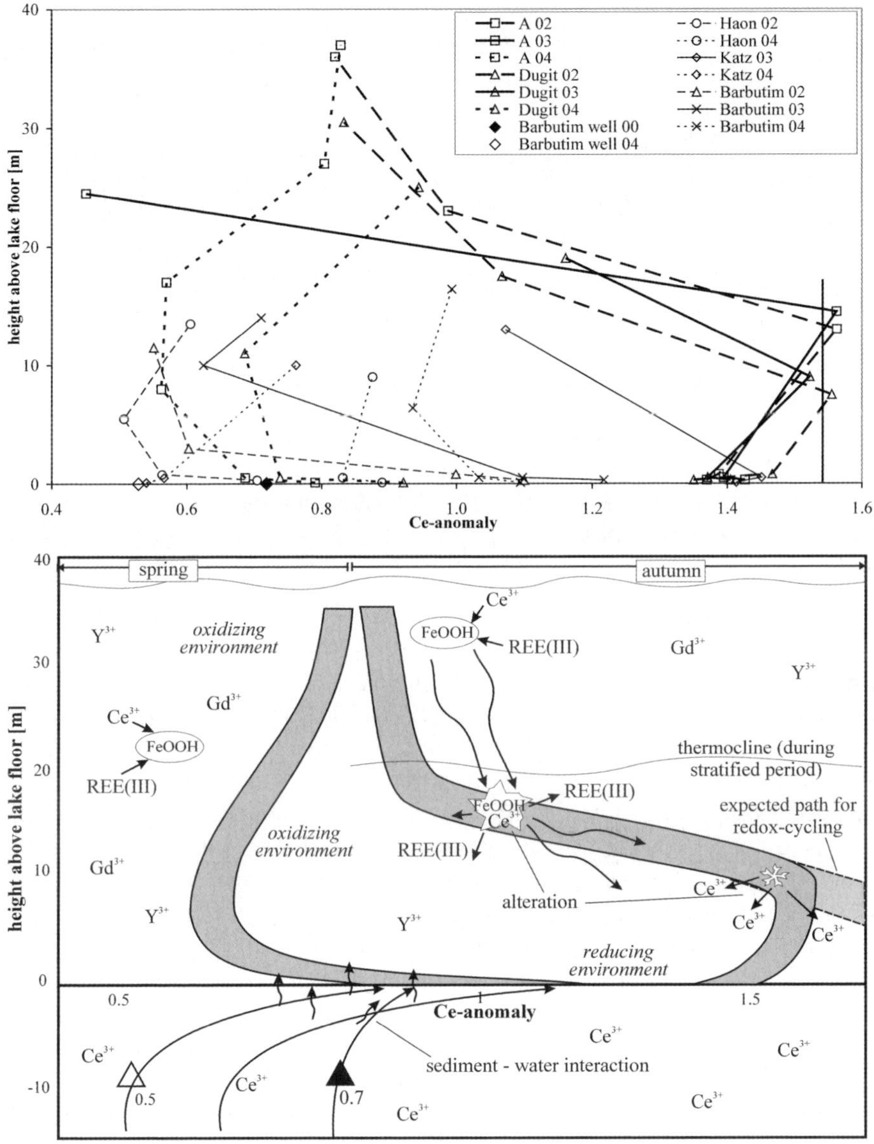

Fig. 3.2.3-4. Distribution of the estimated Ce anomalies in Lake Tiberias. The upper graph shows the estimated Ce anomalies. The bold lines represent the situation in autumn, the thin lines that in springtime. The lower drawing shows the processes in the water column in springtime (left) and in autumn (right). The developing Ce anomalies during different seasons in different water depths are shown as greyish stripes

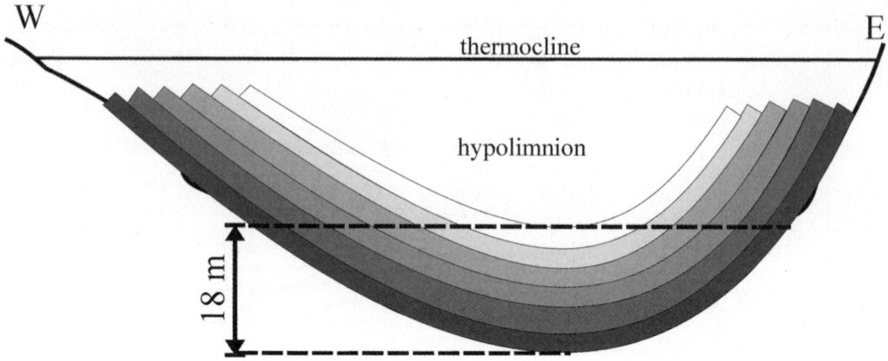

Fig. 3.2.3-5. Schematic section through the hypolimnion showing layers of assumed constant Ce anomalies as applied to estimate an average Ce anomaly in the hypolimnion. Each grey-scaled layer represents a 3 m thick water-body which is used for volumetric weighting of the respective Ce anomaly

$$F_{AD} = -p \cdot D_i \cdot \frac{dC}{dz} \pm v \cdot C_Z p \qquad (3.2.3\text{-}3)$$

Pore water analyses (Stiller 1994; Dror et al. 1999) of different boreholes all over the Lake and accompanying Cl^- fluxes have been classified into four groups (Table 3.2.1-1). These four groups are according to the four classes of specific resistivity measured in 5 m depth of the Lake's sediment (Goldman et al. 2004) (Fig. 3.2.3-6). The areas of equal specific resistivity at 5 m below the sediment/water interface (Goldman et al. 2004) are supposed to be transferable to the upper part of the sediment column. Comparison of measured specific resistivities (which can be converted into Cl^- concentrations) with chlorinity of pore waters at the same depth, differ by ±500–1000 mg/l Cl^-. On the other hand, extrapolated salinities of F5 and U8 (Fig. 3.2.3-1) at 20–30 m sediment depth (18,000 mg/l Cl^-) fit well to the 0–1 Ωm of specific resistivity at that depth. It is most probable, that the above mentioned differences are the result of a temporarily varying salt-water table below the surface of the Lake as the pore water chemistry and specific resistivity were not determined simultaneously. In this study, chlorinity fluxes were weighted by the size of the areas of equal resistivities. The results are given in Table 3.2.3-1.

Gradients of chlorinity in sediments of Lake Tiberias vary strongly (Tab. 3.2.3-1). Therefore, an enhanced concentration gradient dC/dx occurs in borehole E, while locations U8, D4, F5 and A–G (Stiller 1994) (Fig. 3.2.3-6) show lower but similar gradients with depth. Stiller et al. (1975) also reported a very low advective flux of 0.04 m/a (Tab. 3.2.3-1)

Table 3.2.3-1. Groups of pore water, their input parameter and their particular Cl^- fluxes caused by diffusion, advection and the size of the respective area of specific resistivity (Goldman et al. 2004)

Pore water samples	Diffusion (mg/ cm²/a)	Advection (mg/ cm²/a)	v (cm/a)	dC/dx (average of class) (mg/ $cm^4 \times 10^{-3}$)	F_{AD} Eq. (3.2.3-2) (mg/ cm²/a)	Area (km²) /(%)	Mass Cl^- (10^6 kg)
A (B-D, D4, F, F5, U8 G)^S	4.37	1.04	3.8^S	12.2	5.4	117 / 68	6.3
E^S	7.16	0.95	3.8^S	20	7.5	36 / 21	2.7
Fuliya^D	-	64	56^D	771	64	15 / 9	9.6
Tiberias^D	-	340	25^D	1546	340	1-3* / 2	3.5-10

^{D} after Dror et al. (1999): ^{S} after Stiller et al. (1994)
* size of area of equality: 1 km² after Goldman et al. (2004) or 3 km² after Braudo et al. (1970)

in borehole A in the centre of the Lake. In contrast, in boreholes Fuliya and Tiberias, the chlorinity below -0.1 m depth is constant and at a high level (1,150 mg/l and 14,000 mg/l) (Dror et al. 1999). In these areas saline waters enter the Lake by advective flux which is 0.56 m/a in Fuliya and 0.25 m/a in Tiberias (Dror 1998). Fuliya and Tiberias boreholes are drilled next to the shoreline and the chemistry of the pore water is similar to that of the respective on-shore spring cluster, which is hydraulically connected to the Lake (Rimmer et al. 1999). It is most probable that saline pore waters are connected to the hydraulic systems of the respective on-shore spring clusters and the flux is subjected to specific seasonal variations.

From geophysical and hydraulic analyses (Ben-Avraham et al. 1996; Reznikov et al. 2004; Rimmer 2000; Abbo et al. 2003) it is known that about one third of Lake Tiberias (west of longitude 254000 New Israel Grid) is outside the graben *sensu stricto* and is overlying and connected to the aquifers of the Lower Galilee (Abbo et al. 2003). It can be assumed, that in this area, advective fluxes dominate diffusion in this area. Following this line of thought, inside the graben *sensu stricto*, diffusive flux dominates, while advective flux is mostly bound to faults, i.e., zones of enhanced hydraulic conductivity. This is supported by the chlorinity in F5 (2,500 mg/l) and U8 (3,300 mg/l) at 4 m depth outside the graben, which is much higher than in D4 (1,700 mg/l) within the graben fill.

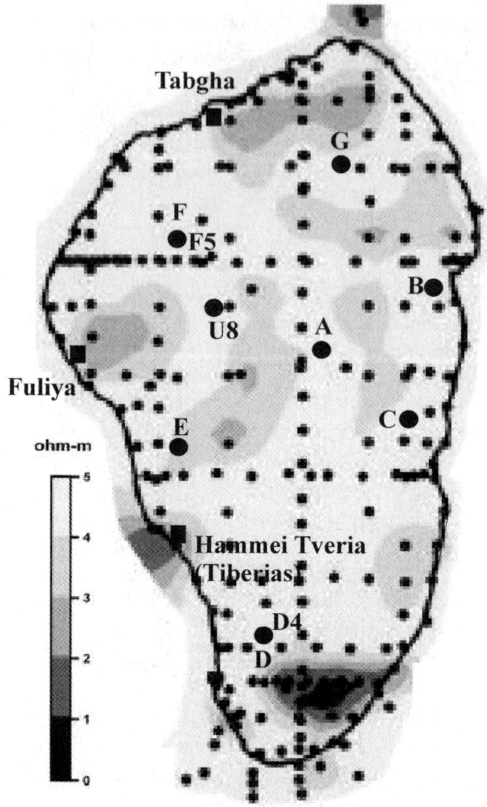

Fig. 3.2.3-6. Areas of equal specific electrical resistivity at 5 m depth of Lake's sediment (Goldman et al. 2004) are indicated by greyscale. Locations of boreholes (thick dots and squares) are taken from Stiller (1994) and Dror et al. (1999)

Assuming only small participation of advection, Stiller (1994) integrated calculated fluxes over the entire Lake bottom, which resulted in an annual input of 9×10^6 kg Cl^-. Our classification and weighting of boreholes of similar Cl^- fluxes with areas of similar specific resistivity yields a minimum input of about 22×10^6 kg Cl^-. This amount equals a mass-input of 5 mg/l or 2% increase of average chlorinity of the Lake water (260 mg/l) with an average volume of 4.2×10^9 m^3. This increase is at the verge of the limit of detection.

The minimum amounts for Cl^- derived from the consideration of Ce anomaly change in the lower water column, and the diffusive and advective transport is based on concentration gradients. Determined flux rates are comparable and both represent minimum inputs. The fault-bound flow is not included in the advective transport, whereas it is indirectly included in the Ce balance. Both approaches are minima for the Cl^- input to the

Lake. We suggest that the input of Cl is definitively >$20 \cdot 10^6$ kg/a without knowing precisely an upper limit.

Mero (1978), Smith et al. (1989) and Nishri et al. (1999) reported results on Cl input of 65–90×10^6 kg/a based on the volume and chemical balances of the Lake, which at first sight are higher than ours. However, they did not evaluate the uncertainties of their results. Considering the uncertainties mentioned by Rimmer and Gal (2003), the above results are within the range of the overall uncertainty of the balance (1σ).

Plate 3.2.3-1: Station "A" on Lake Tiberias. Approach by sampling boat. (Photo: C. Siebert)

Time-bound variations and interaction of hydrochemical patterns in spring clusters and Lake water throughout the year

Rare earth elements including yttrium (REY)

From April/May to December (Eckert et al. 2002) Lake Tiberias is stratified by a thermo- and chemocline. While the Lake is stratified, physicochemical conditions in the upper water body (epilimnion) differ from the hypolimnion (Fig. 3.2.3-7). Only little mixing occurs due to internal movement of the water bodies (Laval et al. 2003).

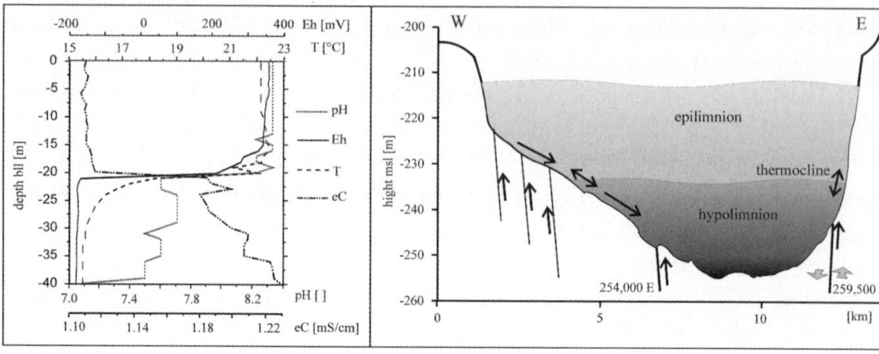

Fig. 3.2.3-7. Trends of physicochemical parameters at station "A" in autumn 2002 and stratification of Lake Tiberias. Epi- and hypolimnion are shown and arrows indicate water movement. The graben faults are situated at 254,000 and 259,500 E (New Israel Grid)

Period of stratification

In autumn, all samples from the epilimnion show soup-bowl like REY patterns with a La_n/Lu_n ratio 1.6–2.1 which resemble freshwaters from aquifers and rivers of the area. Furthermore, negative Ce-, Eu- and positive Gd- and Y anomalies (Fig. 3.2.3-8) are present in all waters down to the thermocline at about 15–20 m below Lake level (bll) (Fig. 3.2.3-6).

Waters, sampled below the thermocline, from the Lake floor and up to 10 m above, show REY-patterns with anomalies of Ce, Gd and Y and La_n/Lu_n of 3.6–4.5. They are enriched in REY up to one order of magnitude compared with waters from above the thermocline (Fig. 3.2.3-8). In shallow locations, in which anoxic conditions do not develop, REY patterns vary in different ways.

In water profiles taken in autumn, the Ce-anomalies change from negative in surface waters to positive below the thermocline and decrease towards the bottom. Y-anomalies are smaller below the thermocline than above. Y/Ho molar ratios at 1 m bll (below Lake level), are about 108, whereas, next to the bottom, the Y/Ho ratio yields 80. During the period of stratification, changing orientation of Ce-anomalies in water column is the result of redox-cycling. In oxygen-rich water of the epilimnion, Fe-Mn-oxyhydroxides precipitate and effectively scavenge REY (Fig. 3.2.3-4), which are partly released during aging (Bau and Dulski 1996; Bau et. al. 1997). In this process the medium REE are more retained than the very light and very heavy ones. Gd and Y preferably stay in solution. In autumn, the REY patterns of the surface Lake water reflect the results of REY scavenging, whereas the deep-water samples represent REY patterns after dissolution of the precipitated oxyhydroxides. Redox-cycling is a common phenomenon in chemically reducing waters in which Fe- and Mn-oxyhydroxides are re-dissolved. The outstanding result of this process is a positive Ce anomaly as seen for example in the bottom water in the Tyro sub-basin (Fig 3.2.3-8) of the Mediterranean Sea (Bau et. al. 1997). These strongly reducing highly saline waters show REY patterns similar to those of the chemically reducing waters in Lake Tiberias, although the latter are not older than one year.

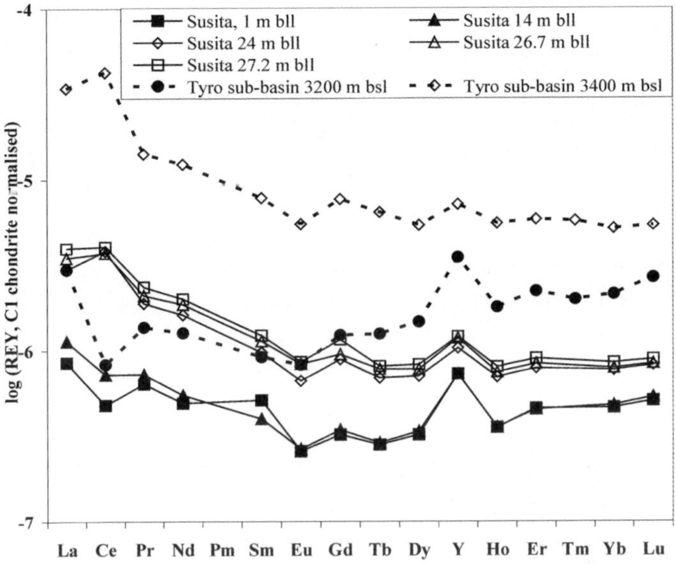

Fig. 3.2.3-8. Rare earth element pattern at the "Susita" point in Lake Tiberias and in the Tyro Basin, Mediterranean Sea (data from Tyro Basin from Bau et al., 1997). bll: below Lake level, bsl: below sea level

Lake site "Barbutim" (henceforth all Lake-sites are given in quotation marks) is situated just above the Barbutim offshore well in the north-western corner of the Lake and in the oxygen rich zone (Figs.3.2.3-1; 3.2.3-7). The Barbutim well is hydraulically connected to the Tabigha area and is supplied by both JGA-aquifers and - as indicated by $\delta^{18}O$ and $\delta^{2}H$ values - probably receives fractions of water which flow through the Korazim block. Most of the studied thermo-saline waters show steeper REY patterns than Barbutim well water, but with similar anomalies, which points more or less to the same origin of these waters. Assuming that all thermo-saline waters represent mixtures of deep-seated brine with recent water (Möller et al. 2003), the change of the slope of the patterns may be caused by mixing of these thermo-saline waters of varied fractions of both types of water.

As for “Barbutim”, the patterns correspond with those of the chemically reducing waters below 20 m. This correspondence is only explicable by assuming that “Barbutim” and waters from anoxic Lake sites originate from the same aquifer. The chemically reducing waters of the well Barbutim are immediately mixed with the oxic waters of the epilimnion. The analysed samples of the Barbutim well, which penetrates 10 m of the sediments, show patterns which are dissimilar to that of “Barbutim”. This indicates that at least at the Barbutim site, waters ascending from depth undergo alteration when passing through the upper sediment layer (Fig.3.2.3-4). In this process a systematic change of REY patterns occurs by which the lightest and heaviest REY are more enriched than the medium ones.

Barbutim well water represents a mixture of thermo-saline waters of the area with water from the bottom of the Lake which is identical with the diffusive Barbutim water (Siebert et al. 2004a; 2004b). As a result of this mixing, the original thermo-saline water has to be very light: $\delta^{18}O$ = -8‰ and $\delta^{2}H$ = -60‰. Such a light water is typical for the recharge from higher elevated areas like Mt. Hermon (Fig. 3.2.3-9).

In the hypolimnion, the positive Ce anomaly is lowest at the water-sediment interface (Fig. 3.2.3-8). The highest positive Ce-anomalies are always found several meters above the bottom of the Lake. A combination of the above arguments suggests that waters with REY patterns similar to those of the Barbutim seepage water also enter the deepest parts of the Lake which are always anoxic during the stratification period in Lake Tiberias. Here, the rising water of the type encountered in the Barbutim well, passes the sediment and changes from strongly negative to slightly positive Ce-anomalies near the interface. If such water enters anoxic parts of the Lake through the bottom sediments and mixes at the water-sediment interface, the high positive Ce anomaly of the latter is diminished (Fig. 3.2.3-4). This process can be observed everywhere in the hypolimnion during the stratified period. Water next to the bottom is a mixture of (1) water with high positive Ce anomalies resulting from redox-cycling and (2) ascending water with low positive Ce anomalies. Internal motions cause fast mixing in the benthic boundary layer (Eckert et al. 2002). Waters of deep origin discharge more or less over the entire bottom into the water body of the Lake. This is supported by time-domain-electromagnetic measurements reported by Goldman et al. (2004). Nevertheless, diffusive inputs may occur locally (bound to faults) and areally. It cannot be excluded that the same type of water even seeps into the shallow, oxygen-rich part of the Lake, which is characterised by slightly varying but always negative Ce anomalies.

Some of the shallow sites show Ce variations occurring next to the bottom, for instance at "Shaldag" (Fig. 3.2.3-1) in 2003. The Lake site "Shaldag" shows a typical pattern of shallow and oxygen rich sites, but with a diminished negative Ce anomaly next to the bottom. In "Shaldag" redox-cycling is impossible, because the water column is completely oxic. Typical thermo-saline waters from the Tiberias cluster (e.g., Main Spring) show a positive Ce-anomaly, which is supposed to be traced in the deep water from "Shaldag". The ascending waters in the "Shaldag" area are in contact with mainly coarse siliceous clastics like quartz and palygorskite ($Al_2((OH)_2/Si_4O_{10})$) (Serruya 1978). Mixing of less than 1% water of the Main Spring type (Tiberias cluster) and more than 99% of Lake water, yields the composition of the deepest water from "Shaldag" with respect to REY, Cl, Br, Na, K, and $\delta^{18}O$ (Siebert 2006).

Period of homogeneity

In March the situation is different because the entire Lake is mixed. Water from the lower water column is enriched in REY. It shows none or, at the most, positive Ce-, Gd- and Y anomalies and La_n/Lu_n about 1.4. The negative Ce- and positive Y anomalies and La_n/Lu_n decrease towards the surface. Only at station "A" and at the "Barbutim" seepage (Fig. 3.2.3-1) where deepest waters become anoxic and redox-cycling is active, negative Ce anomalies are larger at the surface than at the bottom.

In March, discharge patterns of springs and wells show that the saline input to the Lake is mainly dominated by the discharge of the Tabigha and the Fuliya cluster (Rimmer et al. 1999). In springtime 2004, Barbutim well showed a much more negative Ce anomaly than in autumn 2000, which points to an enhanced fraction of water from the Upper Judea Group aquifer during the spring. Ce anomalies in well water are between 0.52 (2004) and 0.72 (2000) (Fig. 3.2.3-3). In March, discharge of Barbutim is more intense, which can be observed by a much higher plume of saline water at the "Barbutim" site.

In springtime, the ascending water passes the still anoxic Lake sediment, leading to the same input of water with none, or sometimes positive Ce anomaly as during autumn. In contrast, in the overlying water column negative Ce anomalies occur as the result of scavenging of REY by precipitating Fe-Mn oxyhydroxides in the whole Lake. In spring, inflowing river waters show slightly positive Ce anomalies, effecting the occurrence of no or even positive Ce anomalies in surface waters of the Lake.

On the bottom of the Lake, mixing occurs in spring as well as in autumn, but in spring the water from most locations is oxic and shows therefore negative Ce anomalies. The ascending water still owns no or only a small positive Ce anomaly, leading to traction of the negative Ce anomalies in Lake water towards more positive anomalies next to the bottom (Fig. 3.2.3-3).

Stable isotopes: δ18O; δ^2H; $\delta^{13}C$

For the search of saline inflows to the Lake, stable isotopes seem not to be helpful, because the signatures of $\delta^{18}O$ and δ^2H of the Lake water are controlled by high evaporation and by an average residence time of 5.5 years (Nishri et al. 1999). They are therefore higher than all known waters in the Kinneret Basin (Fig. 3.2.3-9). The Lake is a mixture of many types of waters.

$\delta^{18}O$ and δ^2H plot below the Local Meteoric Water Line (MWL), with a clear trend along the Paleo- and Global MWL. This may be due to water-rock interaction with marly strata in the Avedat and Judea Groups.
The lightest waters always occur next to the elevated recharge areas in the eastern Galilee, Golan and Hermon Massif. A strong positive correlation between increasing salinity and high isotopic values is noticeable (Fig. 3.2.3-9). On closer examination, waters of different clusters differ in this behaviour. At the same salinity, waters from the Tabigha area are lighter than waters from Fuliya or Gofra area (Fig. 3.2.3-10). The reasons are (1) the elevated recharge area (950 m msl supplying water to Tabigha; max. 522 m msl to Fuliya; 370 m msl to Gofra), known as the altitude effect (Poage and Chamberlain 2001), and (2) inflow of light water from the Hula Valley. The sequence of wells and springs of the Tabigha cluster may be considered as observation wells for any contribution of water from the north flowing from the Hermon Massif through permeable strata and faults in the Korazim block (Belitzki 2002). Not even one of the flows from west of Tabigha (Huqoq, Qalanit, Hazon) shows the same light isotopic values as En Sheva or Kinneret 7. The recharge shows $\delta^{18}O$ and δ^2H values not less than -5.59 and -3.05‰, respectively, (Bergelson et al. 1999), while En Sheva shows values of -5.8 and -28.1‰. Such values are typical for the Shamir well which is located about 30 km northwards.

In contrast, most saline waters, no matter what cluster they belong to, show high isotope values. Such waters become isotopically heavier with increasing salinity. Only the saline well Haon 1 shows an inverse correlation resulting from mixing with isotopically heavy water of Lake Tiberias.

The isotopic values of Lake Tiberias are dominated by evaporation. At the end of the drought in the beginning of the 21st century water was isotopically heaviest at low stands. The isotopic signatures of Lake samples in November 2002 show the influence of stratification with significant differences between hypo- and epilimnion (Fig. 3.2.3-9) and the high ratio between evaporation and recharge in former years.

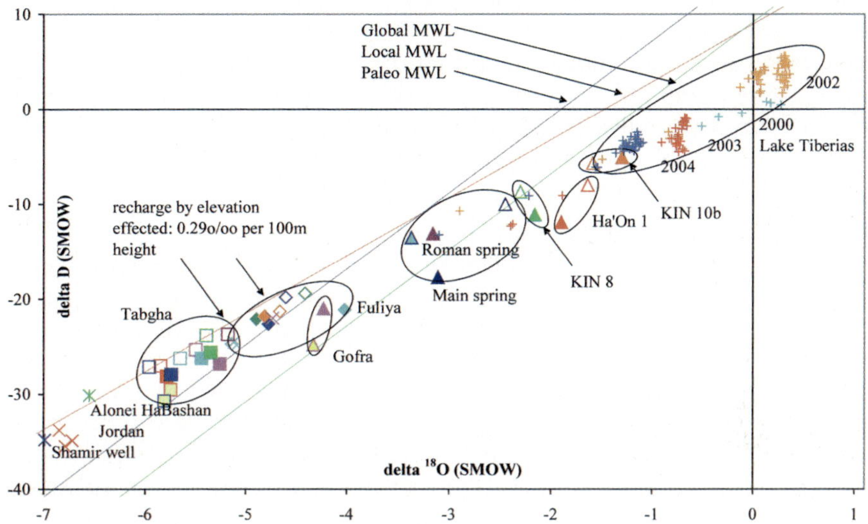

Fig.3.2.3-9. Cross plots of δ^2H vs. $\delta^{18}O$. The Local Meteoric Waterline (LMWL) $\delta^2H=6.1\times\delta^{18}O+9.5$ is based on 52 samples of precipitation (IAEA) from station Har Kna'an in the Upper Galilee (New Israel Grid: 247,000E; 764,000N; 934 m msl), which correlates with the one given by Bergelson et al. (1999). The slope of the LMWL is less than that of the Global MWL ($\delta^2H=8\times\delta^{18}O+10$; Craig 1961) and Paleo MWL ($\delta^2H=8\delta^{18}O+16$; Gat 1981). All isotopic data of this study plot below LMWL. Squares: Tabigha; diamonds: Fuliya; triangles: thermo-saline waters; crosses: Lake water; lying crosses: rivers and distal recharge; Colors indicate individual sampling points, while open and closed symbols represent springtime and autumn, respectively

At location "Tabigha", a $\delta^{13}C$-signature of dissolved inorganic carbon (DIC) of -40.2‰ directly above the bottom strongly points to biogenic conversion of CH_4 to CO_2. "Shaldag", situated above the flat platform in the southern part of the Lake also shows lower values in $\delta^{13}C$ than the average of the Lake. This strongly coincides with depletion of δD towards the deeper parts of most profiles (-8.5‰ in 2002 and -6.3‰ in 2003), while $\delta^{18}O$ remains quite constant.

Mixing of Lake and well water is observed at "Barbutim". Mixing also occurs at point "G", situated about 2 km in front of the inflow of the River Jordan, and is indicated by significant depletion of $\delta^{18}O$ (parallel to Local MWL) and lower Cl/Br ratios in 2002 and 2003. Since the deepest parts of "G" are well within the hypolimnion, absence of any positive Ce-anomaly at the bottom points to the presence of oxygen-rich foreign water because dissolution of settling FeOOH particles does not occur. It is assumed that isotopically light and cold Jordan water flows just above the Lake bottom towards deeper parts of the Lake driven by gravity (Nishri and Koren 1993; Laval et al. 2003). A low temperature, salinity and still oxic conditions in the otherwise anoxic hypolimnion at this sampling location, support this interpretation. Another possibility is seepage of less mineralised or fresh groundwater from the Golan Heights from the bottom.

After the intense rainfall during the winter 2002/03, isotopically light rain and surface runoff mixed with Lake water. Despite of evaporation during the year, the resulting Lake water was still lighter at the end of 2003 (average in Nov. 2003: $\delta^{18}O$ and $\delta^{2}H$ are -0.9 and -2.9‰, respectively). The isotopically lighter recharge is well noticeable in March 2004, when the average was of $\delta^{18}O$ and $\delta^{2}H$ were -1.3 and -3.5‰, respectively (Fig. 3.2.3-9).

$\delta^{18}O$ vs. total dissolved solids (TDS)

The salinity of freshwaters in wells and springs in the periphery of the Kinneret Basin is always very low and does not vary significantly with time. Their isotopic signatures depend only on the signature of recent precipitation (Fig. 3.2.3.-10). Samples, taken at Tabigha in March 2004, were always isotopically lighter and less saline than in autumn, representing the fast component of the uppermost aquifers, which is the dominant fraction of recharge during that time. In the diagram of $\delta^{18}O$ vs. total dissolve solids (TDS) the waters of the Tabigha cluster represent an ideal mixing line between well Kinneret 8 and a fresh and isotopically depleted water of $\delta^{18}O$ of -6‰ or even less. At Fuliya, waters react differently because of a different hydraulic system. In general and similarly to Tabigha, most of Fuliya waters plot on a mixing line between a high-salinity end-member (Kinneret 10b) and a freshwater from distal parts of upper or lower JGA(e.g., Qalanit 2), because all samples are taken from the mixing zone next to the shoreline. In springtime, waters of wells Kinneret 5, D 906, and Kinneret 10b are isotopically depleted and less saline. In these on-shore waters, fresh and isotopically light recharge water dominates in springtime. This trend is similar to that at Tabigha.

As previously mentioned, the freshwater of Fuliya is isotopically heavier than that of Tabigha. Except Kinneret 5, REY patterns of Fuliya waters point to a strong contact with thermally altered limestones. Thus, a $\delta^{18}O$ shift towards enhanced values is explicable, as indicated by the trend line in the plot of $\delta^{18}O$ vs. $\delta^{2}H$ (Fig. 3.2.3-9).

In contrast to on-shore wells, off-shore waters show enhanced (Fuliya A) or no isotopic changes (Fuliya B varies within ±0.1‰) but higher salinities in springtime than in early wintertime. This feature characterises the hydraulic system in Fuliya. During autumn, the Lake level and the discharge of Fuliya springs are low, yielding less salinity in off-shore spring waters due to low discharge from the lower aquifer. At this time, fresh and isotopically light water contributes a major fraction. From the beginning of the recharge period, the Lake level and the discharge of both aquifers increases until the maximum is reached in May (Rimmer et al. 1999). In March, when the antagonistic function of the water-body of Lake Tiberias is not at its maximum, the ascent of water from the lower aquifer is enhanced. Therefore, water from the lower aquifer yields an enhanced fraction in the off-shore mixing zone, which leads to increased salinity and enhanced $\delta^{18}O$. The waters of Fuliya A and B are isotopically enriched, with higher tritium content, and is higher mineralised (enriched in Ca, SO_4^{2-} and Na, Cl) in springtime than in autumn. In general, in all Fuliya waters increasing fractions of saline water are marked by enhanced $\delta^{18}O$ values as proved in the depth profile of Kinneret 10b (Bergelson et al. 1999).

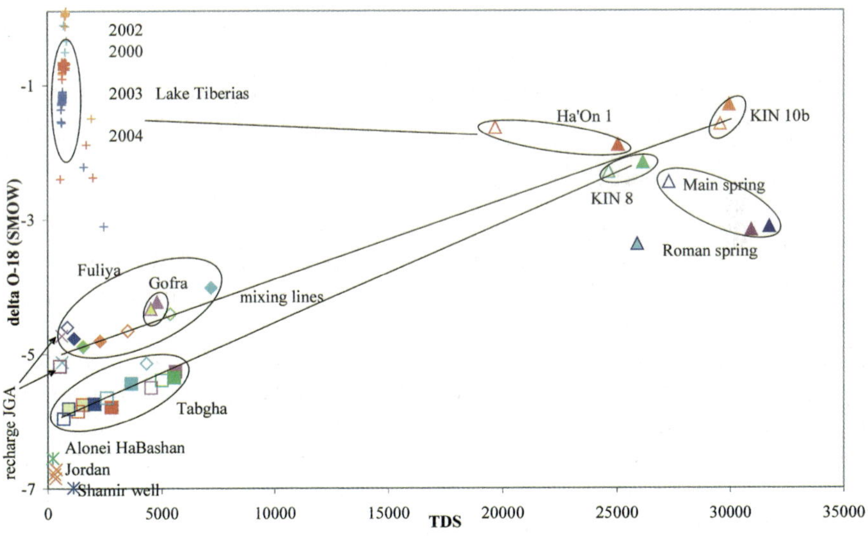

Fig.3.2.3-10. Cross plots of $\delta^{18}O$ vs. total dissolved solids (TDS, mg/l). Squares: Tabigha; diamonds: Fuliya; triangles: thermo-saline waters; crosses: Lake water; lying crosses: rivers and distal recharge. Colours indicate individual sample-points, while open and closed symbols represent springtime and autumn, respectively.

Tiberias Main and Roman Springs do not have a large hinterland and only have an infinitesimal component of young recharge waters, but are isotopically lighter than the other high-salinity fluids of the basin.

Samples from the eastern shore show a different behavior. Gofra spring did not show any variation in salinity or $\delta^{18}O$ between early summer (May, 2003) and late autumn (Nov. 2000). The well Haon 1 differs entirely from all other samples. At a low Lake level in Nov 2000, salinity was high and $\delta^{18}O$ values similar to that of other high-salinity waters such as Kinneret 8 or Kinneret 10b. In March 2004, while the Lake was filled, the salinity in Haon 1 was much less, but $\delta^{18}O$ was slightly enriched. These facts point to mixing with infiltrated, isotopically heavy Lake water. During high stands of the Lake, well Haon 1 contains enhanced fractions of Lake water. At a low Lake level in 2000, the isotopic differences between Haon 1 and other high-salinity waters of the area, were only minor (Fig. 3.2.3-10). Chemical analyses of Bergelson et al. (1999) from the early 90s of water from wells Haon 1 and Haon 2 support this hypothesis. Interaction between Lake and groundwater is expressed by lowered $\delta^{18}O$ values for Lake water. In 2002, while the Lake was strongly divided into hypo- and epilimnion, the lower water body showed a trend to enhanced salinity (TDS ca. 800 mg/l) and $\delta^{18}O$ values around zero because of influx of saline water along the bottom. Samples from shallow locations (TDS 700-750 mg/l) only show this trend in "Shaldag" and "Tabigha". In autumn 2003, the isotopic differences between epi- and hypolimnion are within analytical error and only TDS vary between 800 mg/l in the hypolimnion and 650–700 mg/l in the epilimnion. All points show enhanced TDS towards the bottom (except in O_2 rich or free environments). During the mixing period in March 2004, all profiles became blurred and only Barbutim shows a clear trend.

In 2002 and 2003, the "G" site shows a clear trend towards depleted $\delta^{18}O$ values (-0.94‰ and -2.5‰, respectively) and the lowest TDS (560–630 mg/l) due to the presence of migrating Jordan water or groundwater from the Golan.

Cl/Br vs. $\delta^{18}O$

The Cl/Br wt. ratios of the Lake water (115–140) can trace sources such as the River Jordan (150–250), precipitation (7–12 or much higher) and other known and unknown inputs. The Lake balance reveals unspecified inputs which are hydraulically and most probably spatially connected to the saline spring clusters in the basin. The Cl/Br ratios of the spring clusters vary between 89–111 (Tabigha), 142–200 (Fuliya), 75–78 (Tiberias), 76–87 (Haon), 74–88 (Gofra, Mezar and Hammat Gader) and >>190 for fresh groundwater from the periphery of the watershed. These values are in good agreement with the reported data from the early 90s (Dror et al. 1999; Bergelson et al. 1999), which also show stable ratios for each spring cluster at least during the last decade (Fig. 3.2.3-11).

Precipitation in Israel is considerably enriched in Br. Rainwater analyses from the Golan Heights and Galilee Mts. show Cl/Br ratios of 4.46–16.45 (Herut 1992). In opposite, all freshwaters (< 1,000 mg/l TDS) show Cl/Br ratios >190 (Fig. 3.2.3-11). The increasing Cl/Br weight ratios in fresh groundwater compared with precipitation are the result of leaching evaporites or mixing with brines. Lower ratios (50–150) as observed in groundwater of the Kinneret Basin, represent evaporated seawater or brines derived from such a process (McCaffrey et al. 1987; Starinsky 1974).

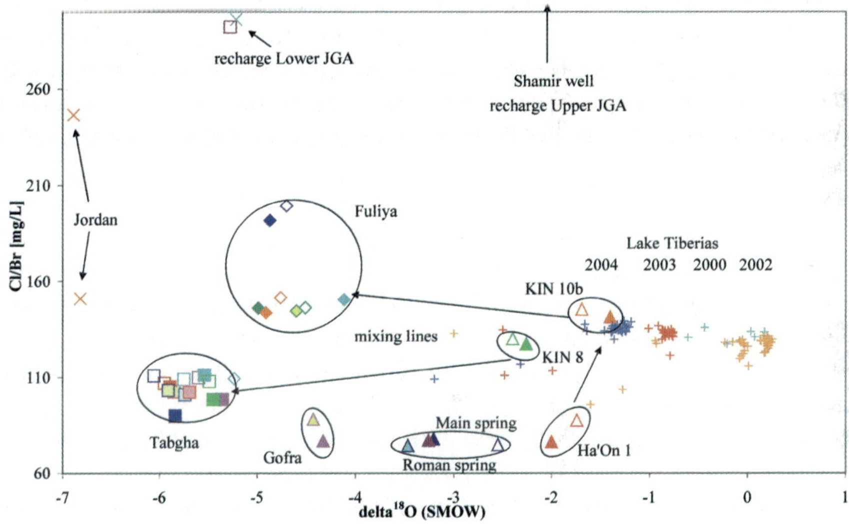

Fig.3.2.3-11. Cross plots of $\delta^{18}O$ vs. weight ratio of Cl/Br. Squares: Tabigha; diamonds: Fuliya; triangles: thermo-saline waters; crosses: Lake water; lying crosses: rivers and distal recharge. Colours indicate individual sample-points, while open and closed symbols represent springtime and autumn, respectively.

The Cl/Br ratios of waters emerging in Fuliya, scatter (140–200), and show a variable behavior during the year. Isotopic signatures and Cl/Br-ratios increase in Fuliya A, B and D 906 in springtime, pointing to an enhanced influence of freshwater deriving from the upper JGA. As result of the lower elevation of the recharge area, recharge water from the upper JGA show enhanced $\delta^{18}O$ values than from the lower JGA. This explains why Kinneret 10b shows lower $\delta^{18}O$ values and higher Cl/Br ratios in springtime than in November.

Well Kinneret 5 shows a different behaviour. Obviously, it gets a lot of fresh, Br-enriched recharge in springtime, because of low Cl/Br ratios and $\delta^{18}O$. During March these parameters approach those of waters in the Tabigha area (Fig. 3.2.3-10).

In the Tabigha cluster, waters from the upper aquifer show lower Cl/Br ratios (90-110). At Tabigha, the waters distinctly differ in the upper and lower aquifers. The observation well Kinneret 8, drilled in the lower aquifer, shows much higher ratios (128) than in the upper aquifer (90-110). The enrichment of Br in the upper aquifer of Tabigha waters is probably not only due to Br-rich saline components; it is additionally due to Br-rich precipitation (Herut 1992). The ratios of Cl/Br in Israel's precipitation vary spatially, but possibly also temporally because of different wind- and cloud systems and elevations (Rosenthal 1987). Hence, in elevated regions of Upper Galilee rainfalls are more enriched in Br than in those from the Lower Galilee. Because of that, Fuliya is fed by precipitation with higher Cl/Br ratios than Tabigha. A second possibility is the successive leaching of residual Br-rich brine pockets from the Avedat Group. The Avedat Group in Fuliya is not as widely present as in the Tabigha watershed.

All waters of Tabigha show a similar behaviour. In March, waters are isotopically lighter and Cl/Br ratios are slightly higher than in November. This indicates increased fractions of water from the recharge areas like in the Qalanit 2 and Huqoq wells (Cl/Br: 296 and 356, respectively) in spring (Fig. 3.2.3-11). At the same time, the pathways of some waters like Sartan springs (fresh and saline) are different. The decreasing chlorinity of Sartan (fresh) spring in March indicates a higher fraction of water from the upper JGA, while decreased Cl/Br ratio indicate a higher influence of brines from the Avedat Group, which probably contains Br-rich brine pockets. Simultaneously, the ratio in the Sartan saline waters increases, as the spring gets less water from the Avedat Group aquifer and more from JGA.

Saline waters from confined deep aquifer show more or less constant ratios, but slightly depleted isotopic values. The Cl/Br ratios range from 74 to 145. While the wells Kinneret 10 b (Fuliya) and Kinneret 8 (Tabigha) show the higher ratios (128 and 141, respectively), Tiberias Hot springs which are nearly free of freshwater, always plot below 80 (74–78) and show enhanced isotopic ratios ($\delta^{18}O$: -3.5 to -2.6‰).

If $\delta^{18}O$ is taken into account, some of the sampling locations in the Lake reveal the origin of inflowing water. In 2002, "Tabigha" was an outstanding sampling point. The next-to-bottom water shows strongly depleted $\delta^{18}O$ values like in the Barbutim well. The ratio of Cl/Br (132) at the sediment surface is at that time, the highest in the Lake. Only at the "Migdal" site (in front of the name-giving village) the bottom waters show similarly high values (131) and indicate an inflow of water with a composition similar to that of Fuliya. The spatial aspect leads to the assumption that "Tabigha" is not fed by the Fuliya cluster system, but from northern, higher elevated areas which show much higher Cl/Br values (e.g., Shamir well). On the other hand, "Barbutim" and "Shaldag" show a clearer trend to lower ratios than all the other patterns. All other Lake samples from 2002 show a trend to lower Cl/Br ratios towards the bottom, even in the highly evaporated epilimnion.

As for well Haon 1, the plot of Cl/Br wt.-ratio vs. $\delta^{18}O$ (Fig. 3.2.3-10) also indicates mixing with saline, Br-rich (ratio: 76–89) water and moderately Br-rich Lake water (ratio: 115–137) in 2004. As the sample of November 2000 does not show effective mixing, low Lake levels seem to prevent it. In contrast, during high water stands in March 2004 and October 1993 (Bergelson et al. 1999), the Lake water gained a larger part of well-water. This is indicated by less salinity and an enhanced $\delta^{18}O$ signature, which is probably caused by evaporation.

3.2.3.4 Conclusions

REY patterns of the water-sediment interface of Lake Tiberias are characteristically influenced by seasonally varying Ce- and Y anomalies and Sm/Lu ratios modified by infiltrating groundwater. By using the changes in Ce anomalies, minimum annual amounts of 44×10^6 m^3 of water with a load of 15.4×10^6 kg Cl^- enter the Lake. An annual minimum of 22×10^6 kg Cl^- is derived from weighted calculations of diffusion and advection. Both results represent distinctly different approaches. These amounts are minima and correspond to an increase of 1.4–2% of the average chlorinity in the Lake which is just at the verge of the analytical limit of detection.

Each spring cluster displays typical Cl/Br ratios which depend on the lithology of the recharge area and their enclosed formation waters. Stable isotopes reflect the average elevation and climate of the recharge area. Thus, different elevations of recharge areas control δ^2H and $\delta^{18}O$ values of individual spring clusters.

The salinization of Lake Tiberias is directly dependent on the hydraulic systems in particular in the tectonic blocks and in the adjoining areas. The interaction between recharge, Lake- and groundwater level and hydraulic head, vary regionally and seasonally (spring and autumn). The unspecified waters flowing to the Lake represent mixtures of different groundwater, such as known from spring clusters and probably from the Hermon Massif and Golan Heights with saline waters residing in the sediments. The mixing processes of fresh and saline groundwater vary seasonally because of alternating flow systems.

Due to their topographical and geological position and to the seasonally varying hydrochemical composition, all waters from the Tabigha cluster plot on an uniform mixing line: Kinneret 8–Barbutim well–Sartan (saline)-Sartan (less saline)-En 7 (En Sheva) –Kinneret 7-(En Tina)-Qalanit 2. The alignment along this mixing line does not represent a flow-path. Water from Fuliya well Kinneret 10b as the most saline member, also plots on this line which probably represents the assumed but diluted Ca-dominated deep-seated brine. However, regional geological differences lead generally to strong hydrochemical differences between Tabigha and Fuliya waters.

3.2.3.5 References

Abbo H, Shavit U, Markel D, Rimmer A (2003) A numerical study on the influence of fractured regions on lake/groundwater interaction; the Lake Kinneret (Sea of Galilee) case. J Hydrol 283:225–243

Arad A, Bein A (1986) Saline versus freshwater contribution to the thermal waters of the northern Jordan rift Valley, Isr J Hydrol 83:43–66

Assouline S (1993) Estimation of lake hydrologic budget terms using the simultaneous solution of water, heat and salt balances and a Kalman filtering approach: Application to Lake Kinneret. Water Resources Res 29:3041–3048

Bau M, Dulski P (1996) Scavenging of dissolved yttrium and rare earths by precipitating iron oxyhydroxides: experimental evidence for Ce- oxidation, Y/Ho fractionation, and lanthanide tetrad effect. Geochim Cosmochim Acta 63:67–77

Bau M, Möller P, Dulski P (1997) Yttrium and lanthanides in eastern Mediterranean seawater and their fractionation during redox-cycling, Mar Chem 56:123–131

Belitzky S (2002) The Structure and Morphotectonics of the Gesher Benot Ya'aqov Area, Northern Dead Sea Rift, Israel. Quat Res 58:372–380

Bergelson G, Nativ R, Bein A (1999) Salinization and dilution history of ground water discharging into the Sea of Galilee, the Dead Sea Transform. Isr Apl Geochem 14:91-118

Braudo CJ, Mero F, Mercado A (1970) Detection of upward percolation of saline lake water. J Hydraulics Div Poc Am Soc Civil Engng 96:1795–1802

Craig H (1961) Isotopic variations in meteoric waters. Science 133:1702–1703

Dror G 1998. In situ measurements of diffusive and advective fluxes from bottom sediments in Lake Kinneret. MSc Thesis. Department of Environmental Sciences and Energy Research, Weizmann Institute, Rehovot, Israel, pp 1–112

Dror G, Ronen D, Stiller M, Nishri A (1999) Cl/Br ratios of Lake Kinneret, pore water and associated springs. J Hydrol 225:130–139

Eckert W, Imberger J, Saggio A (2002) Biogeochemical response to physical forcing in the water column of a warm monomictic lake. Biogeochem 61:291–307

Flexer A, Yellin-Dror A, Kronfeld J, Rosenthal E, Ben-Avraham Z, Artsztein PP, Davidson L (2000) A Neogene salt body as the primary source of salinity in Lake Kinneret. Arch Hydrobiol Spec Issues Adv Limnol 55:69–85

Frieslander A, Bartov Y, Garfunkel Z (1997) The structure of the Arava – new results from geological and geophysical studies. Terra Nostra 4/97

Gat JR (1981) Paleoclimate conditions in the Levant as revealed by the isotopic composition of paleowater. Isr Meteorol Res Pap 3: 13–28

Ginzburg A, Ben-Avraham, Z (2001) Geophysics. In: Horowitz, A (2001) The Jordan Rift Valley. Balkema Publ, pp 1–730

Goldman M, Gvirtzman H, Hurwitz S (2004) Mapping saline groundwater beneath the Sea of Galilee and its vicinity using time domain electromagnetic (TDEM) geophysical technique. Isr J Earth Sci 53:187–197

Gvirtzman H, Garven G, Gvirtzman D (1997a) Thermal anomalies associated with forced and free ground-water convection in the Dead Sea rift valley. Geol Soc Am Bull 109:1167–1176

Gvirtzman H, Garven G, Gvirtzman D (1997b) Hydrogeological modelling of the saline hot springs at the Sea of Galilee, Israel. Water Resource Res 33:913–926

Herut B (1992) The chemical composition and sources of dissolved salts in rainwater in Israel. PhD Thesis, Hebrew University of Jerusalem (in Hebrew, English abstract)

Hurwitz S, Garfunkel Z, Ben-Gai Y, Reznikov M, Rotstein Y, Gvirtzman H (2002) The tectonic framework of a complex pull-apart basin: seismic reflection observations in the Sea of Galilee, Dead Sea transform. Tectonophysics 359: 289-306

Kashai EL (1988) A review of the relations between the tectonics, sedimentation and petroleum occurrences of the Dead-Sea – Jordan Rift System. In: Manspeizer W (ed.) Triassic-Jurassic rifting, Elsevier, Amsterdam, pp 883-909

Laval B, Imberger J, Hodges BR, Stocker R (2003) Modeling circulation in lakes: Spatial and temporal variations. Limnol Oceanogr 48 (3):983–994

Marcus E, Slager J (1985) The sedimentary-magmatic sequence of the Zemah-1 well (Jordan-Dead Sea rift, Israel) and its emplacement in time and space. Isr J Earth Sci 34:1–10

McCaffrey MA, Lazar B, Holland HD (1987) The evaporation path of seawater and the coprecipitation of Br^- and K^+ with halite. J Sediment Petrol 57:928–937

Mercado A, Mero F (1984) The flow of saline springs to Lake Kinneret. TAHAL Consulting Israel, Peport 01/84/48, pp 1-144

Mero F (1978) Hydrology. In: Serruya C (ed) Lake Kinneret, Monographiae Biologicae, 32, Dr. W. Junk Publishers The Hague-Boston-London, pp 87–103

Michelson H, Flexer A, Erez Z (1987) A comparison of the eastern and western sides of the Sea of Galilee and its implication on the tectonics of the northern Jordan Rift Valley. In: Ben-Avraham Z (ed) Sedimentary basins within the Deda Sea and other rift zones. Tectonopysics 141:125–134

Möller P, Rosenthal E, Dulski P, Geyer S, Guttman Y (2003) Rare earths and yttrium hydrostratigraphy along the Lake Kinneret - Dead Sea - Arava transform fault, Israel and adjoining territories. Appl Geochem 18:1613–1628

Nishri A, Koren N (1993) Sediment transport in Lake Kinneret. Verh Intern Ver Limnol 25:290–292

Nishri A, Stiller M, Rimmer A, Greifman Y, Krom M (1999) Lake Kinerat (Sea of Galilee): The effect of diversion of external salinity sources and the probable chemical composition of the internal salinity sources. Chem Geol 158:37–52

Poage MA, Chamberlain CP (2001) Empirical relationship between elevation and stable isotope composition of precipitation and surface waters: considerations for studies of paleoelevation change. Am J Sci 301:1–15

Reznikov M, Ben-Avraham Z, Garfunkel Z, Gvirtzman H, Rotstein Y (2005) Structural and stratigraphical framework of Lake Kinneret. Isr J Earth Sci53:131–15

Rimmer A (2000) The influence of lake level on the discharge of the Kinneret saline springs. Arch Hydrobiol Spec Issues Adv Limnol 55:5–67

Rimmer A, Gal G (2003) The saline springs in the Solute and Water Balance of Lake Kinneret, Israel. J Hydrol 284:228–243

Rimmer A, Hurwitz S, Gvirtzman H (1999) Spatial and temporal characteristics of saline springs: Sea of Galilee, Israel. Ground Water 37:663–673

Rosenthal E (1987) Chemical composition of rain and groundwater in replenishment areas of the Bet Shean-Harod multiple aquifer system, Israel. J Hydrol 89:329–352

Rosenthal E (1988) Hydrochemistry of groundwater at unique outlets of the Bet She'an-Harod multiple aquifer system. J Hydrol 97:75–87

Serruya C (1978) Sediment Chemistry. In: Serruya C (ed.) Lake Kinneret, Monographiae Biologicae, Vol 32. Dr W Junk Publishers, The Hague-Boston-London, pp 205–216

Siebert C (2006) Saisonale chemische Variationen des See Genezareth, seiner Zuflüsse und deren Ursachen. PhD thesis, Freie Universität Berlin, pp 1–226

Siebert C, Möller P, Geyer S (2004b) Investigation of mechanisms of salinization in the Dead Sea Graben, at the example of Lake Tiberias (Sea of Galilee), Israel. Schriftenreihe der Deutsch Geolog Gesellschl 34:319 (in German)

Siebert C, Möller P, Geyer S, Guttman Y, Berger D (2004a) Tracing saline water entries in Lake Kinneret. Proc 5th Intern Symp Eastern Mediterranean Geology, Thessaloniki, pp 1568–1571

Simon E, Mero F (1992) The salinization mechanism of Lake Kinneret. J Hydrol 138:327–343

Smith SV, Serruya S, Geifman Y, Berman T (1989) Internal sources and sinks of water, P, N, Ca and Cl in Lake Kinneret, Israel. Limnol Oceanogr 34:1202–1213

Starinsky A (1974) Relationship between Ca-chloride brines and sedimentary rocks in Israel. PhD thesis, The Hebrew University of Jerusalem Israel (in Hebrew, English abstract)

Stiller M (1994) The chloride content in pore water of Lake Kinneret sediments. Isr J Earth Sci 43:179–185

Stiller M, Carmi I, Münnich KO (1975) Water transport through Lake Kinneret sediments traced by tritium. Earth Plant Sci Lett 25:297–304

3.2.4 Retreat of the Dead Sea and its effect on the surrounding groundwater resources and the stability of its coastal deposits

E. Salameh, H. El-Naser

3.2.3.1 Introduction

In order to quantify the inflows of groundwater into the Dead Sea as a result of its dropping level, this study illustrates the water balance of the Dead Sea system during the predevelopment period and at present. The salt/freshwater interface was taken as the variable to balance the system.

The discrepancy between the actual moderate drop in the level of the Dead Sea and that which should have resulted as a consequence of the extraction of huge amounts of water from its feeding sources (and dedicated for the development of water resources) can only be explained by induced groundwater flows as a result of interface readjustment.

The hydrograph records of the Dead Sea level fluctuations were also analyzed, since their pattern of rise and drop incorporates different mechanisms of surface and groundwater inflows, outflows, bank storage and interface migrations. The configuration of the interface was also elaborated by using Ghyben-Herzberg and Glover approaches and by geoelectric soundings.

The collapses and subsidence in the shore areas are related to dissolution processes caused by freshwater flowing into regimes formerly occupied by salt-saturated waters of the Dead Sea and by flushable fine sediments.

3.2.4.2 Water balance of the Dead Sea

In a previous study (Salameh and El Naser 1999), the water balance of the Dead Sea (Fig. 3.2.4-1) was presented for the time period which preceded the recent development. That water balance represented the natural system of the Dead Sea water, where its level only showed seasonal fluctuations and where evaporation losses were fully compensated by the inflows. The other part of the balance dealt with the present-day water balance of the Dead Sea taking into consideration all the possible effects on that balance which might have been caused by various water projects undertaken within its catchment. These Dead Sea water balances are shown in Tables 3.2.4-1 and 3.2.4-2.

Fig. 3.2.4-1. Satellite image of the Dead Sea and its surroundings (Landsat TM, 2001)

The pre-development water balance

Table 3.2.4-1 summarizes the amounts of water which used to flow into the Dead Sea prior to water resources development within its catchment area. From this balance it occurs that the actual average addition to the water level of the Dead Sea equals 1,980 Mm^3/yr. With a surface of 984 km^2 an evaporation loss of 2,013 mm/yr is calculated.

Table 3.2.4-1. Summary of amounts of water flowing into the Dead Sea prior to the water-resources developments within its catchment area

Source		Amount (Mm^3/yr)
I.	surface water	
	- Lake Tiberias (outflows)	542
	- East Side Wadis	607
	- West Side Wadis	58
	- Dead Sea eastern catchment	219
	- Wadi Araba Basin	
	eastern side	31
	western side	50
subtotal surface water		*1670*
II.	ground water	
	- eastern side	90
	- western side	100
	- northern & southern basins	30
subtotal ground water		*220*
III.	precipitation	
subtotal direct precipitation over the DS		*90*
Total		1980

Source: JVR, WAJ open files, Salameh and Udluft (1985); Kafri et al.(1997)

Current water losses from the Dead Sea

Formerly, the incoming amounts of water to the Dead Sea were only balanced by evaporation losses. However, since the establishment of the Dead Sea Works and Arab Potash Co (1998) on both sides of the Sea, water has been extracted and exposed to evaporation in salt pans to extract the different valuable salts. This process is presently consuming an average of 290 Mm^3/yr (pumping minus return flows). The net amount of water reaching the Dead Sea at present equals therefore (617–290) Mm^3/yr = 327 Mm^3/yr. Evaporation from the Dead Sea (667 km^2) currently amounts to 1334 Mm^3/yr.

Evaporation minus net inflow equals 1000 Mm^3/yr. This amount divided by the total Dead Sea area of 667 km^2 results in a Dead Sea level decline

Table 3.2.4-2. Current water balance

Source	Amount (Mm^3/yr)
Average outflow from Lake Tiberias	40
Average surface runoff of the Western Jordan River catchments	10
Average discharge of the Yarmouk River	50
Eastern Side of the Jordan River	42
Eastern Side of the Dead Sea (surface)	147
Eastern side of the Dead Sea (ground)	90
Western side of the Dead Sea (surface)	13
Western side of the Dead Sea (ground)	50
Western Wadi Araba	10
Eastern Wadi Araba	10
Irrigation return flows + subsurface flows and salt water diversions into the Jordan River.	85
Precipitation over the Dead Sea. (750 km^2)	70
Total inflows	617

of 150 cm/yr. However, the actual drop in the Dead Sea level during the last decade averaged only 80 cm/yr. The difference of 70 cm/yr in water height is compensated for by additional outflows of groundwater from the surrounding groundwater bodies into the Dead Sea. This effect is caused by a seaward interface migration which aims at a hydrodynamic equilibrium state with the dropping Dead Sea level (Fig. 3.2.4-2). The additional quantity of groundwater which is discharged into the Dead Sea to achieve a new equilibrium state amounts to (70 cm/yr × 667 km^2) = 461 Mm^3/yr.

This means that the riparian countries of the Dead Sea contribute 461 Mm^3 of freshwater yearly to the saltwater regimes in their subsurface to compensate for evaporation losses and to consequently alleviate the drop in the level of the Dead Sea. Table 3.2.4-3 shows the project area, the yearly consumption quantities, the project period of consumption, diverted quantities per project area, and the grand total of diversions.

Table 3.2.4-3 also shows that from the early sixties to 1997, the Dead Sea was deprived of around 37.6×10^9 m^3 of water. Assuming an average area of the

Dead Sea of 850 km^2, it amounts to a hypothetical drop of the sea level of around 44 m, although the actual drop is of only 16 m. These conceptual calculations and the resulting discrepancy can only be explained by inflows into the Dead Sea from the groundwater resources of its surrounding areas.

The hydrodynamic disequilibrium brought about by the dropping Dead Sea level must result in a seaward migration of the salt/freshwater interface

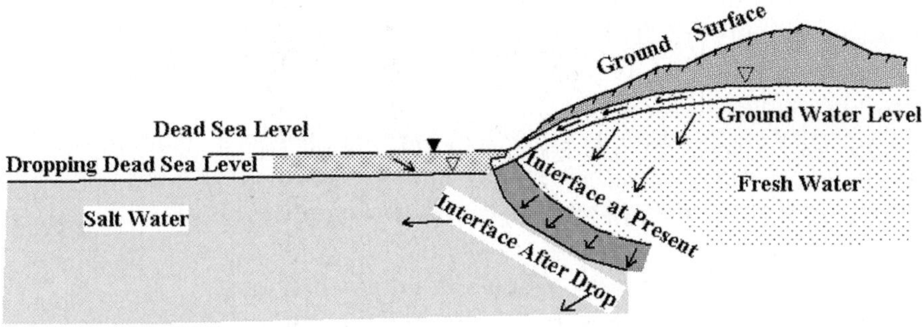

Fig. 3.2.4-2. Schematic pattern showing the displacement (arrows) of the interface due to dropping Dead Sea level

Table 3.2.4-3. Project areas, yearly consumption quantities, period of consumption, total consumed quantities per project area and the grand total consumption in the Dead Sea catchment area of the water amounts which used to flow into the Dead Sea.

Project, Area and Period	Average Quantity Mm^3/yr	Period of Consumption (No of years)	Total Mm^3
National Water Carrier Israel (1964-1997)	500	33	16500
Yarmouk river			
Israel 1967 - 1987	50	20	1000
1987 - 1997	100	10	1000
Syria 1967 - 1988	90	21	1890
1988 - 1997	180	9	1620
Jordan 1964 - 1997	110	33	3730
		20	1200
Wadi AlArab dam (Jordan)	20	16	320
Dams in Wadis Ziqlab, Shueib, Kafrein (Jordan)	8	20-30	200
Jordan Wadis (West)	gradual development		875
West of Dead Sea (Surface)	gradual development		1220
East of Dead Sea (Surface)	gradual development		1080
East of Dead Sea (Ground)	gradual development		400
West of Dead Sea (Ground)	gradual development		1000
Wadi Araba (East)	gradual development		400
Wadi Araba (West)	gradual development		200
		Sub-total	32635
Net use Potash Works			5000
		Total	37635

and by the discharge of saltwater into the Dead Sea. The fresh groundwater must accordingly fill the space which becomes free of saltwater due to the seaward migration of the interface. The substituting fresh groundwater hence becomes salinized (Fig. 3.2.4-2).

As a result, it can be concluded that the Dead Sea as a water body is not only what it appears on the surface, but that it extends widely under the surrounding areas and mountains.

3.2.4.3 Configuration of the fresh/salt-water interface

The high-density Dead Sea water (1.235 g/cm^3) forms a special interface configuration with the fresh groundwater resources of its surrounding aquifers. The fresh groundwater column beneath its surroundings is around one tenth of its length compared to oceanic water (Fig. 3.2.4-3). This fact alone indicates the vulnerability of the fresh groundwater resources to the impacts of changes in the Dead Sea level and to saltwater migration.

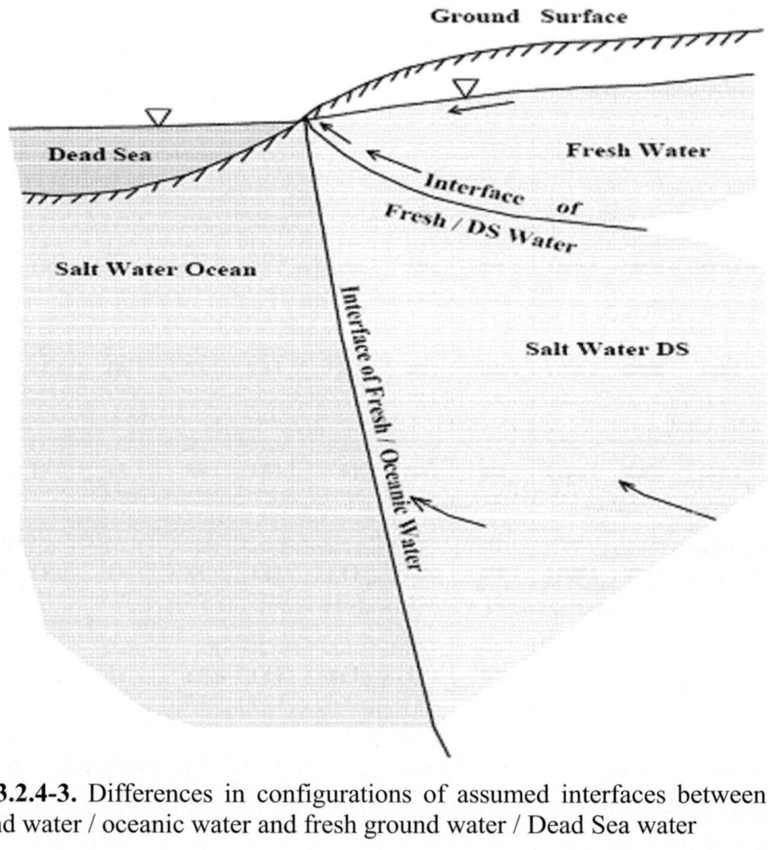

Fig. 3.2.4-3. Differences in configurations of assumed interfaces between fresh ground water / oceanic water and fresh ground water / Dead Sea water

The Ghyben (1988) - Herzberg (1901) and Glover (1959) equation (Eq.(3.2.4-1)) is used to calculate the volumes of water in coastal aquifers which were replaced by freshwater due to seaward migration of the interface as a result of the drop of the Dead Sea level. The dynamic equation of the Glover approach has been integrated to accommodate this type of interface readjustment (Salameh and El-Nasser 2000).

The calculated volume of aquifer, where freshwater substituted the salt water of the Dead Sea due to the migration of interface, is 3.21×10^{11} m^3, corresponding to the drop of the Dead Sea level of -392 m to -411 m msl.

Static interface configuration (Ghyben-Herzberg)

$$\delta_1 \times \gamma_s = (\delta_1 + \delta_{11}) \times \gamma_f \qquad (3.2.4\text{-}1)$$

where

- γ_s : density of Dead Sea water (1.235 g/cm^3),
- γ_f : density of freshwater (1 g/cm^3),
- δ_1 : depth of the interface below Dead Sea level,
- δ_{11} : height of the groundwater level for point x_i relative to the Dead Sea level.

Equation 3.2.4-1 describes a static groundwater condition. Supposing δ_{11} is the groundwater level which lies 1 m above Dead Sea level, then δ_1 becomes (see Eqs.(3.2.4-2) and (3.2.4-3)):

$$\delta_{11} = \frac{\gamma_r - \gamma_{ii}}{\gamma_s - \gamma_f} \qquad (3.2.4\text{-}2)$$

$$\delta_1 = \frac{1}{1.235 - 1} m = 4.255\,m \qquad (3.2.4\text{-}3)$$

This means that for each meter of fresh groundwater level above the Dead Sea, 4.255 m of fresh groundwater are found beneath to compensate for the higher density of the Dead Sea water. This is a ratio of 1: 4.255.

Dynamic interface configuration and effects of Dead Sea level changes on its position

Glover (1959) and Kremer (1977) approximated the dynamic interface configuration by the following equation (Eq.(3.2.4-4)):

$$Z = \sqrt{\frac{2qX}{(\gamma_s - \gamma_f)k} + \frac{q^2}{(\gamma_s - \gamma_f)^2 k^2}} \qquad (3.2.4\text{-}4)$$

where

Z: depth of the interface at a landward lying point (X) in respect to sea water level,

q: quantity of fresh groundwater flowing to the sea per unit length (m^2/s),

k: permeability of the aquifer for freshwater (m/s).

The integration of this equation gives Eq. 3.2.4.5:

$$\int Zr\,dx = \frac{2(\gamma_s - \gamma_f)k}{6q}\sqrt{\left(\frac{2qX}{(\gamma_s-\gamma_f)k} + \frac{2qXr}{(\gamma_s-\gamma_f)^2 k} + \frac{q^2}{(\gamma_s-\gamma_f)^2 k^2}\right)^3} + XrR \qquad (3.2.4\text{-}5)$$

where

Xr: shift of the coastline in meters (landward or seaward),

R: the rise or drop in the Dead Sea level in meters (Fig. 3.2.4-4)

For the case at hand and by using the following parameters:

k	=	10^{-6} m/s
q	=	3×10^{-5} m^2/s
Xr	=	2000 m
R	=	21 m
γ_s	=	1.235 g/cm^3
γ_f	=	1.000 g/cm^3

The difference between the two integrations of the present and former sea levels of -390 m bs1 and -411 m bs1 yields:

$$A1 = \frac{2(\gamma_s - \gamma_f)k}{6q}\sqrt{\left(\frac{2q\times2000}{(\gamma_s-\gamma_f)k} + \frac{q^2}{(\gamma_s-\gamma_f)^2 k^2}\right)^3} - \frac{2(\gamma_s - \gamma_f)k}{6q} \times \sqrt{\left(\frac{2q\times2000}{(\gamma_s-\gamma_f)k} + \frac{q^2}{(\gamma_s-\gamma_f)^2 k^2}\right)^3} + 2000m \times 21m \qquad (3.2.4\text{-}6)$$

After Eq.(3.2.4-6) $A1$ yields 1.72×10^6 m^3/m. This is the volume of freshwater-substituting Dead Sea water per meter length of the Dead Sea shoreline. With a shore line length of 150 km and an average Dead Sea area of 800 km^2 the total volume of inflowing freshwater is about of 3.21×10^{11} m^3.

The amount of freshwater which substituted Dead Sea water is equivalent to the amount of saltwater which joined the Dead Sea due to interface readjustment, both equal 9,450 Mm^3 of water (Eq.(3.2.4-6)). The average porosity of the aquifer then equals (Eq.(3.2.4-7)):

$$\frac{9450\times10^6}{3.21\times10^{11}} = 2.94\% \qquad (3.2.4.7)$$

which agrees well with the calculations carried out according to the Ghyben-Herzberg equation of 2.8% (Eq.(3.2.4-6)).

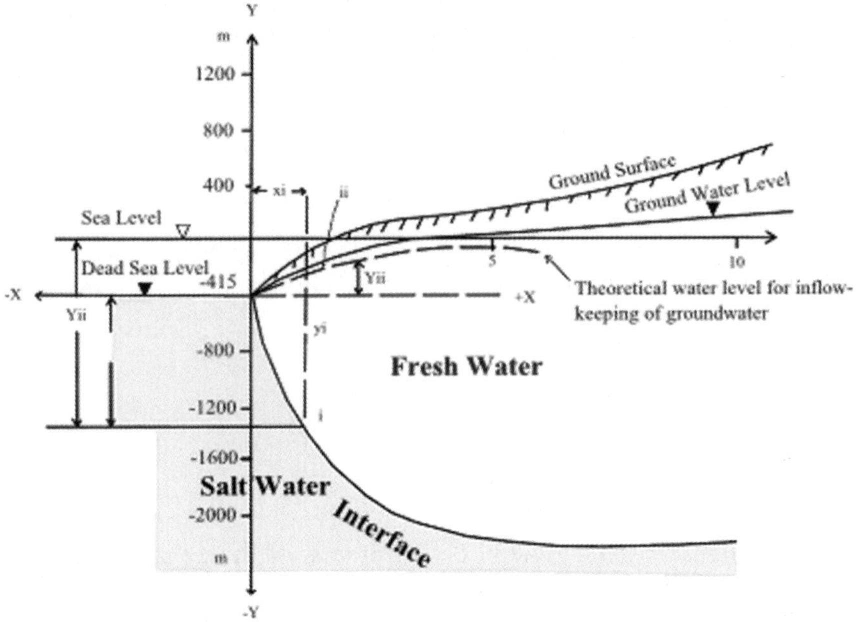

Fig. 3.2.4-4. Configuration of the interface between fresh groundwater and Dead Sea water

3.2.4.5 Hydrograph analyses of the Dead Sea

When analyzing the water level records of the Dead Sea (Fig. 3.2.4-5) it was observed that the higher the rise in the level of the Dead Sea during the winter season, the higher is the rate at which it drops in the dry season. Moreover, the lower the amount of water reaching the Dead Sea during the winter season, the lower is the drop of the Dead Sea level (Salameh and El-Naser 2000).

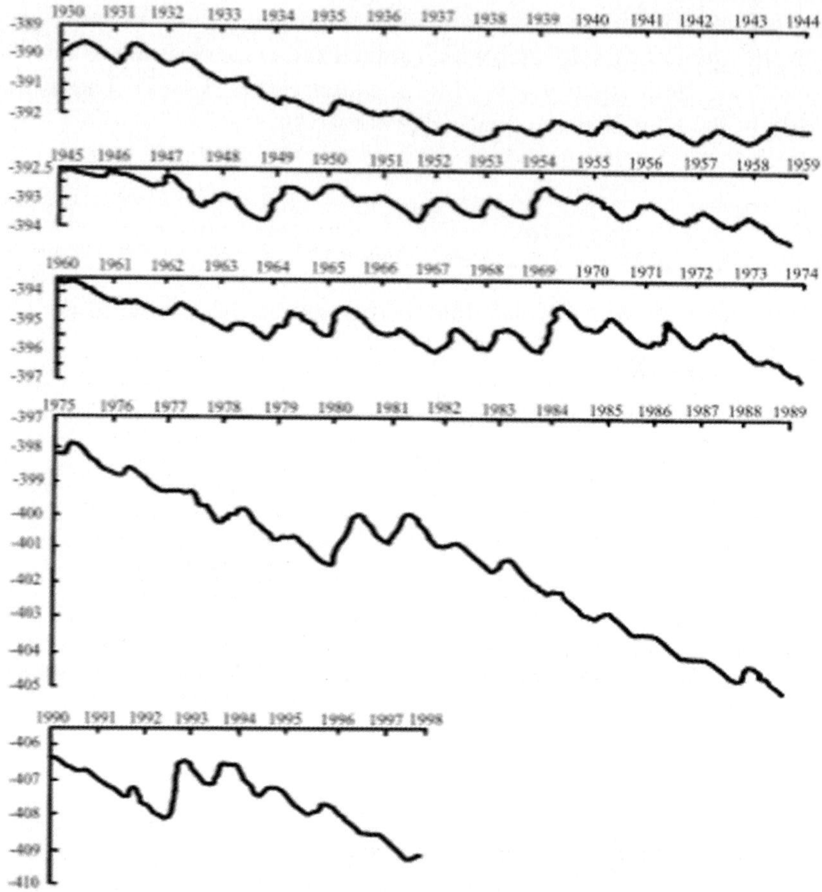

Fig. 3.2.4-5. Drop in the level of the Dead Sea in the period 1930-1998

Eventual evaporation differences due to varying salinities cannot explain these discrepancies. The only explanation for this behavior is that the Dead Sea as a water body is not what we observe at the surface, but also what extends beneath its shorelines. Figures 3.2.4-6a–f explain this situation. This results in an amount of drained water of 1.72×10^6 m^3/m of shore line in the last four decades.

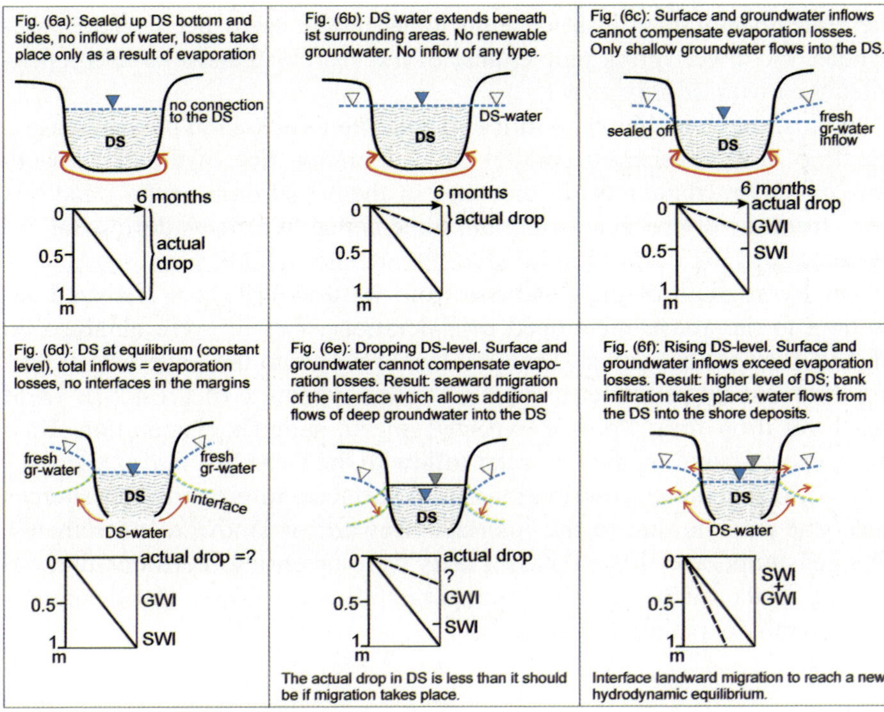

Fig. 3.2.4-6. Possible declines in the level of the Dead Sea during the dry season as a function of influencing factors.

The hydrographs

Considering the hydrographs of 1989/90 and of 1991/92, the following differences can be found (Salameh and El-Naser 2000):

- The 1991/92 summer slope of the Dead Sea level is steeper than that of 1989/90 (1991/92: 65°, 1989/90: 46°).
- During 1988/89 the Dead Sea water level dropped by around 0.9 m compared to 1987/88.
- During 1991/92 the Dead Sea level rose by 1.95 m compared to 1990/91.

Surface water inflows during the dry seasons of 1990 and 1992 were 153 and 392 Mm^3, respectively. The extractions from the Potash Plants in both years were almost the same, since there were no changes in their activities and no new lines were introduced. Evaporation losses in both years were considered equal (water density 1.23 g/cm^3). Although the surface layer of the Dead Sea in the rain-rich years of 1991/92 was diluted and therefore evaporation losses in this time should have been slightly enhanced, the lower spring temperature of the year 1992 must have compensated the lower salinity effect.

In those cases in which no inflows reach the Dead Sea in the dry season, the drop in its level must parallel the evaporation line of the dry season. Any deviation (flattening) in the slope of the actual drop of the Dead Sea level from the dry season evaporation line is due to inflows during the dry season.

The hydrograph of the Dead Sea from 1930 to 1995 was analyzed according to the above mentioned considerations. Details were obtained on highest rises in the winter Dead Sea level relative to the lowest level during the preceding summer as well as on the deviation of the actual Dead Sea level drop from the one expected due to summer evaporation. Data were also gathered on the net water inflow to the Dead Sea during each dry season, the net water contributions from surface and groundwater sources and contributions due to the interface seaward or landward movements. The consumptions of the Potash Plants were taken into consideration according to their actual average pumping and return flows. The deviations are generally explained in Fig. 3.2.4-7.

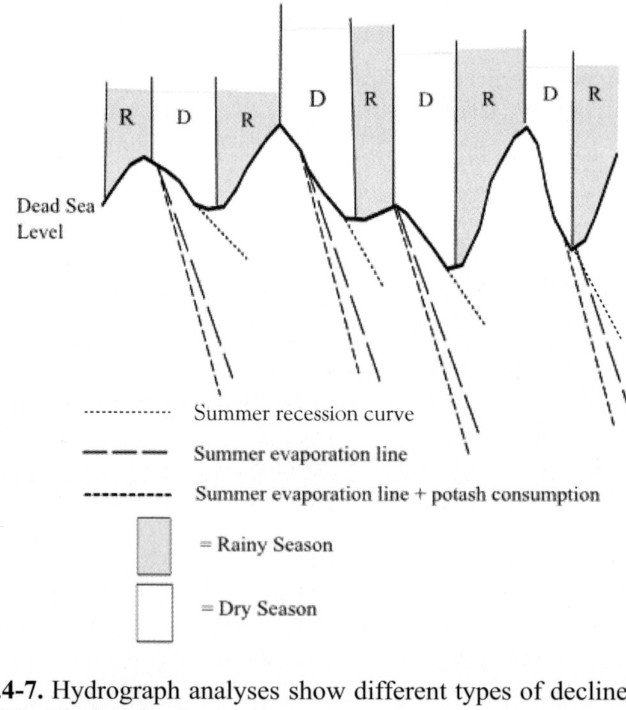

Fig. 3.2.4-7. Hydrograph analyses show different types of declines in the level of the Dead Sea during the dry season.

Table 3.2.4-4 lists the results of hydrograph analyses for Dead Sea levels from 1931 to 1995 to show the contributions of the interface migration (sea- or landward) of Dead Sea water. It clearly shows that no contribution from the Dead Sea to the interface or the contrary took place during periods of more or less constant sea levels. It also shows that in the rain-rich years 1945-46, 1979/80, and 1991/92, the Dead Sea contributed to push

Table 3.2.4-4. Contributions to and from the Dead Sea to the groundwater body due to interface readjustment as a result of a rising or dropping Dead Sea level

Hydrologic Year	Contribution to the Dead Sea due to interface movement Mm^3	Hydrologic Year	Contribution to the Dead Sea due to interface movement Mm^3
1931/32	263	1963/64	0
1932/33	65	1964/65	0
1933/34	159	1965/66	0
1934/35	214	1966/67	0
1935/36	372	1967/68	0
1936/37	162	1968/69	0
1937/38	0	1969/70	0
1938/39	0	1970/71	0
1939/40	0	1971/72	0
1940/41	0	1972/73	154
1941/42	0	1973/74	85
1942/43	0	1974/75	51
1943/44	0	1975/76	65
1944/45	0	1976/77	141
1945/46	-50	1977/78	236
1946/47	0	1978/79	160
1947/48	0	1979/80	-67
1948/49	0	1980/81	67
1949/50	0	1981/82	388
1950/51	0	1982/83	367
1951/52	0	1983/84	283
1952/53	0	1984/85	210
1953/54	0	1985/86	285
1954/55	0	1986/87	457
1955/56	0	1987/88	492
1956/57	0	1988/89	850
1957/58	263	1989/90	626
1958/59	160	1990/91	356
1959/60	188	1991/92	-139
1960/61	369	1992/93	72
1961/62	345	1993/94	157
1962/63	406	1994/95	205

the interface landward, whereas in periods of dropping Dead Sea levels, hundreds of millions of cubic meters of groundwater reached the Dead Sea to compensate its dropping level and decreased its pressure on the interface.

Although the amounts of summer surface inflow in 1992 were higher than those of 1990, the slope of the Dead Sea level decline in 1992 was higher than that of 1990 and it resembled that of the evaporation line. This can only be explained by outflows from the Dead Sea water body during the dry sea- son. The only explanation in this case is the infiltration of the Dead Sea water into the coastal aquifer due to the high rise in the Dead Sea level during the rainy season of 1991/1992 (Fig. 3.2.4-7).

The additional load due to the rise in the Dead Sea level in 1991/92 must also have produced an increasing pressure on the salt-water side of the interface and pushed that interface landward. Accordingly, water from the Dead Sea must have entered the aquifer to compensate for the interface migration in a landward direction. The coastal infiltration and the infiltration to compensate for the interface migration are estimated at: (392 + 195) Mm^3 = 585 Mm^3 (392 Mm^3 is the inflow in the dry season and 195 Mm^3 is the net Potash Plant extraction).

The year 1991/92 was extreme as to encroachment of Dead Sea water into the surrounding aquifers to readjust the interface and account for the rise in the Dead Sea level. Contrary to 1992, in the summer of 1990, the inflows into the Dead Sea were of 193 Mm^3 and the potash net extraction was 195 Mm^3. Therefore, the actual drop in the Dead Sea level must coincide with the summer evaporation line. However, this is not the case and additional water must have flown into the Dead Sea by the seaward movement of the interface. According to the hydrographgraph of 1989/90, this amount must have equaled 626 Mm^3. The deviation between the actual Dead Sea level at the end of the dry season from the evaporation line slope, was calculated as 0.94 m x 666 km^2 = 626 Mm^3 of water. This is of course a very extreme year concerning outflows from the aquifer due to interface adjustment. Applying the same procedure to the different periods of drops in the level of the Dead Sea, gives the amounts contributed by the groundwater to the waters of the Dead Sea as a result of interface readjustment.

3.2.4.5 Environmental impacts of the retreat of the Dead Sea

The riparian states of the Dead Sea are currently loosing an average of 420 to 461 Mm^3 of their fresh groundwater resources to the Dead Sea with each lowering of the Dead Sea level by 1 m. The current annual drop in the level of the Dead Sea averages 80 cm, which means that every year,

Jordan, Israel, and Palestine loose an average amount of 370 Mm^3 of fresh groundwater to compensate for the declining Dead Sea level and the corresponding fresh-salt water interface readjustment.

It is worth mentioning here that the coastal rocks, which are saturated with Dead Sea water (below the interface), are somehow cemented by salts deposited from the oversaturated Dead Sea water (Neev and Emery 1967) (Fig. 3.2.4-7). Lowering the Dead Sea level allows these rocks to be percolated by fresh-or brackish water, which causes the dissolution of the salts in the underground of the coastal areas.

Before the drop of the Dead Sea level, groundwater was flowing within the highly permeable alluvial deposits overlying the thick, low permeability Lisan marls (Fig. 3.2.4-8). At that time, the relatively small amount of water discharge allowed only for a very flat interface to form, with groundwater flowing only in the upper parts of the saturated alluvial deposits. The continuous drop in the Dead Sea level gradually caused the groundwater to flow on the interface between the alluvial deposits and the Lisan marls. The newly created high groundwater gradients along the shorelines, resulting from the drop in the Dead Sea level, caused an increase in the groundwater velocity and of the discharged amounts of groundwater. This fresh, fast flowing groundwater at the top of the soft and partly salty Lisan marls, resulted in the dissolution of salts and the erosion of particles, which were transported into the Dead Sea with the groundwater flow leaving behind cavities at the top layer of the Lisan marls underlying the alluvial deposits.

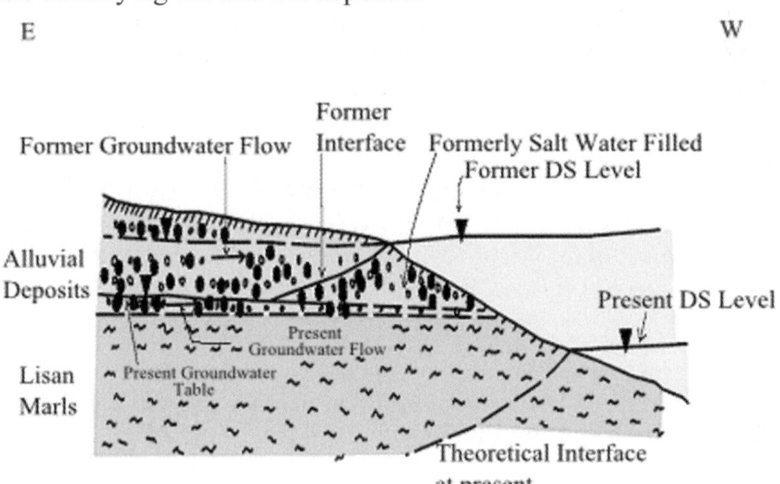

Fig. 3.2.4-8. Dead Sea level-groundwater level relationships and the effects of the dropping groundwater level on salt-dissolution and particle erosion at the gravel/Lisan marls interface

The formation of the cavities seems to have been enhanced by the pressure of the saline Dead Sea water on the coastal rocks (22 m of Dead Sea water level decline are equivalent to 2.7 kg/cm^2 of pressure). Such a release of pressure results in uplifts of the pressurized rocks, which in turn results in widening of joints and enhancement of water movement and hence, to the formation of cavities. At a certain stage the roofs of these cavities consisting of unconsolidated coarse-grained sediments collapsed (Fiog. 3.2.4-9), thus forming sinkholes at the surface of up to 30 m in diameter and 25 m in depth (Fig. 3.2.4-10). In some places along the southern shore lines of the Dead Sea (Ghor Mazraa) the collapses took place within only a few hours.

An additional impact of the retreat of the Dead Sea is the shrinkage of its surface area, which results in less evaporation amounts. Since the prevailing wind direction in the area is towards east and southeast, areas in Jordan which used to receive the moist air passing over the Dead Sea, are now receiving drier air masses. This may also have contributed to the decrease in precipitation over these areas.

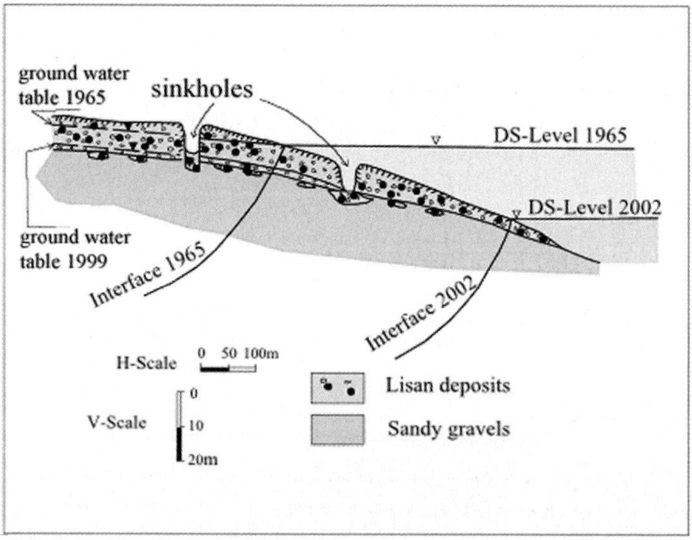

Fig. 3.2.4-9. Dissolution and removal of salts due to the flow of freshwater at the interface of the sandy gravels at the salty Lisan deposits and the accompanying cavity, collapse and sinkhole formation

Plate 3.2.4-1. Sinkhole at Ghor Haditha, approx. 6 m in diameter and 4 m in depth

3.2.4.6 Conclusion

As a result of the retreat of the Dead Sea, since the early sixties of the last century the interface between the fresh groundwater and the Dead Sea water has steadily migrated seaward in order to reach a new hydrodynamic equilib-rium. This has been achieved through the lowering of the groundwater levels by groundwater discharges towards the Dead Sea. The riparians of the Dead Sea are currently loosing an average of 425 to 461 Mm^3 of fresh groundwater to the Dead Sea per meter drop in the sea level.

The induced and accelerated flow of the groundwater towards the Dead Sea resulted in the dissolution of salts and the erosion of friable, fine-grained, non-consolidated Lisan marls and their removal to the Dead Sea, leaving behind cavities in the underground of the coastal areas, which, when reaching a critical size, collapse and form huge sinkholes at the ground-surface. The shrinking Dead Sea resulted in lower total evaporation amounts from its surface and hence, lower humidity of the surrounding areas, and thus increasing aridity.

All these and other impacts support the idea of raising the Dead Sea level to the altitude it had in the 1960s. Bringing additional water from the Red Sea to the Dead Sea through the Red Sea/Dead Sea canal is expected to restore the Dead Sea and its environment and, as a result of the elevation difference between the two seas which presently is 418 m, to additionally produce vast amounts of hydroelectric energy.

References

Arab Potash Company Jordan (1998) Open Files. Amman, Jordan

Ghyben WB (1888) Nota in verband metde voorgenomen putboring nabij Amsterdam-Tijdsch. Van Kouinglijk Instituut Van Ingenieurs 5:8–22

Glover RE (1959) The pattern of freshwater flow in coastal aquifers. J Geophys Res 64:457–459

Herzberg A (1901) Die Wasserversorgung einiger Nordseebäder. J. Gasbeleucht. Wasserversorgung 44:815–819

Kafri U, Goldman M, Lang B (1997) Detection of subsurface brines, freshwater bodies and the interface configuration in-between by the time domain electromagnetic method in the Dead Sea Rift, Israel. Environ Geol 31:34–48.

Kremer R (1977) Ausbildung und Bewegung der Interface Süss-zu-Salzwasser in Grundwasserleitern unter dem Einfluß natürlicher und künstlicher Anreicherung. Dissertation, Rheinisch-Westfälische Technische Hochschule Aachen, Aachen, pp 1–118

Neev D, Emery KO (1967) The Dead Sea depositional processes and environments of evaporites. Ministry of Development, Geol Surv Isr, Jerusalem, pp 1–147

Salameh E, El-Naser H (1999) Does the Actual Drop in Dead Sea level Reflect the Development of Water Sources within its Drainage Basin? Acta Hydrochim Hydrobiol 27:5–11

Salameh E, El-Naser H (2000) Changes in the Dead Sea Level and their Impacts on the Surrounding Groundwater Bodies. Acta Hydrochim Hydrobiol 28:24–33

Salameh E, Udluft P (1985) The hydrodynamic pattern of the Central Part of Jordan. Geol Jahrb, Reihe C 38:39–53

3.3 Aquifers in the western Jordan Valley

3.3.1 The uppermost aquifer in the Wadi Qilt area

M. Toll, C. Messerschmid, J. Wolfer, H. Hötzl, W. Ali, M. Sauter

3.3.1.1 Introduction

The occurrence of this shallow aquifer is limited to a small area in the eastern part of the West Bank (Fig. 3.3.1-1). Although this aquifer is of major importance, it has neither been described nor characterized elsewhere. The aim of this study was to develop a detailed hydrogeological understanding of this local aquifer and its significance for the aquifer system of the Jordan Valley.

The study area has been previously mapped on a large scale by Rofe and Raffety (1963), Roth (1969), Roth (1970), Begin (1974), Arkin (1976a, b), while Wolfer (1998) and Toll (2001) concentrated on geological details of the upper and lower Wadi Qilt area. The main hydrogeological investigations were carried out by Rofe and Raffety (1963), Arad and Michaeli (1967), Kroitoru et al. (1985), and Kroitoru (1987).

3.3.1.2 Geography

The study area extends from the mountainous crest along the axis Jerusalem and Ramallah in the west to the city of Jericho in the east. The area can be subdivided into three different morphological units: the Central Hills, the Eastern Slopes, and the western Jordan Valley. The Central Hills are forming the apex of the mountainous crest in the West Bank and stretch along an approximately N-S axis. The highest point is located around 15 km NE of Ramallah, between Deir Jarir and Silwad at Tel A'sour with an elevation of +1,036 m msl. From the Central Mountains, slopes descend towards the Jordan River Valley in the east.

Most built-up areas concentrate on the Central Hills, where the climate is more suitable for human life than the very dry and hot climate in the Jordan Valley.

Agriculture is hardly developed on the Eastern Slopes of the study area because of very thin soils. Only in the surroundings of Palestinian villages located close to the Central Hills, some olive trees grow in morphological depressions and the Bedouin families keep their livestock.

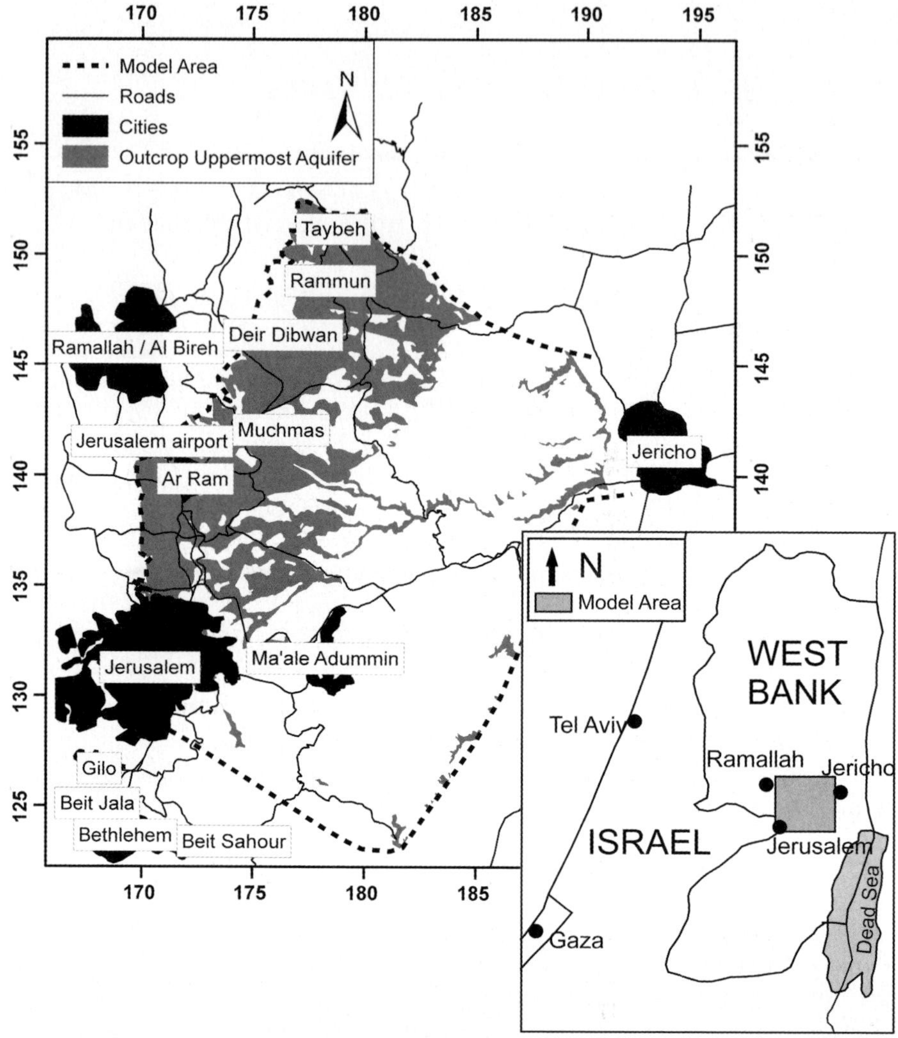

Fig. 3.3.1-1. Overview of the study area, showing the outcrop of the uppermost aquifer (outcrops were digitized from geological maps, scale 1: 50,000, Geological Survey of Israel)

Different surface drainage basins can be distinguished in the study area (Fig. 3.3.1-2):

Auja basin	Ghazal basin
Nueima basin	Muqalek basin
Mafgir basin	Qumran basin
Qilt basin	Nar/Qidron basin

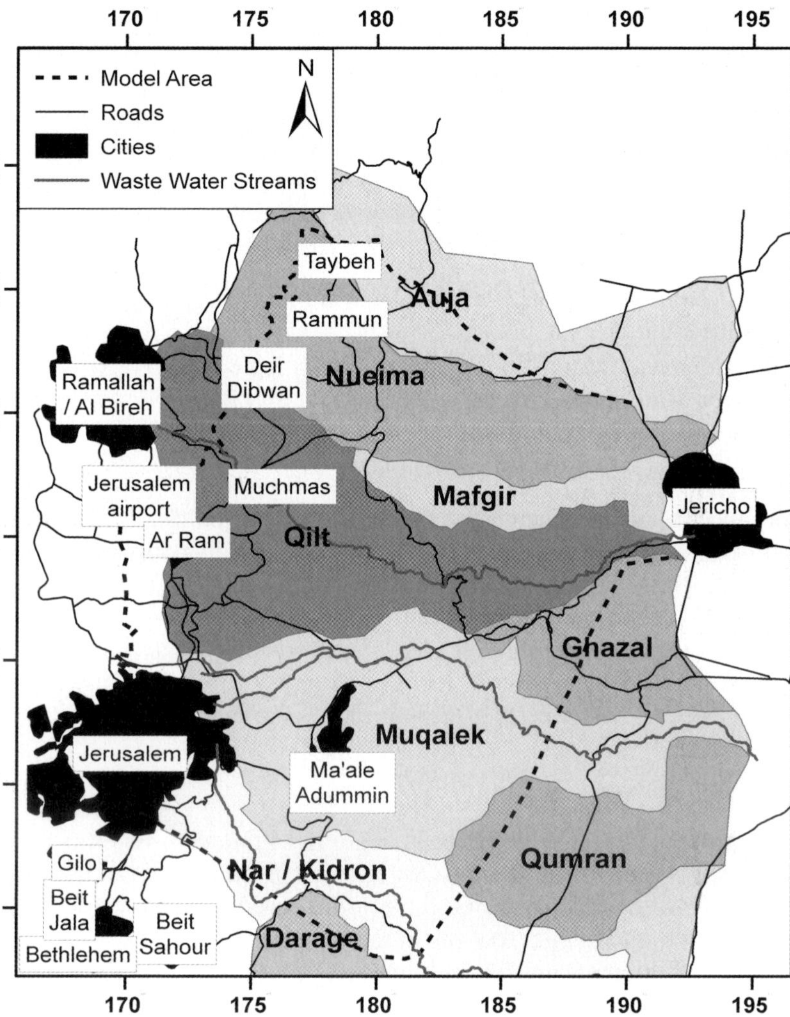

Fig. 3.3.1-2. Surface drainage basins and wadis contaminated by waste-water streams

None of them show perennial flow. However, during and shortly after rain-storm events, they drain floodwater down to the Jordan Valley. The floodwater carries sediments and sometimes even rocks forming fan deposits in the Jordan Valley.

3.3.1.3 Climate

Since the morphology has the largest influence on the prevailing climate, a division into three different climatic zones can be made: the Central Hills, the Eastern Slopes and the Jordan Valley. The Eastern Slopes act as a transition zone between the sub-humid Central Hills in the west and the semi-arid to arid Jordan Valley in the east.

The climate in the Central Hills is of Mediterranean type. Hence, it is characterized by long, hot, dry summers and short, cool, rainy winters. Towards the east, the climate undergoes a rapid change to semi-arid and arid climate inside the Jordan Valley. According to Thorntwaite and Mather (1955) the different zones can be classified as:

- Dry sub-humid mesothermal zone with moderate winterly water excess (Central Hills)
- Semi-arid mesothermal climate with no or little winterly water excess (transition zone, western part of the Eastern Slopes)
- Arid mesothermal zone (eastern part of the Eastern Slopes)
- Arid mega-thermal climate (Jordan Valley).

Rainfall only occurs during the winter months when the climate changes abruptly from dry hot conditions to humid, cold, and stormy ones. Usually it starts to rain in November and rainfall continues until the end of April, whereby 70% of the annual precipitation falls between November and February. The average precipitation rate at the eastern ridges of the Central Hills is around 700 mm/yr and decreases rapidly towards the Jordan Valley. In the Jericho area the long-term average is around 150 mm/yr.

3.3.1.4 Hydrogeology

Karstified rocks of Upper Cretaceous age and related to the Judea Group and commonly known as the eastern aquifer or eastern mountain aquifer (Chapt. 2.2) are characteristic of the aquifer system and represent one of the most important water resources in the region.

All aquifers located on the Eastern Slopes consist of carbonate rocks. On a regional scale, in the Eastern Slope area of the West Bank, two aquifers can be distinguished. They are known as the upper and lower aquifer (Fig. 3.3.1-3) and both consist of Upper Cretaceous limestones and dolostones.

The lower aquifer is almost entirely composed of Albian to Cenomanian carbonates. The upper aquifer is of Cenomanian to Turonian age. Both aquifers consist of various sub-aquifers, the distribution of which varies on a local scale. These sub-aquifers are interbedded with marly and sometimes chalky units which inhibit hydraulic connections to various degrees.

This study focuses on the uppermost member of the upper aquifer (which is called uppermost aquifer from here on) which is limited to the study area. The uppermost aquifer consists of limestones and dolomites of

Period	Epoch	Group (Israeli terminology)	Formation (Israeli terminology)			Formation (Palestinian terminology)	Hydrogeological Unit		Thickness [m]
Quaternary	Holocene Pleistocene	Dead Sea	Alluvium		Q	Lisan			Base unexposed
			Lisan		Qli				
Teritary	Pliocene		Samra		Qs(a)				
Upper Cretaceous	Campanian	Mt. Scopus	Mishash		Kumi	Abu Dis	Aquiclude		100- 400
	Santonian-Campanian		Menuha		Kum				
	Turonian	Judea	Bina	Nezer	Kub	Jerusalem	Aquifer	Aquifer	90-100
				Shivta				Aquifer	
				Derorim				Aquiclude	
	Upper Cenomanian		Weradim		Kuw	Upper Bet-Lehem	Aquifer		90-100
			Kefar Shaul		Kuks	Lower Bet-Lehem	Aquitard		30-40
	Lower Cenomanian		Aminadav		Kua	Hebron	Aquifer		110-140
			Moza	Ein Yorqe'am	Kuey	Yatta	Aquiclude	Aquitard	10
			Beit Meir				Aquitard		120-140
Lower Cretaceous	Albian		Kesalon		Kuke	Upper Bet Kahil	Aquifer		30
			Soreq		Kus		Aquitard		110-170
			Givat Yearim		Kugy	Lower Bet Kahil	Aquifer		20-40
			Kefira		Kk		Aquifer		160-180

Lower Aquifer   Upper Aquifer   Uppermost Aquifer (part of the upper aquifer)

Fig. 3.3.1-3. Stratigraphic Table of rock units in the study area. The Bina Fm. divides into three formations in the eastern part of the study area, namely: Netzer, Shivta, and Derorim Formations (Begin 1974). Yatta Formation in the West comprise the Moza and Beit Meir Formations. Towards the east, the Moza Formation thins out and the Beit Meir unit ends as a tongue within the Ein Yorqe'am Fm. (Begin 1974).

Cenomanian to Turonian age (Upper Bet-Lehem and Jerusalem Formations or Weradim and Bi'na Formations, respectively, Fig. 3.3.1- 3) and is hydrogeologically separated from the upper aquifer by an impermeable marly to chalky unit (Lower Bet-Lehem or Kefar Shaul Formation, Fig. 3.3.1-3). The uppermost aquifer is mainly exposed in the Central Hills (Fig. 3.3.1-1). Further outcrops occur in the deeply incised valleys of Wadi Mukhmas, Wadi Fara, and Wadi Qilt, in the lower part of Wadi Makuk, close to the western boundary fault and in some small areas of the core of the Marsaba anticline. The thickness of the uppermost aquifer ranges between 170 m in the Jericho area and 200 m in the Jerusalem area.

The uppermost aquifer (believed to feed the springs in the study area) has been explored by two deep wells. The first one, Ma'ale Adumin-2 (Fig. 3.3.1-4), encountered the water level of the uppermost aquifer (Bi'na Formation) between 51–54 m below ground level (49–46 m a.s.l.) (Shahnai 1981). The water level of the upper aquifer was tapped at 253.2 m (Aminadav Formation) and the lower aquifer at 391 m below ground level (Beit Meir Formation). At present the well is abandoned and capped. The second well, Jericho 4, is located in the northeastern part of the study area, close to the Western Boundary Fault (Fig. 3.3.1-4). The uppermost water level was encountered at 17 m below ground level (-129 m msl; Guttman 1980) within the Bi'na Formation. The water level of the upper aquifer was tapped at 193 m below ground level (Aminadav Formation). No water is abstracted by these wells from the uppermost aquifer.

In the Jordan Valley, east of the Western Boundary Fault, another shallow aquifer unit is distinguished and consists of clastic sediments deposited in the Jordan Valley since the Pleistocene.

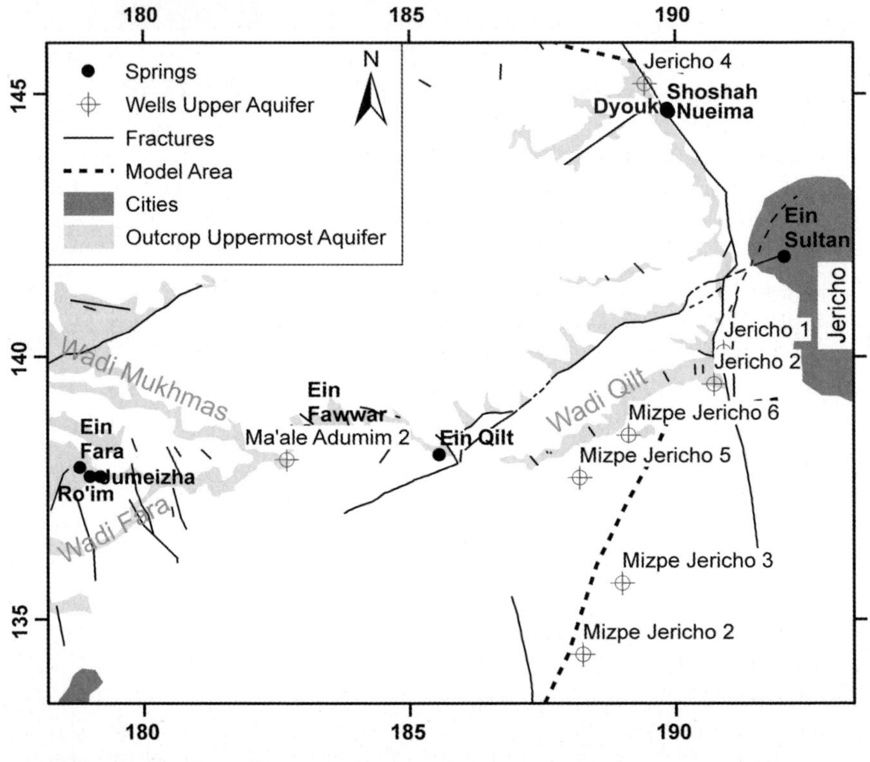

Fig. 3.3.1-4. Springs and wells within the eastern part of the study area

3.3.1.5 Springs

The springs in the Wadi Qilt discharge from the uppermost aquifer. All of them discharge freshwater with 8 to 14 meq/l total dissolved ions. However, in all springs of the study area, faecal and coliforms are found. Consequently, the water is not suitable for drinking water without prior purification.

According to their geographic distribution and discharge behavior, the springs can be divided into two groups: The Eastern Slope springs (Ein

Qilt, Ein Fawwar, Ein Fara, Ein Ro'im, and Ein Jumeizha) and the Jordan Valley springs (Ein Shoshah, Ein Nueima, Ein Dyouk, and Ein Sultan). The words "Ein" or "En" refers to the Arabic and Hebrew words for spring. In this chapter the more common Arabic names are used and the Hebrew names are given in parentheses.

Eastern Slope springs

This group of springs is located in Wadi Qilt and in its tributaries. These wadis are deeply incised canyons and only here the Turonian aquifer is exposed and springs discharge. The springs are characterized by an immediate response to precipitation and by a highly fluctuating discharge pattern.

In Fig. 3.3.1-5 the relationship between rainfall and discharge is shown. The precipitation values of the rain gauge located at the Jerusalem airfield was chosen because it is located near the estimated recharge area. It is apparent that the delay between rainfall and increasing discharge is less than one month. This discharge pattern is characteristic for springs developed in a karstic environment. In the studied aquifer it is characterized by high flow velocities due to high karstification and an adjacent recharge area. The rapid decrease in discharge reflects next to high transmissivities, low-storage features which characterize karst systems.

Ein Fara (En Perat)

Ein Fara is the most western spring. This spring discharges into Wadi Fara at an elevation of +290 m msl. Approximately 4 km downstream, Wadi Fara merges with Wadi Mukhmas forming Wadi Qilt.

In the winter months a continuous surface water stream leads from Wadi Fara towards Wadi Qilt, where it joins the spring outflow of Ein Fawwar and Ein Qilt.

Until 1967, the water of Ein Fara was used for the domestic water supply of Jerusalem. The remains of pumping facilities and pipelines can still be seen. Today, only Bedouins use the spring to supply their livestock.

The outflow mechanism of the spring is strongly related to the Auja Monocline (Fig 3.3.1-6). Ein Fara discharges at the bottom of the wadi, at the base of the Shivta Formation, and at the intersection of the wadi with the Auja Monocline. In the vicinity of the spring outlet, the Shivta Formation is underlain by the Derorim Formation (Fig. 3.3.1-3), which usually consists of chalk to chalky limestone. However, the Derorim Formation is of a reduced thickness of only 10 m and its composition varies

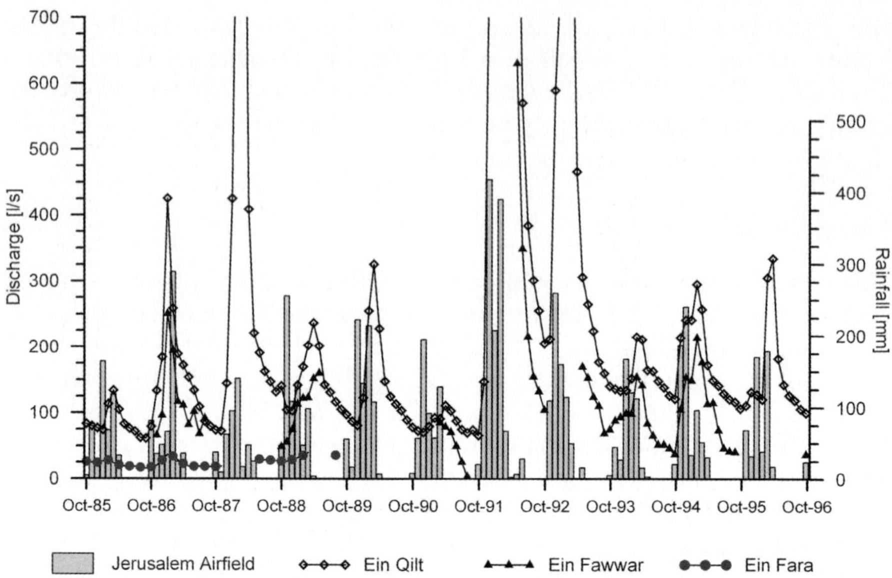

Fig. 3.3.1-5. Response of the Eastern Slope springs to precipitation (data provided by the Hydrological Survey of Israel, HSI)

within the study area from chalk to limestone. Therefore, the Derorim Formation may act as an aquiclude at Ein Fara. Consequently, the spring can be classified as a contact spring.

Ein Fara has an average annual discharge of 940,000 m^3/yr (Israeli Hydrological Survey data base (HSI) 1984-1999). The years with the lowest annual discharge rates coincide with the years of low precipitation, but the spring never dries out not even during drought periods (Rofe and Raffety 1963).

"Small" springs

Approximately 500 m downstream of Ein Fara, two minor springs emerge from the uppermost aquifer. Ein Jumeizha and Ein Ro'im discharge at minor faults and are not directly related to the Auja Monocline. Like Ein Fara, these springs respond very sensitively to precipitation. The available spring data, however, are too sparse to draw further conclusions about the hydrogeological characteristics. The annual discharge of Ein Jumeizha and Ein Ro'im ranges from 150,000–500,000 m^3/yr and from 200,000–600,000 m^3/yr, respectively (HSI data base 1984-1999).

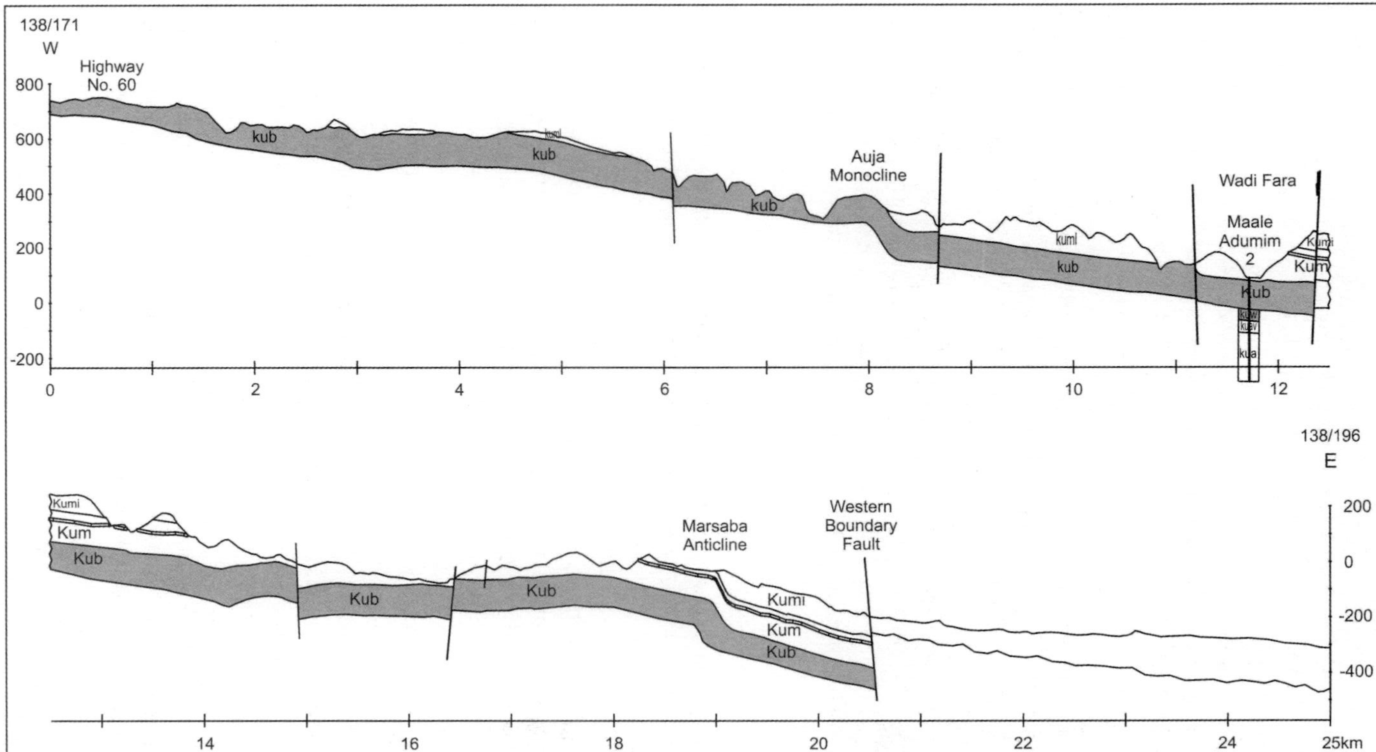

Fig. 3.3.1-6. W-E cross section through the area of interest; the upper part of the uppermost aquifer in gray; kub = Bi'na Formation (includes Derorim, Shivta, and Nezer Formation), kum = Menuha Formation, kumi = Mishash Formation (modified after Wolfer 1998)

Ein Fawwar (En Mavua')

This spring is located a few hundred meters downstream the confluence of Wadi Fara and Wadi Mukhmas into Wadi Qilt and around 4.5 km east of Ein Fara at an elevation of +75 m msl.

The spring was equipped with a pump and used for water supply during the British Mandate and Jordanian times. A spring pool was erected above the spring mouth. A cylindric concrete structure of around 4 m diameter located above the cave, housed the pump. The water discharges turbulently at high hydrostatic pressure. Hence it was given the local name 'Fawwar' (boiling up, bubbling up). Because of its vulnerability to contamination the pump is out of order since years. Groundwater is collected in the pool around the pump house and directed into an irrigation channel draining towards Ein Qilt.

Ein Fawwar is the biggest and - with regard to its mechanism - the most interesting Eastern Slope spring. Wolfer (1998) assumes that Ein Fawwar is fed by NW-SE trending normal faults and joints, located 200 m to the west with an offset of around 10 m to SW. The faults and joints may represent a zone of higher permeability and consequently preferential groundwater flow path which is tapped by an old karstic shaft in the wadi bottom which now forms the spring outlet (Wolfer 1998). Consequently, Ein Fawwar can be classified as a sinkhole karst spring (Fetter 1994).

The average annual discharge of Ein Fawwar is around 4 Mm^3/yr, but it varies highly between 1.25 and 11.8 Mm^3/yr, depending on the annual precipitation (HSI data base 1984–1999).

Due to high flow velocities and the proximity of the recharge area, Ein Fawwar discharges continuously only for a few months during winter and spring. Shortly after the rainy season, it turns into an intermittent or siphon-type spring with discharge cycles, i.e., filling the surrounding spring pool, approximately every 20 minutes. In autumn, the duration of the cycles extends to 40 minutes and more. In very dry years, the spring dries out completely (Rofe and Raffety 1963). However it should be noted, that, due to the siphon mechanism, the water volume discharged during each discharge cycle remains constant throughout the year.

Ein Qilt

Ein Qilt is located approximately 2.5 km further to the east, at the northern side of the canyon. The spring discharges from several NW-SE-striking joints. The erosional cut of the Nezer Formation in the deeply incised canyons along with the damming of groundwater further eastwards is believed to cause a backwater effect that results in a local rise of groundwater-level which allows spring outflow. The dam is built by a ENE-WSW striking normal fault, placing the impermeable Senonian chalks on the downthrown northeastern block against the series of the uppermost aquifer in the southwestern block. The fault is located approximately 800 m to the southeast and has a vertical offset of around 80 m towards SE (Fig. 3.3.1-4).

A clear-cut classification for Ein Qilt is not possible. Hence the spring mechanism can be described as a combination of discharged backwater and of a joint outflow (Fetter 1994).

The spring outlet is located very close to the concrete channel coming from Ein Fawwar. Since the outflow of the spring is located close to the channel, the flow was measured in the channel as a combination of the Ein Qilt outflow and the water flowing in the channel. Because the outflow of Ein Fawwar and hence the water flowing in the channel (Ein Fara contributes only a small amount compared to Ein Fawwar) exceeds the discharge of Ein Qilt, the discharge pattern is controlled by the much larger spring of Ein Fawwar (Fig. 3.3.1-5). This can be seen in a similar trend in the spring hydrographs. Normally the difference between the flow measured in the channel and the discharge of Ein Fawwar is around 40 to 80 L/s. However, in certain years, the measured flow in the channel exceeds substantially the discharge of Ein Fawwar and vice versa.

The mean annual flow in the concrete channel (water coming from Ein Fawwar and Ein Fara together with the outflow of Ein Qilt) is around 7.5 Mm^3/yr. Minimum flow was measured in the dry year of 1978/79 with around 2 Mm^3/yr and the maximum was measured in 1991/92 with an annual flow of around 28.5 Mm^3/yr (HSI data base 1984–1999).

The concrete channel continues further downstream, then passes St. Georges Monastery and ends near Jericho. On its downward way along Wadi Qilt, the water is used by the monks for drinking and irrigation, as well as by Bedouins for drinking and to feed their livestock.

Jordan Valley springs

The springs located in the Jordan Valley are related to the Western Boundary Fault – as is the case of Ein Dyouk, Ein Nueima and Ein Shoshah - or to a major normal fault as in the case of Ein Sultan. Due to various step-faults close to the Jordan Valley, the base of the uppermost aquifer is exposed. The springs located in or near the Jordan Valley discharge at a constant rate. The fluctuations in the discharge pattern match only indirectly with precipitation records available for the working area. This observation can be explained by multiple porosity and hence different flow paths in the karstified aquifers. Very slow-, slow-, medium-, and fast-flow paths are creating a mixing effect and hence cannot be easily interpreted and attributed to one cause. Also, the distance of the recharge area and large size of the catchment contribute to this difficulty. It should be noted that the discharge of the Jordan Valley springs follows to some extent a similar pattern. It is apparent from Fig. 3.3.1-7 that the spring hydrograph of Ein Sultan and Ein Dyouk show almost the same minima and maxima, though separated by a distance of 3.7 km. This strengthens also the hypothesis that they are fed by the same aquifer, i.e. the uppermost aquifer, because travel times in it are higher than in the underlying upper aquifer.

Ein Sultan (En Elisha)

Ein Sultan is located in the Jordan Valley, around 2 km NW of Jericho and discharges at an elevation of -215 m msl. The spring has been used for more than 10,000 years for domestic and irrigation water supply of Jericho, but since a couple of years the pumping facilities were abandoned.

The outflow of the spring is structurally controlled by a large SW-NE trending normal fault (Wolfer 1998). The spring can hence be classified as a fault spring. Because the spring outflow is located 50 m above the local groundwater table of the Cenomanian-Turonian "upper aquifer", the spring is believed to be fed by the uppermost aquifer.

The mean annual discharge of Ein Sultan is around 5.6 Mm^3/year. Maximum annual discharge was measured in 1994/95 (5.96 Mm^3/year) and minimum in 1970/71 with 4.95 Mm^3/yr (HSI data base). As stated above the discharge pattern does not coincide with any precipitation record in the area neither on a yearly nor on a monthly basis, and can be classified as nearly constant. The only exception is the hydrological year 1991/92, the year with the most intensive rainfalls ever measured at the rain gauges in the study area. Fig. 3.3.1-7 shows that the hydrograph of Ein Sultan reflects clearly the extremely wet year of 1991/92. Discharge response is also very quick, i.e., within two months.

Ein Dyouk, Ein Nueima, and Ein Shoshah

Another spring system discharges at the margin of the Eastern Slopes 3.7 km NW of Ein Sultan. The system consists of three springs emerging at -110 m msl at the coordinates 190,000 PGE and 144,700 PGN.

The big vertical offset of various step faults near the Western Boundary Fault exposes the uppermost aquifer and allows spring outflow.

The outflow of Ein Dyouk, Ein Nueima, and Ein Shoshah is collected in a concrete channel and water is used by farmers for irrigation. Due to the good water quality, especially the minor contents in chloride, the water is very well suited for irrigation. Like the water emerging from the Eastern Slope springs, the water is potable only after previous purification from microbiological contamination.

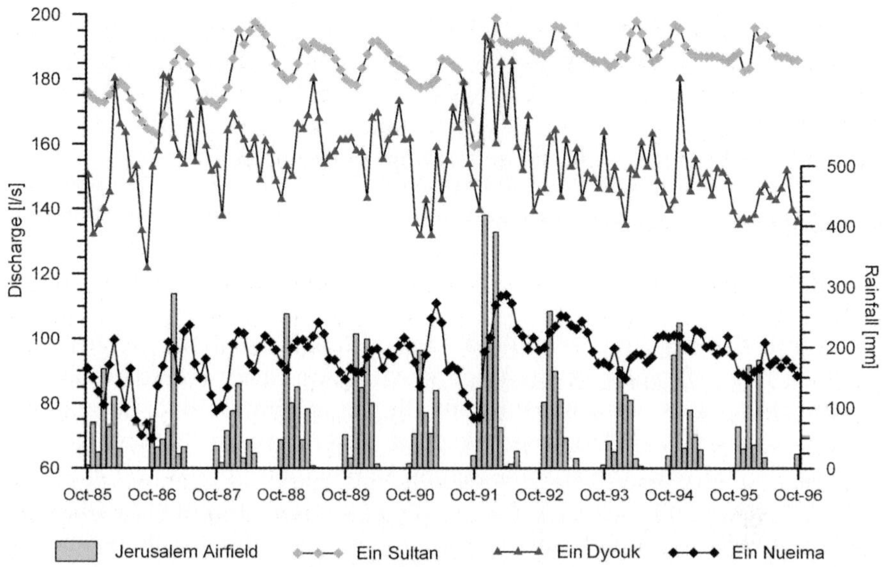

Fig. 3.3.1-7: Response of the Jordan Valley Springs to precipitation (HSI-data base)

In Fig. 3.3.1-7 the hydrographs of the springs can be seen. Because of low discharge rates Ein Soshah is not included in Fig. 3.3.1-7. Although the springs are located very close to each other, discharge patterns are not the same and differ slightly from each other. The maximum discharge of Ein Dyouk and Ein Nueima was measured during the intense rainy season 1991/92. The response of Ein Dyouk coincides with Ein Sultan and is quicker than the response of Ein Nueima (three months delayed with respect to Ein Dyouk). The maximum discharge of Ein Shoshah was recorded in June 1993 and thus with more than one year delay after the intensive rainfall of the winter 1991/92. In general, the discharge pattern of Ein Shoshah differs from the other two for most of the recorded period.

The largest spring is Ein Dyouk. The average annual discharge of Ein Dyouk is 5 Mm^3/yr. The maximum annual discharge was measured 1980/81 with 5.67 Mm^3/yr and the minimum was measured 1997/98 with 4.43 Mm^3/yr (HSI data base 1984-1999).

The second largest spring is Ein Nueima. The average annual discharge of Ein Nueima is around 2.7 Mm^3/yr. The maximum annual discharge was measured 1992/93 with 3.19 Mm^3/yr and the minimum was measured 1972/73 with 2.12 Mm^3/yr (HSI data base 1984–1999).

The smallest spring is Ein Shoshah. The average annual discharge of Ein Shoshah is 668,000 m^3/yr. The maximum annual discharge was measured

1992/93 with 900,000 m^3/yr and the minimum was measured 1978/79 with 480,000 m^3/year (HSI data base).

3.3.1.6 Recharge

The main source of recharge to the uppermost aquifer is direct rainfall onto the outcrops of the aquifer, infiltration and subsequent downward percolation. Most of the recharge occurs in the Central Hills with their very high precipitation rate and over 90% of the uppermost aquifer is outcropping (Fig. 3.3.1-1).

Towards the east, the rainfall decreases rapidly and the outcropping limestone and dolomites of the uppermost aquifer are covered by almost impermeable chalks, reducing substantially the infiltration rate. Groundwater flow is structurally determined and the general dip of the watershed is from the Central Hills towards the Jordan Valley.

In semi-arid to arid areas such as in the greater Wadi Qilt area, high annual variations in rainfall are common. For example at the Ramallah weather station, the average annual rainfall for the last twenty years (1980–99) was around 715 mm/yr. However, the values for individual years lie in a wide range, between 307 mm (1998/99) and 1,591 mm (1991/92), varying by a factor of more than 5. Similar ranges are found in the Jericho area.

Besides the annual variation, there is a strong variation on a monthly and even daily basis. Almost 95% of the annual rainfall occurs between November and March and more than 60% in two months between December and February. Usually, no rainfall is observed between May and September. In the winter months, daily rainfall range between 5 mm and 20 mm, but values of up to 50 mm are frequently recorded. During extreme storm events daily values can reach even 100 mm. Spatial rainfall distribution also varies highly. However, within the study area, there is a clear pattern of decreasing rainfall towards the east. In the rainy season of 2000/01, for example, 492 mm/yr of rain were recorded in Jerusalem, while only 122 mm/yr of rain was measured in the Jericho area (variation by a factor of 4).

In addition, the above runoff-recharge relationship is largely a function of geological parameters. For a large part of the study area, the Senonian chalks inhibit recharge but enhance runoff. In some parts of the study area, runoff is generated on impervious Senonian outcrops and further accumulates in the canyons with permeable alluvial and limestone beds. At these contact zones, runoff infiltrates to form recharge. The epikarst zones in the Turonian and Cenomanian limestone formations are the main contributors to infiltration recharge. The highly irregular pattern of karst channels, caves, sinkholes etc., further complicates the picture and, as a consequence, infiltration may vary significantly (by order of magnitude) within the same Turonian strata over distances of just a few kilometers.

Another source of uncertainty is the disparity between groundwater- and surface-water catchment areas. The western groundwater divide usually follows the axis of the anticline. The surface-water divide starts 2–5 km further west of the groundwater divide. Hence, rainfall over the area west of the study area generates runoff to the east and by this partly contributes to the groundwater recharge of the uppermost aquifer.

Finally, waste-water return flows have to be considered as well. Three large waste-water streams were identified within the study area, in Wadi Mukhmas, Wadi Muqalek, and in Wadi Nar (Qidron Valley) (Fig. 3.3.1-2).

- Wadi Mukhmas flows from the Palestinian cities Ramallah and El-Bireh towards the treatment plant of El-Bireh. After treatment it flows through Wadi Mukhmas towards Wadi Qilt. In early 1998, prior to the opening of the treatment plant, traces of faecal contamination were found in Ein Fawwar spring that were interpreted as a connection with El-Bireh waste-water effluents. Due to the high infiltration rates, surface water reaches Wadi Qilt only in the rainy season. In the summer, the stream infiltrates totally within a few kilometres. The daily flow during periods of dry-weather is 5,750 m^3/d (El-Bireh Municipality/GTZ 1999). Dry-weather hourly peak-flow is 480 m^3/h, while rainy season hourly peak-flows reach 780 m^3/h. Hence, it can be concluded that during the dry season, around 1.0 Mm^3/yr, and during the rainy season around 1.6 Mm^3/yr of discharge occur. This amounts to an estimated average waste-water return flow in Wadi Mukhmas of around 2.62 Mm^3/yr.
- Wadi Muqalek discharges untreated water from northeast Jerusalem into Wadi Muqalek, located East of Jerusalem. The yearly amount of waste water was estimated at 3.6 Mm^3/yr (Israeli Ministry of Agriculture 1989).
- Wadi Nar (Qidron Valley) carries untreated waste-water from East-Jerusalem. The amount of waste water was estimated at around 5.5 Mm^3/yr (Israeli Ministry of Agriculture 1989).
- Other sources of waste water are the settlements located in the study area, but no data concerning their effluents are available. An estimation was made using the average Israeli per capita domestic consumption of 100–110 m^3/yr (Shuval 1999). Around 90,000 Israeli settlers live within the study area of the uppermost aquifer system (Statistical Yearbook of Israel 1999). Approximately 83% of the water usage can be considered as waste water (Anonymous 2000). Hence, the total amount of waste-water return flow is around 8.5 Mm^3/yr.

The contribution of the Palestinian villages is small, because the average Palestinian per capita domestic consumption (in the West Bank) is around

26 m^3/yr and thus too small to be considered as a significant contribution to recharge.

3.3.1.7 Conceptual hydrogeological model of the uppermost aquifer

The findings of this chapter are based mainly on the hydrogeological interpretation of geological maps published by the Geological Survey of Israel. The maps were published at a scale of 1: 50.000 and incorporate the maps of Jerusalem (Arkin 1976a, b), Ramallah (Shachnai 2000), Wadi Qilt (Roth 1970), and Jericho (Begin 1974). In addition various borehole logs provided by the Israeli National Water Company Mekorot and the Deutsche Gesellschaft für Technische Zusammenarbeit (GTZ) GmbH were used.

Aquifer boundaries

The boundaries of the aquifer system are mainly determined by structural features. The doming of the central part of the West Bank during Senonian to Eocene times and subsequent weathering resulted in the exposure of the aquitard underlying the lower aquifer in areas near the axis of the Ramallah anticline. Consequently, in the upper highlands, the uppermost aquifer is eroded. In those areas in which it is not eroded, i.e., near the Ramallah-Jerusalem road (Fig. 3.3.1-6), the western border of its occurrence forms a physical boundary which is the western boundary of the model (Fig. 3.3.1-8).

The Ein Samia Fault zone forms the northern boundary of the uppermost aquifer. This zone is composed of a series of mostly down-throwing faults extending over a width of around 1 km. The total throw of the faults is several hundred meters (Fig. 3.3.1-8).

The Marsaba anticline builds most of the eastern boundary (Fig. 3.3.1-8) and strikes SW- NE with its axis plunging to the NE before the anticline is cut off by the Western Boundary Fault. Before the Marsaba anticline is cut off, it is transformed into a monocline.

The amplitude of the Marsaba anticline has not been exactly determined but was estimated at around 500 m in the southwestern parts of the structure (Roth 1970). Towards the northeast the left flank of the Marsaba anticline flattens towards a monocline facilitating groundwater flow in the uppermost aquifer. Hence, the Marsaba anticline prevents groundwater flowing in the uppermost aquifer from flowing further eastwards. The existence of this uppermost aquifer was not reported in the drilling logs of wells Mitzpe Jericho 2, Mitzpe Jericho 3, Mitzpe Jericho 4, and Mitzpe Jericho 5 (Fig. 3.3.1-4). At first sight, the existence of an independent uppermost aquifer unit in this area could be questioned; but after a closer look at the geological map it occurs that all Mitzpe Jericho wells were drilled in the core or on the flanks of the Marsaba anticline. This not only explains why no independent uppermost aquifer ground-water level was encountered but also confirms the hypothesis that the groundwater of the uppermost aquifer does not flow across the Marsaba anticline.

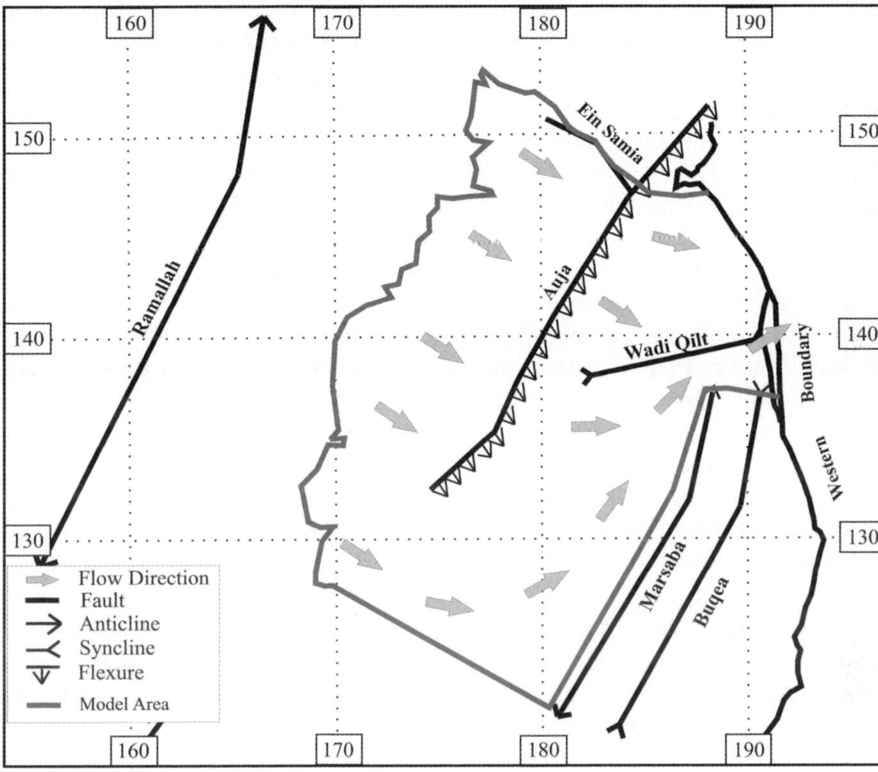

Fig. 3.3.1-8. Flow paths of the uppermost aquifer, main structural features are emphasized

For the rest of the eastern border, it is assumed that all the water from the Eastern Slopes discharges at the springs located near the margins of the Jordan Valley. This assumption is based on water budgets prepared by Wolfer (1998) and Toll (2001).

The Marsaba anticline controls also indirectly the southern border of the uppermost aquifer. The flow path is mostly determined by the regional NW-SE dip of the strata. The southern border can be described as a no-flow boundary or as a subsurface water divide. Water flowing to the north is directed to the northeast because of the Marsaba anticline. Consequently, water flowing to the south of the southern water divide continues to flow southeast to a different aquifer system.

Groundwater flow paths

Generally, groundwater flows from the Central Hills in the west, where the water levels lie at around 700 m above m.s.l., down to the Jordan Valley in the east, where water levels are at around 300 m below m.s.l. The regional flow path is mostly controlled by the regional dip of the strata, which is on average 3° NW-SE (Fig. 3.3.1-6). However, on a smaller scale, flow directions depend largely on folding, faulting, and fracturing of the rocks. In addition, well-developed karst conduits, which cannot be traced, determine the flow pattern.

Four major structural features dominate the groundwater flow in the uppermost aquifer: The Auja (or Uja) Monocline, the Marsaba anticline, the Ein Samia fault strip, and the Wadi Qilt syncline (Fig. 3.3.1-8).

The large SW-NE striking Auja monocline is located approximately in the centre of the study area with its strike more or less perpendicular to the flow direction. Thus, the Auja monocline does not alter the flow direction. However, due to the strong dip of the monocline (with an average of around 22°), the flow gradient increases and, by applying Darcy's law, up to seven times higher velocities can be expected. The Auja monocline also represents a zone of strong fracturing and consequently, even higher flow velocities are expected in the immediate surroundings of the monocline.

Further southeast, the Marsaba anticline influences to a large extent ground-water flow directions. As stated above, the amplitude of the Marsaba anticline is too large and consequently the water of the uppermost aquifer is directed towards northeast and follows the dip of the anticlinal axis until it is transformed into a monocline and crossing of groundwater flow becomes possible.

Ein Samia fault zone is about 1 km wide and is composed of a series of mostly down-throwing faults. The total throw of the faults is several hundred meters. Wells on both sides of the fault, drilled into the lower aquifer at Ein Samia, show different water-table elevation of almost 100 m. This large difference can be considered as evidence for an efficient sealing and redirection of the ground-water flow along the faults, especially, where the faults juxtapose aquifers to aquicludes. Occasional upthrow and block tilting add to this trend. Consequently, the regional southeastern flow is redirected and forced to change to south-southeast.

Another syncline extends northeastwards and almost parallel to Wadi Qilt. This syncline was named by Roth (1969) as the Wadi Qilt syncline. It directs the water, coming from the western flanks of the Marsaba anticline and the Ein Samia fault zone eastwards, towards to the Jordan Valley.

The impact of faults

Research by numerous authors throughout the world has documented the critical importance of fractures in controlling groundwater movement in carbonate aquifers of Mesozoic age. Normal (down-throwing) faults which are the result of extensional forces, tend to create openings, whereas reverse (upthrowing) faults, are more likely to be impermeable, as they are the result of compressional processes. Consequently, faults can represent zones of higher permeability or can act as dams when impermeable chalks or marls are displaced and juxtaposed against the aquifers in the opposite fault block. Because of their nearly planar nature, normal faults, if they stay open, often provide inhomogeneous secondary porosity in carbonate aquifers. On the other hand, when clogged with impermeable material, they can play an important role as inhibitors to flow. As a first approximation, faults of a compressional regime are considered to be rather sealing off while extensional faults are assumed to show increased permeabilities. Groundwater movement along open fault planes will gradually enlarge the fracture aperture by dissolving the carbonates and hence, permeability increases with time. Closer look at the most important faults will be given below.

Three major fault systems dominate the geology and hydrogeology of the study area. The first one strikes SW-NE and prevails in the west and east. Due to its orientation, it may represent zones of preferred horizontal flow. The second system also occurs in the west and east, with strike directions of around NNW-SSE. This fault system includes large faults such as the western boundary fault. The third fault system is found only in the northern part of the study area and strikes WNW-ESE. Due to its nearly parallel orientation to the regional dip, it could also be a zone of preferred horizontal flow.

Just a few hundred meters downstream of the Ein Qilt spring, faults are believed to be responsible for the outflow of the Ein Qilt spring. Faults with a vertical offset of several meters cause a sealing effect.

A much larger fault was mapped in the eastern part of the working area. The fault strikes SW-NE and was traced from Wadi Abu Arabi to the western boundary fault at coordinates 190,900 PGE and 141,500 PGN and may represent a preferential flow direction (Fig. 3.3.1-4). It has a vertical offset of around 80 m and a connection with the underlying upper aquifer is possible. A contribution to the uppermost aquifer is unlikely because the water of the upper aquifer has a different isotopic fingerprint. If the uppermost aquifer would receive a contribution from the upper aquifer, water emerging from Ein Sultan should show significant lower tritium concentrations because the spring is located at the expected continuation of the fault in Jericho. However, the Ein Sultan spring water has the same isotopic fingerprint as the other springs located further westwards.

Aquifer geometry and characteristics

Based on a number of well logs, geological maps and cross-sections, a model for the studied aquifer is proposed.

The total outcrop of the aquifer was estimated (by using the commercial ARCVIEW software package) to be 127.8 km^2. The average thickness of the uppermost aquifer is around 150 m.

It should be noted that there is a problem of the interpretation of the drill cores and geophysical well logs. In some wells, stratigraphic data are very poor due to frequent loss of circulation that occurred during the drilling process. The cross-sections provide an overview of the formation thickness.

The characteristics of the uppermost aquifer are estimated on the base of field observations, discharge behavior, chemical analyses, and hydrogeological considerations.

From field observations, several conclusions can be drawn. There is a facies transition between the Shivta and Nezer Formation. These formations consist of limestone in the Jerusalem- Ramallah area and were mapped by Toll (2001) as such in Wadi Mukhmas and Wadi Fara. They turn into dolomite further eastwards in the Wadi Qilt and Jericho area (Wolfer 1998). All formations of the uppermost aquifer are highly karstified, whereby the coarse grained limestone of the Shivta Formation underwent most intensive karstification. A large number of karst conduits and large caves can be found in the deeply incised canyons of Wadi Mukhmas and Wadi Fara. Caves are not only found in these valleys but also in various other areas (e.g., Wadi Taqua' with the famous Chareitun Cave south of Bethlehem).

Kroitoru et al. (1985) proposed flow velocities for the uppermost aquifer based on arrival of the bomb tritium at Ein Sultan spring. A tritium value of 25±5 T.U. was observed in the Ein Sultan spring in 1968, indicating a minimum flow velocity of 1.45 km/yr (4 m/day).

In most springs of the study area, tritium was measured continuously. The results indicate young (post-bomb) water, even in the samples taken in the late sixties.

Rapid flow velocities are also indicated by the discharge pattern of the springs. The outflow of the wadi springs is very sensitive to precipitation and responds quickly to rainfall events.

3.3.1.8 Summary

The uppermost aquifer is a local, shallow aquifer located in the Central Hills and Eastern Slope of the West Bank. The aquifer is confined to the Wadi Qilt area and bordered by the large cities of Jerusalem in the Southwest, Ramallah/Al Bireh in the Northwest, and Jericho in the East. The uppermost aquifer represents a major groundwater drainage system, conveying recharge from the Central Hills down to the Jordan River Valley.

The uppermost aquifer in the Wadi Qilt area consists of limestones and dolomites of Cretaceous age (Upper Cenomanian-Turonian). The carbonatic rocks can be subdivided into the Weradim, Shivta, and Nezer Formation (Israeli terminology) or Hebron and Jerusalem Formation (Palestinian terminology). Four stages of karstification resulted in a well-developed karst system. Detailed geologic mapping identified the carbonates of the Shivta Formation (Lower Jerusalem Formation) as the most productive part of the uppermost aquifer. The total thickness of the aquifer is around 150 m. The size of the total outcrop is 127.77 km^2.

The outcrop and recharge area of the aquifer lies predominantly in the semi-humid Central Hills and in addition, in the deeply incised valleys on the Eastern Slopes and at the margin of the Jordan Valley. Several springs emerge from the uppermost aquifer. They can be divided into two groups: Eastern Slope and Jordan Valley springs. The Eastern Slope springs are characterized by intensive seasonal and annual variations, while the Jordan Valley springs show only minor seasonal and annual variations.

Since the eighties, a decrease in groundwater quality was recorded, reflecting an increasing population growth in and near the outcrop of the aquifer. Because of microbial contamination groundwater is not suitable for drinking without treatment. Three raw or partially treated domestic waste-water streams infiltrate the uppermost aquifer in the Wadi Qilt area. The only functioning treatment plant is El-Bireh treatment plant, opened in 1999 in upper Wadi Ain near El-Bireh. Most Palestinian villages and Israeli settlements have no waste-water treatment.

Recharge of the uppermost aquifer is mainly derived from two sources: infiltrating precipitation and infiltrating wastewater. The amount of recharge depends highly on the amount of rainfall during the rainy season.

References

Anonymous (2000) Environmental Protection of the shared Israeli-Palestinian Mountain Aquifer; University of Michigan, Department of Environmental and Industrial Health in collaboration with the Hebrew University and the Palestine Consultancy Group; Final Report, pp 1–88

Arad A, Michaeli A (1967) Hydrogeological investigations in the western catchment area of the Dead Sea. Isr J Earth Sci 16:181–196

Arkin Y (1976a) Explanations to the geological map of the Jerusalem- Bet Shemesh area. Isr J Earth Sci 16:46– 47
Arkin Y (1976b) Geological Map of Israel, Jerusalem and Vicinity, 1: 50,000. Geol Surv Isr, Jerusalem.
Begin ZB (1974) Geological Map of Israel, Jericho, Sheet 9-III, 1:50,000, with Explanatory Notes. Geol Surv Israel, Jerusalem.
Fetter CW (1994) Applied Geology, Prentice-Hall, USA pp 1–691
Guttman J (1980) The Jericho 4 and 5 wells, summary of findings. TAHAL Consulting Engineers Ltd., Tel Aviv, Israel [in Hebrew] pp 1–12
Kroitoru L (1987) The characterization of flow systems in carbonatic rocks defined by the groundwater parameters: Central Israel; PhD thesis for the Weizman Institute of Science, Rehovot, Israel, pp 1–120
Kroitoru L, Mazor E, Dror G (1985) Hydrological characteristic of the Wadi Kelt and Elisha springs. In: Scientific Basis for Water Resources Management, IAHS Publ., Wallingford, 153:207–218
Rofe and Raffety Consulting Engineers (1963) Geological and hydrological report, Jerusalem and district water supply. Hashemite Kingdom of Jordan, Central Water Authority, London, pp 1–79
Roth I (1969) The geology of the Wadi Qilt region. M. Sc. Thesis, Hebrew University, Jerusalem, Israel [in Hebrew] pp 1–52
Roth I (1970) The Geological Map of Israel 1: 50000, Sheet 12: Wadi Qilt
Shachnai E (1981) Ma'ale Adumin 2 well, summary of findings. TAHAL Consulting Engineers Ltd., Tel Aviv, Israel [in Hebrew] pp 1–35
Shachnai E (2000) The Geological Map of Israel 1: 50000, Sheet 8: Ramallah
Thorntwaite CW, Mather JR (1955) The water balance, publ. In: Climat. VIII, No.1; Drexel Inst. Laboratory of Climatology, Centerton, N.J.
Shuval H (1999) Sharing the wealth. In The Jerusalem Post. http//www.jpost.com/com/Atchive/16.Nov.1999/Features/Article-1.hmtl
Toll M (2001) Hydrogeological characterization of the Uppermost Aquifer in the Wadi Qilt Area, Diploma Thesis; Department of Applied Geology, Tübingen University, Tübingen, Germany, pp 1–137
Wolfer J (1998) Hydrogeological investigation along the Jerusalem- Jericho Transect (Wadi Qilt), West Bank/ Israel,-Diploma Thesis, Department of Applied Geology, Karlsruhe University, Karlsruhe, Germany, pp 1–106

3.3.2. 3-D Hydrogeological model of the Marsaba-Feshkha region

A.Yellin-Dror, J. Guttman, A. Flexer, H. Hötzl, W. Ali, J. Bensabat

3.3.2.1 Introduction

This study summarizes an investigation in the Marsaba-Feshkha area with the aim of improving exploitation strategies of groundwater resources. Natural discharges of the catchment are the Feshkha springs (Fig.3.3.2-1).

The 3-D hydrogeological model developed in the study area allows the identification of groundwater flow patterns (Fig.3.3.2-1), testing various exploitation schemes prior to their implementation and assessing the hydrological potential of the area, i.e., the amount of water that can be safely extracted.

Its geographical area covers a surface of approximately 700 km^2 located on the eastern slopes of the Judea anticlinorium (Fig.3.3.2-2) bordered in the west by the water divide running along the Hebron and Ramallah anticlines, the Jordan Valley and the Dead Sea in the east, the latitude line 140 in the north and the water divide between Ein Feshkha and Qane-Samar springs in the Judea Desert in the south (Fig.3.3.2-2).

The study area can be defined as semi-arid to arid and suffers from acute shortage of freshwater. It is a part of the eastern drainage basin of the Judea and Samaria hills, and is one of the few locations where water resources can be utilized by capturing the freshwater flows before they reach the Dead Sea. Rainfall sharply decreases from 500–700 mm/yr along the western margins of the area to less than 100–150 mm/yr in the east, along the Dead Sea shore. Previous studies indicated that substantial amounts of freshwater penetrate the aquifer in the relatively rainy replenishment zones on the western part of the basin (the eastern margins of the Judea Mountains). The ground water flows east of the Dead Sea where it mixes with local brines (see chapter 3.1.4). An indication to this is the relatively high discharge of saline water in the Feshkha springs, which constitute the main outlet of the Cenomanian aquifers in this area. There is probably also some leakage of freshwater through the Rift fault into the basin-fill sediments in the area north to the Dead-Sea (Fig.3.3.2-2).

Little information is available on the area as most of the wells that penetrate the relevant aquifers are located in two clusters separated by great distances (Jerusalem-Herodion area in the west and the Jericho–Mizpe-

Jericho area in the east (Fig.3.3.2-2; Table.3.3.2-1). The geology and hydrogeology of this area is complex and suffers from a serious lack of reliable and geographically spread data. Therefore, the work done in the construction of the model relies heavily on soft data (data not directly measurable) inferred from available information of the regional geology, on accepted concepts regarding the most probable flow patterns and on the main components of the water budget in the area.

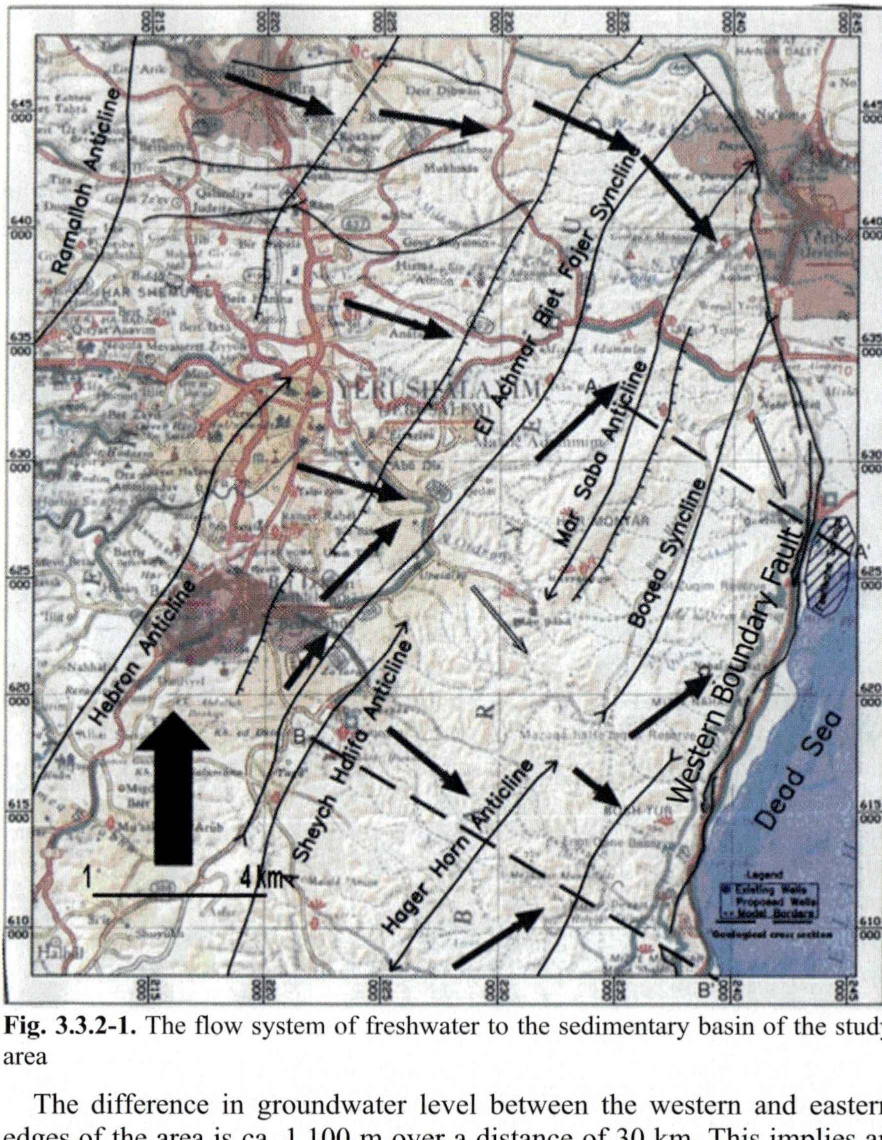

Fig. 3.3.2-1. The flow system of freshwater to the sedimentary basin of the study area

The difference in groundwater level between the western and eastern edges of the area is ca. 1,100 m over a distance of 30 km. This implies an average hydraulic gradient of about 3%, which is obviously very unlikely. In order to cope with this apparent discrepancy geological and structural elements were considered. The area was divided into a number of secondary basins, separated by low permeability flexures through which large drops in the groundwater head occur. Within each basin, the hydraulic gradient is relatively moderate. This is equivalent to assuming the existence of a number of groundwater "cascades" in the area.

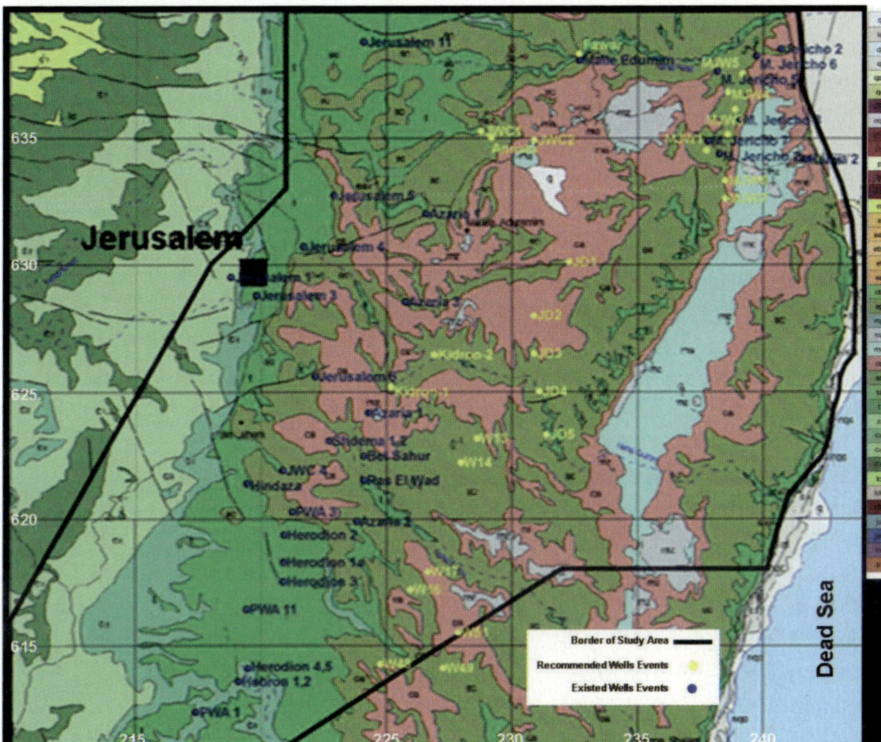

Fig. 3.3.2-2. Geology of the study area (after Sneh et al. 1998)

Obviously, a computational model that aims to simulate groundwater flow processes in such an area suffers from the acute lack of information. However, it may also help in screening the various concepts of flow patterns by testing the resulting flow fields that it produces.

Keeping in mind the strong limitations imposed by the lack of information, the work carried out in this study included: structuring of the geological model, provision of hydrological data and new classification of the aquifer layers, determination of the 3-D computational grid in order to take into account the geological features, preparation of initial conditions (in the Lower and Upper Cenomanian layers), determination of the boundary

Table 3.3.2-1. List of wells in the study area

Name	E coord New Israel Grid	N coord New Israel Grid	Drilling Year	Total Depth m	Elevation m
PWA-1	17406.00	12150.00	1990	601.00	744.60
Hebron-2	16837.00	11311.00	1990	387.00	710.00
Hebron-1	16835.00	11310.00	1990	704.00	710.00
Herodion-4	16946.00	11408.00	1984	691.00	680.63
Herodion-5	16946.00	11408.00	1985	326.00	685.00
Beit-Fajjar-1	16960.00	11510.00	1965	237.00	728.00
Beit-Fajjar-1A	16960.00	11510.00	1990	300.00	692.19
Beit-Fajjar-3	16930.00	11630.00	1965	320.00	750.00
PWA-11	16916.00	11630.00	1998	851.00	751.50
Herodion-3	17090.00	11750.00	1949	742.80	616.78
Herodion-1a	17090.00	11830.00	1992	351.00	569.47
Herodion-2	17092.00	11933.00	1973	770.00	562.99
Azariya-2	17386.00	11988.00	1998	793.00	589.00
PWA-3	17125.00	12025.00	1998	741.00	608.30
Hanza	16938.00	12140.00	1997	672.00	627.20
JWC-4	17087.00	12189.00	1997	787.50	599.80
Ras El Wad-1	17406.00	12150.00	1997	661.00	484.75
Ras El Wad-2	17406.00	12150.00	1970	455.00	484.75
Shdemah-1	17273.00	12305.00	1990	910.00	553.00
Shdemah-2	17273.00	12305.00	1990	440.00	553.00
Jerusalem-6	17217.00	12559.00	1972	820.00	593.90
Jerusalem-2	16980.00	12878.00	1971	227.00	715.61
Jerusalem-3	16980.00	12877.00	1972	778.00	715.72
Jerusalem-1	16879.00	12949.00	1971	600.00	692.19
Jerusalem-4	17170.00	13070.00	1975	999.90	694.95
Jerusalem-5	17291.00	13271.00	1975	741.00	718.67
El Azariyah-1	17660.00	13200.00	1982	827.00	379.31
Mizpe-Jericho-2	18826.00	13434.00	1982	493.00	-18.15
Mizpe-Jericho 1	18810.00	13460.00	1965	305.00	5.09
Kaliya-2	19170.00	13413.00	1979	289.00	-165.58
Mizpe-Jericho-3	18911.00	13567.00	1985	370.00	-60.00
Mizpe-Jericho-5	18821.00	13762.00	1990	515.00	32.16
Jerusalem-11	17411.00	13876.00	1976	759.00	575.00
Maale-Adomim-2	18270.00	13805.00	1982	296.20	100.00
Mizpe-Jericho-6	18912.20	13850.50	1992	417.00	-27.00
Jericho-2	19072.00	13948.00	1976	324.00	-155.48
Jericho-1	19090.00	14080.00	1975	243.50	-170.72

conditions; selection of the calibration period, preparation of the pumping and rainfall data for the selected calibration period, running of the model for the calibration period and assessment of model performance. This sequence has been simulated many times. In the light of the results the process stopped when an adequate and satisfactory fit between measured and simulated ground-water head elevations was achieved.

At the end of this process, the results obtained were believed to be a relatively reliable computational model that reproduces the aquifer response to meteorological events and to pumping regimes. Groundwater exploitation scenarios for this area were elaborated, comparisons between scenarios were made and finally, recommendations were suggested regarding the enhanced exploitation suggesting alternatives for the sustainable exploitation of the groundwater resources.

Such development is of considerable importance considering the water agreements formulated between Israel, the Palestinian Authority and Jordan.

3.3.2.2 Geological background

The stratigraphic profile exposed in the study area varies from the Lower Cenomanian to Holocene Formations (Table 3.3.2-2). The thickness of the Judea Group is about 800–850 m (Arkin et al. 1965; Arkin and Hamaoui 1967). The exposed layers are formations of the Judea Group, composed mainly of limestone layers in the vicinity of the Hebron anticline in the west, formations of the Mt. Scopus Group exposed near the Jordan Valley and throughout the Judea Desert, and young formations of Pleistocene-Holocene age located in the Jordan Valley and along the shores of the Dead Sea and associated wadis.

The predominant elements governing the geological structure and flow regime are the anticlinal and synclinal structures crossing the entire study area. West-east geological cross-sections illustrating these structural features are presented in Figs. 3.3.2-3–5.

The Hebron anticline is the principal structural element in the study area. Its axis runs from the Hebron-Halhul area to Jerusalem in a general SW-NE direction and dips towards Jerusalem. This anticlinal axis represents the maximum elevation in the area. To the east and west there are minor folds, representing the descent of the structures towards the plain in the west and the Dead Sea in the east. The anticlines descend towards the Jordan Valley and the Dead Sea in a series of undulations, forming secondary structures. The anticlines and synclines are asymmetric, with steepest inclinations to the eastern part of the anticlines. The principal fault structure in the area is the western fault of the Dead Sea-Jordan Valley. This fault runs along the western shore of the Dead Sea up to Jericho City and its direction is N-S. Around Jericho the fault direction turns to N-W, continuing as the Ein Samiya fault zone (Fig.3.3.2.1) (Roth 1970; Begin 1976).

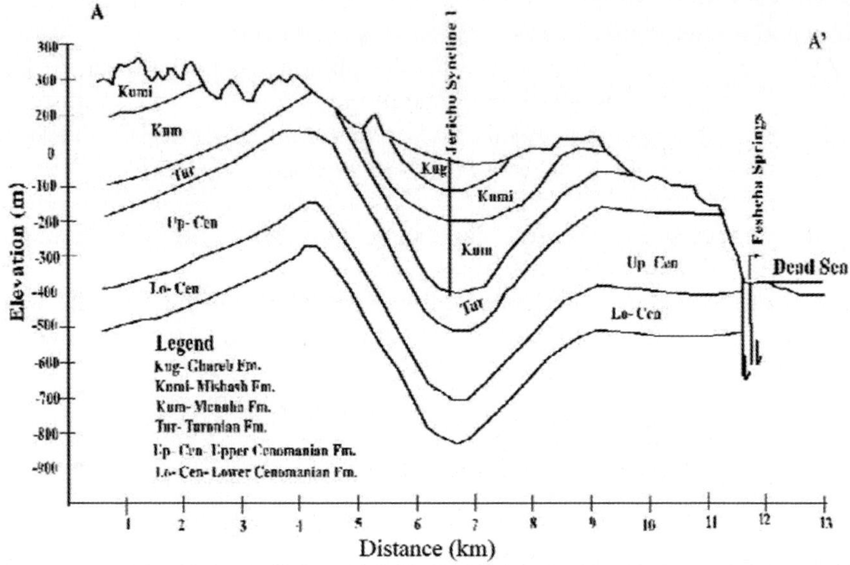

Fig. 3.3.2-3. West-East Geological cross-section A–A' in Fig. 3.2.2-1

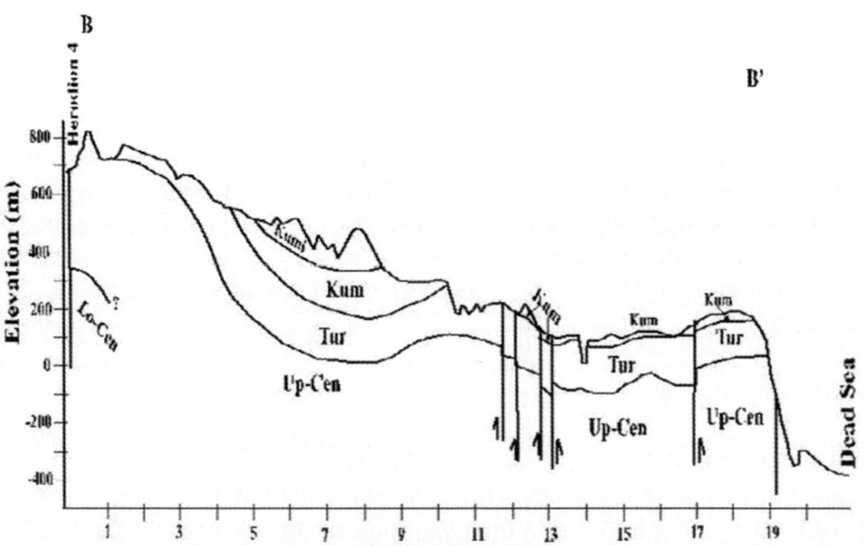

Fig. 3.3.2-4. West-East Geological cross-section B–B' in Fig. 3.3.2-1

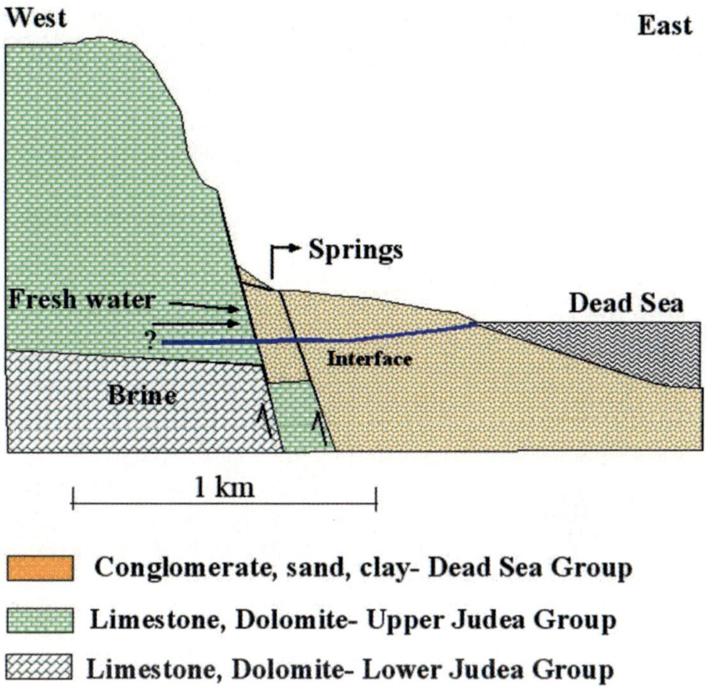

Fig. 3.3.2-5. Schematic geological cross section of the Ein Feshkha area

3.3.2.3 Hydrogeological background

Precipitation

The study area is characterized by sharp fluctuations in precipitation rates. On the high mountains in the west, the average rainfall ranges from 500–700 mm/yr. Towards the east and southeast there is a sharp drop in precipitation over a relatively short distance. In the Jordan Valley, the average rainfall is less than 200 mm/yr and precipitation decreases to less than 100 mm/yr along the Dead Sea shore.

Aquifers

The Judea Group aquifer is the main aquifer in the study area with a thickness of ca. 800–850 m and comprises several sub-aquifers (Table 3.3.2-2). Two sub-aquifers are present in most of the study area: the upper and the

Table 3.3.2-2. Stratigraphic table of rock units in the study area

Age	Group	Formation	Lithological composition	Thickness [m]	Hydrological properties	Remarks
Plio-Pleistozän	Dead Sea	Lisan	Clays, silt, conglomerate, gypsum, chalk			Pebble horizons serving as local aquifers
Paleocene	Mt. Scopus	Taqiye	Marl, chalk, clay, limonite concentrations		Aquiclude	Thickness of Mt. Scopus varies from anticline margins to synclinal center
Maastrichtian		Ghareb	Yellowish chalk		Aquiclude	
Campanian		Mishash	Chalk with black chert layers	100-400	Aquiclude	
Santonian-Campanian		Menuha	White chalk, some calcareous chalk		Aquiclude	
Turonian	Judea	Bina [(1)]	Fine grained limestone and dolomite	90-100	Aquifer	
Upper Cenomanian		Veradim	Hard gray dolomite, porous	90-100	Aquifer	
Lower Cenomanian		Kefar Shaul	Chalky limestone, chalk, marl	30-40	Aquitard	
		Aminadav	Hard gray dolomite, porous	110-140	Aquifer	
		Moza [(2)]	Marl, clay	10	Aquiclude	Not always present
		Bet Meir [(2)]	Chalky limestone, bluish clay, chert, dolomite in Bet Meir Lissan	120-140	Aquiclude	Bet Meir Tongue constitutes an aquiferous layer
Albian	Kurnub	Kesalon	Limestone and dolomite rich in fossils	30	Aquifer	Poor aquifer due to small thickness
		Soreq	Limestone and marly dolomite, marl and some chert	110-170	Aquitard	Limestone and dolomite horizons constitute an aquifer
		Givat Yearim	Hard dolomite, at times also limestone	20-40	Aquifer	
		Kefira	Limestone alternating with marl	160-180	Aquifer	Profile at the base is marly
		Qatana	Marl with some limestone	50	Aquiclude	
Aptian		Ein Qinya	Limestone	60-70	Aquifer	
		Tammun	Limestone, marl and some sand	Base not penetrated	Aquitard	

[(1)] The Bina Fm. divides into three formations in the eastern part of the study area, namely: Netzer, Shivta and Derorim Fms;
[(2)] In the eastern part of the study area the Moza and Bet Meir Fms. are referred to as the Ein Yorqe'am Fm (Begin 1976).

lower aquifer. The upper aquifer generally includes the Bina (Shivta and Netzer on the eastern side), Veradim and Aminadav Formations, and at times the aquiferous horizons in the Bet Meir tongue. Its total thickness varies between 350–370 m. A local aquifer, the uppermost one, is found in the Turonian layers (in the Netzer and Shivta Formations) in the Wadi Qilt region. This aquifer is perched on impermeable layers of the Derorim Formation and its existence is known from the appearance of various springs: Gihon and Jericho, and in Wadi Qilt. This aquifer is phreatic in most parts, with the exception of a small section of the Boqea Syncline (Jericho Syncline). This aquifer is separated hydraulically from the previously described upper aquifer and has no influence on the flow regime towards Ein Feshkha and therefore, it has been excluded from the model. The lower aquifer includes the Givat Yearim and Kefira Formations and the aquiferous horizons in the Soreq Formation. Its total thickness varies between 300 m and 320 m.

According to the well logs, the hydraulic separation between the two sub-aquifers is primarily in the mid-lower section of the Bet Meir Formation, composed of blueish-greenish clay, marl and chalk (middle aquitard). The total thickness of this aquitard varies between 150 and 200 m. The separation is manifested by the difference in water levels in the two sub-aquifers, the level in the upper aquifer being higher than that in the lower aquifer in most of the study area.

Wells

Most of the wells in the study area are concentrated in two clusters (Fig. 3.3.2-2): in the area of Mizpe-Jericho and Jericho and around the Herodion area. Overall, there are 37 wells (17 in the Upper Cenomanian and 20 in the Lower Cenomanian). Most of the wells are production wells and they are located in the western cluster.

Pumping has begun during the early 1970s. Since then, pumping rates have been steadily increasing due to the drilling of new wells. Most of the water is extracted from the lower aquifer of the Judea Group. The depths of the wells range from 350–900 m depending on their geological location and the depth of the lower aquifer.

Ein Feshkha springs

The Feshkha springs are located along the Dead Sea shore (Fig. 3.3.2-1). They represent the outlet of the Judea Group aquifer (upper and lower aquifers). Various attempts have been made to monitor the discharge of these springs. Unfortunately, they were not continuously monitored because the springs are spread along 3–4 km of the Dead Sea shoreline and because of complicated topographical conditions it is extremely difficult to measure their discharge. Measurements vary from 32–40 Mm^3/yr (Tahal, 6 measurements during the period 1984–1986) to 80–85 Mm^3/yr (Israel Hydrological Service, 1992–93). This discrepancy may also be due to the effect of the uncommonly wet year of 1991–92. There are also model evaluations of spring discharge that vary from 53–61 Mm^3/yr.

Since the year of 2003 the Israel Hydrological Service (IHS) currently records the springs' discharge (Rofe 2003). The total discharge varies between 62–67 Mm^3/yr.

Groundwater flow patterns

Most of the natural replenishment in the study area is from precipitation on the outcrops in the western part of the basin. Accordingly, groundwater flow direction is mainly from west to east (see the flow pattern in Fig. 3.3.2-1). Groundwater head varies from +450 m msl along the Hebron and Ramallah anticlinal axis to -300 m in the Jordan Valley and -400 m msl in the Dead Sea area. This drop of groundwater head of 850 m occurs over a distance of approximately 25–30 km. However, this flow pattern is far from being uniform. There are zones with a relatively low groundwater head gradients (Herodion, Ma'ale Edumim-Mizpe-Jericho wells) while there are others in which the gradient is rather steep (e.g., between Jerusalem 5 and Azariya 1). This variability in the hydraulic gradient is indicative of either a strong heterogeneity of the aquifer or significant changes in the layer geometry.

Groundwater flow is governed by the geological structure of the area, i.e., uplifting of the secondary anticlinal axes, as well as by flexures and fault systems (Rofe and Raffety 1963; Bensabat et al. 2004; Guttman et al. 2004). Groundwater flow bypasses low conductivity zones, as shown by the fact that groundwater flow from Herodion and Shdema is mainly northeastward - parallel to the Biet Fajer Syncline axis, and subsequently eastwards.

Water salinity

In the western part of the area, both aquifers contain freshwater. The chloride remains low up to the Mizpe-Jericho well field there the lower aquifer contains saline water (up to 13,000 mg/l Cl^- in well Jericho 2). The possible explanation is that the freshwater in the lower aquifer in the west, flows vertically into the upper aquifer in the Jericho-Mizpe-Jericho region. The upper aquifer in the Mizpe-Jericho well-field contains freshwater that has the same chemical and isotopic composition as the freshwater in the lower aquifer in the western part of the aquifer (e.g., Jerusalem and Azariya wells).

The transition zone in which the freshwater apparently flows from the lower aquifer to the upper aquifer is somewhere west of the Mizpe-Jericho well field. However, it is difficult to find field evidence for this assumption. The whole region is covered by thick formations of the Mt. Scopus Group. It is difficult to identify the nature of the tectonic element (a fault and/or a flexure) responsible for the hydraulic connection between the lower and upper aquifer in Jericho-Mizpe-Jericho region. The assumption in this case is that the lower aquifer is blocked towards the Dead Sea. Otherwise, the salty water of the lower aquifer would have been flushed out during a relatively short period of time and replaced by freshwater. In that case, freshwater should have been found in the lower aquifer, similar to the upper aquifer. In the present case, the lower aquifer contains only saline water.

Geochemical and isotope studies support this assumption. Rosenthal (1978) used uranium isotopes and found that the activity ratio ($^{234}U/^{238}U$) and ^{238}U concentrations in the water of wells Jericho 4 and 5 that penetrate into the lower aquifer, are similar to the values measured in the water of wells Jericho 1 and 2 that pump from the upper aquifer.

The findings of Mizpe-Jericho wells show that the upper aquifer layers (Aminadav and the upper part of Bet Meir Formations (Table 3.3.2-2) in the Marsaba anticline are apparently above the regional water table. Hence groundwater flow originating from the west shifts to the north (and apparently also to the south) towards the Mizpe-Jericho well field located on the saddle inside the Marsaba anticline, from where the flow is channeled towards the Feshkha springs. The Feshkha springs flow out from the upper aquifer and the water is a mixing product of freshwater deriving from the upper aquifer with the saline water from the lower aquifer.

The possible origin of the saline water is residual water from the Lisan Lake (Begin et al. 1974) that extended in the Jordan Rift Valley and rose up to an elevation of 180 m below sea level (ca. 230 m above the present Dead Sea level. This ancient lake was the precursor of the modern Dead Sea and existed from 60,000 until 12,000 years ago. As mentioned before, it covered large areas in the Jordan Valley. The saline water could also be relicts of the ancient Dead Sea when it was at higher stands. The salty water penetrated into the aquifer at the contact zones between the aquifer strata and the Lisan Formation strata. Over the last 12,000 years freshwater (that flows to the outlets in the Dead Sea) flushes out the salty water from the upper aquifer.

3.3.2.4 Conceptual model

The model area is located on the eastern slopes of the Judean anticlinorium, which slopes eastwards by several undulations creating a series of parallel anticlines and synclines with axes generally striking northeast-southwest. The anticlines are asymmetric, with steep dips of their eastern flanks. The asymmetric structures are characterized by flexures generally directed northeast-southwest.

The major western fault zone of the Dead Sea Rift Valley is the dominant tectonic element in the area acting as a hydrological border between the Judea Group aquifer strata and the basin-fill strata of the Dead Sea-Jordan Rift Valley. The vertical movement of the fault in the Feshkha springs area reaches an offset of hundreds to thousands of meters. The Feshkha springs emerge along this fault zone. Other fault systems of tens of meters offset are located further westwards. These relatively small faults are of minor hydrological importance and were neglected.

The aquiferous formations of the Judea Group are exposed on the western part of the anticlinorium. Eastwards, younger formations of the Mt. Scopus Group composed mainly of chalks and marls overlie these layers. Older formations of the Judea Group are exposed along the area of the fault escarpment descending towards the shoreline of the Dead Sea.

The Judea Group aquifer is composed of thick beds of limestone and dolomite interbeded with thin beds of marl, shales and chalk. The total thickness of the aquifer is 800–850 m. In the study area the aquifer is divided into two regional sub-aquifers. A shaly, blueish, highly plastic horizon of 10–20 m thickness occurs in the lower part of the Ein Yorqeam Formation separating the two sub-aquifers. This horizon was identified in many wells in the Herodion-Beit Lehem area in the west and in the Mizpe-Jericho wells in the east.

A relatively high gradient of more then 9% exists between the Jerusalem-5 well (+360 m) and the Azariya-1 well (-15 m). The distance between these wells is about 4 km. On the other hand, a lower gradient of 0.8% was measured between Maale Adumim-2 well and the Mizpeh-Jericho (distance of about 6 km) and the water level difference is about 50 m. Such a steep hydraulic gradient can be explained only by a geological structure in the form of underground "cascades" (Fig. 3.3.2-6).

The upper and the lower aquifers in the west contain freshwater flowing in both aquifers from the west eastwards. In the east (in the area west of the Mizpeh-Jericho wells and up to the eastern boundary of the model) the lower aquifer is not an active aquifer because it contains brine. The freshwater flows only in the upper aquifer, the Lowermost Aminadav and the uppermost Ein Yorkeam Formations.

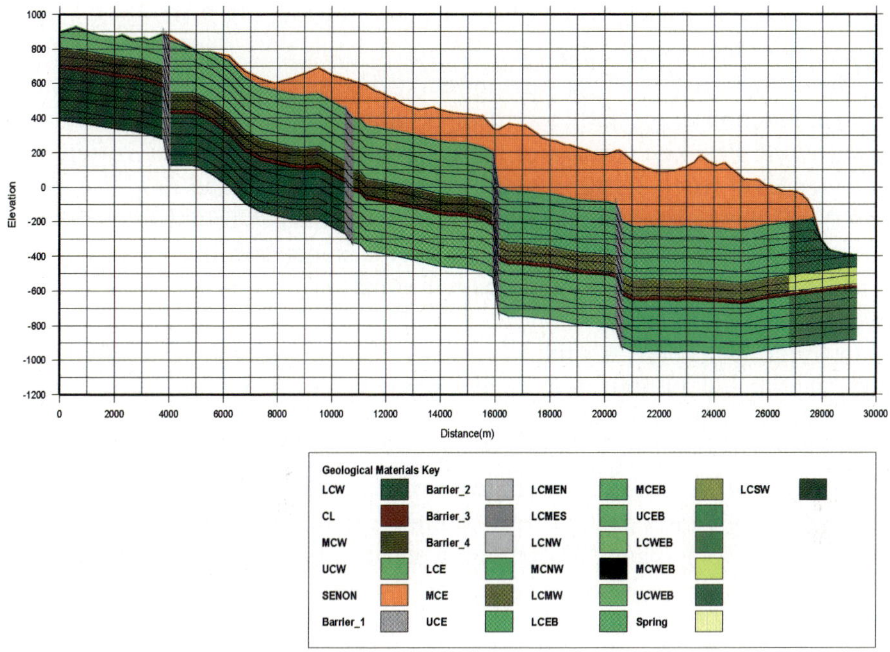

Fig. 3.3.2-6. Hydraulic gradient through geological structure

In this current study the assumption is that the brine found in the lower aquifer acts as a hydrological barrier that prevents groundwater flow from the lower aquifer towards the Feshkha springs. In the regional western aquifer, the flow path of the freshwater is probably west of the previously mentioned "brine obstacle". It flows into the lower part of the upper aquifer and drains (through this aquifer) towards the Feshkha springs. The hydraulic connection between the upper and lower aquifer can be obtained only by using this unique model path which could take place near the Marsaba flexure.

The Marsaba anticline in the east is the most dominant tectonic element and influences the groundwater flow path and the outlet at the Feshkha Springs. The Marsaba structure is an asymmetric anticline plunging northeastwards. In the Mizpe-Jericho area, the lower part of the upper aquifer is saturated because of the geological structure, which deflects this individual flow path eastwards. It is still unclear why the groundwater path is deflected southeastwards and why does groundwater emerge through the Feshkha springs located a few km to the south, instead of following the direct and short path eastwards.

The major western Rift fault is a relatively sealed boundary through which leak only few Mm^3/yr. This sealed tectonic boundary causes the change in groundwater flow path to the south, parallell to the southern boundary of the Feshkha springs.

The characteristics of the flow model can be summarized as following:

The direction of groundwater flow is from the recharge areas in the west towards the Feshkha springs in the east (see Fig.3.3.2-1). The differences in water level reach 1,100 m over a distance of 30 km and the flow gradient is of about 3% (assuming constant gradient). Groundwater flow is through two separated aquifers, namely the upper and the lower aquifer. The lower part of the Ein Yorqeam Formation acts as an aquiclude between the two aquifers.

In order to account for the previously mentioned high-flow gradient, a huge amount of rainfall and large replenishment areas are required. The available rainfall data and the hydrogeological information on the replenishment areas show that such volumes of water do not exist in the area. Therefore, the volume of groundwater within the aquifers cannot account for such a high and constant flow gradient. A series of hydrological barriers is needed to account for the high water levels in the higher areas of the aquifer and the low water levels in the lowermost areas. The flow system from west to east follows probably a geological cascade-like structure. The cascade-like structure (Fig. 3.3.2-4) is the result of the asymmetric geological structures crossing the area. The flexure (cascade) area is of low permeability, which causes a drop of the water level in the eastern lowermost areas of the flexure. The flow gradient in the uppermost areas of the flexure is relatively moderate. The current study defined five flexures crossing the model area from west to east. The flexure structure determines the flow system, the flow gradients and the water levels in the research area. The flow of the Feshkha springs is highly influenced by the water levels in the Marsaba (easternmost) flexure. Pumping in the area of the Mizpeh-Jericho wells rapidly influence the discharge of the springs. On the other hand, pumping in the uppermost areas of the basin (Jerusalem-Beit Lehem) hardly influence the outflow of the Feshkha springs.

The Feshkha springs are the major outlet of groundwater in the area. There is probably some minor leakage of freshwater through the Rift fault into the basin fill sediments in the area north to the Dead Sea. Guttman and Rosenthal (1991) found that according to geological and geochemical evidence, groundwater within the deep basin-fill strata are chemically identical to the saline water measured in the Mizpeh-Jericho wells.

Boundary conditions

The western boundary of the model (see Fig.3.3.2-2) is assumed to match the groundwater divide that separates flow from the western and eastern parts of the mountain aquifer. The northern and southern boundaries are assumed to be streamlines through which no flow takes place. The eastern boundary deserves a much finer treatment. This boundary is composed of the following two segments: (a) the Dead Sea shore and (b) the alluvial aquifer of the Jordan Rift. Along these two segments the subsurface contains brines. A precise representation of this boundary would have required taking into consideration density-dependent flow and salt transport. However, since the research deals only with the flow processes, a boundary condition of type III (mixed type) was assumed, and the cross-section flow was artificially limited in order to represent the presence of the brines (through which no freshwater flow occurs). This type of condition allows a better control of the fluxes crossing the boundary.

The computational model

In order to represent the complex geological features that characterize the level of topography area, the computational mesh was created with a fine node density (Fig. 3.3.2-7).

Mathematical formulation

The mass balance equation solved by the model is:

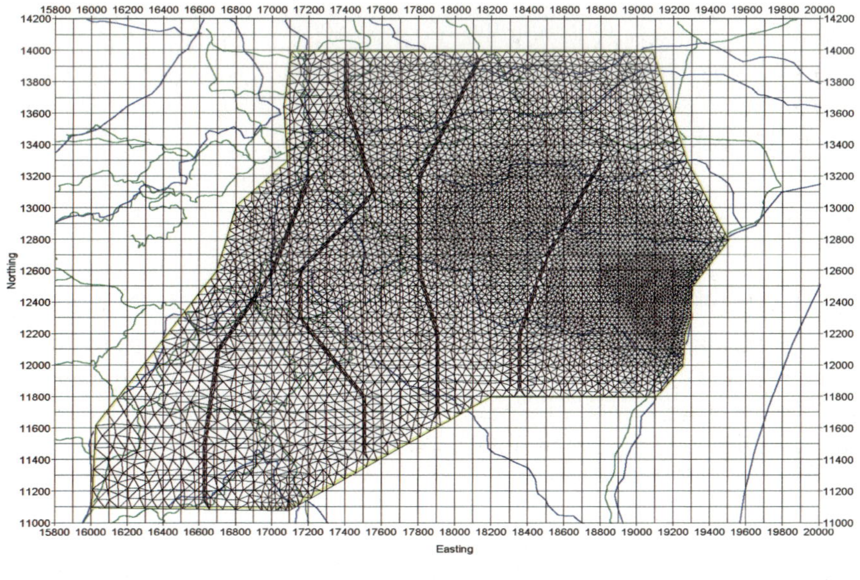

Fig. 3.3.2-7. Computational mesh for the modeled area

$$S_0 \frac{\partial \varphi}{\partial t} = \nabla[K_{ij} \nabla \varphi] - \sum_i P_i(t)\delta(x - x_i) - Q_s(t) \tag{3.2.2-1}$$

with S_0 = storativity coefficient; φ = piezometric head; K_{ij} = hydraulic conductivity tensor; $P_i(t)$ = pumping rate at well i at time t; $Q_s(t)$ = spring discharge at time t.

The spring discharge is determined with the linear approximation:

$$Q_s(t) = \max(0, \alpha_s(\varphi_s(t) - \varphi_{s0})) \tag{3.2.2-2}$$

with $Q_s(t)$ = spring discharge at time t; α_s = spring coefficient; $\varphi_s(t)$ = piezometric head at spring; φ_{s0} = threshold piezometric head, below which the spring dries out.

Boundary conditions:

First type boundary conditions

$$\varphi(x,t) = \varphi_\Gamma(x,t) \qquad \forall x \in \Gamma_I \tag{3.2.2-3}$$

where Γ_I is the portion of the model boundary over which a first type boundary condition is imposed and φ_Γ is a known function of space and time.

Second type boundary conditions

$$q_n(x,t) = q_n^\Gamma(x,t) \quad \forall x \in \Gamma_{II} \tag{3.2.2-4}$$

where Γ_{II} is the portion of the model boundary over which a second type boundary condition is imposed, q_n is the groundwater flux entering (or leaving) the flow domain and q_n^Γ is a known function of space and time.

Third type boundary conditions

$$q_n(x,t) = -\kappa(x)(\varphi_\Gamma(x,t) - \varphi(x,t)) \quad \forall x \in \Gamma_{III} \tag{3.2.2-5}$$

where Γ_{III} is portion of the model boundary over which a third type boundary condition is imposed, q_n is the groundwater flux entering (or leaving) the flow domain, φ_Γ is a known function of space and time and κ is a known coefficient.

Boundary conditions on the phreatic surface

At the phreatic surface two conditions must be satisfied:

Groundwater flux continuity through the surface:

$$q_n = -\left(R(x,t) - S_y \frac{\partial \varphi}{\partial t} \right) n_z \tag{3.2.2-6}$$

The pressure of the water at the phreatic surface is zero, i.e.,

$$\varphi(\mathrm{x},t) = \varsigma_s(x,y,t) \tag{3.2.2-7}$$

where q_n is the water flux entering (or leaving) the flow domain through the phreatic surface, $R(x,t)$ is the replenishment flux (groundwater recharge) entering though the phreatic surface, and n_z if the vertical component of the vector normal to the phreatic surface.

The simulations were performed with the FEASi-3-D-Flow module simulating groundwater flow in three dimensional anisotropic and heterogeneous aquifers. This module takes care of dynamic boundaries using a moving mesh technique. It has been applied in various regional models in Israel (in the Coastal Plain, in the Western Galilee and in the Southern Arava).

Model calibration

Calibration approach

The model was calibrated using a trial and error approach. A calibration period of 10 years (1983–1992) was chosen. This period is interesting since it includes a drought period in 1989 and a period of high rainfall intensity in 1992.

The calibration of the model was extremely tedious due to the high complexity of the geological setup (the presence of flexures with large drops in the layer's topography) and to the resulting drastic changes in the groundwater head elevation (from about +700 m msl in the west to about -400 m in the east).

Since the boundary conditions at the eastern boundary were rather artificial, care was taken to keep the flow of freshwater through this boundary within reasonable limits.

The relatively high value of groundwater recharge in the western side of the flow domain indicates that there may be a flux of freshwater entering the domain from this boundary.

The average water budget

The average water budget at the final stage of the steady state model includes the following components: groundwater recharge of 70 Mm^3/yr, groundwater abstraction of 12 Mm^3/yr, discharge from the Feshkha springs of 35 Mm^3/yr, and leakage of freshwater through the eastern boundary into the basin-fill sediments in the area north to the Dead-Sea of 23 Mm^3/yr. This outflow is too high and is a result of the boundary conditions mentioned before. During the second phase of the research (not mentioned in this paper), the model area was expanded and the influence of the brine on the flow pattern had been taken in consideration.

Calibration quality

Calibration quality was assessed on the basis of comparison between measured and simulated groundwater heads in wells within the flow domain, by drawing the groundwater elevation map in the upper and lower aquifer layers and by verifying the compliance of the obtained flow fields with the conceptual hydrogeological model.

Time series of measured and simulated groundwater heads for wells Jerusalem 1, Herodion 3 and Jerusalem 6 are presented in Fig.3.3.2-8. A satisfactory agreement was achieved between the measured and simulated data.

3.3.2.5 Exploitation scenarios

After calibration, groundwater exploitation scenarios were elaborated, comparisons between scenarios were made and finally, recommendations are suggested regarding the enhanced exploitation of this area suggesting alternatives for the sustainable exploitation of the groundwater resources in the study area.

The baseline scenario is based on the pumping regime that was applied in 1998 (before the drought period). Within the model area there were 24

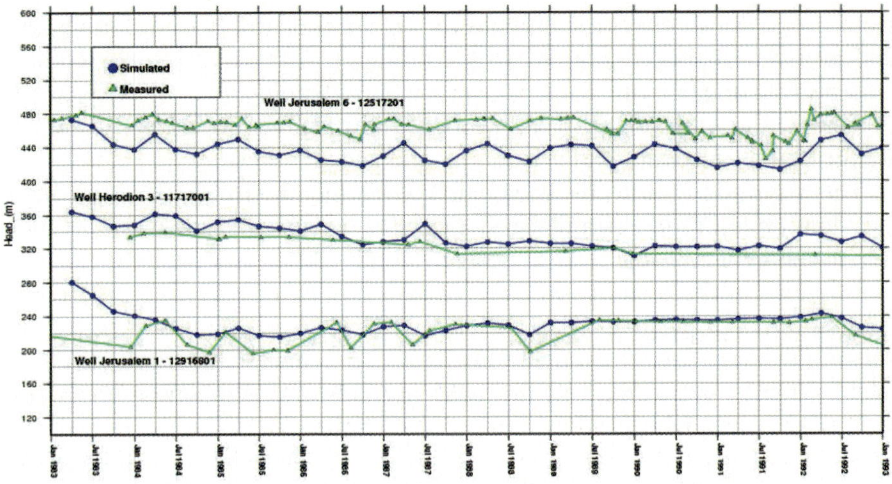

Fig. 3.3.2-8. Time series (1983–1993) of measured and simulated groundwater heads (m below surface) for wells Jerusalem 1, Herodion 3 and Jerusalem 6

wells pumping 23.6 Mm^3/yr. All the exploitation scenarios were based on this baseline scenario.

Rainfall was taken from two rain gauges with relatively long histories of records covering the spatial variability in the recharge zones of the model.

Constraints were put on the pumping wells in the model, i.e., wells will be shut-off once the dynamic head in the well falls below a threshold value. These constraints take into account the technical structure of the well, the specific discharge and a simplified behavior of the efficiency curve of the well pumps. The computational model was modified accordingly so that these constraints are checked. In addition, the constraints may induce an erratic behavior of the well pumping regime, depending on the impact of the pumping on the well itself. If the groundwater head drops on a regional scale, due to regional overexploitation then it may be expected that the well will be shut-off for long periods of time and a smooth behavior is to be expected. On the other hand, a well may be shut-off due to the excessive pumping in the well itself, in which case an erratic behavior is to be expected, i.e., the well is activated and shut off periodically. In this case the water level will rise as a result of the reduction in pumping or after a very rainy year. The well may than be reactivated.

The exploitation scenarios are based on two principles: (a) using the existing infrastructure, i.e., increasing water extraction from existing wells, in order to minimize development and water distribution costs, and (b) developing new well fields so that the expected groundwater head drops are evenly distributed over a larger area. Most of the new wells were taken from proposals that were submitted by the JWC (Joint Water Committee) and in the TWC (Technical Water Committee). A third option is a combination of the former two.

Three new well fields were proposed for the exploitation scenarios: Mizpe-Jericho, Annata-Kefar Edumim and Judea Desert.

The Mizpe-Jericho wells are located along the eastern flank of the Marsaba anticline. The assumption is that this area is the easternmost line where the groundwater is still fresh. It comprises 7 new wells located in the vicinity of the existing wells of Mizpe-Jericho (see Fig.3.3.2-2). They penetrate the Upper and Middle Cenomanian layers, above the clay layer separating these layers from the Lower Cenomanian. Suggested pumping is 2.2 Mm^3/yr per well, i.e., the total groundwater abstraction from this well field is 15.4 Mm^3/yr.

The second new well field is Annata-Kefar Edumim. It is located in the area between the Annata and Kefar Edumim villages. This region is intermediate between the Jerusalem wells in the west and the Mizpe-Jericho wells in the east. It comprises five wells that are located on the eastern flank of Far'a flexure. From each well 1.25 Mm^3/yr are abstracted so that the total groundwater extraction from this well field totals 5 Mm^3/yr.

The third new field contains a series of wells that was proposed to the JWC for the development of the groundwater in the Judea Desert before it reaches the springs along the Dead Sea western shore. Each of the wells will pump 1 Mm^3/yr so the total groundwater extraction from this well field would be 10 Mm^3/yr. Part of the wells pump from the upper aquifer and part from the lower aquifer. All scenarios were tested for a period of 25 years (2005–2024).

A0 is the baseline scenario upon which all the other scenarios are built and compared to. The pumping during 1998 (23.57 Mm^3/yr) was taken as the pumping regime for scenario A0. According to scenarios A1 and A2 the pumping was increased by 20% (scenario A1) and by 10% (in scenario A2).

A3 includes scenario A0 and the implementation of the three well fields. The total model abstraction for this scenario is 53.57 Mm^3/yr. While scenario A4 consists of scenario A1 and the three well fields and the model total extraction result for this scenario is 58.28 Mm^3/yr, scenario A5 consists of scenario A2 and the three well fields and the total model abstraction for this scenario is 55.93 Mm^3/yr. The flexures crossing the model area divide the area into several segments. During the simulations, the regional pattern of the flow regime (groundwater flow from west to east) as well as the structure of the segments remains unaffected. However, local hydrologic depressions were observed probably as a result of local overpumping. The depth and the size of the hydrologic depression depend in each scenario on the pumping regime, the natural recharge (local and regional) and the distance between the wells. The drop of the water level in the lower aquifer is larger than in the upper aquifer because most of the water in the new scenarios is abstracted from the lower aquifer.

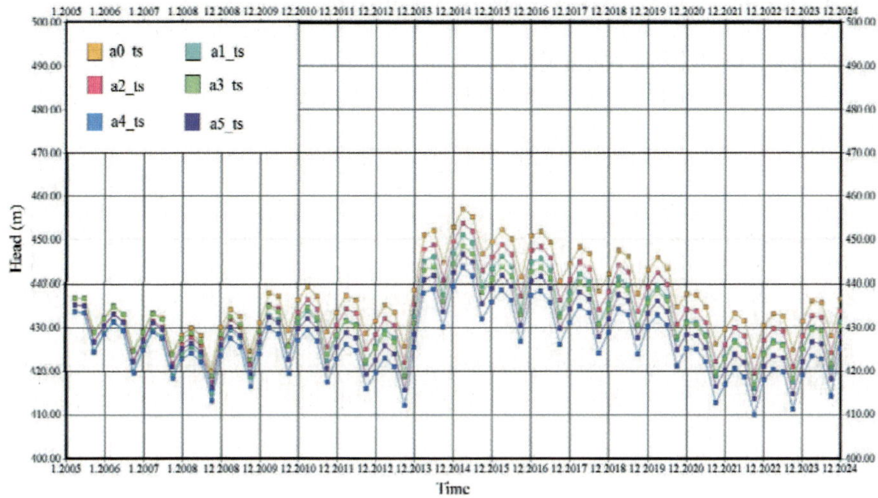

Fig. 3.3.2-9. The impact of different scenarios (see text) on the water level in well Jerusalem 3.

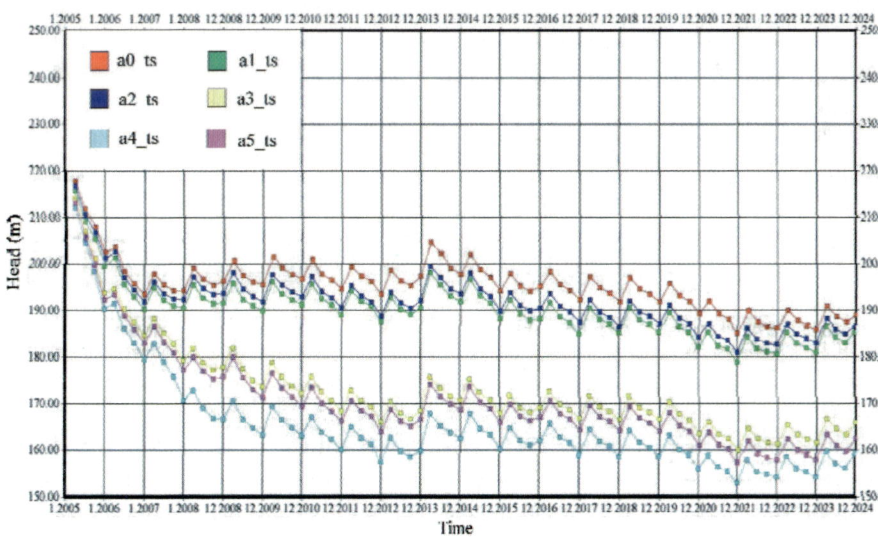

Fig. 3.3.2-10. The impact of different scenarios (see text) on the water level in well Shdema 2.

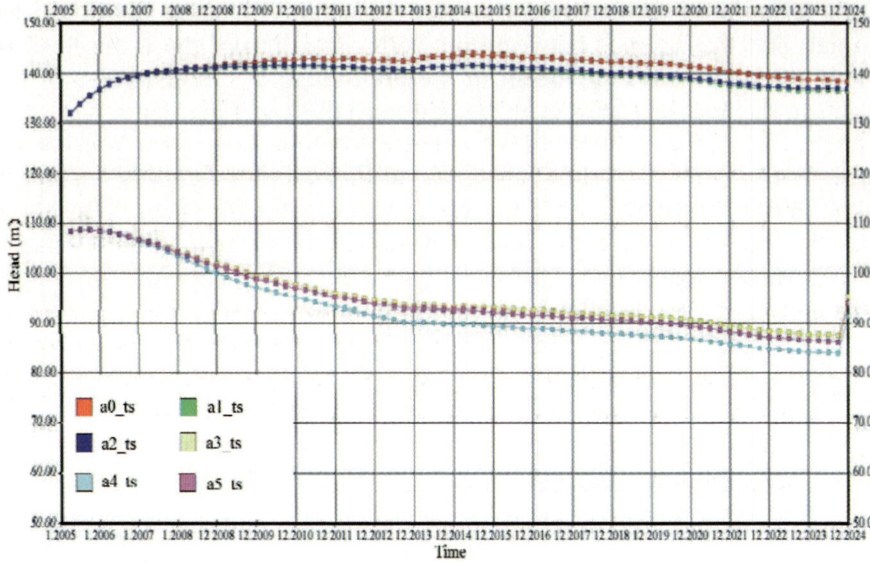

Fig. 3.3.2-11. The impact of different scenarios on the water level in well W16.

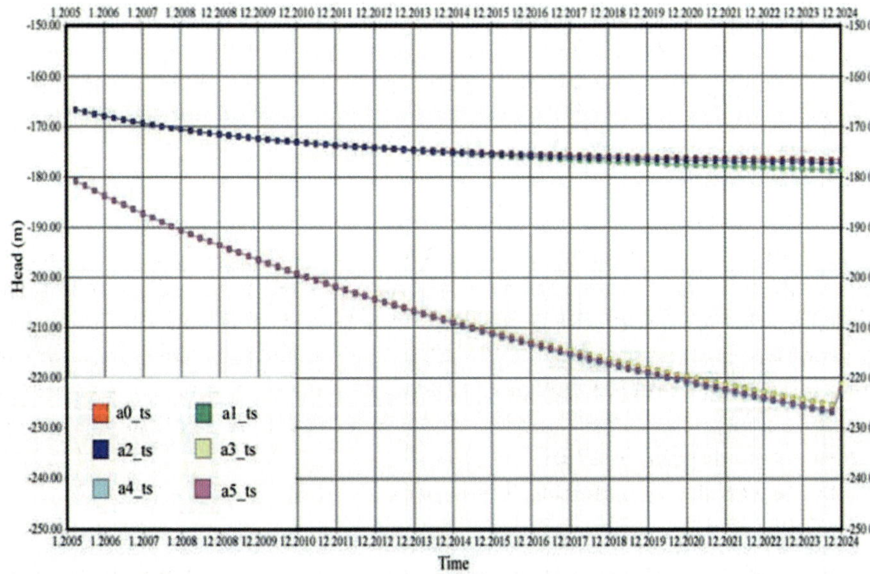

Fig. 3.3.2-12. The impact of different scenarios on the water level in well JD5.

Figs. 3.3.2-9–12 shows the water level at the end of the simulation (year 2024) in various wells along a west-east line. In the wells located in the recharge area (for example Jerusalem 3), the influence of the fluctuations in

the natural recharge is clearly seen. At further distance from the recharge area (Shedema 2 (Fig. 3.3.2-10) and further east- W16 (Fig. 3.3.2-11) and JD5 (Fig. 3.3.2-12)), the influence of the natural recharge decreases and the abstracted water in these wells appears to be derived from storage.
The results of the simulation show that the amount of drawdown in each well depends on the total pumping rate in the well itself, within the whole well field and of course on the annual natural recharge. The results show that in order to reduce the large groundwater drops, it is important to spread the wells over larger areas and to manage the pumping in the single well and in the whole well field according to the yearly recharge. The results show that the amount of pumping that was tested in the some scenarios may be exaggerated, particularly in case of dry years. Therefore, sustainability of the groundwater exploitation may not be achieved.

3.3.2.6 Conclusions and recommendations

This study dealt with an arid area that suffers from an acute shortage of freshwater. Its rationale was based on previous studies, which indicated that substantial amounts of freshwater infiltrate into the aquifer in the relatively rainy replenishment zones in the west and flow eastwards to the Dead Sea where they mix with the brines present there and become saline. A clear proof of this is the relatively high discharge (of saline water) in the Feshkha springs, which are the main outlet of the Judea Group aquifers in this area.

Therefore, the objective was to suggest ways to capture part of this water before it reaches the saline zones.

A three-dimensional geohydrological model and a computational groundwater flow model were constructed and calibrated, scenarios for the exploitation of the freshwater in this area were suggested and the response of the aquifer was evaluated with the help of the model. The design period of time was 20 years (from 2005 to 2024). The results obtained are promising and clearly indicate the possibility to produce additional significant amounts of freshwater during the next 20 years. This could clearly contribute to the development of this area.

The exploitation scenarios chosen in this study were based on the following principles:

- Usage of the existing infrastructure by increasing the water abstraction from existing wells, in order to minimize development and water distribution costs, and
- Development of new well fields so that the expected impact on the aquifer is distributed over a larger area.

Most of the new wells were taken from proposals that were submitted to the JWC (Joint Water Committee) and in the TWC (Technical Water Committee). Taking into consideration these proposals, the current study indicates three well fields: Annata-Kefar Edumim (additional pumping of 5 Mm^3/yr); Mizpe-Jericho (additional pumping of 15.4 Mm^3/yr), and Judea Desert (additional pumping of 10 Mm^3/yr).

The main results are:

The flexures divide the aquifer area into secondary basins (segments). The overpumping in one segment first induces a drop of the water level within this unit. As this drop increases, it can propagate to the down gradient segment.

The size of the drawdown within a segment is affected by natural recharge in the segment, the pumping regime at individual wells, the total amount of extracted water, and the distance between the wells.

The flexures are thin and low-permeability barriers. The major component of groundwater flow is parallel to them. The local depressions caused by the pumping have an elongated form extending along the direction of the flexure.

In the wells that are located in the recharge area (Jerusalem and Herodion well fields), the influence of the fluctuation in natural recharge is apparent. Farther from the recharge area, the influence of the natural recharge decreases and the abstraction in these wells appears to be derived from storage.

The results of the simulation show that in order to reduce drawdown, it is important to spread the new wells over large areas and over large distances between the wells so as to adapt pumping rates in the single well and in the whole well field to the annual recharge rate.

The conceptual model estimates that the brine in the lower aquifer acts as a hydrological barrier that blocks the flow from the lower aquifer towards the Feshkha springs. The regional flow path of fresh groundwater in the western parts of the aquifer probably occurs west of the brine region. It flows into the lower part of the upper aquifer and drains through the upper aquifer towards the Feshkha springs.

One of the recommendations is to adopt scenario A3 (scenario A0 plus additional new well fields) as the leading development scenario for this arid zone, with the following reservations and modifications:

The large drawdown in Annata-Kefar Edumim well field indicates that the nominal extraction of 5 Mm^3/yr may be too high. This may be a real issue or an artifact of the model, due to a severe lack of information in the area. We suggest proceeding by a step-by-step approach. One well, Fawwar, should be drilled, put in operation and monitored for a few years. Depending on the aquifer response, decisions should be made regarding further development of the well field.

The additional extraction in the Mizpe-Jericho area (15 Mm^3/yr) should be reduced to 8 Mm^3/yr in 3 to 4 wells. Here too, further development should be decided depending on the aquifer response.

The Judea Desert well-field seems to be a feasible option for development. In order to better control the drawdown, we suggest spreading the wells over a larger geographical area.

The conceptual as well as the flow model show that the saline water bodies existing in the lower aquifer in the Mizpe-Jericho area play a major role in the calibration. In the second phase a flow and transport model is calibrated taking in consideration also the source and distribution of the saline water.

References

Arkin Y, Braun M and Starinsky A (1965) Type sections of Cretaceous formations in the Jerusalem-Beit Shemesh Area. Stratigraphic Sections, No. 1, Geol Surv Israel

Arkin Y and Hamaoui M (1967) The Judea Group (Upper Cretaceous) in Central and Southern Israel, Geol Surv Israel. Bull, No. 63

Begin ZB (1976) Geological Map - Jericho Sheet (1:50,000), published by the Geological Survey (in Hebrew)

Begin ZB, Erlich A, Nathan I (1974) Lisan Lake, the Representative of the Pleistocene Age in the Dead Sea. Geoll Surv Bull, No. 63 (in Hebrew)

Bensabat J, Guttman J, Flexer A, Hötzl H, Ali W, Yellin-Dror A (2004) Groundwater model of the Marsaba-Feshkha region-Israel, 5th International Symposium on Eastern Mediterranean Geology, Thessaloniki, Greece, pp 1487–1490

Guttman J, Flexer A, Hötzl H, Bensabat J, Ali W and Yellin-Dror A, (2004) 3-D hydrogeological model in the arid zone of Marsaba-Feshkah region-Israel, 5th International Symposium on Eastern Mediterranean Geology, Thessaloniki, Greece, pp 1510–1513

Guttman J, Rosenthal A (1991): Mitzpe Jericho Area - Study of Salinization Mechanisms and Water Potential: Tahal. Rep. No. 01/91/46 (in Hebrew)

Rofe B (2003): Measurement campaign in Feshkah Spring-Summer 2003 (In Hebrew)

Rofe and Raffety Consulting Engineers (1963) Geological and Hydrological report, Jerusalem and District Water Supply. Hashemite Kingdom of Jordan, Central Water Authority, London

Rosenthal A (1978): Non-Equilibrium of $^{234}U/^{238}U$ Isotopes in the Water of the Judea Group Aquifer on the Eastern Slopes of the Judea and Samaria Hills. M.Sc. thesis, Tel Aviv University (in Hebrew)

Roth I (1970) Wadi Qilt, geological map, 1:50,000. Geol Surv Israel

Sneh A, Bartov Y, Weissbrod T, Rosensaft, M (1998) Geological Map of Israel, 1:200,000. GSI

3.3.3 Groundwater flow and transport model of the Jericho area

M. Beinhorn, J. Guttman, O. Kolditz

3.3.3.1 Introduction

The shallow aquifer discussed here is located in the Jericho area. It consists mainly of alluvial deposits, clay, silts, conglomerate, chalk, marl and gypsum layers of the Lisan Formation. Due to the high water demand in the area of Jericho and consequently high groundwater abstraction rates and irrigation back-flow, groundwater quality in the aquifer has deteriorated considerably. Detailed investigations performed by Marie and Vengosh (2001) using geochemical data suggest that high chloride concentrations, particular in the eastern part of the aquifer, are derived from:

- Upconing of deep brines
- Leaching of salts within the aquifer
- Anthropogenic contamination of agriculture return flow and waste water infiltration

Another possible source of salinity is residual water resulting from former highstands of the Dead Sea. Some 15,000 years ago, the water level reached its maximum of about 230 m above the present Dead Sea level. During that time the Lisan Lake, the precursor of the present Dead Sea, covered large areas of the lower Jordan Valley up to an elevation of -180 m msl (Begin et al. 1974) and therefore, the underlying rocks and sediments contain brackish water. The model domain comprises this range of elevation.

A hydrogeological model of the area has been developed based on data available from project partners. A numerical model was used to improve and verify the hydrogeological model. The work carried out in this study does not take into consideration the evaluation of possible sources of salinity. The first aim was to place some constrains on boundary conditions and hydraulic properties of the aquifer in order to improve our understanding of the hydrogeology of the area. The second aim was to extend the capacities of the modeling software so that the hydrogeological system can properly be represented by the numerical model.

3.3.3.2 Methodology

Data availability

Hard copies of topographic and geological maps and satellite images of the study area were available. Relevant information was collected from a groundwater resources report of Golani (1972). Further information was added by personal communication during project meetings. Altogether the following data set was gathered for the modeling study:

- Topographic and geological maps at a scale of 1:50.000
- Satellite images
- Groundwater abstraction rates from wells of the shallow Plio-Pleistocene aquifer for the years 1961/63, 1968, 1969/70
- Yearly spring discharge volumes of the Jericho springs Sultan, Dyouk, Nueima, Shosha and Uja
- Groundwater contour lines for the year 1982
- Chloride concentration contour maps for the years: 1961/63, 1968 and 1969/70
- Pumping test map

Data handling

Topographic and geological maps and satellite images were scanned, geo-referenced and then placed in an ArcGIS project. Groundwater contour maps as well as concentration contour maps were digitized and then also placed in an ArcGIS project. Tables of groundwater extraction and spring discharge were stored and organized using Excel and then partly displayed as ArcGIS point themes.

Simulator used

The numerical model used for the simulations is the FE Simulater GeoSys/RockFlow (Kolditz et al. 2003) which is programmed in C and C++ according to object oriented software concepts (Kolditz and Bauer 2004). The numerical approach is based on the Galerkin finite element method. Time derivatives are evaluated by a finite difference scheme. The iterative coupling between the discretized flow and transport equations is realized by a Picard iteration scheme, i.e., actual densities of a new iteration step are calculated from the field variables of the previous iteration step. The code has been tested against the classical seawater intrusion problem of Henry, the free convection problem by Elder and the salt dome problem (Kolditz et al. 1998).

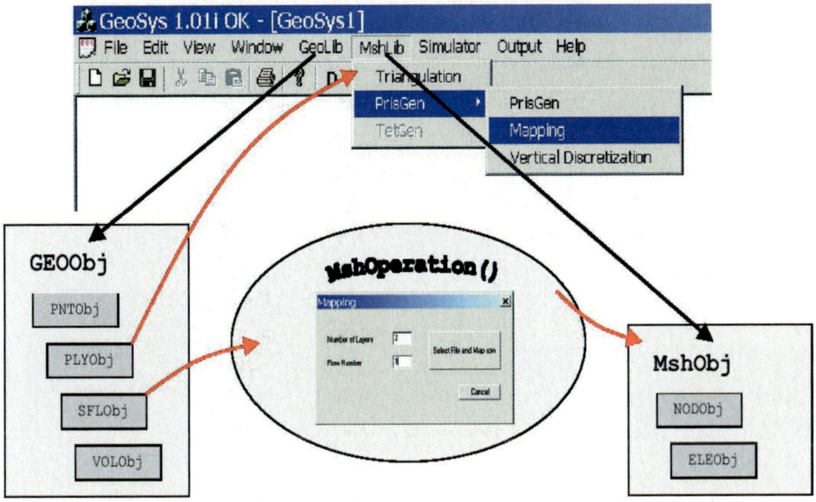

Fig. 3.3.3-1. Links of geometric data and meshing operations for hydrogeological models

Basic modules of GeoSys/RockFlow are the GeoLib, MshLib, and FEMLib, which are related via a graphical user interface (GUI) (Fig. 3.3.3 -1). The interface displays the relationship between GeoLib and MshLib. Geometric objects of the GeoLib are: points, polylines, surfaces and volumes which form domains. These geometric entities are implemented as C++ classes: CGLPoint, CGLPolyline, CGLSurface, CGLVolume, CGLDomain. Instances of these objects can alternatively be stored as vectors and lists for convenience and specific purposes. A detailed description of the implementation of GeoObjects is given in Kalbacher et al. (2005). As an example, Fig. 3.3.3-2 shows geometric information of the hydrogeological system described in the following chapters within the framework of the GUI. The polylines seen in the GeoSys GUI stem from the shape file imported form ArcGIS. The geometric object can now be used to define boundary conditions, source terms or an mesh geometry as seen in the associated example in Fig. 3.3.3-3.

The geometric information seen in Fig. 3.3.3-2 also forms the basis for the mesh generation. Currently, the following build-in mesh generators can be used: PrisGen (Kalbacher et al., 2003); TetMesh (Manabe et al., 2003) as well as open and commercial systems such as gmsh (2004) and gOcad (2004). The resulting mesh geometry can be manipulated using mesh operations similar to the mapping operation. Here a surface object (SFCObj) is being used to relate mesh data to surface data file such as top or bottom of a hydrostratigraphical unit. An additional operation of the MshLib is vertical mesh refinement. All operations of the MshLib are controlled via an GUI as seen in Fig. 3.3.3-1.

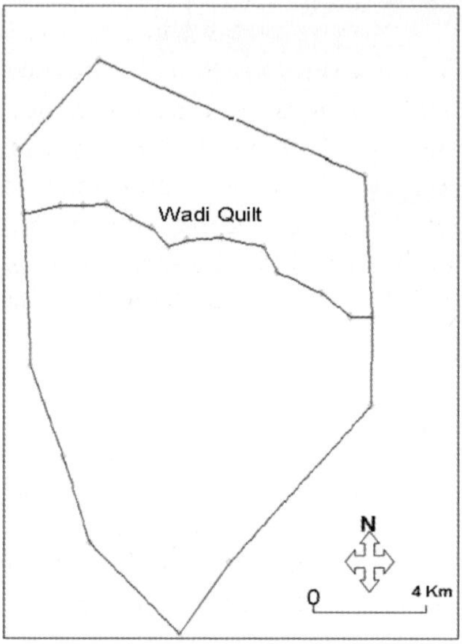

Fig. 3.3.3-2. Geometric units of the model domain.

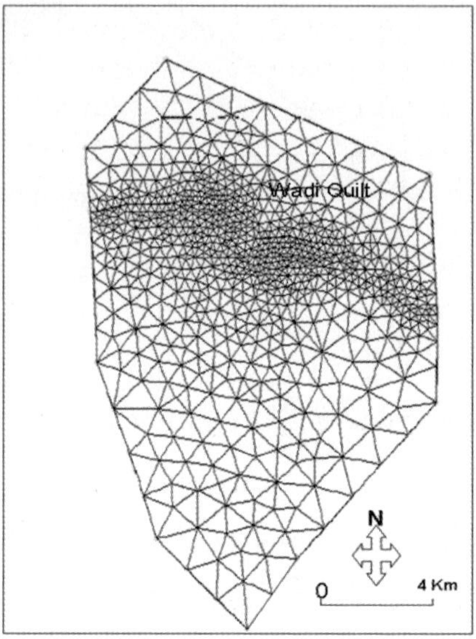

Fig. 3.3.3-3. Triangulated model domain.

The FEMLib contains independent units necessary to set up and solve the finite element problem. As a result of the numerical approximation method described, two sets of algebraic equations were derived. The general form of this equation is (Eq.(3.3.3-1)):

$$A_i\,(t, x, u_i\,, u_j\,)\, u_i = b_i\,(t, x, u_i\,, u_j\,) \tag{3.3.3-1}$$

where i, j denote different processes, with Ai being the system matrix of process i, t= time, x= coordinates, u_i= solution vector (primary variable) of process i, b_i= RHS vector of process i. Dependence of process i on process j means coupling, and the dependence of the system matrix on solution itself means non-linearity. In addition to the standard procedure to evaluate element matrices numerically, analytical solutions for flow and transport problems were implemented for triangular prismatic elements.

Hence, numerical integration is not necessary so that the computational effort is significantly reduced: CPU time needed for calculation of element matrices analytically is 1/5 of the time needed using numerical integration.

The effect on total computation time depends on the problem to be solved but is estimated to be 1/3 to 1/2 of the time needed using numerical integration. Clearly, the reduction of computational effort in the formulation of element characteristics is advantageous, if it comes to large-scale groundwater models or if nonlinearities such as unconfined flow or density-dependent flow are considered. In principle, the technique has been used by Huyakorn and Thomas (1984), there referred to as "influence coefficient" technique. Influence coefficient matrices were presented for linear rectangular elements. At a later stage Huyakorn et al. (1986) presented coefficient matrices for linear rectangular prism elements and linear triangular prism elements for fluid flow. Details of the method are presented in Beinhorn and Kolditz (2003b) and Beinhorn (2005). In these reports the matrix expressions are derived in detail for fluid flow as well as tracer transport.

Geology and hydrogeology of the area

Two Dead Sea Group aquifers are located in the Jericho area, the Holocene or sub-recent alluvial aquifer and the Pleistocene Samra aquifer.

The Holocene sub-recent Alluvial aquifer is distributed and used mainly in the Jordan Valley and neighboring areas. It is built up of sub-recent terrigenous deposits formed along the outlets of the major wadis. These alluvial fans are still under accumulation after large floods and consist of debris from all neighboring lithologies and are deposited according to their transport energy. The transport normally takes place along alternating channels. Thus permeable horizons alternate with impermeable lithologies within the deposits. The total thickness is maximal near the margins of the Rift thinning out towards the centre of the Rift basin. The alluvial aquifer often directly overlies the Pleistocene gravel aquifer and by that is hydraulically interconnected with this aquifer. The two formations that are exposed in the study area are the Samra Formation and the Lisan Formation. The clastic unit in the Samra Formation is the aquifer unit in the study area. The Samra Formation contains also a silt unit. There are lateral facies successions from terrestrial/fluvial, to deltaic/limnic and limnic/brackish lake environments that reflect the Plio-Pleistocene depositional conditions of the Lisan Lake. The Lisan Formation consists of marl, gypsum and silt beds and is generally considered as an aquiclude with no exploitable water. The Lisan Formation extends mainly in the central parts of the graben, where it overlies or interfingers with a silt member. Westwards, the Lisan beds interfinger with a coarse clastic member which consists mainly of gravel, interbedded with clay, sand and marl horizons (see chapter 3.3.1).

The natural recharge by rain is almost negligible. The sources of fresh water sources are from leakage of storm water, especially from Wadi Quilt and overflow and leakage from the Jericho springs. Other sources are irrigation return flow that became an important source in the last three decades, and leakages from the dolomite-limestone aquifer.

Geometry

The model domain is defined using geological and topographical maps and groundwater level information. Based on this information, displayed as ArcGIS themes (see for example Figs. 3.3.3-4 and 5), the model boundaries are:

- The south-western boundary along the main Rift fault, where Upper and the Lower Cenomanian rock-units are adjacent to the Plio-Pleistocene aquifer
- The north-western boundary outlined by a minor fault, where Upper and Lower Cenomanian strata are adjacent to the Plio-Pleistocene aquifer
- The northern boundary outlined by a boundary flow line
- The eastern boundary marked by the Jordan River area
- The south-eastern boundary defined by the Dead Sea and by the Jordan River area

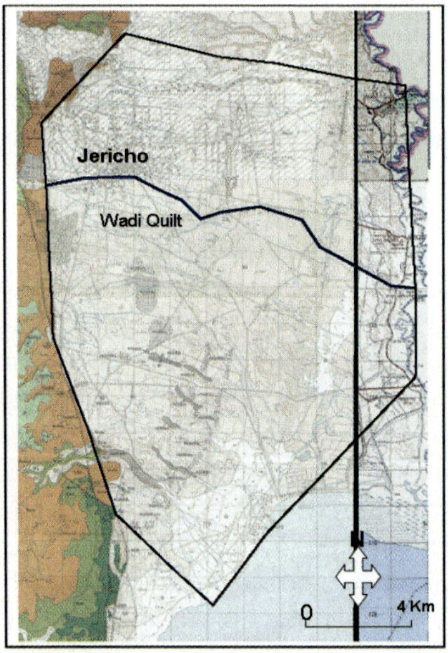

Fig. 3.3.3-4. Model domain geological and topographic maps in ArcGIS.

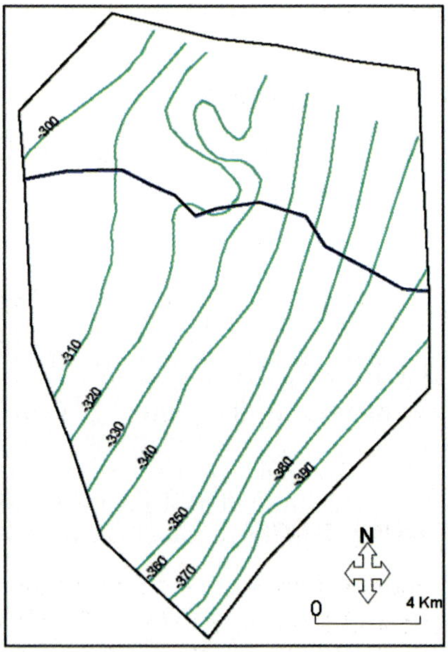

Fig. 3.3.3-5. Model domain and groundwater contours from 1982 in ArcGIS.

Water Balance

Measurements in the area have shown that recharge is mainly due to irrigation and to floods during the rainy season. Irrigated land covers an area of 28 km² as calculated from the satellite image (Fig.3.3.3-6). The water for irrigation is taken from the Sultan, Dyouk and Nueima springs located outside the model area. The volume of spring discharge is presented in Table 3.3.3-1. It is estimated that only 2/3 of the water for irrigation taken from the Dyouk spring is included in the model area. Therefore the estimated contribution to recharge is no more than 2/3 of that spring discharge. The yearly total average spring discharge and hence water used for irrigation is 12 Mm^3/yr. An estimated 10–15% of the water contributes to recharge by canal leakage and irrigation backflow, i.e., 1.2–1.8 Mm^3/yr.

Recharge through floods in the wadis is estimated to be in the order of 2 Mm^3/yr. Recharge due to rain is negligible although additional water enters the domain along its NE-boundary. The irrigated areas and the distribution of wells in the wadis are shown in Fig. 3.3.3-7.

Groundwater is abstracted from ca. 80 wells in the Jericho area. Total abstraction rates for the years 1961/63, 1969/70 and 1968 amount to 2.7, 1.3 and 1.6 Mm^3/yr, respectively. The last value corresponds to a yearly abstraction of about 5 Mm^3.

Hydraulic properties

The values for hydraulic conductivity (summarized in Table 3.3.3-2) are extracted from a map containing well test data. Values of transmissivities of the aquifer range from 25–430 m²/d. The hydraulic conductivities are calculated for each well considering the saturated thicknesses. They range from about 1–9 m/d. Due to the limited information, the data do not allow a detailed resolution of different conductivity zones. The average hydraulic conductivity of 3.4 m/d calculated from these values is not considered as representative for the region. A value of 1 m/d is thought to be more realistic.

Numerical model

The model domain is discretized using triangular prismatic elements (Beinhorn and Kolditz 2003a,b). The total domain is formed by six layers of 525 elements each. The edge length varies from 500 m at the model boundaries to 200 m in the vicinity of the wadi. The allocated conductivity of the uppermost, 20 m thick layer, is of 1 m/d in the west and 0.5 m/d in the east (Fig. 3.3.3-8), corresponding to the distribution of coarse deposits in the west and finer material in the east.

Table 3.3.3-1. Yearly discharge volume of the Jericho springs.

year	Sultan (1000 m³/yr)	Dyouk (1000 m³/yr)	Nueima (1000 m³/yr)
79/80	4753	5460	2403
79/81	5536	5672	2555
79/82	5674	5065	2974
79/83	5910	4955	3136
79/84	5833	4885	3136
79/85	5628	4864	3124
79/86	5465	4673	2920
79/87	5569	5115	2645
79/88	5915	4954	2867
79/89	5873	5051	2939
79/90	5837	5079	3047
79/91	5720	4802	2950
79/92	5856	5219	2959
79/93	5980	4819	3141
79/94	5957	4755	3194
79/95	5965	4768	2972
79/96	5939	4477	3123
79/97	5660	4460	2891
79/98	5870	4430	2960
79/99	5463	3954	2950
79/00	5349	4094	2718
Total mean	5702	4836	2934
Area factor	1	0.66	1
Total mean in area	5702	3192	2934

Table 3.3.3-2. Transmissivities (T), saturated thicknesses (Ds) and hydraulic conductivity (K) values extracted from a pumping test map. (coordinates: Palestine Grid)

Well ID	Well Name	X Coord.	Y Coord.	Elevation m	T m²/day	Ds m	K m/d
13819803	19-13/12	198.98	138.31	-350.00	230	90	2.56
13919503	19-13/50	195.81	139.38	-290.00	160	45	3.56
13919601	19-13/15	196.15	139.50	-292.44	25	30	0.83
14019509	19-14/17	195.92	140.31	-290.00	180	20	9.00
14019603	19-14/71	196.87	140.09	-302.00	120	125	0.96
14219701	19-14/81	197.06	142.31	-302.07	430		

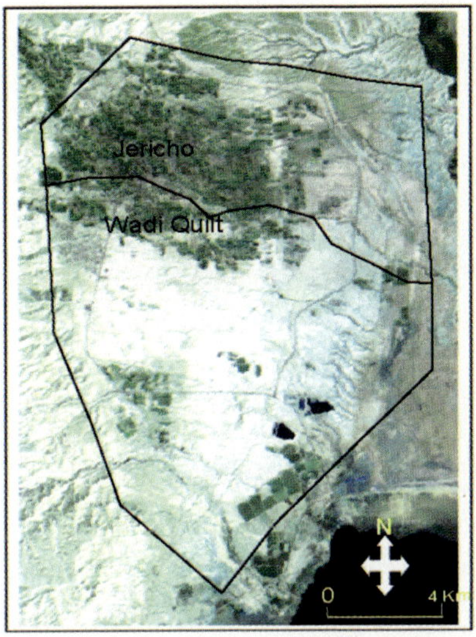

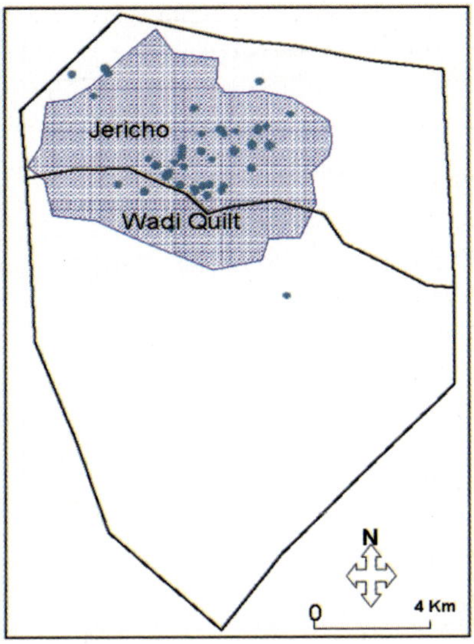

Fig. 3.3.3-6. Satellite image for the determination of irrigated areas in ArcGIS

Fig. 3.3.3-7. Distribution wells and irrigated area in ArcGIS

The second layer with a thickness of 10 m is assigned a value of 0.5 m/d.

The following boundary conditions are assigned to the model domain:

- SE, main Rift fault: no flow boundary condition
- NW, minor fault: constant head boundary condition
- NE, boundary flow line: no flow boundary condition
- E, Jordan river area: constant head boundary condition
- SE, Dead Sea-Jordan river area: constant head boundary condition

The medium abstraction rates in the years 1961/63, 1969/70 and 1968 are assigned to nodes close to active well locations. Along the polyline of the wadi a source term with 2 Mm^3 over the total wadi length is defined. Standard procedures of model calibration cannot be adapted here as reliable measurements of groundwater heads remain elusive. As an alternative it was envisaged to use time series of chloride concentration maps for the model calibration. Concentration contour maps were available from Golani (1972) for the years 1961/63, 1968 and 1969/70. The information available is displayed in Figs. 3.3.3-10 and 11. Using the 1961/63 concentration distribution as initial condition, the model was tested for the time interval of 8 years using boundary conditions and source/sink terms as described above. The results of the concentration distribution after 8 years of pumping are

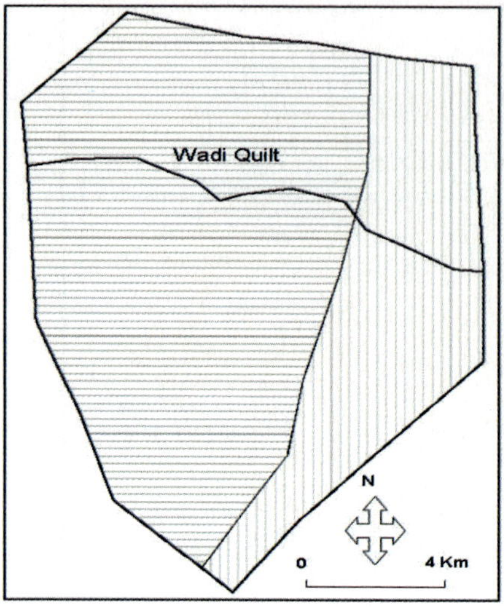

Fig. 3.3.3-8. Material groups in the first layer displayed as ArcGIS themes.

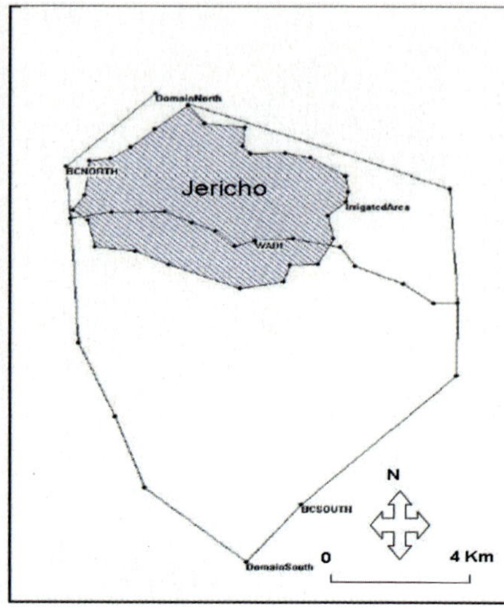

Fig. 3.3.3-9. GeoSys GUI, elements of the GeoLib.

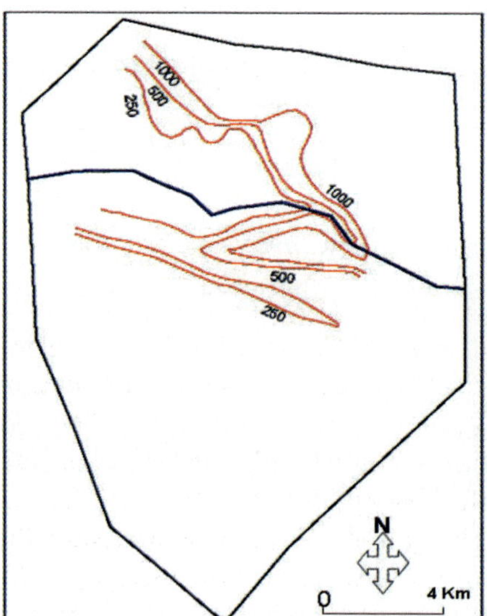

Fig. 3.3.3-10: Concentration contours (1961) in ArcGIS

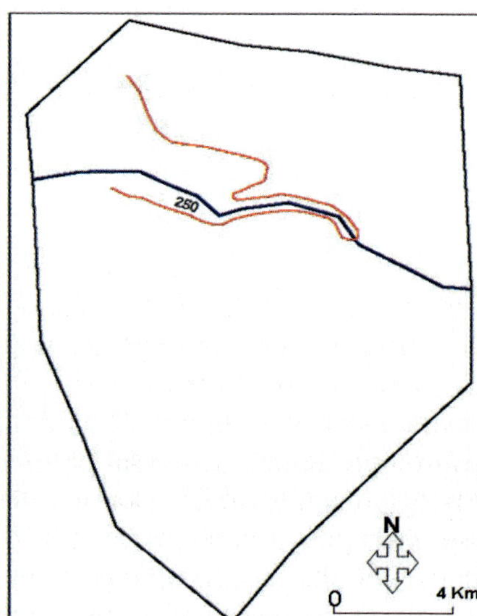

Fig. 3.3.3-11: Concentration contours (1970) in ArcGIS

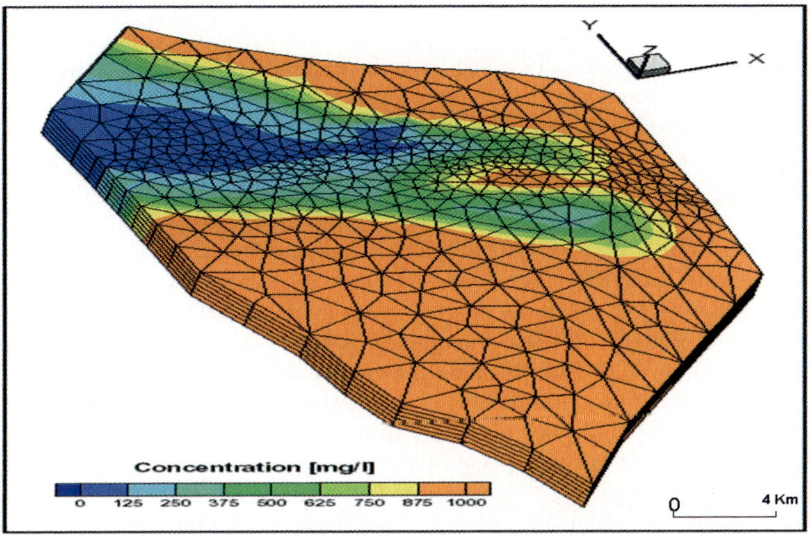

Fig. 3.3.3-12. Resulting concentration distribution after 8 years of pumping.

shown in Fig. 3.3.3-12. The pattern displayed resembles the measured pattern given in Fig. 3.3.3- 10. Therefore, it appears that the numerical model could be used to verify the hydrogeological model to some extent.

3.3.3.3 Summary and conclusions

A hydrogeological model of the area has been developed based on data available from project partners. An ArcGIS project was used to visualize and organize the data.

Extensions of the existing software GeoSys concern the implementation of a new element type, i.e., triangular prismatic elements, a mapping function to relate element thicknesses to a digital elevation model and a function to read in distributed initial conditions. The new model components were necessary to set up the numerical model.

Using time series of chloride concentration contours, the hydrogeological model was verified to some extent. The system is dominated by recharge through flooding in the wadis, irrigation return flow and also the groundwater abstraction taking place in the area. In combination with the hydraulic conductivity of 1 m/d in the west and 0.5 m/d in the east, a groundwater recharge of 2 Mm^3 over the total wadi length produces results comparable to measured data and seems to be in the right order of magnitude.

References

Begin ZB, Erlich A, Nathan I (1974) Lisan lake, the representative of the Pleistocene age in the dead sea (in Hebrew). Geol Surv Isr Bull 63:1–30

Beinhorn M (2005) Non-linear flow processes in subsurface and surface hydrosystems, PhD- thesis, Center for Applied Geosciences, University of Tübingen gmsh (2004) http://www.geuz.org/gmsh/. gOcad (2004) http://www.ensg.inpl-nancy.fr/GOCAD/

Beinhorn M, Kolditz O (2003a) Triangular prismatic elements PART I: Numerical integration. Tech Rep, Nr. 2003-2, Center for Applied Geosciences, Univ Tübingen, http://www.uni-tuebingen.de/zag/geohydrology/literature/, pp 1–11

Beinhorn M, Kolditz O (2003b) Triangular prismatic elements PART II: Analytical integration. Tech Rep Nr. 2003-30, Center for Applied Geosciences, Univ Tübingen, http://www.uni-tuebingen.de/zag/geohydrology/literature/, pp 1–11

Golani U (1972) Groundwater resources in the Jericho region. Tech Rep, Tahal Report HR/72/016 (in hebrew), pp 1–33

Huyakorn PS, Springer EP, Guvanasen V, Wadsworth TD (1986) A three-dimensional finite-element model for simulating water flow in variably saturated porous media. Water Resour Res, 22:1808–20

Huyakorn PS, Thomas SD (1984) Techniques for making finite elements competitive in modeling flow in variably saturated porous media. Water Resour Res, 20:1115–24

Kalbacher T, Beinhorn M, Kolditz O (2003) PrisGen - Mesh generator for prismatic elements. GeoSys-Report 2003-34, Center for Applied Geoscience, University of Tübingen, pp1–17

Kalbacher T, Wang W, McDermott Ch, Kolditz O and Taniguchi T (2005) Development and application of a CAD interface for fractured rock. J Environ Geol 47:1017–1027

Kolditz O, Bauer S (2004) A process-oriented approach to compute multi-field problems in porous media. J Hydroinform 6:225–244

Kolditz O, de Jonge J, Beinhorn M, Xie M, Kalbacher T, Wang W, Bauer S, McDermott C, Chen C, Beyer C, Gronewold J, Kemmler D, Manabe T, Legeida D, Adamidis P (2003) GeoSys - Theory and users manual, release 3.9.11. GeoHydrology/HydroInformatics, Center for Applied Geoscience, University of Tübingen, pp 1–67

Kolditz O, Ratke R, Diersch HJG, Zielke W (1998) Coupled groundwater flow and transport: 1. Verification of variable density flow and transport models. Adv Water Resour 21:27–46

Manabe T, Kalbacher T, Kolditz O, Taniguchi T (2003) TetMesh - Mesh generator for tetrahedral elements. GeoSys-Report 2003-35, University of Okayama and Center for Applied Geosciences, University of Tübingen. http://www.uni-tuebingen.de/zag/geohydrology/literature/

Marie A, Vengosh A (2001) Sources of salinity in ground water from Jericho area, Jordan Valley. Groundwater 39:240–24

3.3.4 Hydrochemistry and isotope hydrogeology in the Jericho area/Palestine

S. Khayat, M. Ghanem, A. Tamimi, M. Haddad, S. Geyer, H. Hötzl,

W. Ali, P. Möller

3.3.4.1 Introduction

The eastern basin of the West Bank with Jericho in the centre comprises the wadis Malih-Buquei'a, Faria, Auja-Fasail, the area of Ramallah-Jerusalem and the eastern desert sub-basin (Fig. 3.3.4-1). The area is part of the Jordan Rift Valley (Chap. 2.1). Its geology and structure of the area is rather complex (Rofe and Raffety 1963). All sub-basins drain eastward and southeastward to the Jordan Valley. Its wells and springs are used for domestic water supply and irrigation. The continuously increasing demand for water leads to rapid degradation of the quality of freshwater resources by salinization and contamination processes (Vengosh and Rosenthal 1994; Salameh 2002). The lack of sufficient water, combined with rapid deterioration of water quality, presents a serious challenge to the people in the region. Therefore, it is crucial to understand the origin and mechanisms of all contamination processes. Hydrochemical monitoring of aquifers is of great importance for public services and farming.

3.3.3.2 Hydrogeology of the eastern basin

The study area of the Jericho plain is outlined in Fig. 3.3.4-2. It is covered dominantly by Pleistocene Samra (or Samara) and Lisan Formations. In general, the Lisan Formation is the major source of soil and of water salinity in the Jordan Valley (Salameh 2002). Details about the sources of salinity are discussed in Chap. 2.3 and 3.1.4. The climate is semi-arid with low monthly precipitation rates of about 140 mm (ARIJ 1995). The main recharge areas are the Jerusalem Mountains on the western side (Fig. 3.3.4-1). The Cenomanian-Turonian aquifer system (Fig. 3.3.4-3) in the eastern basin is subdivided into the shallow and relatively thin Upper Cenomanian-Turonian- and the deep Lower Cenomanian sub-aquifer (Chap. 2.2). Additional sub-aquifers are hosted in Quaternary, Neogene and Eocene Formations. These sub-aquifers are described as follows:

Pleistocene sub-aquifer consists of unconsolidated sand, gravel, cobbles, and boulders of different sizes separated by impermeable layers of

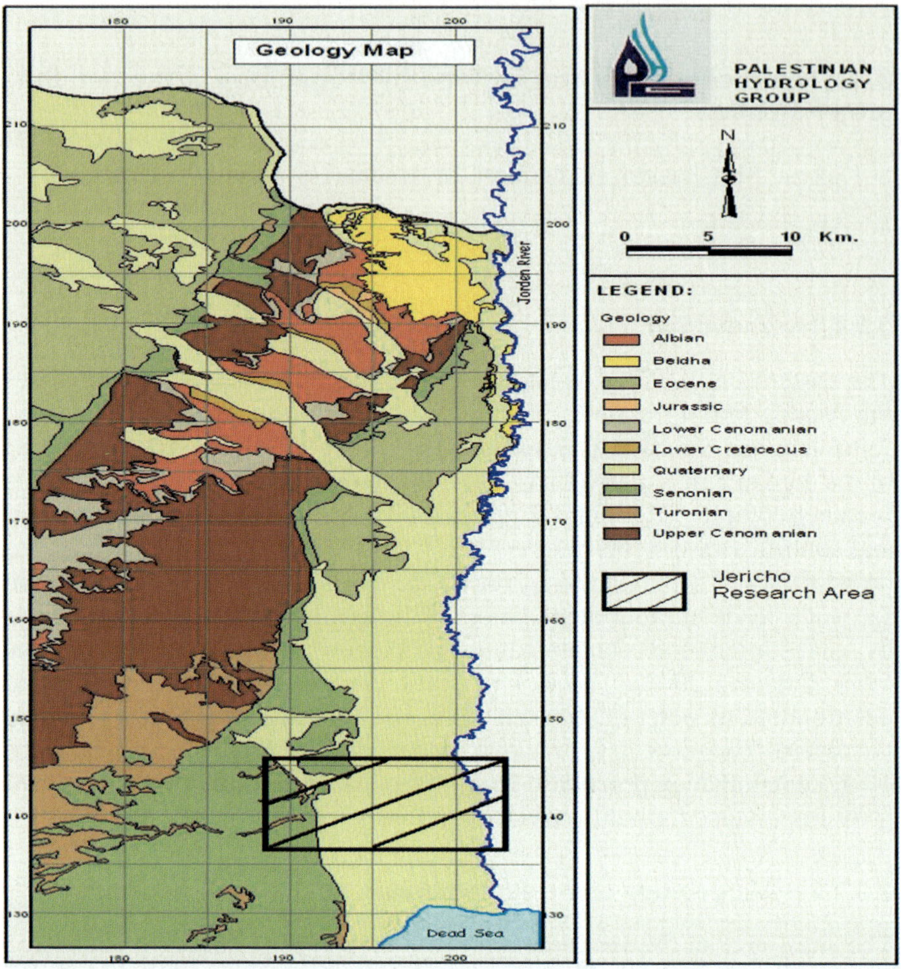

Fig. 3.3.4-1. Geological map of the study area with the mountain aquifer in the western part and Quaternary/Pleistocene sediments of the Dead Sea graben to the east

saline lacustrine marl. The alluvial fans consist of gravel with sand and clay and occasionally sandy layers. The beds of the Lisan Formation (Pleistocene subaquifer) as well as alluvial and gravel fans (Holocene) extend along the Jordan Valley. The groundwater in these alluvial fans differs quantitatively and qualitatively depending on its location within the fan. Freshwater occurs around the apex of the fan, whereas saline water occurs at the fringes. Very steep, deep faults in the Jordan Rift Valley may cause deep circulation of groundwater bringing it into contact with salty formations. Brackish springs occur near the Jordan River.

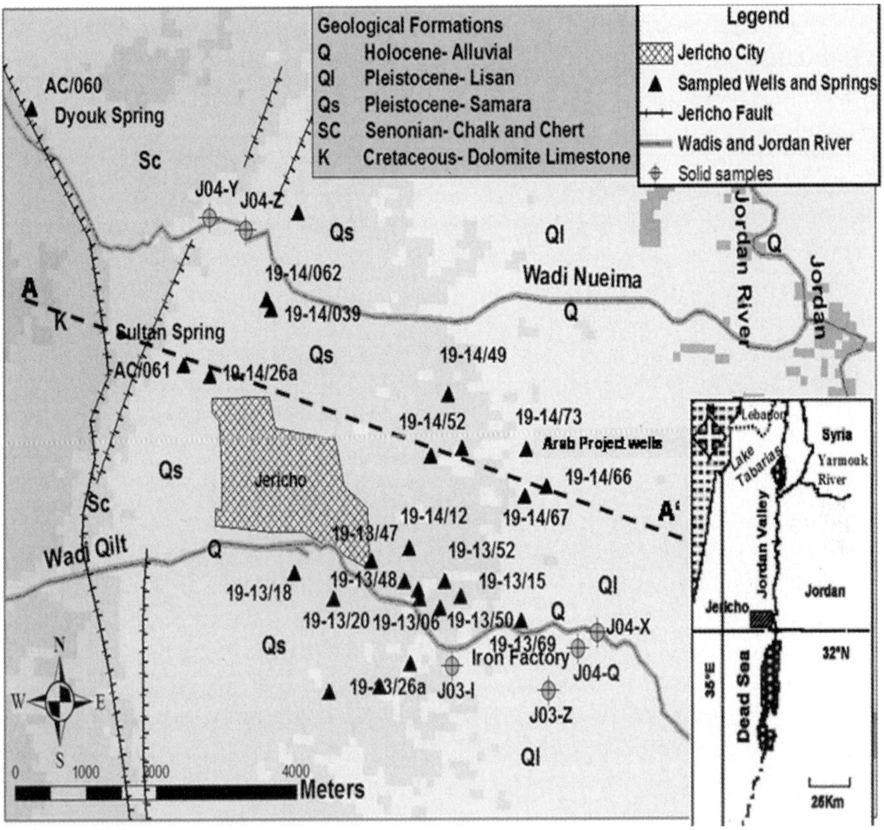

Fig. 3.3.4-2. Study area of Jericho. Geological formations are indicated and locations of sampled wells and springs are shown. Note: Cross section A-A' is given in detail in Fig. 3.3.4-3

The *Neogene sub-aquifer* consists of well cemented conglomerates and contains a small amount of freshwater. It is composed of conglomerate lenses, and marls of the Beida Formation (See Fig. 2.1-7), and Lower Tertiary clays.

The *Eocene sub-aquifer* consists mainly of nummulitic limestone with chalks, chert bands and marl. The limestone is thin-bedded with chalk, chert and marl intercalation.

The *Cenomanian aquifer* (Ajlun Group; Chap. 2.1) is confined. The well heads indicate that there is no hydraulic separation between the Lower and Upper Cenomanian. The Upper Cenomanian aquifer composed of the upper parts of Ajlun series that include the Bethlehem, Jerusalem and Hebron Formations (see Fig. 2.1-7), consists mainly of interbedded dolomites and chalky limestones. The Lower Cenomanian sub-aquifer consists mainly of dolomitic limestone and thick to thin bedded marly limestones.

It is classified as an excellent aquifer due to its thickness of 800–850 m (Guttman 1995).

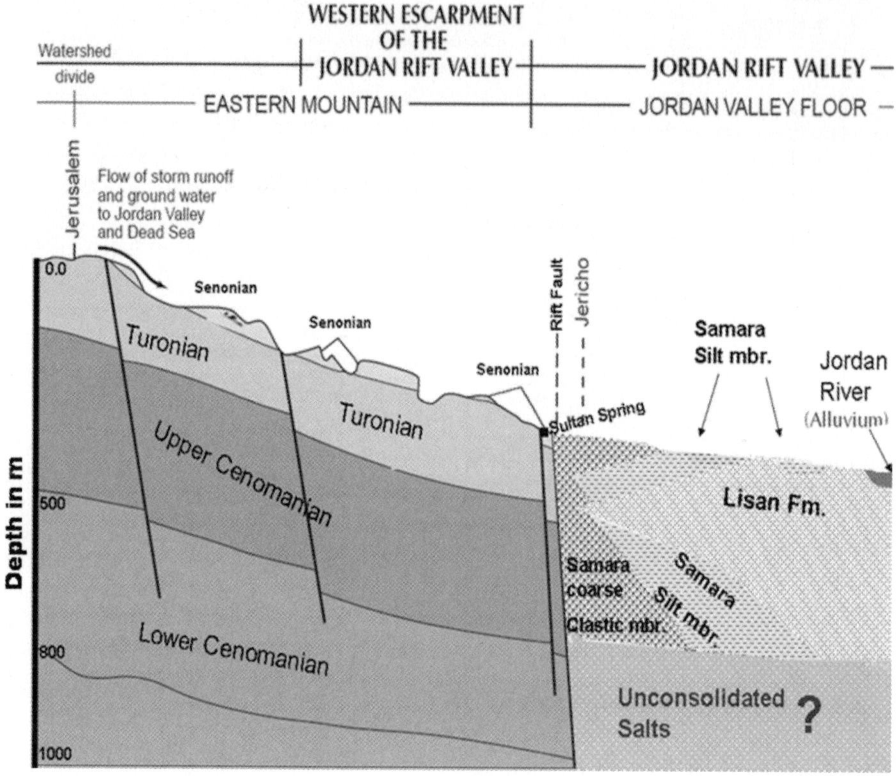

Fig. 3.3.4-3. Schematic geological cross section of the study area

The area of Jericho comprises three major spring systems: (i) Wadi Qilt spring system including Ein Qilt, Ein Fara and Ein Fawwar, (ii) Wadi Makkuk (Nueima) spring system including Ein Dyouk, Ein Nueima and Shosha, and (iii) Ein Sultan spring system. All springs are used for domestic water supply. Ein Sultan is also used for irrigation.

The sub-basin is mainly fed by inflows of excess water from surface runoff in the wadis and from the adjacent aquifers such as the Lower Cretaceous Kurnub sandstone aquifer and the Albian to Turonian limestone and dolomite aquifer, also known as the lower and upper aquifer (ARIJ 1997). The annual precipitation in the main recharge zone of this aquifer extends to the Jerusalem and Ramallah Mountains and is of about 540 mm (Marie and Vengosh 2001).

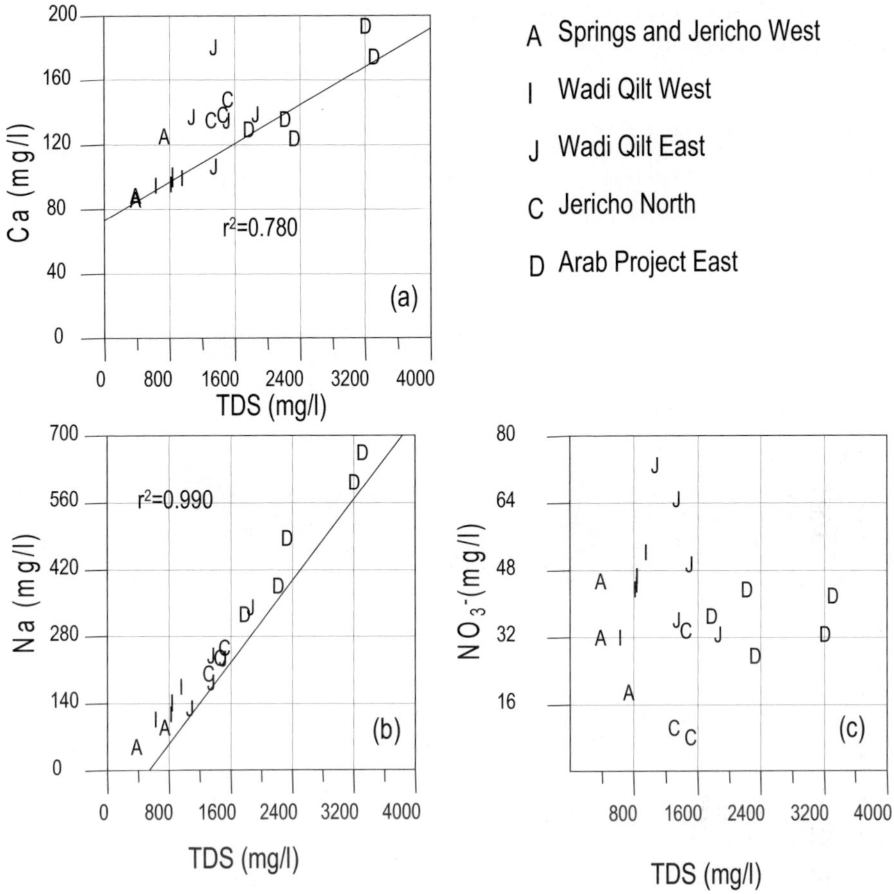

Fig. 3.3.4-4. Major anions and cations vs. TDS for the well- and spring waters of Jericho area (October 2004). Salinity increases from west to east.

3.3.4.3 Results and discussion

Hydrochemistry and tritium

The concentration of total dissolved solids (TDS) in water is higher during the wet than in the dry season. This may be caused by flushing water from an earlier dry season. In both seasons all the major anion and cation concentrations show an increasing trend from west to east (Fig. 3.3.4-4). Mg, K, Na, and Br are strongly correlated with the TDS, while Ca, Sr, B, SO_4^{2-}, HCO_3^- are less correlated. Nitrate shows no correlation at all which indicates a source different from all the other elements. In general, the data reflect distinctive end members of freshwater in the west and saline water towards the east (Fig. 3.3.4-5).

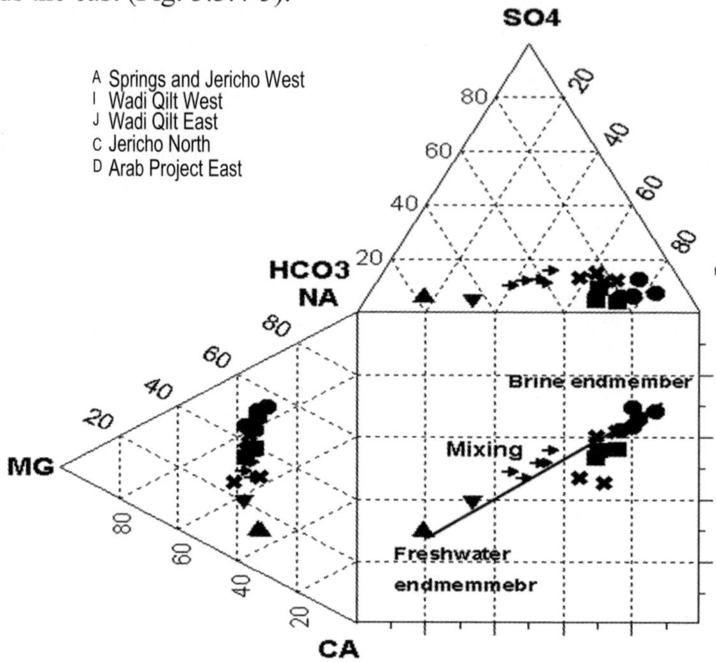

Fig. 3.3.4-5. Durov diagram

Molar ratios of Mg/Cl (~1.2 Fig. 3.3.4-6a), SO_4^{2-}/Cl (~0.18 Fig. 3.3.4-6e) and Sr/Ca ($7.2\text{-}7.4*10^{-4}$) decrease from west to east. The groundwater is over-saturated with respect to dolomite and calcite, while it is under-saturated with respect to gypsum and aragonite (Fig. 3.3.4-7a, b, c, d).

The groundwater from wells in the *western Wadi Qilt* shows intermediate values between the fresh and saline end members (Fig. 3.3.4-5). The chloride concentration is much lower than in the eastern wells, and is sometimes close to those of freshwater emerging from the springs. Tritium units range from 3.3–3.9 T.U. The ionic molar ratios of these well waters plot between those of the fresh and saline end members (Fig. 3.3.4-6). The saturation indices of dolomite and calcite are lower in saline than in spring waters. The saturation indices for anhydrite and gypsum are higher in spring waters than some of the saline ones. The concentrations of carbonate, boron and nitrate in these well-waters are high.

The hydrochemistry of the wells in the *eastern Wadi Qilt* is similar to that in the western part, but with enhanced TDS. The area is characterized by overlapping of Lisan and Samra (or Samara) Formations. Tritium units in the groundwater from these wells are lower than in the western wells

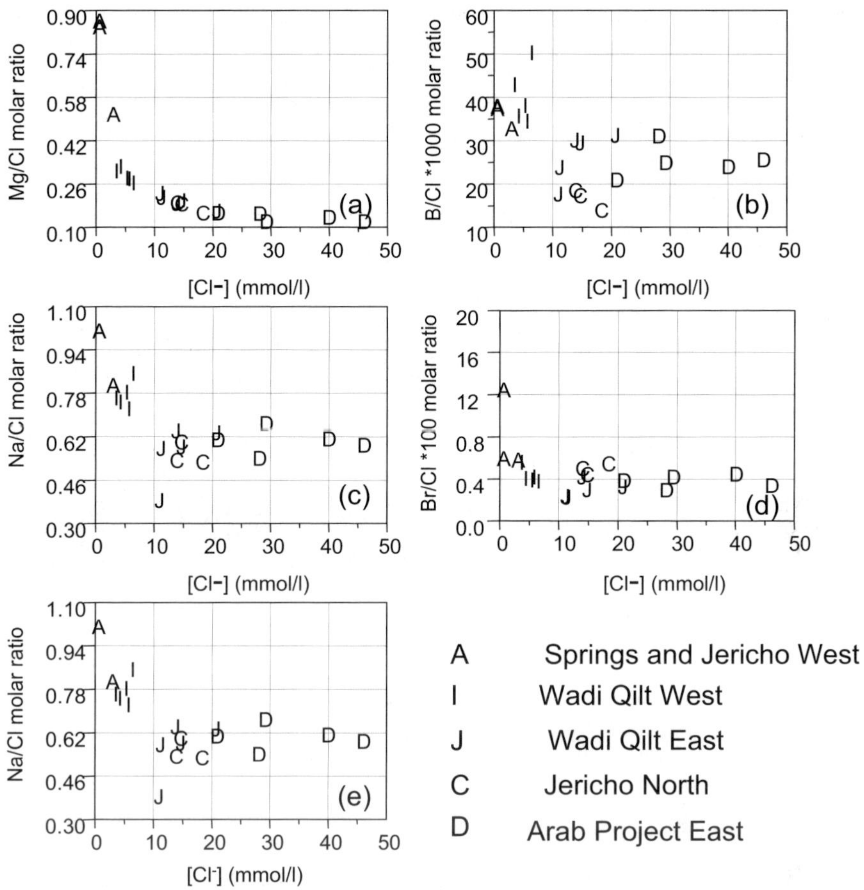

Fig. 3.3.4-6. Ionic molar ratios of well- and spring waters from the Jericho area. The wells in the east show constant ratios.

with values between 2–3.5 T.U. The waters show saturation indices for dolomite and calcite similar to the wells west of the Wadi Qilt but much higher ones for gypsum and anhydrite (Fig. 3.3.4-7). Even though the chloride concentrations of these wells are higher (500–900 mg/l), the various molar ionic ratios in Fig. 3.3.4-6 are much lower.

The well waters in the *Wadi Nueima North* area show higher salinities than those from wells in the west and the western Wadi Qilt. This area is dominated by Samra (or Samara) Formation with high hydraulic conductivity. The Mg-Na-Cl-HCO_3 type of water is composed similarly to that from the wells in the west. These waters are high in Mg concentration which originates from leaching of dolomitic limestone formation adjacent to the west of the area by wadi runoff. All groundwater from these wells

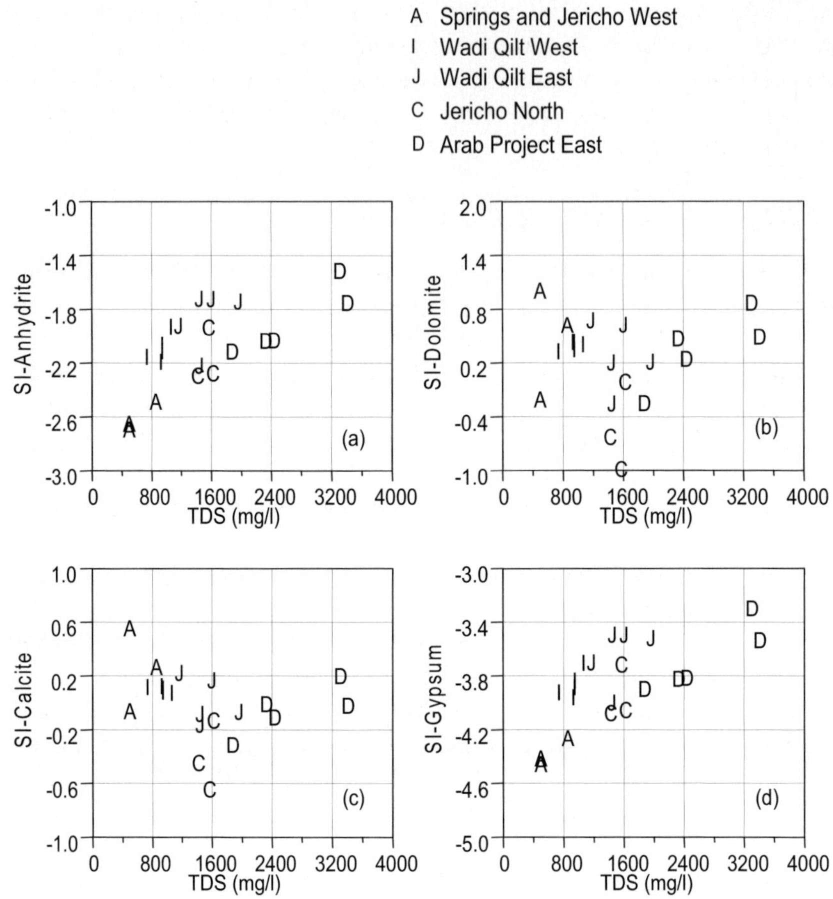

Fig. 3.3.4-7. Estimated saturation indices for dolomite, calcite, gypsum and anhydrite vs. TDS for groundwater samples from the Jericho area. Waters from the eastern wells mostly show SI values increasing with total dissolved solids (TDS). Saturation is reached for SI=0.

are under-saturated with regard to carbonate and sulfate minerals. The various molar ionic ratios (Fig. 3.3.4-6) are close to those in the eastern wells. Tritium units in these waters are around 1.5 TU.

East Arab Project Wells (brackish end-member)

The groundwater from Arab Project wells in the east of Jericho shows the highest salinity in the study area, with Cl^- concentration and total dissolved solids (TDS) in well 19-14/067 exceeding 1,800 mg/l and 3,664 mg/l, respectively. This highly saline water is of the Na-Mg-Cl and Mg-Ca-Cl type. The SO_4^{2-} concentration is 320 mg/l. The groundwater from these wells is over-saturated with respect to dolomite and saturated with respect to calcite, while gypsum and anhydrite are under-saturated (Fig. 3.3.4-6). Carbonate concentration of 440 mg/l is relatively high. The groundwater from these wells shows low T.U. below detection limit.

Overall hydrochemical conditions

The concentrations of major elements in spring- and well waters show a gradual increase from the west towards the east along the flow path of water. The salinity varies from Ca-Mg-CO_3 type of freshwater in the west to highly saline Na-Mg-Cl and Mg-Ca-Cl water in the east (electrical conductivity >5000 µS/cm) (Fig. 3.3.4-6). Some wells in the western Wadi Qilt show less Cl^- contents but high boron and nitrate values. Wells in the north are less affected by anthropogenic activity.

The linear correlation in the Durov diagram (Fig. 3.3.4-5) suggests mixing that increases from the west towards the east and north of the study area. The water type varies along the mixing line. It is of Ca-Mg-HCO_3^- water in the spring water and Mg-Na-Ca-Cl-HCO_3^- water for the pumped wells in the west and the centre of the area, but the water becomes more saline towards the east, where Mg-Ca-Cl and Na-Mg-Cl types of water dominate. The variability in Br^- and nitrate concentrations depends on the location of the well and on the agricultural/domestic activities in the area. The groundwater from wells further to the east (19-14/67) with ^3H below detection limit must have been recharged prior to 1952 and can be considered as sub-modern water. Most of the other groundwater samples show T.U. between 0.8 and 4.5 which reflect modern recharge (Clark and Fritz 1997a). These values could, however, also reflect mixing of old and modern water. The groundwater in the western springs and in wells of the alluvial aquifer comprises mixtures of recharge from the past, i.e. less than 5 years. It seems that the tritium level declines with increased distance from the recharge area along the flow path of water through the wadis. This emphasizes that the mechanism of recharge is not related to direct infiltration of precipitation but of infiltrated runoff along the wadis that drained the Jerusalem-Ramallah Mountains.

Salinization

Salinization of groundwater may be caused by many factors. There is good evidence for evolution of *brackish water* in the east of the study area. It is believed that the over-exploitation of the wells results in upwelling of saline water from underlying aquifer(s). However, the source and location of these saline waters cannot be identified. The analysis of samples of waters from eastern wells shows highly saline Ca-Cl and Na-Cl waters. The Na/(Na+Cl) ratio, which is less than 0.5 in combination with TDS higher than 500 mg/l, suggests reverse softening by mixing of brackish water with freshwater from the west. Moreover, the continuous trend Na/Cl and B/Cl with chlorinity from the west towards the centre of the study area, changes to rather constant ratios for the eastern group (Fig. 3.3.4-6). The low tritium concentrations also emphasize the presence of mixing waters from different sources. T.U. below 1 in the eastern areas may reflect mixing of recent infiltrated water from runoff from the wadis with deep but old saline water. This evidence was proved by studies of the rare earth elements concentration (Chap. 3.1.2).

Another aspect is *dissolution of evaporites and weathering of rocks*. From east to west in the study area, the molar ratios increase (Fig. 3.3.4-6): Na/Cl (0.4 to 0.7), Mg/Cl (0.18 to 0.5), and SO_4^{2-}/Cl^- (0,015 to 0.15), with the exception freshwater emerging from springs. These ratios are even lower than those in the Lisan leachate (Na/Cl=0.54, Mg/Cl=0.23, SO_4^{2-}/Cl^- =0.05) (Salameh 2002) and much lower than in seawater (Na/Cl=0.86, SO_4^{2-}/Cl^-=0.135). The Br/Cl ratio (0.006 in the east and 0.014 in the western parts) is higher than that of seawater ratio (0.0016). Groundwater from many of the studied wells show Br/Cl molar ratios that may be affected by dissolution of halite from the evaporitic Lisan Formation (with Br/Cl molar ratios higher than in brines or seawater) or by fuming of methyl bromide in agriculture.

The Mg/(Ca+Mg) ratios, which in most cases exceeds 0.5, indicate possible Mg-rich leachates by wadi runoff from the dolostone formation in the north-west and south-west of the area. In the eastern well-waters Ca/(Ca+SO_4) ratios are between 0.75 and 0.95 which are due to assumed Ca-Cl brines and to leaching of (i) gypsum or of carbonate from the Lisan Formation to the east and north and/or (ii) of salts from the Samra (or Samara) Formation and of the dolomite/limestone formations further to the west. Fig. 3.3.4-7 shows that the groundwater in the west is over-saturated with respect to dolomite and calcite and becomes less saturated further towards the east of Wadi Qilt and north of Wadi Nueima. This trend changes in the wells in the east, where an additional input of Mg-rich salts raises the saturation indices for dolomite and calcite. Saturation indices increase until they reach the brackish water end-member in eastern wells of the Arab Project. Relatively high SO_4^{2-} concentrations in the east, in parallel with low saturation index of gypsum, indicate gypsum dissolution. Sr/Ca molar ratios between 0.0025 and 0.0155 emphasize recrystallization and probably dolomitization of biogenic carbonates (Starinsky 1997).

The greatest *anthropogenic effect* on groundwater quality in the Jericho area is agriculture in which methyl bromide gas is used as a fumigant and which involves input of pesticides, nitrate from nitrogen fertilizers and animal manure. Fertilizers which contain K, Mg and Ca reach the aquifer by irrigation return flow. The wells in the centre of the study area and west of Jericho are affected by intensive agriculture. This is more evident in the east, even though the high Cl^- concentration is due to the presence of natural evaporites which make this indicator less unambiguous.

Fig. 3.3.4-8 displays three end members. The end member Iron Factory is located in a non-agricultural area but $NaNO_3$ is used for hardening of the steal. These nitrate-rich solutions, together with human sewage, pass septic tanks. Additionally, it is expected that the groundwater in this area also receives the nitrate flushed out from the recharge stream flowing through the whole area. The water encountered in the Arab Projects wells is the other end member. These wells are located in an extensive agricultural area. There are two different trend lines for the two end members. Other groundwater samples are mixing products of the two end members. Nitrate values fluctuate as the result of local human activities. The NO_3^- values increase significantly in the wells located in agricultural and animal farm areas or along the outflow of septic tanks (well Sabiru Rantizi 19-14/06). This well is next to an animal farm and in an agricultural area which necessitates usage of pesticides. Similarly high values are shown by the Iron factory well which exploits the alluvial fans of the Wadi Qilt south-east of Jericho, which are interfingering the sediments of the Lisan Formation. In this area there is no agricultural activity whatsoever. The well is clearly replenished by surface water flowing in the wadi. The main reason for high levels of nitrate is the wide use of septic tanks. Unsuitable location of septic tanks in this area proves to be hazardous, especially when they are located in marly and gypsiferous layers. The Lisan Formation aquiclude contains highly consolidated sands and evaporites with many solution conduits through which pollutants can easily pass and reach the alluvial aquifer. The pesticides also yield nitrogenous compounds from which an additional amount of nitrate can infiltrate into the groundwater. However at present, the pollution of groundwater by decomposition of pesticides yielding nitrate is minor. This may be due to long persistence of pesticides in soils without decomposition by oxidation or by hydrolysis.

The continuous input of K to the groundwater from eastern wells may reflect an agricultural source. K is part of synthetic fertilizers. Another source of K may be ascending brines (Chap. 2.3, 3.1.1 and 3.1.4).

Hydrogen and oxygen isotopes

In the main recharge area precipitation occurs over the Palestinian eastern mountain-chain of Jerusalem-Ramallah. It is represented by a local meteoric water line (LMWL: δD=8.1‰; $\delta^{18}O$= + 19.5‰; Fig. 3.3.4-9) with a

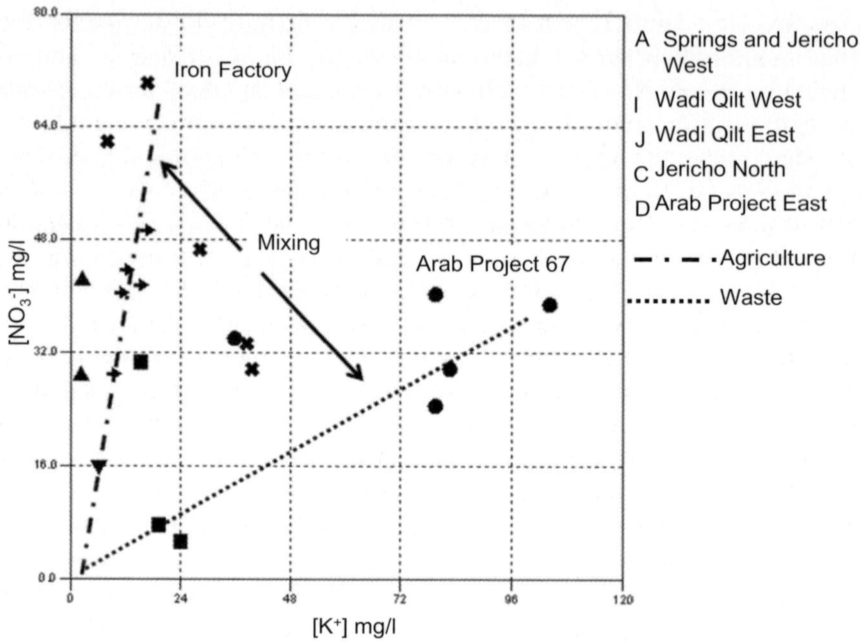

Fig. 3.3.4-8. Potassium versus nitrate in groundwater samples from the Jericho area: the Iron Factory end member is regarded as indicator for sewage; the Arab Project 67 end member represents return flow from agriculture. Most of the wells are mainly affected by sewage.

slope that runs nearly parallel to the Mediterranean Meteoric Water Line ($\delta D = 8\ \delta^{18}O + 22$; Gat and Dansgaard 1972). $\delta^{18}O$ shows no significant variation between dry and wet seasons, whereas δD is enriched in wet seasons. In dry seasons, δD and $\delta^{18}O$ in spring water plot on the local meteoric water line (LMWL), indicating no further evaporation of the recharge water and confirms direct infiltration without any isotopic alteration. The isotope data of waters from the Pleistocene sub-aquifer in the Jericho Plain plot, with considerable scatter, on a mixing line defined by the brackish water of the Arab Project Well 67 which is taken as an end member. Another source is the water from the Samed well located to the west and represents the freshwater end member. The seasonal variation of δD could be explained either by variation within rain or by fractionation processes which only affect deuterium. The first option does not seem to apply because there is no seasonal rain and all the precipitation events occur in winter under similar conditions. The fractionation process was defined by Coplen and Hanshaw (1973) as 'ultrafiltration' (Hoefs 2004) showed that the fractionation of hydrogen isotopes is affected by absorption of H_2O on mineral surfaces due to the tendency of clays and shales to act as semi-permeable membranes. The fractionation of hydrogen isotopes occurs during this process in such a way that the residual water is enriched in deuterium due to the preferential adsorption of lighter hydrogen isotopes. During the continuous recharge in wet season the increasing hydraulic pressure flushes out the previously adsorbed water. Gat (1998) suggested that the isotope transfer function (ITF) is the main cause for modifying the isotopic composition in arid environments and wet lands with stream run-off. The behavior of isotopes shows that rainfall at the top of the mountains and along the wadis, is partitioned into singular fluxes of surface run-off. The water infiltrates through soil and sinkholes at different distances from the main recharge area. The isotopic signature is shifted by evaporation during runoff and selective rainfalls. δ^2H and $\delta^{18}O$ for spring waters in the western area indicate a different recharge and infiltration mechanisms. The springs are outlets of the regional mountain aquifer in which precipitation infiltrates the Upper Cretaceous aquifer.

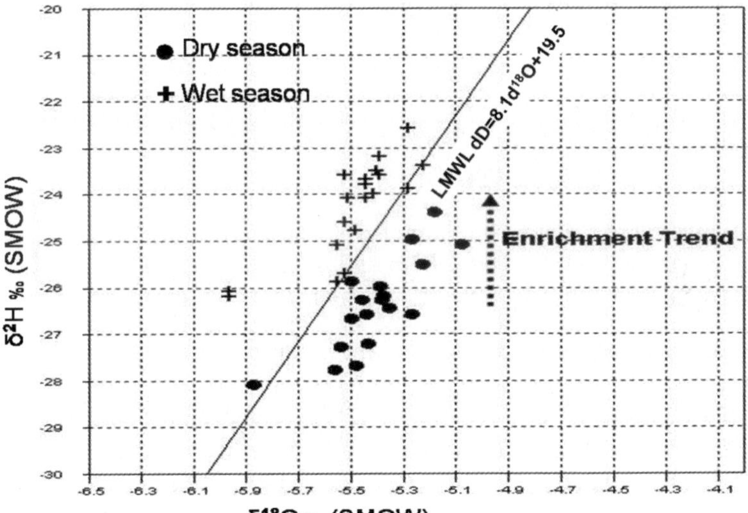

Fig. 3.3.4-9. Seasonal variation of groundwater isotopic signatures around LMWL, the shifted δ^2H values located as a parallel line on the left of LMWL

The wells of the shallow Pleistocene aquifer are mainly fed by infiltrated surface water runoff that drained through wadis. Normally the runoff water of the wadis flow sub-aerially over long distances and infiltrate into the shallow Samra (or Samara) aquifer or into the alluvial fans in the lower plain of the study area. The deviation of the water from the shallow wells in the lower and warmer area of Jericho plain from the local meteoric line

(Fig. 3.3.4-10) is caused mainly by the climate of the area which is characterized by low humidity and high evaporation rates. Thus, the isotopes undergo fractionation during evaporation and are therefore heavier.

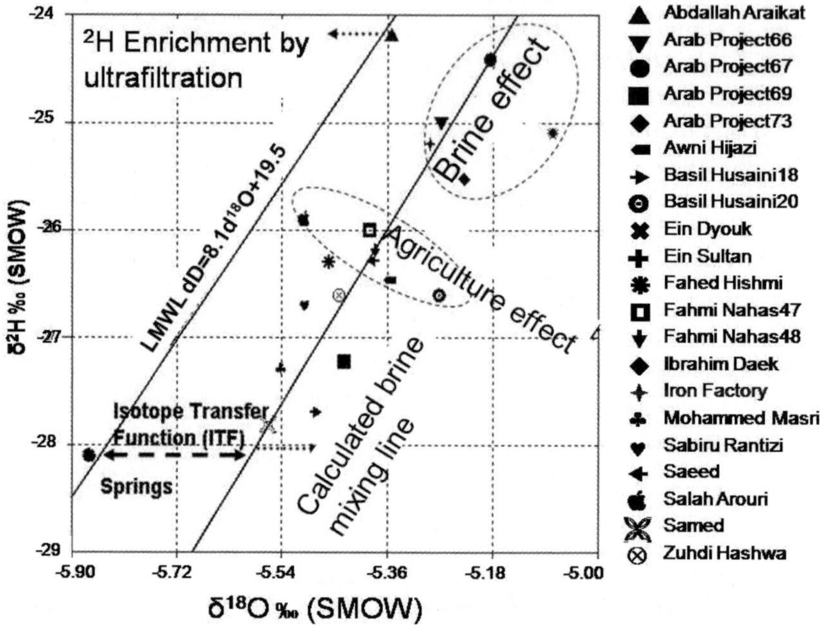

Fig. 3.3.4-10. Isotopic signature of water from wells and springs during the dry season. The signature for the water from springs plot along LMWL, while those from well waters are shifted towards the calculated mixing line of freshwater from the west and brackish water from the east. The isotope shift of this groundwater can be explained by the isotope transfer function (ITF) according to Gat (1998)

The relationship between the molar ratios of Sr/Ca and δ^{18}O reveals the distribution of the groundwater between three end members (Fig. 3.3.4-11). δ^{18}O values are mainly affected by the evolution of brackish water towards the east. The increase in δ^{18}O is associated with increase of Sr/Ca molar ratios, which reach those of brine Sr/Ca ratio of ~$1*10^{-2}$. The freshwater end member of 'Samed' well shows a Sr/Ca ratio of $2.6*10^{-3}$ which is close to the freshwater ratio of $2\text{-}4*10^{-3}$ (Siemann and Schramm 2002).

The anthropogenic effect on the groundwater is shown in Fig. 3.3.4-12. The lower trend is the calculated mixing line of brackish and freshwater. The upper trend reflects the anthropogenic effect. The low Cl^- concentration of these waters evidences of absence of major amounts of brackish water. Thus, the small anthropogenic effects are clearly visible. These wells also show high concentrations of boron, nitrate and bromide. All of these indicators suggest input of isotopically heavy water from sewage and irrigation return flow.

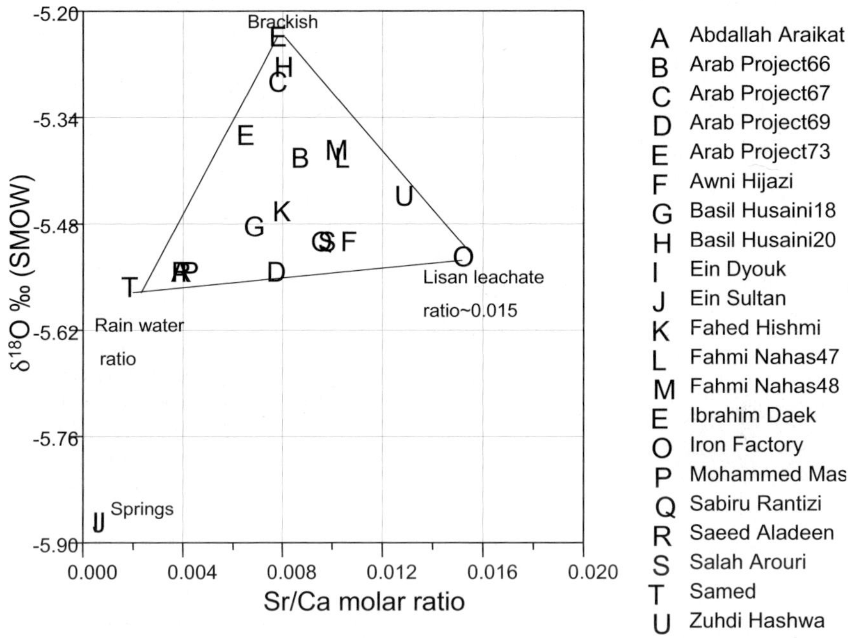

Fig. 3.3.4-11. Sr/Ca molar ratio vs. $\delta^{18}O_{water}$ suggest three inputs to the groundwater, the brackish water are the main significant input that affects the $\delta^{18}O$ signature. Most waters from wells undergo mixing between the three end members.

Nitrogen isotopes

Nitrate concentrations vary depending on the well location and surrounding human activities, ranging between 3–100 mg/l in wet seasons. Nitrate concentrations are more or less the same in dry and wet seasons (Fig. 3.3.4-13). The values of $\delta^{15}N$(nitrate) range from +3.0 to +15.8‰. $\delta^{18}O$(nitrate) and were mostly between +3.3 and +8.5‰, with two of the water samples exceeding this range. The two anomalous samples show very low nitrate concentrations. Nitrate in the environment has various sources such as atmospheric deposition, denitrification of organics in soils, fertilizer, sewage and manure. Nitrate from each of these sources is characterized by distinct isotopic values. Typical ranges of $\delta^{15}N$(nitrate) are -4 to +4‰ in chemical fertilizers; +7 to more than +30‰ in human and animal waste; -10 to +4‰ in soils. $\delta^{18}O$(nitrate) values for atmospheric deposition is > +25 ‰, while it ranges in chemical fertilizers from +18 to +22‰, and in human and animal wastes and in soil nitrate from -10 to +10‰ (Fritz and Clark 1997b). Hence, the isotopic composition of nitrate constitutes a useful tracer for determining nitrate sources provided that alteration of isotopic ratios by biogeochemical reactions is absent. Excepting two wells in the north, most of the isotopic signatures for the sampled wells and springs suggest sewage or manure as a main source for the increased nitrate concentrations, with $\delta^{15}N$ values > +7‰, and $\delta^{18}O$(nitrate) between +3‰ and +6‰.

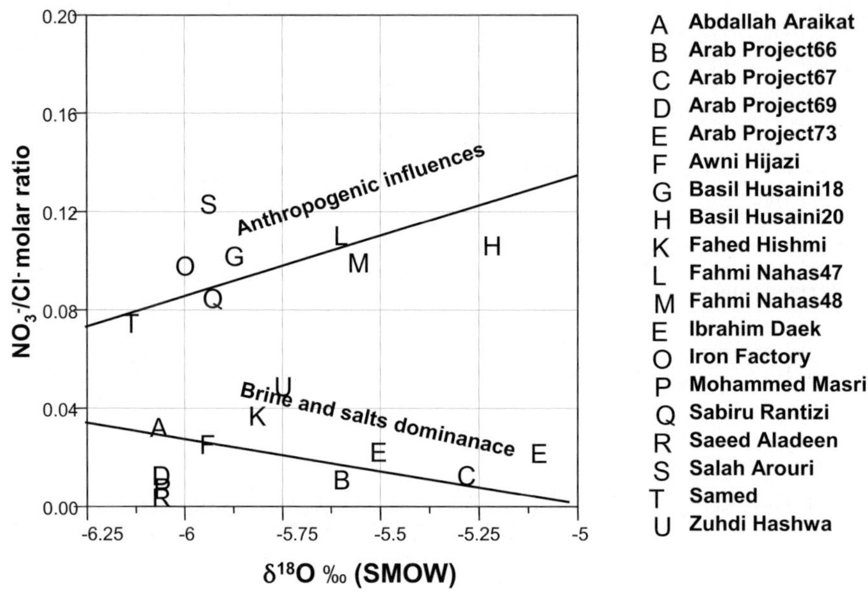

Fig. 3.3.4-12. δ18O vs. NO_3^-/Cl^- molar ratio shows two distinct trends based on either anthropogenic pollutants or mixing with saline waters

Fig. 3.3.4-14 shows increasing in $\delta^{15}N$(nitrate) and $\delta^{18}O$(nitrate) with decreasing [NO_3^-]. These combined trends indicate denitrification in the aquifer, where the heavy isotopes ^{15}N and ^{18}O are preferentially retained in the remaining nitrate (Böttcher et al. 1990). Sewage is considered as a direct source of nitrate because it contains high amounts of urea and other organic and inorganic nitrogenous compounds. Manure in the presence of aerobic condition in shallow aquifers is another source of nitrate. The presence of many animal farms especially in the eastern Wadi Qilt and Ein Dyouk area suggest that animal manure is used as natural fertilizer. The wells in the north of the study area show low nitrate concentrations, but enhanced $\delta^{18}O$(nitrate) values (Fig. 3.3.4-14b). This might indicate either a different nitrate source or represent the residual nitrate from bacterial denitrification of nitrate in sewage or manure, which gives rise of $\delta^{18}O$ values above +12‰.

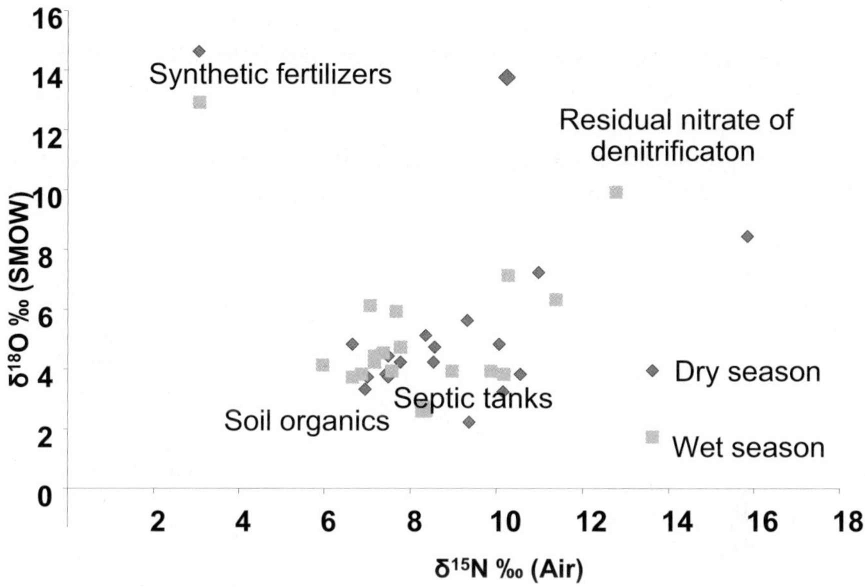

Fig. 3.3.4-13. δ^{15}N(nitrate) vs. δ^{18}O(nitrate) in dry and wet seasons.

Sulfur isotopes

The eastern wells show sulfate and chloride concentrations that vary between 100–320 mg/l and 800–1,800 mg/l, respectively, whereas the wells in the west range from 20–90 mg/l and 100–750 mg/l, respectively. The δ^{34}S values of sulfate eluted from the Lisan samples J04-X, J03-I, J03-Z and J04-Q range from 9.8–16.6‰, while δ^{18}O varies around 16‰. In contrast, the sulfate for the two samples J04-Y and J04-Z originating from the Samra (or Samara) Formation are more depleted with δ^{34}S signatures of 6.9 and 9.1 ‰ and 1.4 and 9.1 for δ^{18}O, respectively. Most of the samples including springs and wells range between -2.2 to +12.5‰ in δ^{34}S and +3.3 to +17.3 ‰ in δ^{18}O which is close to the value of eluted sulfate from the Lisan and Samra (or Samara) Formation (Fig.3.3.4-15). However, the wells lying further to the east and southeast, particularly the Arab Project wells, show a high depletion in δ^{34}S to -10‰ which is paralleled by high sulfate concentrations.

High sulfate concentration in the groundwater from the Pleistocene aquifer is mainly due to leaching of Lisan sediments (Marie and Vengosh 2001). The in-situ water/rock interaction is considered one of the main contributors to salinization. δ^{34}S(sulfate) and δ^{18}O(sulfate) of most of the water samples are similar to those of the solid samples which reflects a

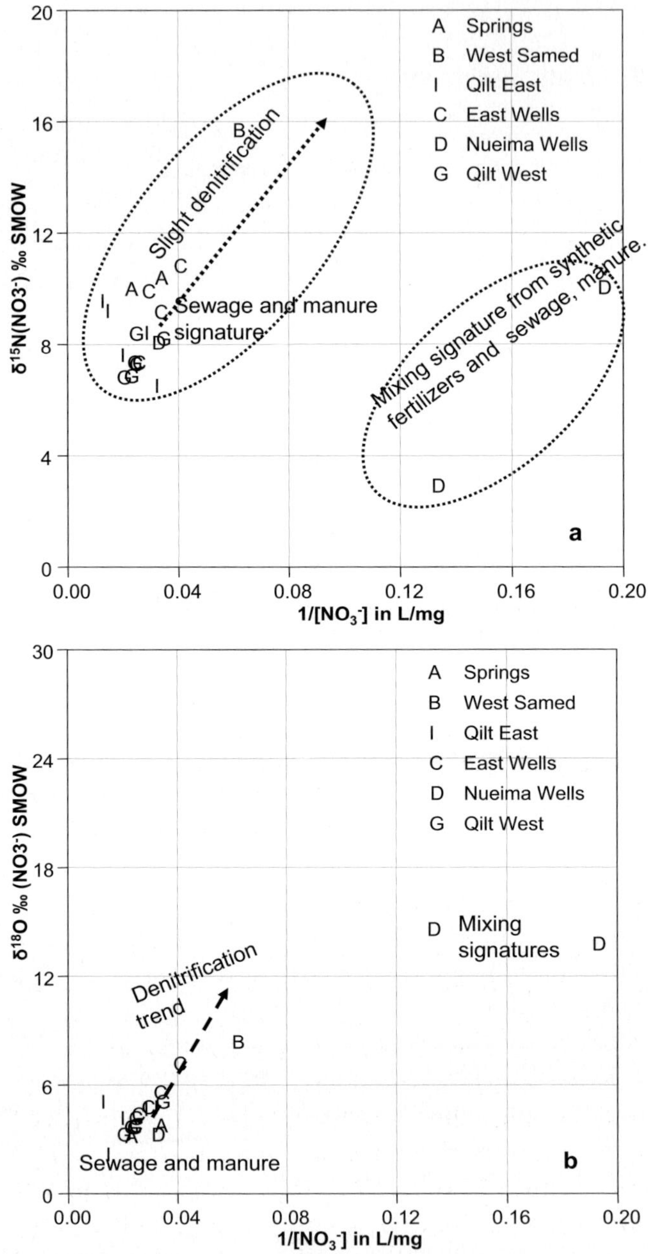

Fig. 3.3.4-14. δ^{15}N(nitrate) and δ^{18}O(nitrate) vs. 1/[NO3]. Most of the wells and springs show insignificant differences within the isotopic signature, except for the Samed well with its low nitrate concentration. Jericho north wells show low nitrate concentration with a higher δ^{18}O(nitrate) signatures which reflect mixing.

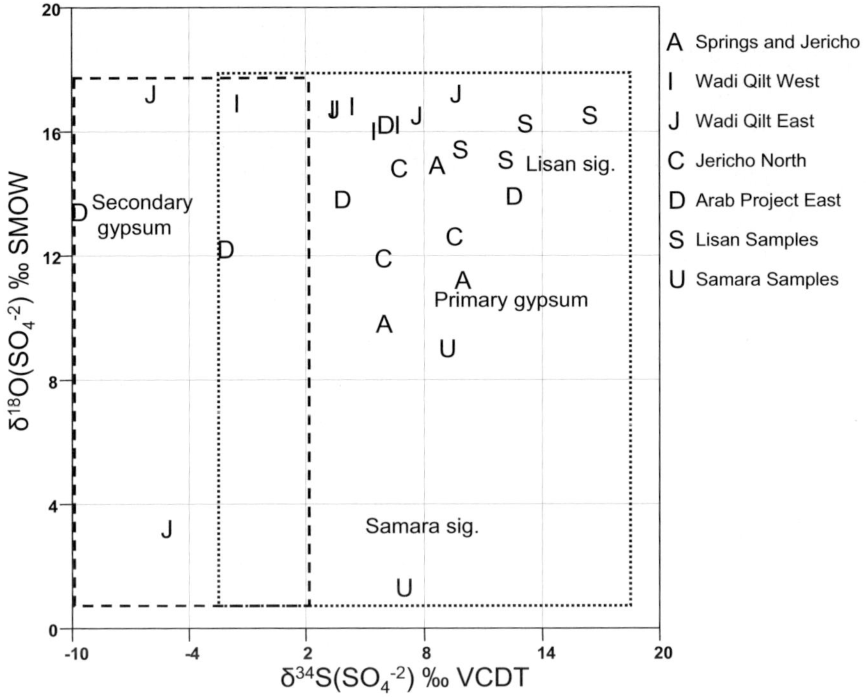

Fig. 3.3.4-15. The isotopic signatures of sulfate in groundwater and in solutions eluted from solid samples. Primary sulfate dissolution from the Lisan and partially Samra (or Samara) Formation are enriched in δ^{34}S and δ^{18}O, while disseminated sulfate from secondary gypsum and oxidized pyrite has more depleted signatures.

dissolution of gypsum (Fig. 3.3.4-15). However, some wells in the eastern part of the study area show lighter δ^{34}S(sulfate) which even reach -10‰. These depleted signatures indicate another sulfate source (Fig. 3.3.4-15). Gavrieli et al. (1998), reported a range in δ^{34}S(sulfate) between 14 and 28‰ for the primary gypsum in Lisan sediments, and up to -26‰ for the disseminated secondary gypsum in the aragonitic layers. Thus, the observed light values could indicate mixing of leachates from the two gypsum sources. Farber et al. (2003) also suggested mixing between two different sulfate sources in the southern Jordan Valley, namely of a hypersaline brine with δ^{34}S of 6–10‰ and a sulfate-rich groundwater of >200 mg/l with δ^{34}S of -17‰ from Wadi el Malih, further to the north of the study area. However, the results on Pleistocene groundwater in Jericho do not reflect any influence or a mixing between such two sources. In the centre and west of Jericho, δ^{34}S values are typical for leachates from Lisan sediments combined with relatively low chloride contents and high Na/Cl, Br/Cl molar ratios (Khayat et al. 2005) which cannot be explained by ascending saline water. Moreover, groundwater from the east with light $\delta^{34}S$ values contain high Cl^- and SO_4^{2-} concentrations of 1800 mg/l and 320 mg/l, respectively. Br/Cl and Na/Cl molar ratios are 0.006 and 0.4, respectively (Khayat et al. 2006). The different isotopic and chemical signatures in the west and the east of the study area, suggest two different sources of sulfate (Fig.5.5.3-16).

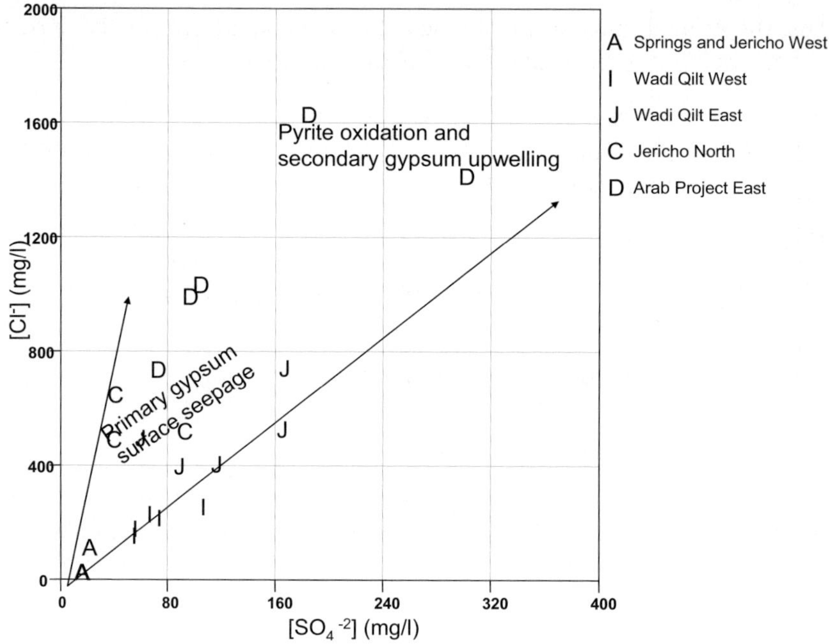

Fig. 3.3.4-16. Chloride vs. $SO4^{-2}$ in mg/l. The figure indicates two different sources for sulfate relatively to chloride content.

The relation between sulfate and $\delta^{34}S$ (Fig. 3.3.4-17) may indicate three sources: the sulfate from Lisan sediments, soils, and a chemically more complex one. $\delta^{34}S$ of water from the wells of the Arab Project are very low but contain high sulfate concentrations. This contradictory relationship probably reflects dissolution and mixing of primary and secondary gypsum, which are leached under heavy abstraction of water. The previous hydrochemical study indicates that the waters from some of the eastern wells are relatively enriched in iron (Khayat et al. 2005), which may result from pyrite oxidation, a process which also yields sulfate (Clark and Fritz 1997a). Contrastingly, the given depleted signatures of $\delta^{34}S_{sulfate}$ and $\delta^{18}O_{sulfate}$ in some wells to the east suggest that sulfate is derived from disseminated secondary gypsum and/or from other sulfide compounds originating from a deep anoxic environment and being oxidized in the shallow aquifer. Both of the above mentioned sources indicate interaction between the upper groundwater layer and a deeper saline layer. On the other hand, the seepage water from the upper surface runoff and irrigation return flow infiltrate within the Lisan and Samra (or Samara) layers forming a saline leachate low in Cl^- concentrations but with positive $\delta^{34}S$(sulfate) values (Fig. 3.3.4-15).

Thus, the high salinity in the shallow Pleistocene aquifer of the Jericho plain area turns out to be a function of three contributing sources.

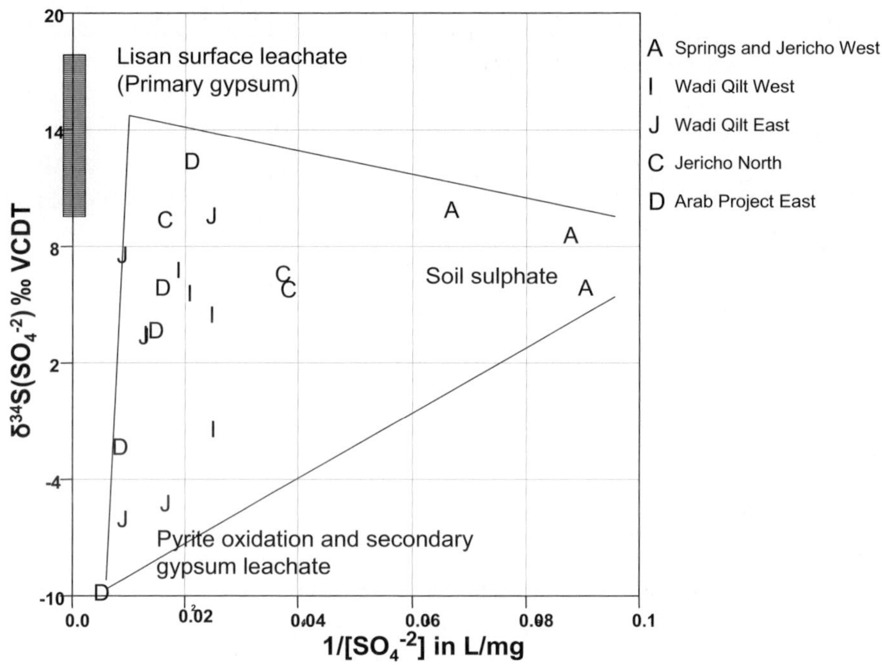

Fig. 3.3.4-17. $\delta^{34}S_{sulfate}$ vs. $1/[SO_4^{2-}]$. The light $\delta^{34}S_{sulfate}$ values indicate pyrite oxidation while the heavy $\delta^{34}S_{sulfate}$ values indicate sulfate from primary gypsum in Lisan and Samra Formations

3.3.4.4 Summary and conclusions

The hydrochemical and isotopic data of water in the study area reveal various aspects:

- The Ca-Cl type of waters in the eastern part of the Jericho plain is enriched in Sr and Fe and shows low element/chloride molar ratios. This is attributed to reverse softening, mixing of brackish water and oxidation of pyrite.

- The continuous change of the element/chloride molar ratios towards the ratios in waters from Lisan sediments, suggest the in-situ dissolution of rock components as an additional source of salinity. Saturation indices of dolomite and calcite decrease for most of the well waters in the west towards the east, in parallel with an increase of Mg, Ca and HCO_3^- concentrations.
- The anthropogenic effect by sewage and agriculture can be observed in low-chloride waters, which are high in Br and nitrate indicating the presence of pollutants. This water may result from irrigation return flow or infiltrating sewage. The lithological formation of the area has a twofold effect on the salinity in the area: in-situ weathering of the rocks, and transmissivity of the formations. The Samra (or Samara) Formation consisting of gravel and marl and the Lisan Formation with fine silts and low hydraulic conductivity but highly fissured structure, serve both as a good media for pollutant transport and storage. The isotopes analyses (δD, $\delta^{15}N$, and $\delta^{34}S$) reveal a clear picture on the possible sources of salinity, and confirm the conclusions of hydrochemical data in tracing the mode, tracks and various sources of salinity and pollutants in groundwater. $\delta^{15}N_{nitrate}$ and $\delta^{18}O_{nitrate}$ values show that most of the well- and spring waters from the Jericho area are dominantly contaminated by nitrate leaking from septic tanks.

δD and $\delta^{18}O$ yield the following results:

- There is no variation between dry and wet seasons except in δD when flushed water is enriched by ultrafiltration.
- Fitting the spring water signatures with LMWL indicate that the recharge area for the spring water is different from that of the wells in the lower Jericho plain with no further evaporation.
- Many wells are fed by different infiltration zones in the wadis. Overexploitation induces mixing with saline waters.
- The waters with scattering light $\delta^{18}O$ values and low salinity may reveal anthropogenic contaminants. Low-salinity waters with heavy $\delta^{18}O$ values indicate irrigation return flow. It has to be mentioned that it is common practice to connect well pumps directly with the fertilizer container in which the pumped water is mixed for use in irrigation. Together with enhanced boron, bromide and nitrate contents, this suggests presence of sewage and manure from animal farms and septic tanks which are widely used in the area. Any rising sulfate concentration in the Pleistocene aquifer of the Jericho area is mainly caused by leaching of gypsum from the Lisan and Samra (or Samara) Formations. There are two different sources or contribution mechanisms for sulfate into the groundwater in the area. These are: (i) surface seepage leachate with the

positive $\delta^{34}S_{sulfate}$ values of primary gypsum, and (ii) upwelling saline water with the negative $\delta^{34}S_{sulfate}$ values and high sulfate concentrations. Such saline waters ascend under heavy abstraction (particularly in the eastern wells). Oxidation of pyrite would yield the same sulfate characteristics. Two main contradictory factors act as an end member that effect the carbonate system and the water quality in the aquifer as a whole. Some freshwater leaks from the Tertiary Formation through the faults bordering the Tertiary and Quaternary Formations, which most probably feed the springs. The wells in the west and north produce mixtures of freshwater and infiltrated water that feeds mainly the Pleistocene aquifer. The anthropogenic influence plays an additive factor that affect the composition of the aquifer system. Thus, it is difficult to differentiate the local anthropogenic- from the brackish effect in the east. More tools are still needed for tracing the different inputs to the groundwater system and differentiate clearly the main participants in the deterioration of water.

As an overall result, the isotope and hydrochemical data suggest two main recharge mechanisms: The aquifer system in the Jericho plain area is fed by the surface runoff that infiltrates across Wadi Qilt and Wadi Nueima, whereas recharge of the Cretaceous aquifer in the west occurs by infiltration along the fault systems and fractured rocks. Water from the Cretaceous aquifer in the west may leak in small quantities along faults into Quaternary formation.

References

ARIJ (Applied Research Institute-Jerusalem) (1995) Environmental Profile for the West Bank, Volume 2: Jericho District. ARIJ, Jerusalem

ARIJ (Applied Research Institute-Jerusalem) (1997) Water Resources. Chapter 8 in: The Status of the Environment in the West Bank. ARIJ, Jerusalem, pp 95–107

Böttcher J, Strebel O, Voerkelius S, Schmidt HL (1990) Using isotope fractionation of nitrate nitrogen and nitrate-oxygen for evaluation of microbial denitrification in a sandy aquifer. J Hydrol 114:413–424

Clark ID, Fritz P (1997a) Environmental Isotopes in Hydrogeology: Lewis Publishers, New York, USA, pp1–328

Clark I, Fritz P (1997b) Nitrogen Cycling in Rural Watershed. In: Clark ID, Fritz P (eds) Environmental isotopes in hydrogeology. Lewis Publishers, Boca Raton, FL, pp 148–151

Coplen TB, Hanshaw BB (1973) Ultrafiltration by a compacted clay membrane. I. Oxygen and hydrogen isotopic fractionation. Geochim Cosmochim Acta 37:2295–2310

Farber E, Vengosh A, Gavrieli I, Marie A, Bullen TD, Mayer B, Holtzman R, Segal M, Shavit U (2003) Hydrochemistry and Isotope Geochemistry of the Lower Jordan River: Constraints for the Origin and Mechanisms of Salinization. Geochim Cosmochim Acta 68:1989–2006

Gat JR, Dansgaard W (1972) Stable isotope survey of the freshwater occurences in Israel and the Northern Jordan Rift Valley. J Hydrol 16:177–212

Gat JR (1998) Modification of the isotopic composition of meteoric waters at the land-biosphere-atmospheres interface. In: Isotope techniques in the study of environmental change. IAEA Vienna, Austria, pp 153–164

Gavrieli I, Stein M, Kolodny, Spiro B (1998) Sulfur Isotopes in Gypsum as indicator for Paleolimnological conditions of Lake Lisan, the Dead Sea Rift Valley. Am. Geophys. U. Ann Fall Meet, San Francisco, December, p F521

Guttman J, Zuckerman CH (1995) A model of the flow in the easter basin of the mountains of Judea and Samaria from the Far'ah to the Judean desert. Water Planning for Israel. Inc (Tahal) Tel Aviv (unpubl.)

Hoefs J (2004) Stable isotope geochemistry. Springer Verlag, Berlin, Heidelberg, New York, pp 1–244

Khayat S, Hötzl H, Geyer S, Ali W (2005) Hydrochemical investigation of water from the Pleistocene wells and springs, Jericho area/ Palestine. Hydrogeol J 14:198–202

Khayat S, Hötzl H, Geyer S, Ali W, Knöller K, Strauch G (2006) Sulphur and oxygen isotopic characters of dissolved sulphate in groundwater from the Pleistocene aquifer in the southern Jordan Valley (Jericho area, Palestine) Isot Environ health Stud 42:289–302

Marie A, Vengosh A (2001) Sources of salinity in groundwater from Jericho area, Jordan valley. Groundwater 39:240–248

Rofe and Raffety Consulting Engineers (1963) Geological and hydrological report, Jerusalem and district water supply. Hashemite Kingdom of Jordan, Central Water Authority, London, pp 1–79

Rosenthal E, Vengosh A (1994) Saline groundwater in Israel: Its bearing on the water crisis in the country. J Hydrol 156:389–430

Salameh E (2002) Sources of water salinities in the Jordan Valley area/Jordan. Acta Hydrochim Hydrobiol 29:329–362

Starinsky A (1997) The evolution of brines along the Dead Sea Rift. In: The Dead Sea Rift as a Unique Global Site. Terra Nostra 4/97

Vengosh A, Rosenthal E (1994) Saline groundwater in Israel: it's bearing on the water crisis in the country. J Hydrol 156:389–430

3.3.5 A GIS based hydrosystem model for the Jericho Plain, Palestine

C. Chen, S. Khayat, S. Geyer, W. Ali, H. Hötzl, O. Kolditz

3.3.5.1 Introduction

The study area around Jericho is located in the lower Jordan Valley (Fig. 3.3.5-1) extending 10 km to the north of the Dead Sea, and 7 km to the west of the River Jordan. Although it has a desert climate, its location on the eastern slope of the Jerusalem and Ramallah Mountains in the west makes it a catchment area that provides abundant water resources. The area is covered by sedimentary sequences of the Lisan and Samra Formations that mainly consist of two alternating lithological sequences of gravel and marl which build a limited part in the profile. By considering the lithological data of twelve well logs obtained from the Palestinian Water Authorities (PWA 2003; Khayat et al. 2004) seven alternative layers of the formations are defined. The δ^2H, $\delta^{18}O$ and $\delta^{13}C$ groundwater data (Khayat et al. 2006) suggest two possible mechanisms of recharge: direct infiltration of surface water runoff along wadis and irrigation return flow of bicarbonate type of water from Sultan spring which is fed by the Upper Cretaceous aquifer system in the western part of the area. The isotope data also indicate leakage from the water of Upper Cretaceous aquifer to the adjacent wells in the west, suggesting mixing of water from two sources. Moreover, the tritium units reflect recent groundwater in the western wells compared with old (prior to 1956) water in the eastern Arab Project wells. The water in wells located in-between, represent mixtures of both extremes (Khayat et al. 2006; see also chapter 3.3.4)). However, the mode of the suggested mixing process between the two possible sources is still not well understood (ARIJ 1995).

A 3-D numerical groundwater model was developed in order to analyze the mixing mode of the different water bodies recharging the aquifer system, i.e., effective infiltration and the effect of the pumping rates on the water level with respect to the abstracted quantity and the season of the year. With the help of the numerical model, the effective infiltration into the aquifer could be estimated. Only 2% of the water from the Sultan spring is contributing to aquifer recharge. Thus, the modeling is aimed to support a sustainable exploitation of the groundwater resources in the study area. The GeoSys/RockFlow (GS/RF) scientific modeling software

(Kolditz et al. 2007) is used to develop a numerical groundwater model of flow within the shallow Pleistocene aquifer system. Using GS/RF a detailed 3-D groundwater model is developed using a comprehensive data base for the study area which is provided as a GIS project (Fig. 3.3.5-2). The model of unsteady flow particularly reflects the hydraulic features of well field exploitation and irrigation pattern. Groundwater systems can be directly coupled to overland flow (Beinhorn 2005).

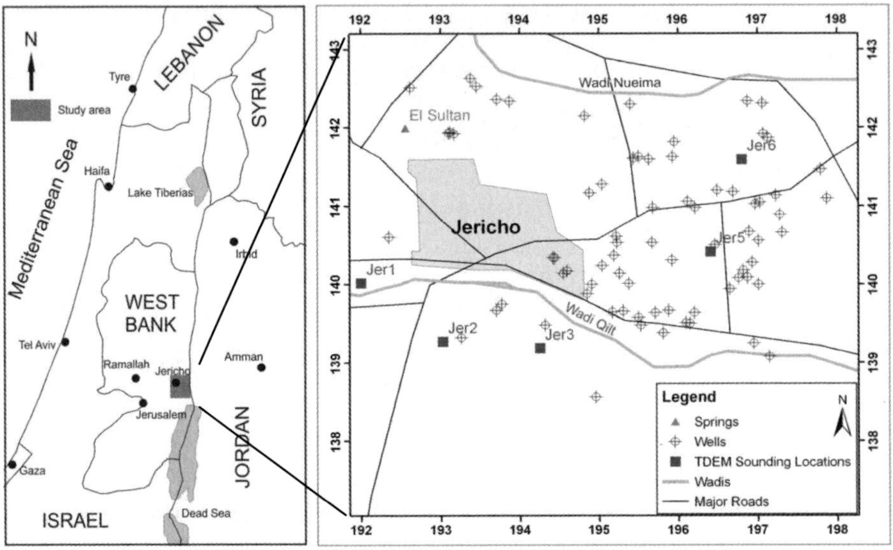

Fig. 3.3.5-1. Geographic location of the study area (Toll et al. 2005)

3.3.5.2 Methodology

The basic concept for the development of the groundwater model for the Jericho area is a combination of 2-D GIS data with hydrogeological information on the structure of the aquifer and the pumping well data base in order to establish a 3-D subsurface flow model (Beinhorn et al. 2004). Fig. 3.3.5-2 depicts the GIS data set, including the model area, catchment boundaries, well field, irrigation area and wadis. Lithological information from the boreholes is used to identify the subsurface structure model.

Conceptual groundwater model

The available data are first interpreted in order to construct a detailed GIS-based model which forms the basis of the conceptual model. The conceptual model is used to create a finite element numerical model. The data

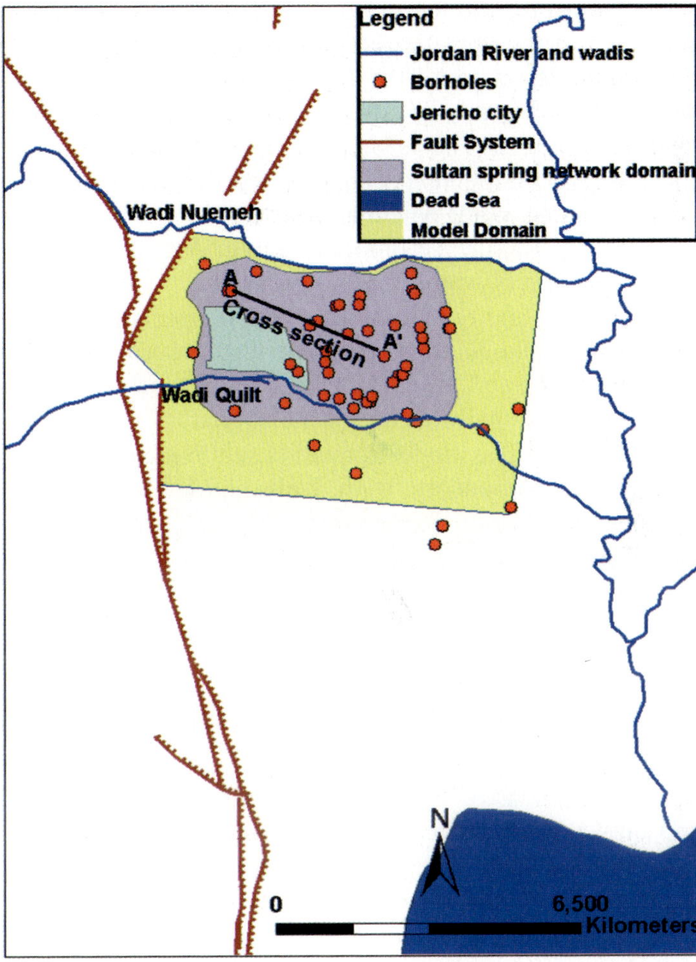

Fig. 3.3.5-2. GIS project and model domain of the study area

from GIS can be directly imported into GS/RF for modeling purposes. There are several GIS data types which can be used for hydrosystem modeling. In general, they are subdivided into raster data and vector data. The fundamental concept of vector GIS is that all real geographic features can be represented as points, polylines or surfaces. The vector data which are involved in the conceptual model are as follows:

- Layers: The alternative layer structure of marls and gravels for the first 160 m depth of the Pleistocene Lisan and Samra formations are converted to shape files, where each shape file represents one layer elevation for twelve wells in the area. The layer elevation describes the geometric information for the top or bottom of each alternative layer structure, and refers to the elevations above mean sea level. The minimum layer elevation was set to be 0.5 m, especially in the area, where the layer forms a closed lense. The 2-D geological profiles are divided into seven layers.

- Wadis: The area is drained eastwards toward the River Jordan by Wadi Qilt in the south and Wadi Nueima in the north. These wadis represent the direct recharge of surface runoff flooding. A river model is created based on the location of wadis.
- Faults: The area is highly affected by faults especially in the western part. These western faults are responsible for the spring system (mainly for the Sultan spring) and also act as a barrier between the Upper Cretaceous aquifer and the shallow Pleistocene aquifer.
- Study area zonation: The model domain is subdivided into four surfaces based on the domain boundary, wadi, faults and Jericho city.
- Borehole locations: More than sixty wells were drilled in the Jericho plain area. The majority of these wells are private wells which have been drilled for agricultural purposes. Twelve of these wells are used to describe the lithological formation of the uppermost 100–170 m of the aquifer.
- Water level: The recordings of water levels of many years are used in constructing the water level contour map in the shallow aquifer system.
- Abstraction rates: Several abstraction rates (m^3/h) are taken from the working wells in the area. The average daily operation of wells is about 8 h/d and thus a long-term estimation for the system output in this aspect could be calculated.
- Sultan irrigation network: Depending on the irrigation network deriving from the Sultan spring, a shape file is created covering the irrigation channel area.
- Boundary condition: Polylines are vector data which are used here to describe the extraction of the boundary conditions which are kept in a shape file. It was assumed that there was no water movement from the north to the south or vice versa. Therefore, the only active model boundaries are assumed to be those in the west and in the east.

In addition to the raster model, the vector type model is also utilized for hydrosystem modeling purposes which include:

- Interpolated raster maps for layer elevation information: The construction of a solid 3-D model depends on the layer models. For the layer model of each geological formation, a regular raster is created from sparse and irregular data using available interpolation techniques: inverse distance weighted, spline, kriging, and natural neighbors. For the subsurface modeling, the data from the underground study is always limited. Based on the interpolation of the selected twelve wells, a seven-layer model is created for the description of the lithological information. The raster maps from all the layers are used to generate a 3-D geometric structural model.

- Wells for surface topography: A computer-generated map is created for the surface topography using the interpolation techniques, which are implemented in ArcGIS. The observation wells with their corresponding elevation are used as sparse points and are interpolated to create the raster map of the surface. The exported ASCII file of the raster map is used for mapping purpose.

Data processing for groundwater modeling

For the generation of the numerical model, GIS based data from the conceptual site model is imported directly in ESRI shape file format via a GIS data interface - Shp2GLI (Fig. 3.3.5-3) implemented in the GS/RF code (Chen 2006). This enabled the automated generation of the GEO objects (points, polylines, surfaces, volumes and domains) requires availability of data from the structural model for the numerical model. In this way, data for the following aspects of the numerical model could be imported from the conceptual model:

- Domain properties: The initial geometric input for the domain discretization
- Node properties: Location of initial conditions, boundary conditions and source sink term
- Element properties: Spatial distribution of the material properties to be assigned to the finite elements (Fig. 3.3.5-4)

The main concept of GS/RF is the separation of geometrical, topological and physical data as GEOLib, MSHLib, and FEMLib. Because of the data format difference between ArcGIS shape files and GS/RF, there is a GIS interface for data conversion. In addition to that, the significant weakness of GIS for subsurface modeling is that GIS does not support solid 3-D modeling. Based on the object-oriented technology, the GEOLib has the ability to describe 3-D data structure. Therefore, by integrating GIS and GEOLib, a structural 3-D model could be built. The hydrogeological conceptual model of the area is created using the ArcGIS software. All required data for modeling, such as the modeling domain, geometric data converts the shape file into the geometric data structure GLI which is of aquifer layers, boundary conditions, initial conditions, source terms and

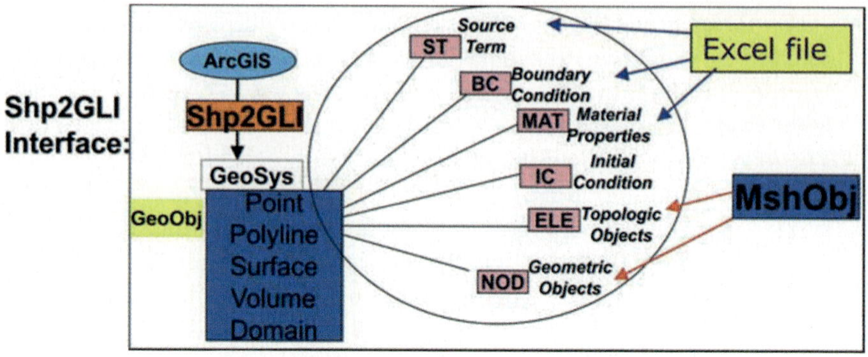

Fig. 3.3.5-3. GIS-based concept of a GeoSys/RockFlow project

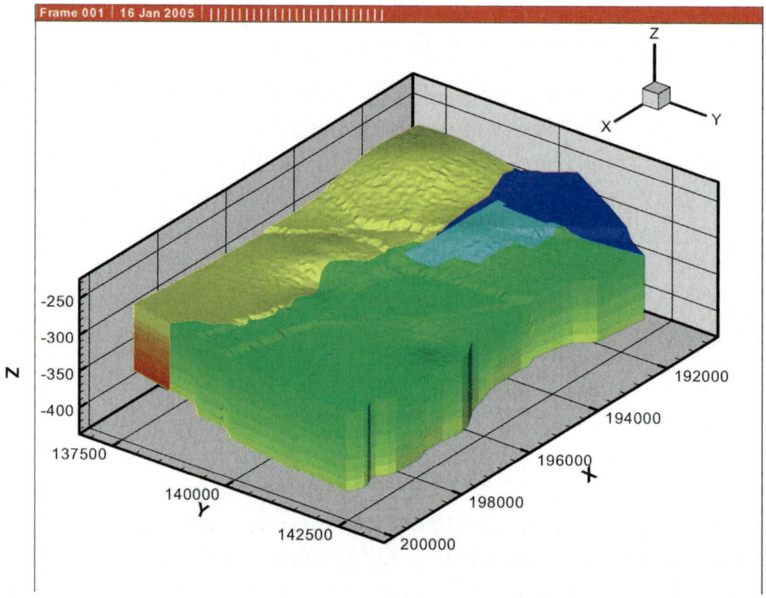

Fig. 3.3.5-4. Distribution of material properties (hydraulic conductivity) in the study area. The different color indicate volumes with different material properties

spatial distribution of material properties are saved in the GIS project. After importing the GIS shape file into GS/RF, the interface the file extension of the GEOLib data. 2-D geometric objects such as points, polylines, and surfaces together with an external Excel data base, which describes the temporal information (i.e., well pumping), can be used not only for locating source terms (ST), boundary conditions (BC), material properties (MAT), initial conditions (IC), but also for creating nodes and elements that form a finite element mesh (Kalbacher et al. 2005, Chen et al. 2005).

3-D structural model

By implementing GIS tools within GS/RF, a 3-D structural model (Fig. 3.3.5-5) is built for analyzing the hydrogeological system. A 7-layer model is built and displayed using the implemented visualization tools in GS/RF. Each layer has its own color. The generation of the 3-D model relies heavily on data from boreholes and geologic cross sections. This part deals with the finite element components used in GS/RF and how these elemental shapes link to the concept of layer models. Such layer models are made up of layer polylines, layer surfaces (TINs), and layer volumes, all of which are required for the construction of a 3-D structural model. Based on the surface triangulation, structured prism elements are generated for each layer, and then adjusted to 3-D structures based on layer elevation raster maps. Thus, the prism elements define the 3-D volume objects used in the construction of the 3-D structural model. At first, an unstructured 2-D triangular mesh is generated for a polygonal 2-D surface representing the model domain. This was then extended to a structured 3-D prismatic mesh based on the 3-D layer interface elevations from the geological model using mapping methods. The mapping process on the prism layers is a procedure to vertical height and modulate the prism elements according to topographic and stratigraphic information, which is provided by TIN (Triangular Irregular Network) or grid files.

In order to build a 3D structural model, the shape file of the model domain is imported into GeoSys and converted into GeoSys geometric data type GeoLib - GLI. One GLI file is composed of hierarchically ordered points, polylines, surfaces and volumes. Then a finite element mesh is created consisting of triangles building the 2D model. Based on this 2D model, a 7-layer 3D model is constructed by prismatic elements. After that the mesh is adjusted to stratigraphic irregularities using raster grid files of 7 layers. The result is saved in a MSH file, which is the file format of MSHLib. Finally, volumes are built for the entire model. Each volume consists of a top and a bottom surface embracing the prismatic elements.

Groundwater model

To create a groundwater model, the assignment of process (PCS) data is the final step before numerical simulations can be performed. PCS data can be subdivided into node and element related properties. The source terms, initial conditions and boundary conditions have to be assigned to mesh nodes. Material properties have to be assigned to mesh elements.

- Boundary conditions: Geometrically, boundary conditions are defined by points, polylines and surfaces. Here, the boundary conditions are defined by polylines, which are used to detect the mesh nodes along the-polylines to which boundary conditions should be assigned for the groundwater flow model.

- Source terms: In general, source terms are fluxes into or out of the model domain. Here, source terms are pumping wells, the Sultan irrigation network, line recharge along wadis, area recharge by precipitation and evaporation. The well data are imported directly from an Excel file. The value of the source term is assigned to the mesh nodes.
- Initial conditions: Water level data from the years 2002–2003 are used as initial reference head data for comparison with the model results.
- Material properties: Values for material properties are assigned to the elements of the mesh according to the spatial distribution of the materials. The material properties used for the different layers are stored in an Excel data base. By connecting the material data with the mesh data, each volume is then assigned a material property.

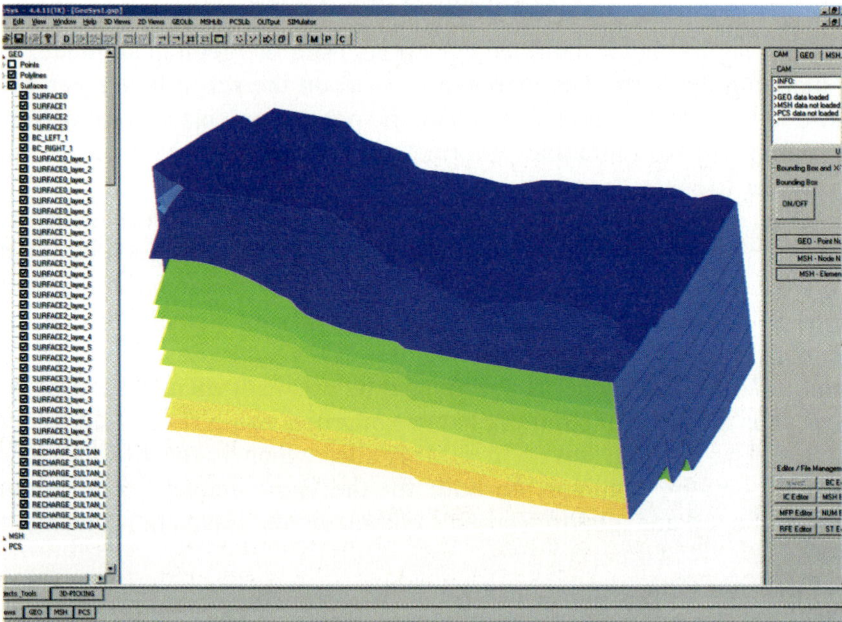

Fig. 3.3.5-5. 3-D structural model of the Jericho area and visualization tools

Results and discussion

The first simulation result of the Jericho area is the regional groundwater flow model showing the hydraulic distribution of the first aquifer layer (Fig. 3.3.5-6) without the effect of pumping wells and irrigation network.

This simulation is then used as initial condition for the transient groundwater simulations.

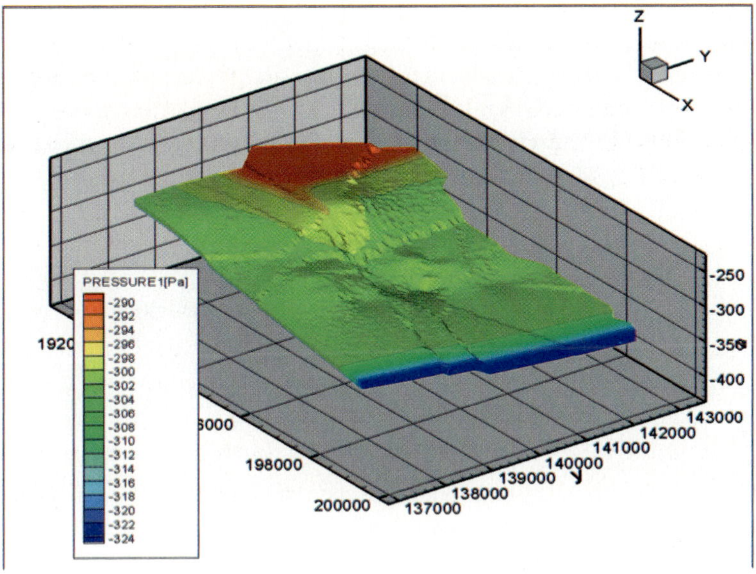

Fig. 3.3.5-6. Hydraulic head distribution in the upper aquifer layer (undisturbed situation without pumping and irrigation)

Successively, data of all pumping wells and irrigation from Sultan spring are then included in the simulation. The overall irrigation from Sultan spring is calculated to 3.024 Mm^3/yr. Irrigation is modeled as a homogeneous infiltration rate because there is no information available for the distribution of irrigation rates. The numerical modeling is used to highlight the hydraulic functioning of each contributor, i.e., irrigation, pumping and combination of both. Fig. 3.3.5-7a depicts the groundwater level, if only pumping is taken into account. Local lowering of hydraulic heads is observed. Fig. 3.3.5-7b depicts the ground-water level only if homogeneous irrigation is taken into account. A rise of ground-water level results from this water source. Figs. 3.3.5-7c illustrates the distribution of hydraulic heads for different assumed infiltration rates (2 and 10% of irrigation rate). In the central-northern part, the groundwater level is lowered but rises in the central -southern part of the investigated area. The modeling results fit to the measured data, if an infiltration rate of only 2% of the irrigation rate is assumed. This is the most important information which is provided by the groundwater model. However, the absolute distribution of hydraulic head under undisturbed conditions is still overestimated. Therefore, further studies have to be conducted in the future in order to improve the groundwater model.

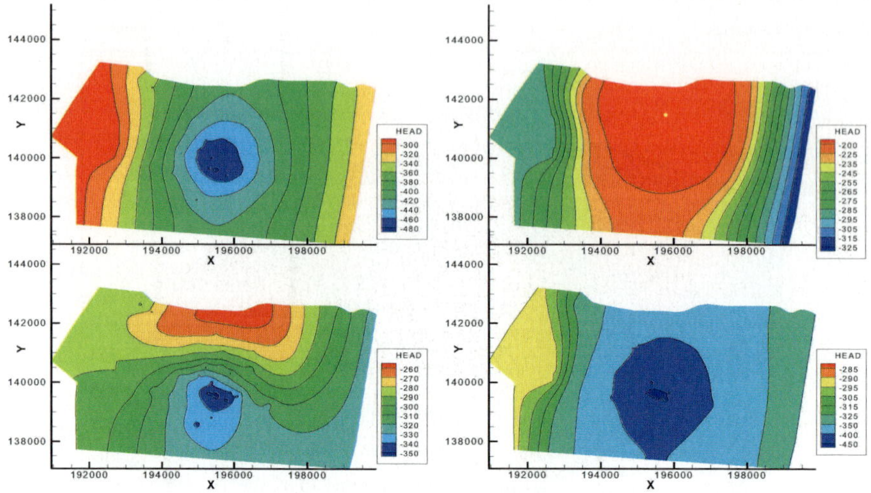

Fig. 3.3.3-7. Hydraulic head distribution in the upper aquifer taking into account: (a) pumping wells only, (b) irrigation from Sultan spring only, (c) pumping and irrigation (10% infiltration rate) and (d) pumping and irrigation (1% infiltration rate)

The above mentioned results are based on high recharge of water from the spring irrigation network to the regional water level. Moreover, it confirms the initially proposed model implying mixing of spring water with the flood water recharge in the wadis (Chap 3.3.4). The water level gradually decreases from the spring recharge area in west towards the area of heavy abstraction by wells in the eastern parts of the study area.

Finally, the effect of pumping rate in a specific season of a year is considered. The ground water level in Jericho is highly affected by the seasonal pumping rate from the agricultural wells. This effect is most conspicuous during the dry season by the end of October (Fig. 3.3.5-8). The effect of the seasonal water pumping on the water head of the shallow aquifer in Jericho is quite obvious. The water level dropped during the first agricultural season, immediately after the replenishment of the aquifer. Pumping during the second agricultural season (immediately after the dry summer season) cause further decrease of the water head.

The model visualized for a certain extend the effect of major extensive pumping at different times of the year. However, this model is still subjected to further refinement as to mixing water from two different sources. Moreover, different expectation scenarios should be considered after refining the model to foresee the changes depending on assumed future environmental conditions.

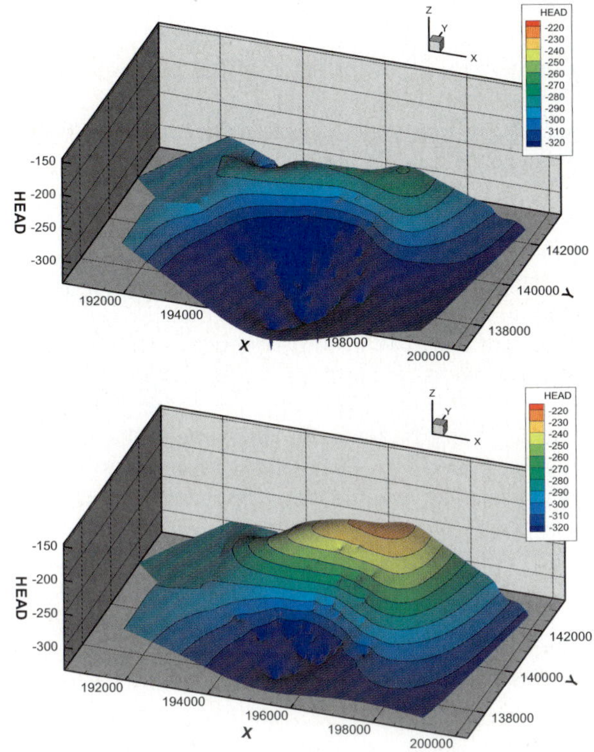

Fig. 3.3.5-8. Groundwater head during the recovery phase (February) [bottom] and in the late dry season (October) [top]. Effective infiltration rate is assumed to be 2% of irrigation rate

References

ARIJ (Applied Research Institute-Jerusalem) (1995) Environmental Profile for the West Bank, Vol 2: Jericho District. ARIJ, Jerusalem

Beinhorn M (2005) Contributions to computational hydrology: Non-linear flow processes in subsurface and surface hydrosystems. PhD thesis. Reihe C Hydro-, Ingenieur- und Umweltgeologie der Tübinger Geowissenschaftlichen Arbeiten (TGA), Uni Tübingen: pp 10–14, 51–52

Beinhorn M, Guttman JS, Sauter M, Toll M, Kolditz O (2004) Groundwater modeling of the shallow aquifer in the Jericho area. In: 5th Int Symposium on the Eastern Mediterranean Geology, Thessaloniki, Greece, pp 1483–1486

Chen C (2006) Integrating GIS Methods for the Analysis of GeoSystems. PhD thesis. Reihe C Hydro-, Ingenieur- und Umweltgeologie der Tübinger Geowissenschaftlichen Arbeiten (TGA), Uni Tübingen: pp 1-157

Chen C, Sawarieh A, Kalbacher T, Beinhorn M, Wang W, Kolditz O (2005) A GIS based 3D hydrosystem model of the Zarqa Ma'In-Jiza areas in Central Jordan, J Environ Hydrol 13:11–13

Kalbacher T, Wang W, McDermott C, Kolditz O, Taniguchi T (2005) Development and application of a CAD interface for fractured rock. J Environ Geol 47:1017–1027

Khayat S, Geyer S, Hötzl H, Ali W, Ghanem M (2004) Hydrochemical investigation for the Pleistocene wells and springs from Jericho Area. – In: Chatzipetros A and Pavlides S (eds): Proc 5th Intern Symp Eastern Mediterranean Geology, Thessaloniki, Greece, pp 1529–1533

Khayat S, Hötzl H, Geyer S, Ali W (2006) Hydrochemical investigation of water from the Pleistocene wells and springs, Jericho area, Palestine J Hydrogeol 14:192–202

Kolditz O, Kalbacher T, Wang W, Bauer S, McDermott CI, Chen C, Gronewold J, Beyer C, Walsh R, Park CH (2007) GeoSys/RockFlow manual. Center for Applied Geosciences, Univ Tübingen, http//www.uni-tuebingen/zag/geohydrology/index.html

PWA (Palestine Water Authority) (2003) Wells profiles lithological data.

Toll M, Gropius M, Beinhorn M, Guttman J, Chen C, Kolditz O, Khayat S, Tamimi AR, Sauter M (2005) Groundwater investigation of the shallow aquifer in the Jericho Area, Israel-Germany Scient Conf Water is Life, Jerusalem, Israel 7-8 December 2005, 5p

3.4 Aquifers in the eastern Jordan Valley

3.4.1 Hydrogeology and hydrochemistry of Wadi Waleh and Wadi Zarqa Ma'in catchment, central Jordan

A. Sawarieh, H. Hötzl, E. Salameh

3.4.1.1 Introduction

The study area covers the northern part of Wadi Mujib groundwater basin. It consists of the Wadi Waleh sub-basin and the Zara-Zarqa Ma'in thermal springs area which is a small sub-basin that lies to the northwest of the Wadi Waleh sub-basin. It extends from the Zara hot springs at the eastern shore of the Dead Sea through the Zarqa Ma'in thermal springs to the Jiza region in the east (Fig. 3.4.1-1). The entire study area has an irregular shape with a total area of about 2,300 km^2 lying between the coordinates 200,000–280,000E and 95,000–145,000N (Palestine Belt Grid, PBG).

A sequence of sedimentary rocks of more than 1000 m thickness consists of the Paleozoic Ram sandstone Group, the Permo-Triassic Zarqa sandstone Group, the Lower Cretaceous Kurnub sandstone Group and the Upper Cretaceous Ajlun and Belqa limestone Groups. The area is dissected by numerous Paleozoic or earlier faults of several trends, covered by Paleozoic and Cretaceous sediments. The NW–SE compressional faults were re-activated during Late Cretaceous to Tertiary and Quaternary times and led to new movements along old faults and weakness zones. These tectonic forces were also responsible for creating new faults, fractures and discontinuities. The Wadi Zarqa Ma'in fault zone is one of the major E-W trending structural features.

3.4.1.2 Hydrogeological setting

The lithological sequence in the study area is divided into two major aquifer complexes separated by an aquitard composed of about 400 m of marl and marly limestone of the Upper Cretaceous ($A_{1–6}$; see Table 3.4.4-1). The upper aquifer complex (B_2–A_7; see Table 3.4.4-1) consists of Upper Cretaceous limestone, chert and marly limestone and is the major source of fresh water for domestic uses. The lower aquifer consists mainly of Lower Cretaceous and older sandstone hosting brackish and thermal water in most areas.

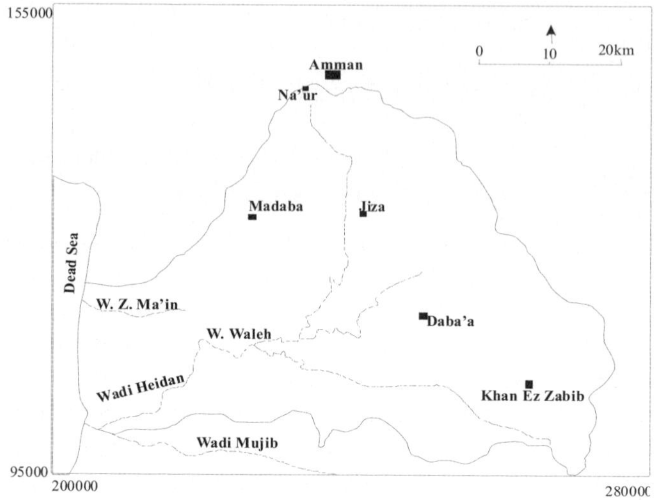

Fig. 3.4.1-1. Location map of the study area

Infiltration into the upper aquifer complex takes place mainly in the western highlands. With the groundwater divide lying 20–30 km to the east of the Dead Sea escarpment, groundwater flows either westwards or eastwards (Fig. 3.4.1-2). Most of the groundwater in the upper aquifer flows to the west. The easterly-directed groundwater flow leaks into the lower aquifer complex passing the aquitard (A_{1-6}) in the eastern parts of Jordan. The infiltrated water is heated up according to the local normal geothermal gradient. The water flows westwards to the Dead Sea and discharges as thermal water in the Zara and Zarqa Ma'in springs. In the lower aquifer all groundwater is flowing westward towards the Dead Sea.

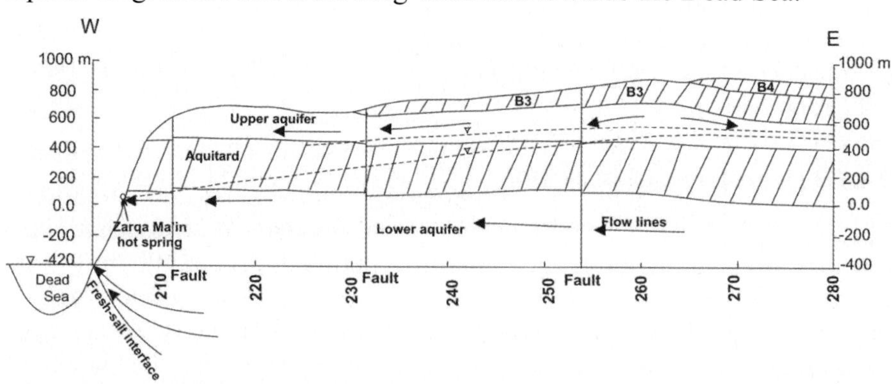

Fig. 3.4.1-2. The hydraulic pattern within the study area (modified after Salameh and Udluft 1985; Sawarieh 2004)

3.4.1.3 Classification of groundwater

A total of 49 chemical analyses were used to characterize the different water types (Table 3.4.1-1).

In the Cl^{-}-SO_4^{2-}-HCO_3^{-} triangular plot (Giggenbach 1991) for thermal water classification (Fig. 3.4.1-3), all samples from the Zara and Zarqa Ma'in thermal springs are located in the chloride corner, whereas most of the samples of the thermal wells fall in the high bicarbonate part. Only two samples (16, 21) can be classified as chloride-type groundwater. Another two samples (9, 10) are located in the sulfate water type. The position of samples 8, 11, 31, and 32 show a shift to the sulfate and chloride corners, but are still within the bicarbonate water range. The SO_4^{2-} increase may result from oxidation of sulfides and dissolution of gypsum/anhydrite occurring in several horizons in the aquifer matrix. The figure suggests that mixing of Cl^{-} and HCO_3^{-} type of waters, originating from two aquifers. Schoeller diagram (Fig. 3.4.1-4) also shows that the cold water from wells is of bicarbonate type (blue circles) and the water of the thermal springs from the lower aquifer is of chloride water type (red stars). Water samples from the thermal wells (green triangles) show mixing of the two types.

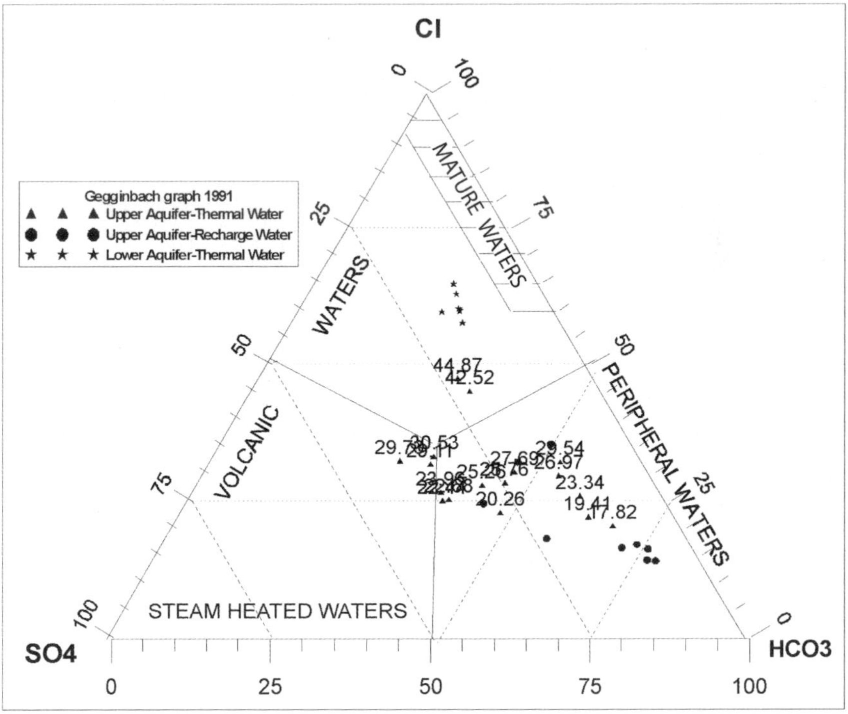

Fig. 3.4.1-3. Giggenbach graph showing the water types and the mixing trend

Table 3.4.1-1. Chemical and isotope analyses of water samples from recharge areas, lower aquifer thermal springs and thermal wells of the upper aquifer. Samples (1–32) were taken from the shallow thermal wells in the upper aquifer (Jaser 2002). Samples (33–43) were taken in the course of the present study and samples (44–49) were taken from a previous study (Sawarieh and Massarweh 1996). All samples were analyzed in the Water Authority Labs, Amman. Out of these samples, 24 samples were also analyzed for stable isotopes in the same laboratory.

No.	$\partial^{18}O$	$\partial^{2}H$	Na	K	Mg	Ca	Cl	HCO_3^-	SO_4^{2-}	NO_3^-	SiO_2	F	Br	B
	±0.2	±1.0						mg/L						
1	-5.6	-26	37	2	35	94	59	398	52	4.5	14	0.5	0.1	0.1
2	-6.6	-32	73	3	55	110	158	419	101	0.4	18	1.1	0.7	0.2
3	-6.1	-31	112	5	50	106	179	338	177	0.5	22	1.3	0.8	0.3
4	-6.4	-32	83	3	51	123	205	386	103	0.4	17	1	1.0	0.2
5	-5.6	-27	64	15	53	77	164	270	119	0.5	19	1.7	0.9	0.2
6	-5.1	-25	43	2	30	100	71	365	50	3.3	9.7	0.6	0.1	0.2
7	-5.7	-30	121	7	66	159	208	445	289	0.3	11	1.6	0.9	0.5
8	-6.3	-33	213	11	100	211	402	465	451	0.4	20	1	1.7	0.9
9	-6.4	-34	245	8	133	223	424	432	570	1	11	1.3	0.1	0.9
10			239	8	80	177	376	397	431	0.5	20	1.4	1.6	0.9
11	-6.6	-33	103	8	55	220	344	420	419	0.3	17	1.2	1.3	0.5
12	-6.1	-32	71	3	41	124	120	469	85	3	14	1.3	0.7	0.1
13	-6	-31	87	3	58	133	151	506	121	0.2	15	1.5	0.8	0.0
14	-5.4	-28	79	5	56	140	124	479	190	0.6	11	1.5	0.6	0.3
15	-5.1	-25	38	2	31	79	59	312	56	4.9	9.5	0.5	0.4	0.1
16	-6.2	-32	198	7	70	145	433	307	225	0.8	12	1.5	0.1	1.5
17	-6.4	-33	174	7	79	187	263	476	433	0.8	11	1.2	0.1	1.0
18	-6.6	-34	174	6	65	171	254	420	386	0.6	11	1.2	0.1	0.9
19	-6.7	-35	134	5	50	140	172	431	246	0.3	13	1.2	0.9	0.2
20	-6.7	-34	170	6	78	204	267	488	422	0.5	14	1	2.7	0.4
21			175	7	60	138	372	305	198	0.6	12	1.4	0.1	1.5
22	-5	-26	81	6	35	113	206	327	92	5.6	13	0.8	1.3	0.0
23			153	7	62	158	225	420	312	0.5	20	1.5	0.7	0.5
24	-6.4	-35	154	8	58	141	226	406	263	0.5	21	1.4	0.8	0.4
25	-5.4	-27	45.8	3	32	85	75.4	427	50		20.6	1.36	1.0	
26	-5.3	-26	41	5	25	71	65	452	51	14	12	0.8	0.3	
27			54	3	26	100	78	451	43		9.7	0.9	0.9	
28			190	7	66	182	250	459	411	0.1	17	1	0.8	
29			187	9	70	174	259	445	385	0.1	14	1.2	1.0	
30			125	4	42	110	174	293	240	0.1	17	1.1	0.5	
31			207	8	69	182	338	401	387	0.1	18	1.1	1.4	
32			161	8	60	145	330	372	186	0.4	16	1.3	1.5	
33	-4.9	-24	44	3	24	72	70	260	35	21				
34	-5.1	-26	44	3	24	72	70	261	35	55				
35	-5.3	-26	45	6	25	73	65	284	26	37				
36	-5.5	-25	26	3	21	77	41	300	14	33				
37	-5.3	-27	40	6	22	100	70.3	320	18	59				
38	-5	-23	38	2	17	66	45	238	35	21				
39	-5.2	-24	31	3	22	59	47	218	31	30				
40	-5.2	-23	45	3	30	60	73.1	249	36	28				
41	-5.2	-26	97	5	51	116	146	413	145	15				
42	-6	-31	148	7	80	153	275	489	229	2.7				
43	-6.2	-33	147	5	67	144	251	527	153	2.7				
44/Zm8	-3.9	-32	382	47	27	132	667	262	174	4.4				
45/Z25	-5	-38	208	27	24	89	374	186	119	0.2				
46/Z8			275	38	21	137	515	232	150		24	0.4	4.7	
47/Z22			195	30	18	115	408	180	118		30	0.4	4.0	
48/Zm21			500	49	34	157	799	288	196		20	0.3	7.7	
49/Zm53			410	63	35	170	740	299	254		25	0.5	6.7	
R. Rabba	-5.6	-22												
R. Waleh	-5.4	-22												

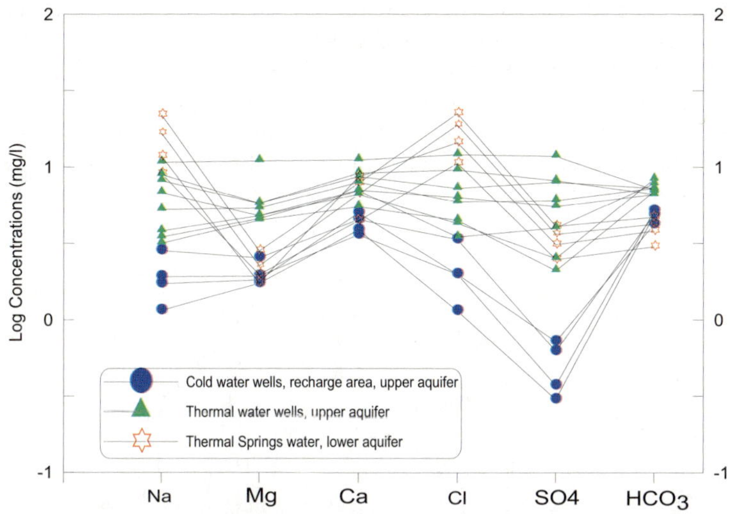

Fig. 3.4.1-4. Schoeller diagram for samples from water of the thermal springs deriving from the lower aquifer and for cold well water from the upper aquifer

3.4.1.4 Thermal water and heat sources

The natural geothermal activity in Jordan is manifested by a large number of thermal springs associated to the Jordan Rift system and other main faults like the Zarqa Ma'in fault. Fumarolic activity and boiling mud pools are unknown. The Zara hot springs (> 40 springs) at the Dead Sea shore and Zarqa Ma'in hot springs (> 60 springs), about 4 km to the NE of Zara, form the main manifestations of geothermal activity in Jordan. The thermal springs emerge from the Lower Cretaceous and older sandstone formations with temperatures ranging from 30–63°C. The average discharge rate of all springs exceeds 700 l/s.

The thermal waters of Zara-Zarqa Ma'in have been subjected to many studies regarding their chemistry, heat source, therapeutic application, and potential energy source (Truesdell (1979), Myslil (1988), Sawarieh (1990) and Salameh et al. (1991). Contrarily, only limited work has been done on the thermal wells further east. The Zara–Zarqa Ma'in thermal springs are fed by water circulating deep within the Paleozoic sandstones. The Na-K geothermometer the maximum temperature yields about 110°C for the water at depth (Truesdell 1979). The lower field temperatures are caused by cooling down during ascend and by mixing with shallow cold water before discharging.

Beside the thermal springs along the Jordan Rift, there is another geothermal field in the Jiza region (Fig. 3.4.1-1), where many wells produce

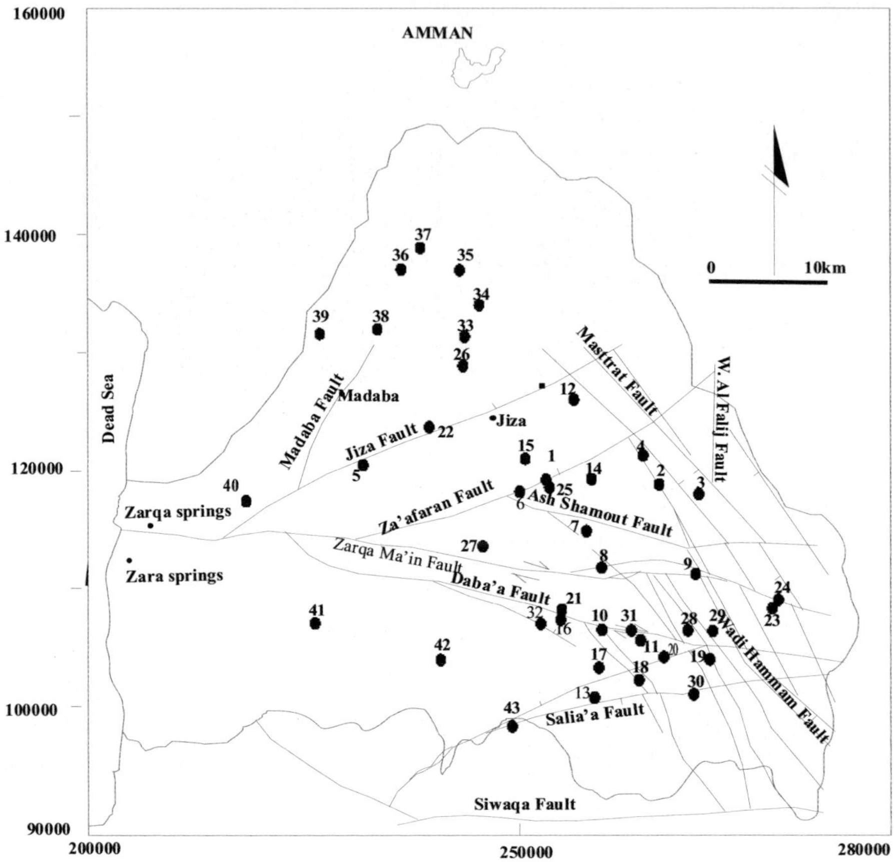

Fig. 3.4.1-5. Location of thermal well water and of the major faults

thermal water (30–46°C) from the Upper Cretaceous limestone. These wells are generally located within the eastern extension of Wadi Zarqa Ma'in Fault Zone and of the other faults in the area such as the Daba'a, Masattarat and Wadi Hammam faults (Fig. 3.4.1-5). The temperature distribution at well heads shows that there are two heat anomalies in the upper aquifer (Fig. 3.4.1-6). One is located around well 5 east of Madaba City, the other occurs along the Daba'a and Masattarat faults.

The dense network of faults striking WNW-ESE and NW-SE strongly suggests that the two aquifer systems are hydraulically interconnected, especially in the eastern part of the study area. This allows the thermal water from the lower aquifer to ascend along faults into the upper aquifer raising the groundwater temperature in the vicinity of these faults.

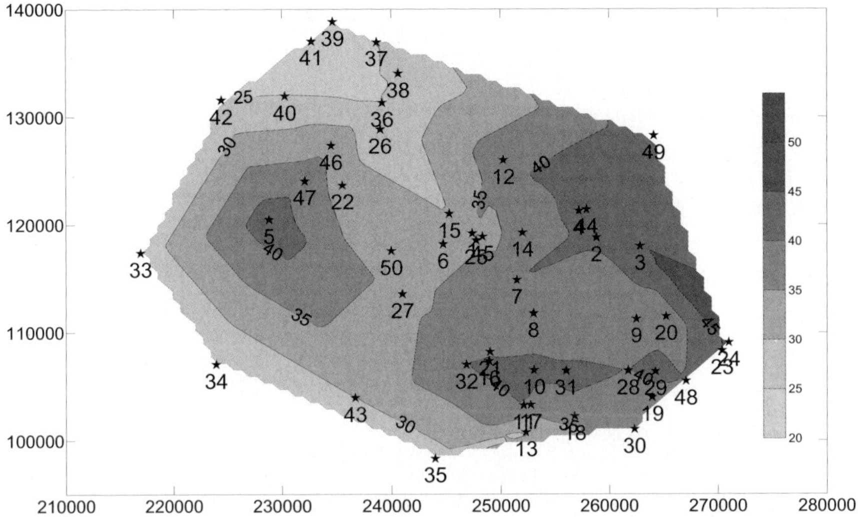

Fig. 3.4.1-6. Isothermal contour (°C) map in the upper aquifer

3.4.1.5 Isotopes content and origin of thermal water

The water isotopes (δ^2H and δ^{18}O) enable differentiation between thermal waters of magmatic, oceanic, and meteoric origin. δ^{18}O (V-SMOW) of all samples from the area range from -6.71 to -3.88‰ with averages of -5.59‰ and -4.43‰ for the upper and the lower aquifer, respectively. δ^2H of the upper aquifer ranges from -34.9‰ to -22.9‰ with an average of 29.18‰, whereas in the lower aquifer it ranges from -32 to -37.8‰ with an average of -34.9‰. In magmatic water, δ^{18}O and δ^2H range from +6 to +9‰ and -40 to -80‰, respectively (Pearson et al. 1980). In seawater δ^{18}O and δ^2H are about 0‰. Therefore, the thermal waters are of meteoric origin.

Cross plots of δ^{18}O vs. δ^2H of all samples from cold and thermal wells from the upper aquifer, plot between the Mediterranean Meteoric Water Line (MMWL: $\delta^2H = 8\ \delta^{18}O + 22$; after Dansgard 1971) and the Global Meteoric Water Line (GMWL: $\delta^2H = 8\ \delta^{18}O + 10$); after Craig 1961) (Fig. 3.4.1-7). Water samples from the Zara-Zarqa Ma'in thermal springs (lower aquifer) fall below the local meteoric water line (LMWL) with an oxygen shift that could be explained by assuming ^{18}O exchange with hot rocks.

In the graph of δ^{18}O vs. δ^2H (Fig. 3.4.1-7), the thermal water samples of the shallow wells of the upper aquifer plot on a line with a slope of less than 8 indicating that these waters are either subjected to evaporation or mixing with water of different origin. Any evaporation process is ruled out by the negative relationship between δ^{18}O and Cl^- and the absence of the

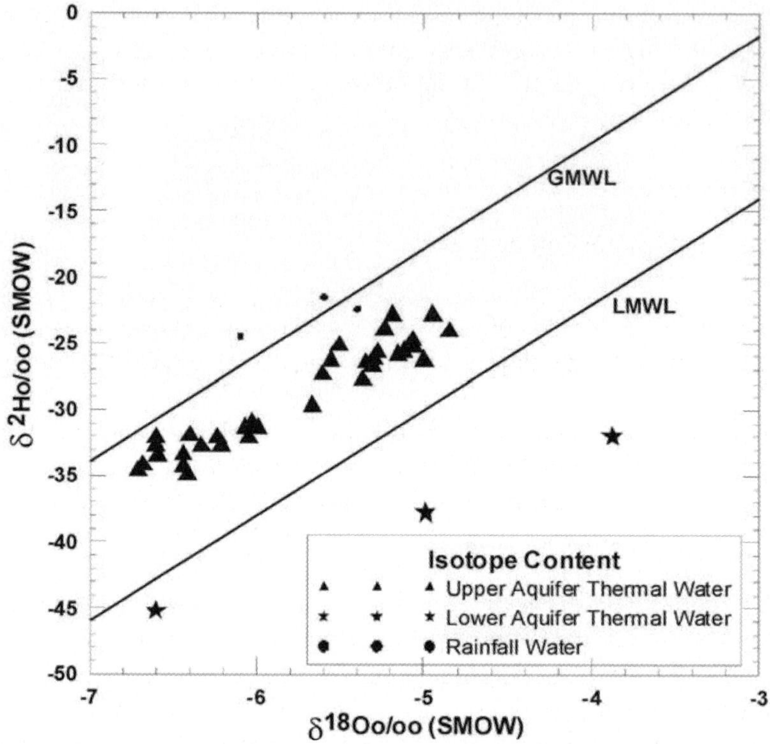

Fig. 3.4.1-7. Stable isotopes content of rain water, thermal and cold water from both the upper and lower aquifers. LMWL=local meteoric water line; GMWL= global meteoric water line

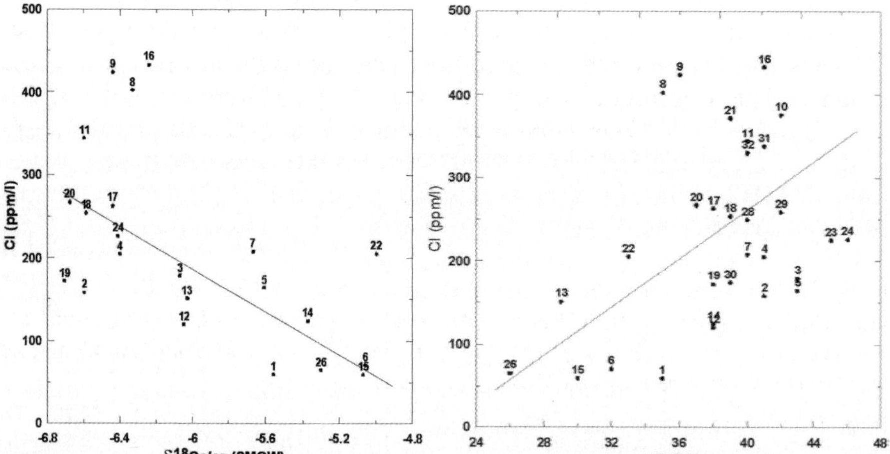

Fig. 3.4.1-8. Cross plots of Cl^- concentration vs. $\delta^{18}O$ and temperature

negative relationship between $\delta^{18}O$ and temperature (Fig. 3.4.1-8a), which indicates a mixing process. The positive relationship of temperature and Cl^- and temperature indicate mixing between the two aquifers in the study area (Fig. 3.4.1-8b).

3.4.1.6 Conclusions

Following conclusions can be drawn:

- The study area which is located in the central part of Jordan belongs to the western part of the regional hydraulic pattern. It hosts two major aquifer complexes, the upper and lower aquifers, both discharging towards the Dead Sea.
- Chemical analyses show that groundwater in the upper and lower aquifers are of the bicarbonate- and chloride type, respectively.
- The isotope content of the thermal water from the two aquifers in the study area indicates that the water is of meteoric origin.
- The chemical analyses and isotope ratios of the thermal well waters in the Jiza region show that this water is the result of mixing of two different types of water, i.e., water from the upper and lower aquifers.
- In the Cl^--SO_4^{2-}-HCO_3^- diagram (Fig. 3.4.1-3) and in the Schoeller diagram (Fig. 3.4.1-4) the thermal water samples indicate mixing between the Na-Cl type of thermal water from the lower aquifer and the Na-HCO_3^- type of water from the upper aquifer.
- Deeply circulating water is heated up in the Paleozoic sandstones. The distribution of water temperature at well heads shows that two heat anomalies exists in the upper aquifer. These anomalies are close to faults such as the Jiza, Daba'a and Masattarat faults suggesting ascent of thermal water from the lower aquifer into the upper aquifer along these faults. The occurrence of high Cl^- concentrations along the faults of Daba'a and Wadi El Hammam suggests that these faults are the main upflow zones of thermal water from the lower into the upper aquifer.

References

Craig H (1961) Isotope variation in meteoric waters. Sci 133:1702 –1703

Dansgard W, Johnson SJ, Clausen HB, Langway CC (1971) Climatic record revealed by the camp century ice core. In Turekian K (ed) The Late Cenozoic Glacial Ages. New Haven Yale Univ. Press, pp 37–56

Giggenbach WF (1991) Chemical techniques in geothermal exploration. In: D'Aniore, F. (coordinator), Application of geothermal reservoir development. Rome UNITARI UNDP publication, pp 119–142

Jaser D (2002) Preliminary results of the geology and structures, geophysics, geochemistry and hydrological works in the course of the NRA-Geothermal project-Central Jordan, pp 1–165

Myslil V (1988) Report on evaluation of geothermal potential of Jordan. Strojexport prague, Gzechoslovakia, Geological Survey, Prague, pp1–42

Pearson FJ Jr, Rightmire CT (1980) Sulphur and Oxygen isotopes in aqueous Sulphur compounds. In: Fritz P and Fontes JC, Handbook of Environmental isotope Geochemistry, Vol. 1, Amsterdam, Oxford, New York, Elsevier Scientific Publ, pp 227–257

Salameh E, Rimawi O, Hamed KH (1991) Curative thermal and mineral water in Jordan. Water Resources Center Issue. No.-15. Jordan University-Amman, pp 1–195

Salameh E, Udluft P (1985) The hydrodynamic pattern of the central part of Jordan. Geol. Jahrb 38:39–53

Sawarieh A (1990) Hydrogeology of the Wadi Waleh Sub-basin and Zarqa Ma'in Hot Springs. MSc thesis, University of London, UK, pp 1–60

Sawarieh A (2004) Hydrogeology of thermal water from Zara-Zarqa Ma'in and Jiza areas, Jordan. Proceedings of the 5th Intern Symp Eastern Mediterranean Geology, Thessaloniki, Greece, Astract 3:1564–1567

Sawarieh A, Massarweh R (1996) Geothermal water in Zara–Zarqa Ma'in Area. Natural Resources Authority, Subsurface Geology Division Amman-Jordan, Internal Report 8, pp 1–36

Truesdell AH (1979) Final report on the chemistry and the geothermal Energy possibilities of Zarqa Ma'in–Zara hot springs-Jordan. USGS, Menlo Park, California, USA (unpubl)

3.4.2 GIS based model for groundwater flow and heat transport in Zarqa Ma'in and Jiza area, central Jordan

A. Sawarieh, C. Chen, M. Beinhorn, J. Gronewold, O. Kolditz, W. Ali, E. Salameh, H. Hötzl

3.4.2.1 Introduction

The study area is located in the central part of Jordan (Fig. 3.4.2-1) extending from the Dead Sea in the west to the Jiza region in the east. It covers Wadi Waleh (northern third of Wadi Mujib groundwater catchment) and Wadi Zarqa Ma'in sub-catchments with an area of about 2,300 km^2 (refer to Chap. 3.4.1).

The area consists of two aquifer systems, an upper and a lower one, separated by one major aquitard (A1-6; see Table 3.4.4-1) of Upper Cretaceous marls and marly limestone. The upper aquifer (B_2–A_7: see Table 3.4.4-1) comprises limestones and cherts of Upper Cretaceous age, and is mainly recharged along the northern and northwestern parts of the study area, where it is exposed and receives the highest rainfall averages of 250–400 mm/yr. Another recharge reaches the aquifer in the southeastern part of the study area and flows from the southern part of Wadi Mujib basin (Fig. 3.4.2-2). In Wadi Waleh sub-catchment the long-term (1975–2003) infiltration average amounts to 21.5 Mm^3/yr being 6% of the average total rainfall (357 Mm^3/yr) (Rimawi 2004). The direct and indirect recharge to Wadi Zarqa Ma'in sub-catchment amounts to 6 Mm^3/yr (Abu Ajamieh 1980). The groundwater within this aquifer flows mainly from north (recharge area) to south, then to the west, and from south to north joining the flow to the west. Part of the water flows further to the east towards the Azraq groundwater catchment (Fig. 3.4.2-2). The main natural outlets from the aquifer are the Haidan springs and the Zarqa Ma'in spring with average annual discharges of about 15.09 Mm^3/yr and 0.7 Mm^3/yr, respectively (WAJ open files).

The lower aquifer consists mainly of sandstones of Lower Cretaceous and older ages. The major contribution to the lower aquifer recharge comes from the upper aquifer by infiltration, which mainly occurs in the eastern parts of Jordan. The temperatures of the infiltrated water raise due to a normal to slightly elevated geothermal gradient. This groundwater flows westwards towards the Dead Sea. It reaches the Dead Sea as subsurface flow or discharges to the ground surface as thermal springs such as

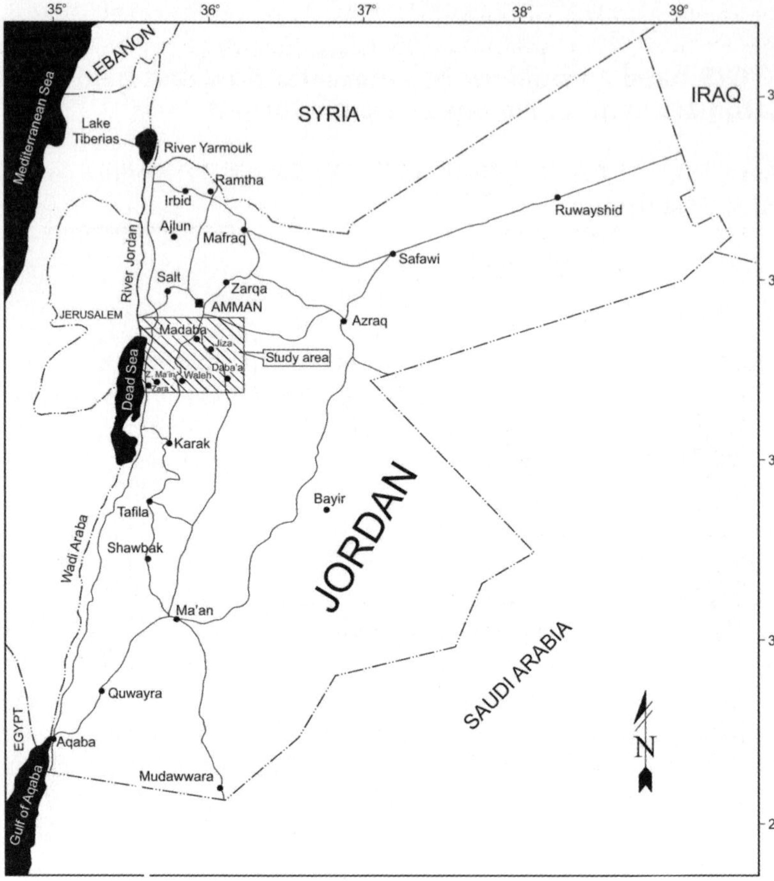

Fig. 3.4.2-1. Location map of the study area

the Zara and Zarqa Ma'in springs. The average discharge of the Zara- and Zarqa Ma'in thermal springs amounts to 20 Mm^3/yr (Salameh and Udluft 1985; Sawarieh 1990). About 5 Mm^3/yr discharge from the lower aquifer along the lower part of Wadi Haidan, increasing the total discharge in the wadi to about 23 Mm^3/yr at the confluence with Wadi Mujib.

During the last three decades in the study area about 300 wells were drilled into the upper aquifer. The private sector drilled most of them for agricultural purposes and the rest were drilled by the government for drinking purposes. Wells drilled mainly in- and east of the of Jiza-Madaba region discharge thermal water ranging in temperatures from 30–46°C. The average annual discharge of all wells is about 35.6 Mm^3/yr (Sawarieh, 2005).

The dense network of faults affecting the area may imply that the two aquifer systems are hydraulically connected, especially in the eastern parts of the study area. This allows the thermal water in the lower aquifer to ascend via some faults into the upper aquifer.

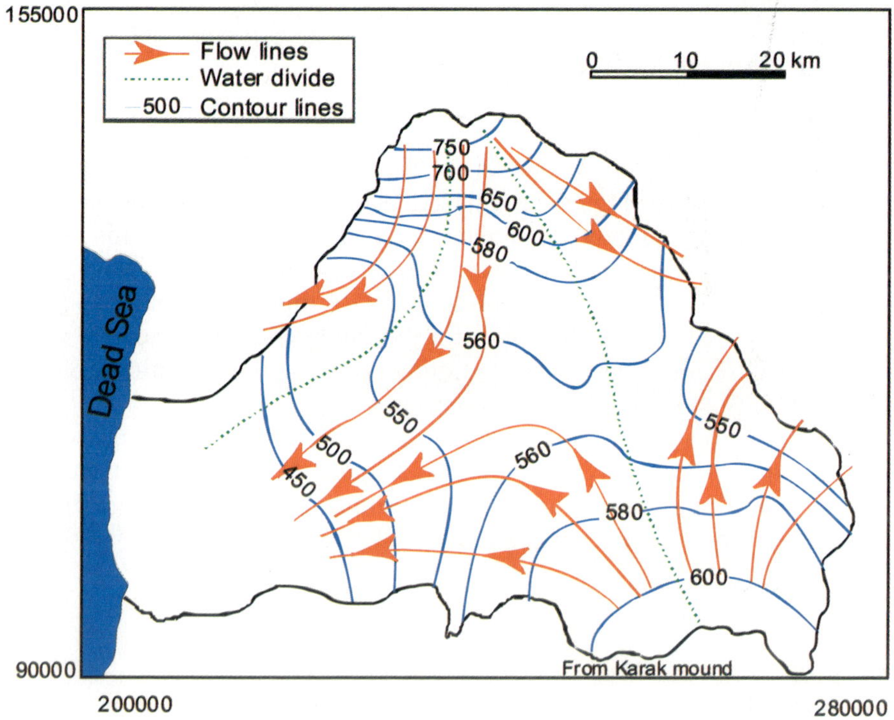

Fig. 3.4.2-2. Water level contour map of the study area

The isotope distribution indicates that the thermal water from both aquifers is of recent meteoric origin (Chap. 3.4.1). The chemical analyses and the isotope contents of the thermal waters from the upper aquifer indicate that it is a mixture of groundwaters derived from the two aquifers (Sawarieh et al. 2004). The detailed hydrology and hydrochemistry of the investigated area are described in Chap. 3.4.1.

In the course of this study, the GeoSys/RockFlow scientific modeling software (Kolditz 2002, Kolditz et al. 2007; Chaps. 3.3.5 and 3.3.3) is used in two applications: firstly to model (2-D) the water flow and heat transport across the Daba'a fault and secondly, to develop a 3-D numerical groundwater flow and heat transfer model for the whole investigated area. This is the first application of using the GeoSys/RockFlow (GS/RF) software for modeling groundwater flow and heat transport based on GIS Project (Chen 2006). The hydrogeological conceptual model of the study area is converted to shape files (GIS Project) using the ArcGIS software. All required data for modeling, such as aquifer geometry, boundary conditions, material properties, well positions and pumping rates etc. are kept in the GIS based data management project (Gronewold 2006) as described in the following section in more detail (refer also to Chap. 3.3.5).

3.4.1.2 Hydrogeological model

GIS Project

The first step in setting the model for the study area is creating the GIS project, where the geological data such as the geological formations, faults, wadis and the hydrogeological data such as aquifers, aquitards, wells, springs, water table, top and base of aquifers and boundary conditions were converted to shape files using the ArcGIS digitizing feature and kept in the GIS project. The GIS Project for this study consists of several shape files (layers): faults, wadis, springs, study domains, well locations, boundary conditions and water level contour maps.

The study area drains westwards to the Dead Sea by two major wadis: Wadi Zarqa Ma'in and Wadi Waleh/Haidan and their numerous tributaries such as Wadi Ez Za'afaran and Wadi Es Sufuq. The tributaries are dry except for short periods after rainfalls.

Fig. 3.4.2-3. Layers of major wadis and springs in the model domain

Perennial base flow occurs only on the lowest reaches, downstream of Wadi Waleh/Haidan, between elevations of 450 and 250 m. The Haidan and Zarqa Ma'in thermal springs are located within a short distance at elevations of 350 m and at about 200 m, respectively. Fig. 3.4.2-4 shows the model domain with the layers of the major wadis and springs.

The whole study domain is subdivided into 29 areas (basic surfaces) by adding the major faults and wadis to the hydrological units (Fig. 3.4.2-4). 22 surfaces represent the upper aquifer system (0–21), 3 surfaces form the aquitard (22–24), and 4 represent the lower aquifer system (25–28). The hydraulic conductivity values computed from all wells located in each surface are averaged and given as a hydraulic conductivity value for that hydrogeological unit.

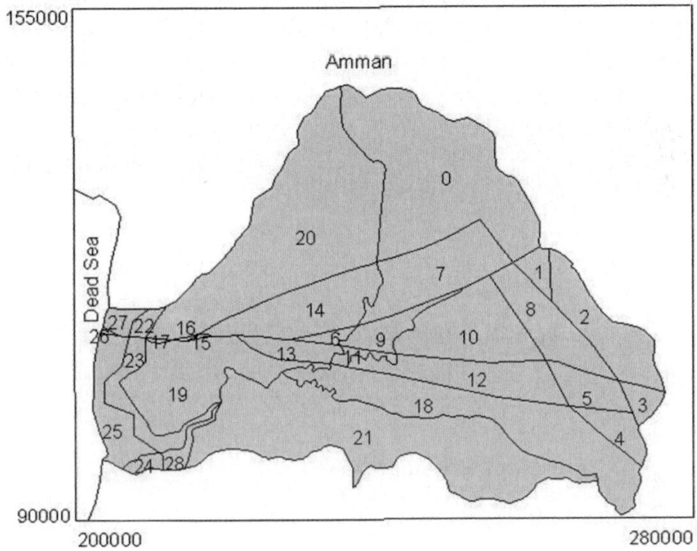

Fig. 3.4.2-4. Basic surfaces of the modeling domain (basic surfaces). The Daba'a Fault corresponds to the boundary of the arrays 12 and 18. See also Fig. 3.4.1-5

For the upper aquifer, the recharge boundary is set along the northern and western highlands, where the major recharge points are located. A no-flow boundary is set along the western border except for the main discharging outlets (Haidan springs and Zarqa Ma'in spring) in the lower reaches of the wadis. An outflow boundary is set along the eastern boundary, where the water is leaving the system to the Azraq groundwater basin in the east. The southern border is divided into a recharge boundary in the south-eastern corner and no-flow boundary in the western part (Fig. 3.4.2-5). For the lower aquifer, no-flow boundaries were set to the northern and southern borders, inflow (recharge) boundary to the eastern border, and an outflow boundary for the western border.

Modeling project

All hydrogeological and hydrological data introduced above can be conveniently imported into the GS/RF software. The data reference for the model such as geometry data, meshing data and processing data are stored in this project file to keep the connection between the project and its data components.

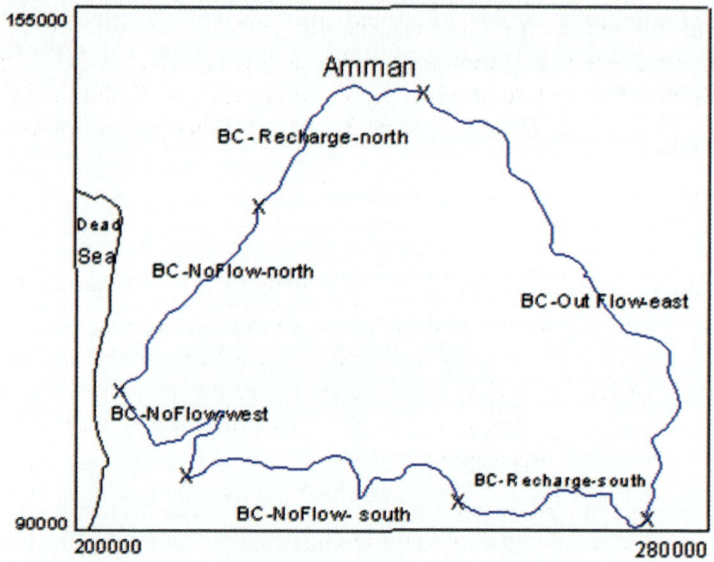

Fig.3.4.2-5. Boundary conditions layer of the upper aquifer system

Geometry

Shape files or the GIS project are imported and converted into the corresponding geometric data structures (GeoLib) as basic geometric data for the modeling area (Chen 2006). Geometric data are composed of points, polylines, surfaces, volumes and domains. A "point" is the basic element of all geometry types which are used to define 1-D geometry elements such as source terms (springs and wells) and boundary condition. A polyline consists of "points" and it is used to define 2-D geometric elements such as boundaries of study domains, boundaries to assign boundary conditions and linear structures (faults). The surface is composed of one closed polyline, the 2-D boundary domain. The volume is composed of top and bottom surfaces and it builds the 3-D component of the domain.

The whole study domain is subdivided into 29 areas due to faults and wadis located in the study domain which are treated as basic surfaces in a

GS/RF project. Each surface is created from a polygon. After all basic surfaces are created, the mesh is generated by using the external program (GMSH) which creates triangulations for all basic surfaces. The results are stored in a mesh data file, which consists of geometric information of nodes and elements of the triangulations. The existing triangulation is used to create TINs (triangulated irregular networks) for the discrete geometric description of surfaces and of volumes. Each TIN object has the same base name as the corresponding surface. The conceptual structure model of the study domain consists of three layers: the upper aquifer, the lower aquifer and the aquitard in between. Therefore, three copies of the surfaces are needed to create the 3-D layer model based on TINs for the original 2-D model.

Meshing

The hydrological/hydrogeological system of the Zarqa Ma'in and Jiza area consists of several hydraulic features, such as wadis, faults and the aquifer as well as aquitard. In order to develop a numerical model including all hydraulic important elements, a hybrid meshing procedure was necessary which couples different types of elements. The general steps for the hybrid meshing procedure in this application are:

- Creation of a triangulation based on surfaces for the 2-D surface model. The surfaces must be topologically consistent
- Creation of the line elements for the 1-D Wadis model based on the triangulation
- Creation of the prismatic elements for the 3-D subsurface model by extending the triangles to the vertical direction
- Creation of the quadrilateral elements for the 2-D fracture/faults model

Mapping

The last step to create the structure model is mapping the finite element mesh based on a digital elevation model (DEM). Surface mapping is the process of making the mesh conform to stratigraphic irregularities, i.e. the thickness and orientation of the mesh slices can be deformed. The types of mapping data, which can be used, are grid data created by either Surfer or ArcView. Four grid files have been created for the upper aquifer (B2-A7) and the lower aquifer (Ks). Fig. 3.4.2-6 shows the final result of the meshing and mapping procedure, the mapped (inside the hybrid) finite element mesh including line, triangle, quadrilateral and prismatic elements to represent the wadis, surfaces, faults as well as aquifers and aquitards.

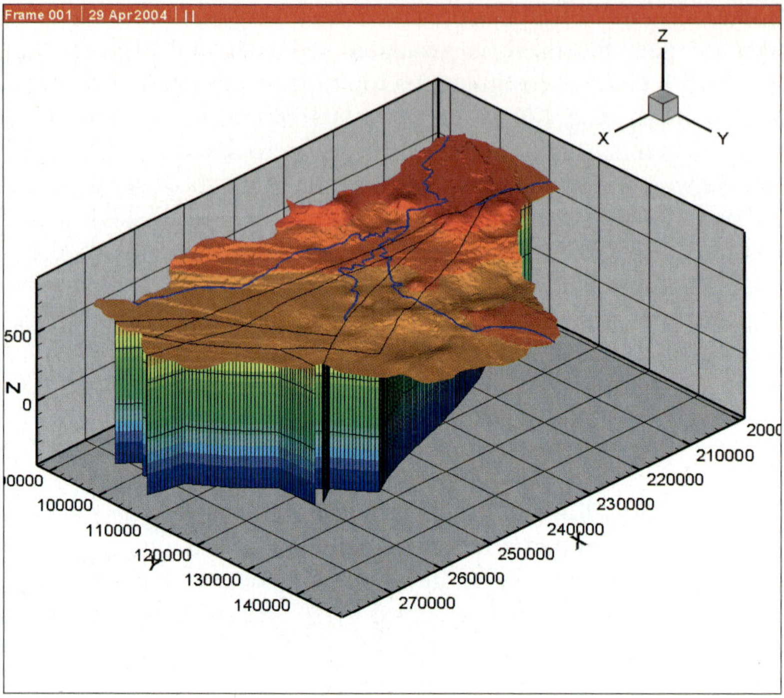

Fig. 3.4.2-6. Inside the hybrid finite element mesh representing the 3-D structure model of the Zarqa Ma'in and Jiza area. In the X-Y plain, elevation increases from brown to orange to red. In Z direction, the upper aquifer is marked in yellow; the lower aquifer is given in blue; the aquitard is marked in green

Process data

After having finished the required geometric and meshing operations, the physical processes are specified (i.e., groundwater flow and heat transport).

Boundary conditions and source terms

The *boundary conditions* (Fig. 3.4.2-5) and source terms (i.e., wells and springs) of the hydraulic system are defined by the recharge and discharge to or from the model area as well as by discharges from springs. The heat flow from the base (100 mW/m^2) is presented by source terms distributed at the bottom surface.

Three types of boundary conditions are given for the upper and the lower aquifer by importing their shape files as polylines. The boundary conditions at surfaces were assigned as identical to those polylines. The water table measurements of the year 1985 were used as initial boundary condition. Hence hydraulic heads at the study domain boundary are used to simulate the hydraulic conductivity within the study area. Points are used to specify the springs as source terms.

Material properties

Material properties have to be specified for fluids, solids and the porous medium (i.e., the aquifers and aquitards). These material data are connected with the finite elements by a material group number. For convenience, the material data have been prepared as Excel tables, which can easily be imported. Fig. 3.4.2-7 shows the geometric distribution of material groups in the model domain based on the hydrogeological data. In the 3-D model all material groups are linked to volume objects. In a finite element model all units belonging to a specific geometric object inherit data from the material group connected to this volume.

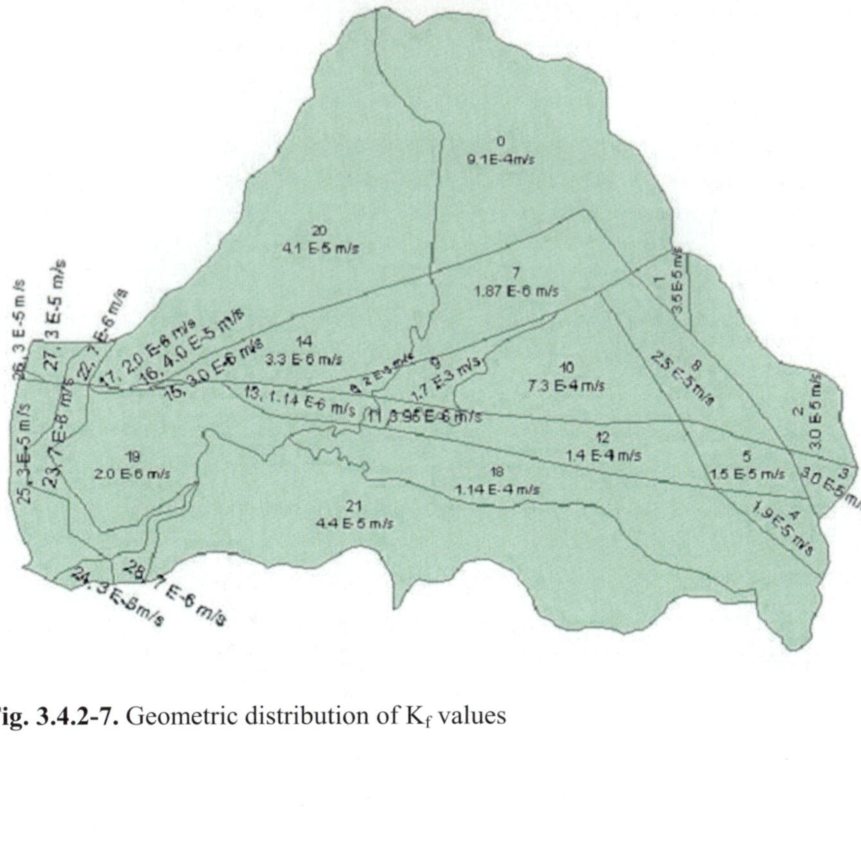

Fig. 3.4.2-7. Geometric distribution of K_f values

3.4.2.3 Simulation results

Groundwater flow modeling

Steady state

The hydraulic system in the area is controlled by recharge and discharge conditions to or from the model area as well as by discharges from Haidan springs. Fig. 3.4.2-8 shows the simulation results of hydraulic heads under steady-state conditions. It can be seen from the figure that high heads are found in the recharge areas in the north and in the southeast. It also shows the effect of the Haidan spring on the groundwater flow. The simulated heads compare well to the water heads measured in 1985.

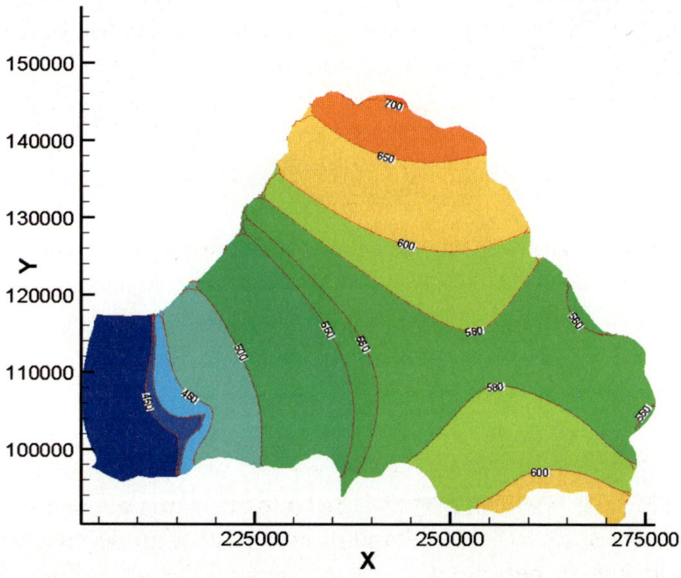

Fig. 3.4.2-8. Well head distribution (steady state) in the upper aquifer

Transient state

In the transient state, well data have to be involved in the hydro-system model from the available Excel data base. This data, i.e., well positions and pumping rates are directly imported by the software. Fig. 3.4.2-9 shows the resulting distribution of hydraulic heads.

Although this is the first implementation of the software based on GIS data in modeling the groundwater flow both for the steady and transient states, the simulation results gave reasonable results. The model was not calibrated for the transient state simulation. Therefore, the simulation results form a solid base for further work to develop a 3-D calibrated flow model for the area using the present GIS Project.

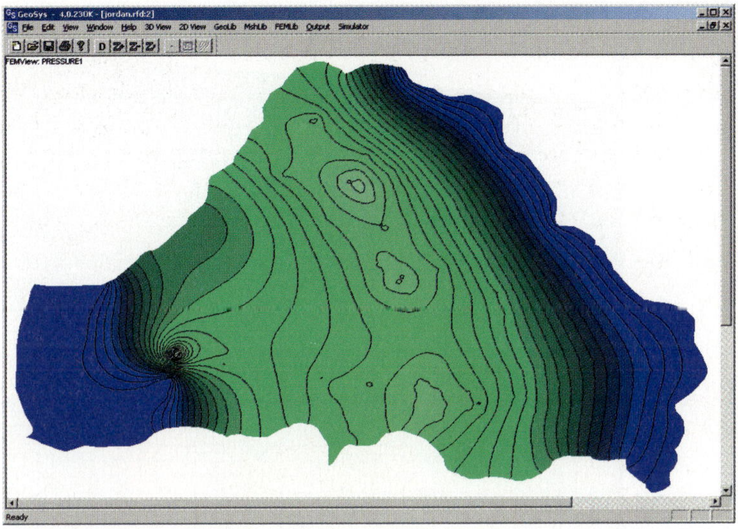

Fig. 3.4.2-9. Transient state hydraulic head in the upper aquifer including pumping data (Green to blue indicate high to low water levels)

Heat transport modeling

Two applications were carried out to model heat transport in the study area. In the first application we used a vertical N-S cross section, 12 km long, crossing the Daba'a Fault, which runs E-W through the study area. Aquifer layers are represented by 2-D rectangular elements while the fault zone is represented by 1-D line elements and given high permeability. The heat transport had to be modeled and coupled to the water flow process. The temperature distribution in the aquifer influences some material parameters, like fluid density and fluid viscosity. Warm water has lower density and thus rises relatively to cold water. The upconing effect of warm water at the fault can be clearly seen by the higher temperatures along the vertical fracture (Fig. 3.4.2-10). Vertical heat transport in the model area without fractures is much less pronounced, which shows the relative importance of heat transport by water movement compared to heat conduction in the aquifer material and stagnant water.

Based on the 3-D model of the groundwater flow regime in the model domain, a 3-D heat transport model is set up to investigate the long-term thermal regime in this region. The heat flow from the base is assumed to be 100 mW/m^2, and is represented by source terms distributed at the bottom surface.

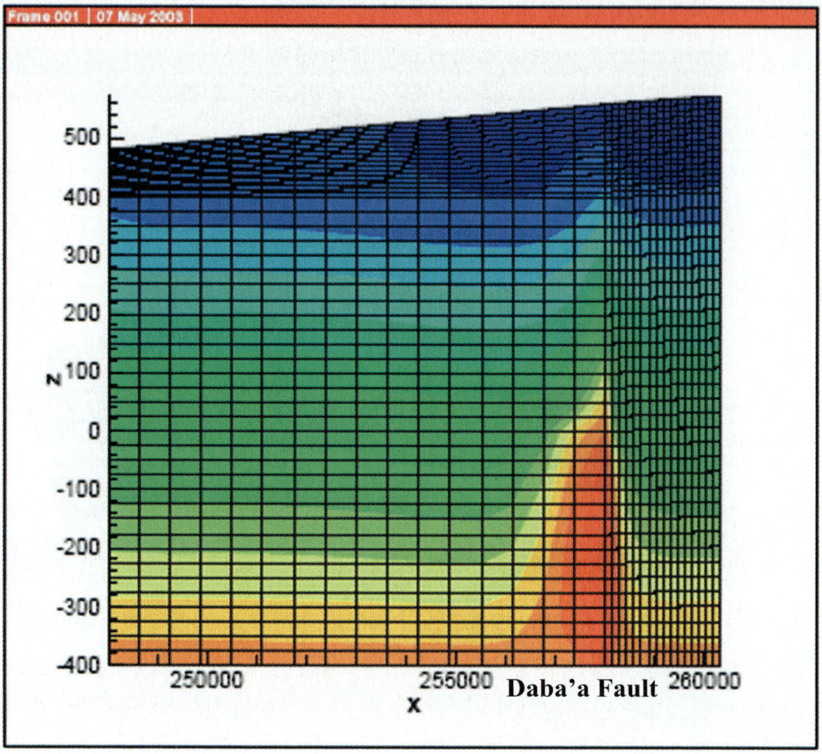

Fig. 3.4.2-10. Vertical N-S distribution of temperature in the fractured aquifer. High temperatures are shown in red when crossing the Daba'a Fault running E-W (see Fig. 3.4.1-5)

Fig. 3.4.2-11 shows a first long-term simulation (30,000 years) of the thermal system based on the hydraulic model presented before. It shows that there is a clear temperature increase along the faults, but it also shows a permanent increase of temperatures in the whole system. In other words, groundwater entering the system is not equilibrating the base heat flux. This points to possible defects in the current model which could be due to the underestimation of groundwater recharge to the domain or to overestimation of the base heat flux. This means that by involving thermal data in the simulation, the hydraulic model should be improved.

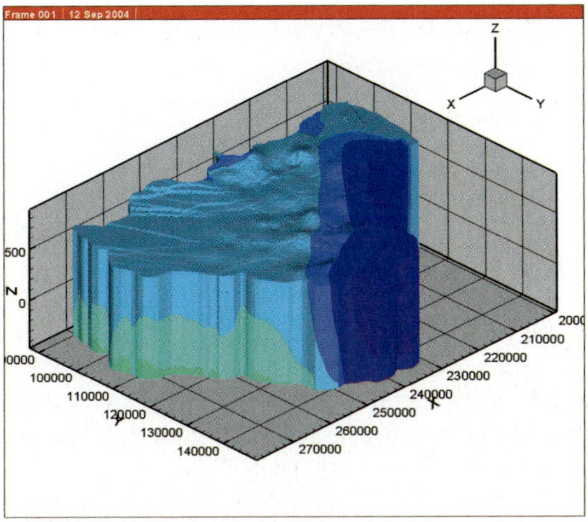

Fig. 3.4.2-11. Temperature distribution in the model area. Light green color indicates enhanced temperatures across fractures

3.4.2.4 Conclusions

In this work the development of a catchment-scale 3-D groundwater flow and heat transport model for the Zarqa Ma'in and Jiza area is presented. This is the first numerical model for this area comprising the aquifer/aquitard, fault system as well as the wadi network. The data processing for the model set-up is completely based on GIS projects and can be therefore be easily adopted. The calibration of the steady-state groundwater model is based on groundwater heads measured in 1985. The groundwater flow model shows the influence of transient effects due to groundwater abstraction in the area. A further calibration of the groundwater model is proposed for future work in order to make the modeling tool applicable for water resources management.

The heat transport simulation results (vertical cross section, Daba'a Fault) show the clear upconing effect of warm water at the fault, i.e., higher temperatures along the vertical fracture compared with other parts of the section. This shows the relative importance of heat transport by water movement compared to heat conduction in the aquifer material and stagnant water.

The long-term thermal simulation shows that there is a clear temperature increase along the faults but it also shows a permanent increase of temperatures in the whole system. In other words, groundwater entering the system is not equilibrating the base heat flux. This indicates a possible defect in the current model most likely due to the overestimation of the base heat flux. This means that by involving thermal data in the simulation, the hydraulic model might be improved.

References

Abu Ajamieh M (1980) The geothermal resources of Zarqa Ma'in and Zara. Report of the geothermal energy project in Jordan (phase-1), Natural Resources Authority, Amman-Jordan

Chen C (2006) Integrating GIS Methods for the Analysis of GeoSystems. PhD thesis. Series C Hydro-, Ingenieur- und Umweltgeologie der Tübinger Geowissenschaftlichen Arbeiten (TGA), Uni Tübingen, pp 1-157

Gronewold J (2006) Development of an internet information system for the modelling of natural attenuation processes in groundwater (in German). PhD Thesis, Universität Tübingen, Germany, pp 1-72

Kolditz O (2002) Computational methods in environmental fluid mechanics. graduate textbook, Springer Publisher, pp 1–104

Kolditz O, Kalbacher T, Wang W, Bauer S, McDermott CI, Chen C, Gronewold J, Beyer C, Walsh R and Park CH (2007) GeoSys/RockFlow manual, Center for Applied Geosciences, Uni Tübingen, http//www.uni-tuebingen.de/zag/geohydrology/index.hmtl

Rimawi O (2004) Groundwater Flow Modelling of Wadi Waleh Water Resources (Amman-Wadi Sir Aquifer System). Draft Final, University of Jordan, Amman, Feb.2004. 89- Salameh E. and Khudeir, K. (1983): Thermal water system in Jordan, Neues Jahrb Geol Paläntol Monatsh 4:249–256

Salameh E, Udluft P (1985) The hydrodynamic pattern of the Central part of Jordan. Geol. Jb 38:39–53

Sawarieh A (1990) Hydrogeology of Wadi Waleh Sub-basin and Zarqa Ma'in Hot Springs. MSc. thesis, Department of Geology, University of London, UK, pp 1–57

Sawarieh A (2005) Heat Sources of groundwater in the Zara-Zarqa Ma'in-Jiza area, Central Jordan. PhD thesis, AGK, University of Karlsruhe, Germany, pp 1–165

Sawarieh A, Hoetzl H, Salameh E (2004) Hydrochemistry and Origin of Thermal Water from Zara-Zarqa Ma'in and Jiza Areas, Central Jordan. 5th Intern Symp Eastern Mediterranean Geology, Abstracts, Thessaloniki, pp1560–1563

3.4.3 Groundwater conditions of the Jordan Rift escarpment northeast of the Dead Sea

W. Ali, J. Glaser, H. Hötzl, S. Lenz, E. Salameh, M. Thiel, H. Werz

3.4.3.1 Introduction and objectives

Though in a first view the general geomorphologic and geological characteristics of the escarpments on both side of the Jordan Rift Valley and the Dead Sea show many similarities, the hydrogeologic conditions differ due to the lithologic variety of the Cretaceous rock-sequences. The deeply lowered Dead Sea forms the base level for the mountain ranges extending along both sides of the Valley. However, while on the western side the two main Cretaceous karst aquifers are dewatered by several large springs (see Chaps. 2.2 and 3.3.1-3.3.3), only small springs discharge along the foothills on the Jordanian side. Most of the water there results from surface flow in the wadis, seeping into the alluvial fans extending at the emergence of the wadis towards the Jordan Valley. Along the shoreline these wadis discharge directly into the Dead Sea, some of them even during the dry summer season.

In order to understand the different discharge patterns and to evaluate the groundwater resources, specific studies were carried out in the hinterland of Suweimeh, particularly in Wadi Udeimi at the north-eastern corner of the Dead Sea, and in Wadi Kafrein and Wadi Shueib further to the north. The hydrogeological conditions of Wadi Shueib are discussed with regard to groundwater vulnerability in Chap. 3.4.4. In this study, the general aquifer conditions are discussed by considering an escarpment section from Suweimeh up to Mount Nebo and in Wadi Kafrein (Fig. 3.4.3.-1)

The main objectives of this study are:

- Compilation of a geological cross section from the area of Amman down to the Jordan Valley at a scale of 1:10,000 based on detailed geologic mapping
- Characterization of the different existing aquifers and their structural relationships
- Hydraulic evaluation of aquifers and pumping tests in the study area.
- Surveying of the springs and hydrochemical analysis of spring water
- Classification and interpretation of spring waters by hydrochemical parameters and stable isotopes

- Water storage in reservoirs, seepage investigation including a tracer test and water level monitoring of wells.

This area is covered by geological maps of 1:50.000 published by the Jordan Natural Resources Authority. The general hydrological properties of the Mesozoic sequence are described by Salameh and Bannayan (1993),

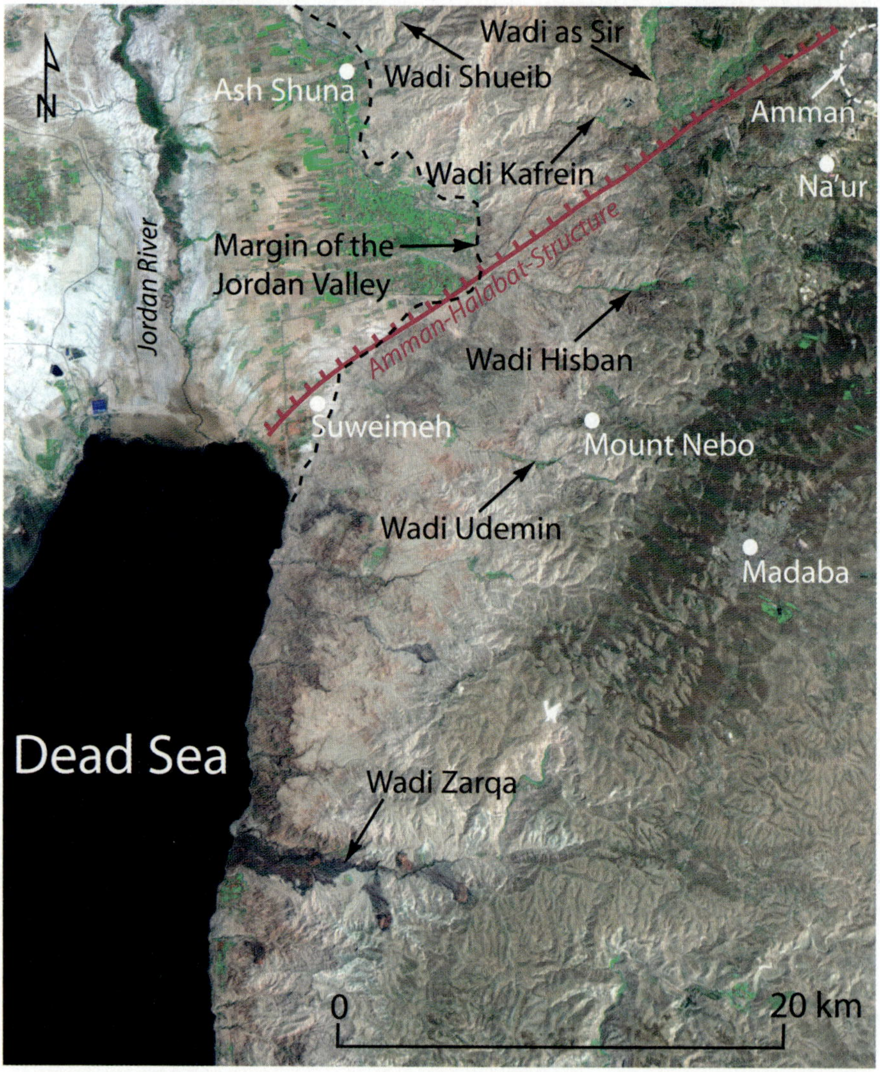

Fig. 3.4.3-1. Satellite image of the study area with location of the main wadis and settlements (source: Landsat ETM^{+} data from 21st of May 2001, kindly provided by the Natural Resources Authority Amman)

Margane et al. (2002) and others. Information on precipitation, discharge of Wadi Kafrein, and waste-water discharge were kindly made available by the Jordan Ministry of Water and Irrigation.

3.4.3.2 Geomorphology and geology

The study area extends over 20x20 km² on the Jordan Rift escarpment starting from the north-eastern part of the Dead Sea near the village of Suweimeh, northwards to the village of Kafrein at the margin of the Jordan Valley. Along the mountain crest it includes part of Mount Nebo (west of Madaba) up to north of Amman. The south-western suburbs of the city are dewatered by Wadi es Sir, a tributary of Wadi Kafrein (Figs. 3.4.3-1–3).

In the southern part of the study area, the escarpment forms steep cliffs overlooking the Dead Sea shoreline (-415 m msl) and reaching at Mount Nebo an altitude of +900 m msl. At Kafrein the foothills start around -380 m msl and west of Amman, the mountain crest reaches altitudes of +1000 m msl.

The slope relief differs according to the geologic structure. South of Wadi Hisban, the slope is of a step-like character. As shown in the digital elevation model in Fig. 3.4.3-4, normal faulting and gravitational block rotation. The lithologic differences in the hardness of the rock sequence cause repeated changes from steep rock walls to flat-slope angles. In the northern part along Wadi Kafrein, the escarpment slope is determined by the layering of the rocks which is dominated by a flexure-like subsidence of the Cretaceous sedimentary sequence starting at the crest with nearly horizontal layering, then followed by increased inclination and reaching locally steep reverse stratification at the margin to the Jordan Valley.

The escarpment is deeply incised and dissected by wadis leading to extreme relief conditions. In their upper parts, the wadis show a wide trough-like shape but become narrow in their lower part, where they form steep gorges in the sandstones. From north southwards, the three main wadis are (i) Wadi Kafrein with its two tributaries, Wadi es Sir and Wadi Naur, (ii) Wadi Hisban (in the central part) which follows partially an important fault system oblique to the rift structure and opens out in the Jordan Valley with an wide triangle shaped alluvial fan and (iii) finally Wadi Udeimi in the south.

The escarpment itself is a product of geological processes forming the Jordan-Dead-Sea Rift System (Horowitz 2001). Though it resembles in its morphologic form an extensional tectonic graben, the Rift system is one of the most impressive transform systems by which the Arabian plate on the Jordanian side moved along a left lateral shear zone more than 100 km northward relatively to the Levant-Sinai block on the western side (ref.

Fig. 3.4.3-2. View from the mountain crest south of Mount Nebo down the escarpment, over the southern part of the study area, towards the Dead Sea (Photo: Hötzl 2006)

Fig. 3.4.3-3. View towards northeast on Wadi es Sir, one of the two main tributaries of Wadi Kafrein. The City of Amman is located on the highest background ridges (Photo: Hötzl 2006).

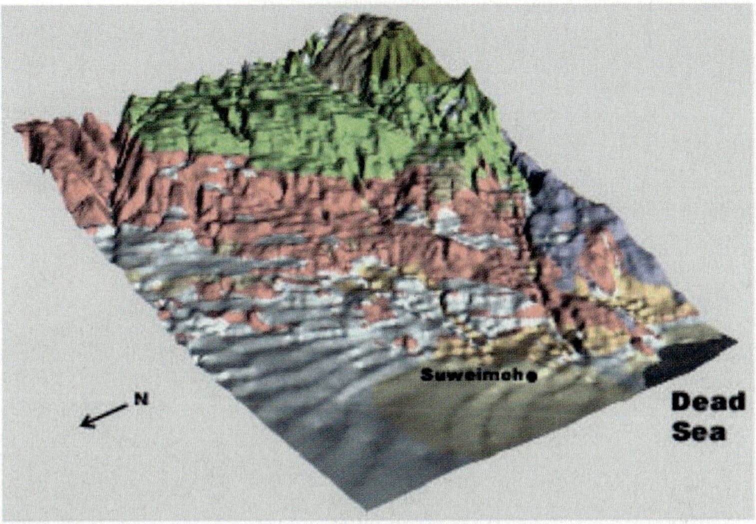

Fig. 3.4.3-4. Digital elevation model (DEM) of the southern study area with the geological coverage, view from north-west (Thiel 2002; Glaser 2002). Legend: red= Kurnub Group; green =Ajlun Group; dark blue = Quaternary basalt; grey = Lisan Formation

Chap. 2.1). A slight uplift of the Arabian plate was connected with the shear movement and, together with lateral displacement, is responsible for the stratigraphic differences on both sides of the Rift. This uplift movement associated with a downward dragging of uplifted blocks, could be responsible for the flexure-like structure of the bedding planes in the Wadi Kafrein area and can be observed from there along the escarpment northward.

The differences relatively to the step-shaped configuration of the escarpment south of Wadi Hisban and further southwards along the Dead Sea, can also be explained by the specific tectonic situation in this section. Responsible for that is the opening of the elongated south-north directed pull-apart basin which is related to the left-lateral shear movement, which formed the Dead Sea. Geomechanically, the long sides of this deep, open pull-apart basin is completely instable and caused gravitational mountain- and slope-sliding which are recognizable now in en-echelon normal fault patterns on both sides of the Dead Sea. Even in recent times, the persistent slope instability is manifested by continuous mass movement along rather shallow lystric planes (Fig. 3.4.3-5).

The recession of the rock slopes at the northern end of the Dead Sea and the following wide indentation of the alluvial fan of Wadi Hisban are caused by the Amman-Halabat structure. Its northeast strike direction and its structure dissects the escarpment from the Dead Sea close to the village

Fig. 3.4.3-5. Example for the instability of the Rift escarpment: block built of the Ajlun Group with its original continuation on the slope top and subsided along a lystric plane. Location: main road from Mount Nebo (Photo: Hötzl 2005)

of Suweimeh up to Amman. This fault zone is 1–5 km wide and consists of many structural elements including faults and folds controlled by the same stress system, which belongs to the pre-Rift system of the Syrian arch (see Chap. 2.1). The north-west down-throw of this fault causes a vertical displacement up to 400 m of the lower Wadi Kafrein area compared with the adjacent southern part (Shawabekeh 2001).

The rocks forming the escarpment belong to the Palaeozoic to Cainozoic sedimentary sequence (Chap. 2.1). In the study area, the slopes are mainly build by Cretaceous sediments (Fig. 3.4.3.-6). The older Palaeozoic sandstones and siltstones (Ram Formation) are only present in the deeper underground as it is known from oil and water drilling, and from outcrops further south in the Wadi Zarqa and Wadi Mujib (Chap.3.4.1), where they reach a thickness of 700 m and more.

The Permian to Jurassic sequence (Zarqa Group), which reaches a thickness of nearly 1,000 m in the southern wadis, is strongly reduced in the area of investigation due to erosional processes before sedimentation of the Lower Cretaceous. Thus, the slopes at the lower part of the escarpment are

are mainly built up by the Lower Cretaceous Kurnub Group. The sequence consists of multi-colored medium- to coarse-grained quartzose sandstones.

Fig. 3.4.3-6. The lower part of the Rift escarpment southeast of Suweimeh. The picture shows the superposition of the Lower Ajlun Group on the upper part of the Kurnub sandstones (Photo: Thiel 2002)

The upper part shows gradual increase in fine-grained sandstone with proportions of marl, clay and siltstone. The formation is exposed along a wide zone from the north-eastern corner of the Dead Sea along the NE-striking Amman flexure zone towards the Wadi Hisban. North of Wadi Hisban along the ridge parallel to the wadi, the Kurnub Group is covered by superposed limestones of the Ajlun Group, which also crops out along the lower parts of the slopes in Wadi Kafrein.

The Middle Cretaceous Ajlun Group (A1–A7; ref. Fig. 3.4.4-1) overlies the Kurnub Group. It contains marls and limestones with intercalated chert and dolomite (Fig. 3.4.3-6). In the study area, the whole sequence has a thickness of more than 400 m. Vertical displacements along normal faults causes even larger thicknesses. The Coniacian-Campanian Belqa Group is at the highest stratigraphic position in the study area, and is built of chalk and silicified limestones.

Only on the southern margin of the study area the sedimentary sequence is partly covered by Quaternary basalts. The eruption centre is situated southeast of Suweimeh at the crest of the escarpment, from where lava spread down-slope unto the shoreline of the Dead Sea. Contrastingly, the main part of the scarp-foot is covered by the Pleistocene marly Lisan Formation and by Holocene talus deposits. The floors in the upper parts of the wadis are frequently formed by eroded bed-rocks only partly covered by gravels or older terrace remnants. Downstream the thickness of gravels increases and passes into the wide alluvial fans mouthing into the Jordan Valley.

3.4.3.3. Hydrogeology

In the sedimentary sequence of Jordan (Chap. 2.2), the following aquifers are present in the study area (from base to top):

- The Cambrian to Ordovician Ram Group aquifer ('Disi aquifer') of sandstones (not outcropping and only present in the deep underground)
- The Lower Cretaceous Kurnub Group sandstone aquifer
- The Upper Cretaceous aquifer including members of the Ajlun and Belqa Groups. In the Ajlun Group the formations consisting of thick limestones and dolomites (A2, A4; ref. Fig.3.4.4-1) form aquifers of local importance, whereas the marly formations (A1, A3, A5/6) form aquicludes. The Wadi es Sir-Amman system (A7/B2) is the most important aquifer in the country (Parker 1970) due to its wide extension and high thickness.
- The Upper Tertiary/Quaternary sedimentary complex of the Jordan Valley includes the Shaquar (JV1), Ghor el Katar (JV2) and Lisan (JV3) Formations as well as the Holocene alluvium (Qal). It forms the major aquifer in the Jordan Valley area. These primarily gravel- and sand-containing sediments in the marginal parts of the valley have generally high permeabilities. The adjacent silts, marls and evaporites in the more central part of the Valley floor are of low permeability and sources for highly saline water.

Groundwater recharge

The recharge conditions along the Rift escarpment on the Jordanian side vary considerably due to the distribution of precipitation, conditions of evapotranspiration, slope angles, and infiltration conditions of the rock. The main factor of influence is precipitation which varies from less than 100 mm/yr along the coast of the Dead Sea (arid climate) and nearly 600 mm/yr in the highlands near Amman (Mediterranean climate). In order to demonstrate the relationship between the yearly rate of precipitation and the altitude above the Jordan Valley, three stations with long term measurements are selected in the study area. The stations are Wadi es Sir (in the highland) at an altitude of +720 m msl with 550 mm yearly average precipitation, Addasieh (on the escarpment) at +530 msl with 260 mm, and South Shuneh (on the foothills) at an elevation of -230 msl with 180 mm (Fig. 3.4.3.-7).

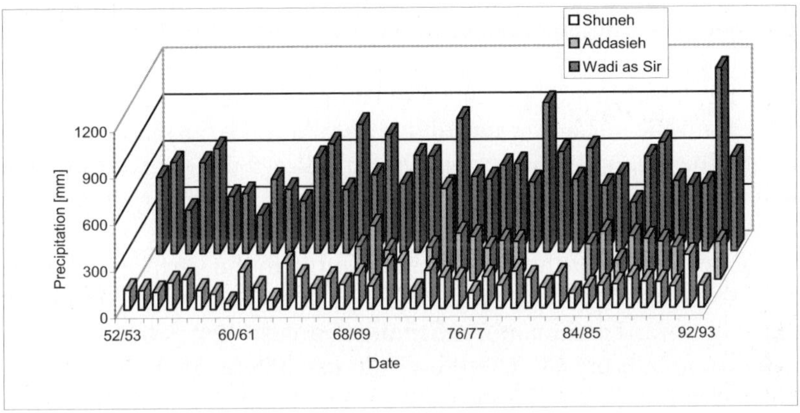

Fig. 3.4.3-7. Long term mean annual rainfall distribution along the catchment area at the stations of Wadi es Sir (+720 m msl), Addasieh (+530 m msl) and South Shuneh (-230 m msl). Data are supplied by the Water Authority of Jordan (2000).

The distribution of rain during the year is strictly seasonal with rainfall in the winter months. The diversity of rainfall is enhanced by evapotranspiration. In the Jordan Valley the highest temperatures coincide with highest evapotranspiration rates and with the smallest amount of rainfall.

Another inappropriate condition are the steep slopes which favor surface runoff. The outcropping area of the deeper aquifers and their catchment is restricted to some very small stripes or to cliffs along the escarpment. The rocks in the crest region from Madaba to Amman belong to the Upper Cretaceous aquifer and, due to the described flexure, are exposed in the northern part down to the Jordan Valley. Moreover, the limestones of the upper aquifer are highly karstified and facilitate fast infiltration of rain water. In comparison with other karst regions in arid areas, the recharge to groundwater could be 60% of the yearly rainfall. Along the Jordanian Rift escarpment, recharge is reduced to 50% due to the steep slopes. Considering the average precipitation of 500 to 600 mm/yr in the upper part of the escarpments the recharge of groundwater could be up to 250 mm/yr. In their lower parts, groundwater recharge decreases rapidly so that in the lower third it is more or less zero because of the low and rare precipitation and the high potential evaporation. Thus, the Lower Cretaceous Kurnub Group aquifer, which is exposed in the deeper part of the escarpment, is not directly recharged by rainfall infiltration. The replenishment of groundwater is caused mainly by leakage from the upper aquifer through marls along faults and fractures.

Groundwater flow

Groundwater flow in the rocks of the Rift escarpment is conditioned by the slope inclination and by the deep base level in the Jordan Valley and finally, of the Dead Sea (Fig. 3.4.3-8). The main question is how far backwards beyond the surface-water divide can the underground catchment be extended. The original dewatering system follows the eastward inclination of the bedding planes. Though the limestones of the Upper Cretaceous aquifer are karstified, the backward underground erosion in this system has not really developed further to the east relative to the surface-water divide. This is also proved by wells located east of Madaba or by the big spring in the old city-core of Amman confirming the eastward ground-water flow from the mountain crests. Therefore, the catchment of the escarpment in the study area is restricted to the westward inclined slopes. This is different from areas further to the south and north, where wadis such as Wadi Mujib (Chap. 3.4.1) cut through the escarpment eastward into the subsequent plateau, tapping the eastward groundwater flow.

The westward flow of groundwater in the escarpment area is restricted to the outcrops of the aquifer base along the slope (Fig. 3.4.3-8). The deepest intersection of the base is naturally located in the wadis and their tributaries where numerous small springs indicate groundwater outflow. For the Upper Cretaceous aquifer the intersection of the aquifer basis with the surface occurs at an altitude between +500 and +700 m msl depending on the tectonic displacement. In the wadis these springs emerge approximately from Ayun Musa north of Mount Nebo, to the village of Adassya (Wadi Naur) and further on to Iraq al-Amir (Wadi es Sir). Beside these main springs emerging in the wadis, numerous small and only temporal springs discharge from small notches and depressions along the protruded ridges at the base of the aquifer or along fault zones.

Only in the main wadis and at least during the wet season, the flow from these springs reaches down to the Jordan Valley. The flow from small springs disappears after a short distance in gravels or seep away into intercalated limestones of the Lower Ajlun Group.

In the lower parts of the escarpment, where marls of the Ajlun Group and sandstones of the Kurnub Formation build the slopes, only few springs occur. As previously mentioned, the low precipitation rate and the high potential of evaporation prevent nearly completely groundwater recharge. The only natural outflow from the lower aquifer (Kurnub Group) occurs directly south of Suweimeh, close to the Dead Sea shore, where warm water with a temperature of 28°C discharges - probably along a N-S fault. However, these springs with a total discharge of 3–5 l/s cannot represent the total groundwater discharge from this section of the escarpment. In addition to these springs, there are few wells in the foothills of the escarpment drilled to test the yield of the lower aquifer. These wells partly penetrate the deeper Zarqa Group. The temperature and the chemical composition of this confined and artesian water (Fig.3.4.3-9) is similar to that of the warm springs south of Suweimeh.

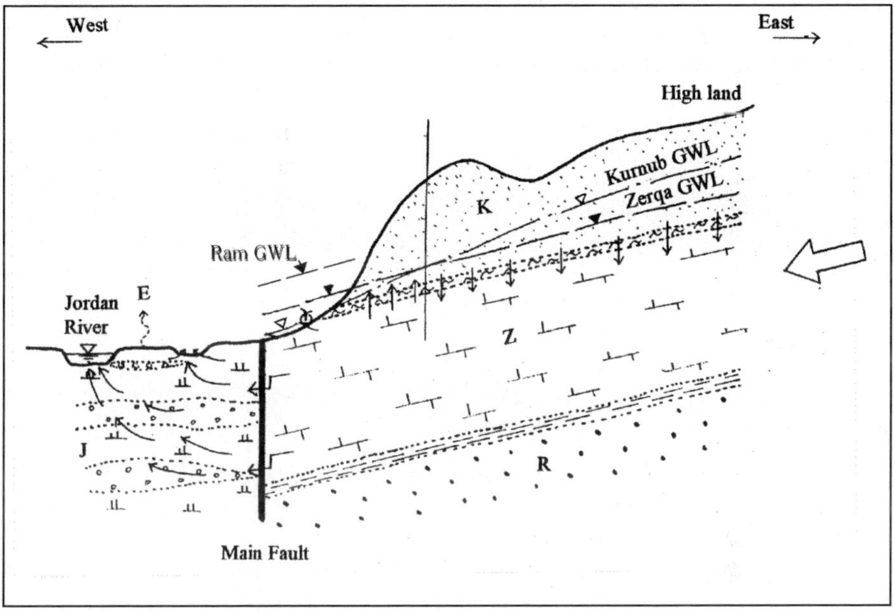

Fig. 3.4.3-8. General scheme of groundwater flow in the lower part of the study area (modified and complemented after JICA 1995). Explanation of the legend: R-Ram Group, Z- Zarqa Group, K- Kurnub Group, J- Tertiary and Quaternary sediments of the Jordan Valley, GWL- piezometric level, E -Evaporation

Fig. 3.4.3-8 shows a schematic graph of the hydraulic pressure in the Kurnub Group as well as in the deeper Zarqa and Ram Groups as it was originally drawn by JICA (1995) on the base of data from wells from different locations. The origin of the water in the different sandstone aquifers is still unclear. Some authors assume seepage from the Upper Cretaceous aquifer of the Azraq basin in the east into the sandstones below. From there the water flows according the hydraulic gradient towards the Dead Sea as the deepest base level (Salameh and Udluft 1985; Margane et al. 2002; Sawarieh 2004). Another scenario suggests flow from the outcrops of the sandstone in Wadi Disi and Wadi Ram in the south of Jordan. This could explain the discharges along the southern part of the Dead Sea (Salameh 2006). The Zarqa Group is regarded as an aquitard. Its many water-bearing intercalations of sandstones, marly beds, and even intercalation of anhydrite in its lower and middle part, act as hydraulic barriers so that the groundwater of the Ram and Kurnub Groups are characterized by different piezometric levels. In the study area, the recharge of groundwater of the Kurnub Group aquifer seems to originate from downward percolation from the upper aquifer and by lateral flow from the east.

Fig. 3.4.3-9. Artesian well situated about 700 m south of the Kafrein dam was drilled 500 m into the Kurnub and Zarqa Group for testing the aquifer. The original closed well, which is under high pressure, was partially opened by sheepherders. The aggressive reduced water has H_2S and high iron content. Water temperature is 31°C. (Photo: H. Hötzl)

Considering the geological situation, it can be assumed that in the northern part of the study area, the groundwater from the sandstone aquifers discharges beyond the main transform fault into the thick young sediment sequence of the Jordan Valley basin (Fig. 3.4.3-8) consisting (especially along its the margins) of intercalations of coarse gravel. Further to the south direct groundwater upflow into the Dead Sea exists along parallel faults.

The resources of groundwater in the study area are limited. The catchment includes more or less only the westward inclined escarpment. The underground water divide resembles closely the surface divide. Along the mountain crest in the highly karstified limestone of the Upper Cretaceous aquifer, the annual groundwater recharge could attain up to 250 mm. This water discharges through numerous springs emerging at the base of the aquifer at elevations between 500–800 m msl, thus generating the flow in the wadis. As the result of evaporation or seepage into the wadi fill, most of the water in smaller channels completely disappears after a short flow distance forming an underground flow towards the large alluvial fans, which mouth into the wide Jordan Valley. By contrast there is nearly no rainfall infiltration into the outcropping Lower Cretaceous aquifer. The small amount of groundwater in this aquifer seems to be supplied mainly by leakage from the upper aquifer. There are no significant springs dewatering this aquifer, therefore it can be assumed that water from the lower aquifer flows directly across the main fault into the thick fill of the Jordan Valley.

Pumping-test in Wadi Kafrein

During the period of October–December 2001 three new wells were drilled in Wadi Kafrein on behalf of the Ministry of Water and Irrigation (MWI) in order to provide the hotels at the Dead Sea with potable water. These wells are the Kafrein Project wells 8A, 9A and 10A. They are located upstream of the Kafrein dam. The distance between the three wells ranges between 500–1,000 m. The aim of the MWI was to drill down to the sandstones of the Kurnub Group because this formation was assumed to provide large volumes of high-quality potable water. The final depths were -530, -500, -500 m msl in wells 8A, 9A and 10A, respectively.

The analysis of the drawdown–time curves of the three wells (Fig.3.4.3-10) indicate that the sandstone of the Kurnub Group does not behave like a fractured rock formation. All wells behave similarly following the exponential function of a relative homogeneous aquifer.

The different graphs do not show a typical drawdown curve which follows the double-porosity concept. Tab. 3.4.3-1 shows the transmissivity and the K-values of the three wells. The saturated thickness for the Kurnub Group is estimated at 177 m for well 8A, 203 m for well 9A and 275 m for well 10A.

The calculated hydraulic parameters for the wells in Wadi Kafrein differ from the ones modeled by Margane et al. (1997) for northern Jordan and agree with the value of 3.0×10^{-5} m/sec published for the Kurnub sandstone (Margane et al. 2002).

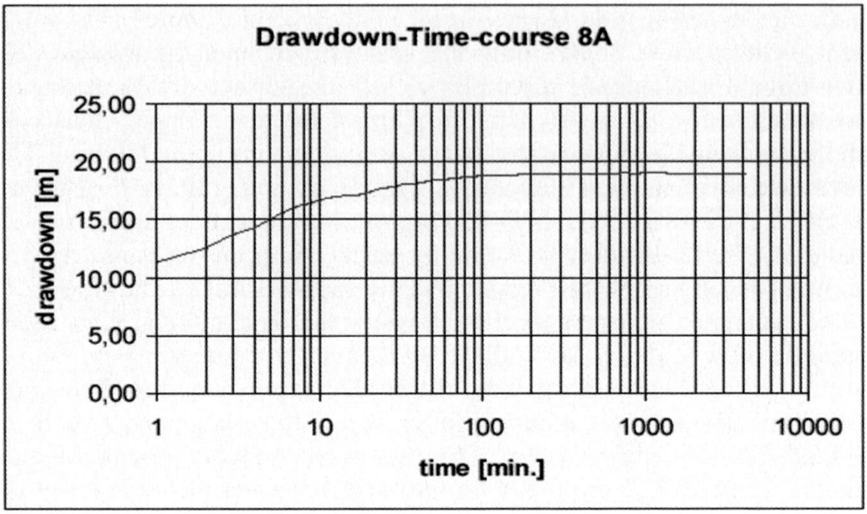

Fig. 3.4.3-10. Drawdown–time curve of well 8A of the Kafrein Project

Tab. 3.4.3-1. Transmissivity and hydraulic conductivity of the Kurnub Group aquifer from pumping test in wells in Wadi Kafrein

Hydraulic Parameters	Well 8A	Well 9A	Well 10A
T_{Theis} [m²/s]	$6.01 \cdot 10^{-3}$	$2.21 \cdot 10^{-3}$	$1.81 \cdot 10^{-2}$
K_{Theis} [m/s]	$3.39 \cdot 10^{-5}$	$1.09 \cdot 10^{-5}$	$6.59 \cdot 10^{-5}$
$T_{Cooper\text{-}Jacob}$ [m²/s]	$5.07 \cdot 10^{-3}$	$1.34 \cdot 10^{-3}$	$7.97 \cdot 10^{-3}$
$K_{Cooper\text{-}Jacob}$ [m/s]	$2.86 \cdot 10^{-5}$	$6.60 \cdot 10^{-5}$	$2.90 \cdot 10^{-5}$

Wadi Kafrein groundwater flow

In the study area the system of deeply incised wadis has been proven as the most important drainage system for the runoff from the escarpment. This stands also true for other parts of the Rift escarpment in Jordan. The wadis are not only collecting the surface runoff, especially the flash floods in the wet winter season, but also the groundwater flow which, in contrary to the escarpment on the western side of the Jordan Valley, is discharging through springs emerging at higher positions on the slope. At present, the wadi system is also the collector of sewage. This began in the past with the fast increase of urbanized areas on the mountain or on the margins of the adjacent plateau which produced an ever increasing amount of (at first untreated) effluents from the sewer systems. At present the wadis carry more

and more effluents from waste-water treatment plants. By infiltration and seepage of the surface wadi flows into the alluvial fans at the foothills, the latter become important aquifers recharging groundwater towards the Jordan Valley.
The outlet of Wadi Kafrein serves as an example for such outflows through wadis and was therefore studied in more detail. It dewaters the Wadi es Sir, which hosts one of the big treatment plants of the Wadi Naur and the city of Amman (Fig. 3.4.3-11). A dam was built at the head of the wadi opening towards the Jordan Valley (Fig. 3.4.3-12). At first the reservoir capacity amounted to 3.8 Mm^3, later the dam was raised up to a capacity of 8 Mm^3.

Fig. 3.4.3-11. Wadi es Sir waste-water treatment plant. The treated effluents are discharged to the wadi and are collected downstream in the Kafrein reservoir

The average rainfall in the catchment area of Wadi al Kafrein is 410 mm/yr (Salzgitter and ICE 1993). In the highlands precipitation may occur as snow once or tw a year but usually as rain (Salameh and Bannayan 1993). As rainfall decreases towards the Jordan Valley, temperature increases. The average annual temperature ranges from 17.5°C in the highlands to 23.5°C in the Jordan Valley, where the average daily temperature rises from 15 and 17°C in the winter months to 31 and 33°C in the summer. Due to the hot temperatures, evaporation losses from the dam are extremely high. For the period February till May 1999 a loss of nearly 300,000 m^3 was calculated (Lenz 1999).

Fig. 3.4.3-12. Wadi Kafrein reservoir with the raised dam has a storage capacity of 8 Mill m^3. It is used to store water in the wet season to facilitate an all-season water supply for the irrigation of the Kafrein-Hisban area in the Lower Jordan Valley

The groundwater contour map (Fig. 3.4.3-13) downstream of the dam in the opening alluvial fan shows a general flow direction of the shallow aquifer towards the Jordan River. The water surface has an average gradient of 3.9%. Kafrein Reservoir is considered a positive boundary with a water level at -135.4 m msl.

The Pleistocene Lisan Formation (JV 3) outcrops at the margin of the alluvial fan with well bedded laminated grey-white silt facies intercalated by gypsum. The alluvial deposits consist of poorly to well rounded gravel. The dam obstructs the continuous water flow downstream into the alluvial fan of the opening wadi. From the dam the water is deviated by channels and distributed to the adjacent Jordan Valley by the Jordan Valley Authority for irrigation. In spite of this obstruction of the through-going flow, the groundwater contour map (Fig.3.4.3-13) indicates a large amount of groundwater discharges under the dam which is used by several wells for irrigation. Upwelling water below the dam shows that the reservoir is not tight. Part of this upwelling water is collected in a pond and pumped back into the reservoir. In addition, sinkholes can be identified in the reservoir area under low water level conditions.

A tracer test was carried out to find the flow path of water from the sinkholes: the deep karst system or a fault zone in the hard rocks passing the dam and then continuing towards the alluvial fan. Another aim was to find out, whether artificial groundwater recharge can be initiated from the reservoir areas in order to infiltrate large amount of surface water and thus avoid extreme evaporation losses from the reservoir.

For the injection of tracers, two areas were identified in which several smaller and larger sinkholes occur. These areas are named zone I and zone II in Fig 3.4.3-14. Zone I is situated along the right abutment at the transition from the sediment surface to the solid rock. The reservoir floor consists of fine material such as silt and clay. Sediment thickness does not exceed 0.5 m covering the rocks of the Naur Formation. Zone II is located in the centre of the reservoir about 600 m north of the dam there the floor is covered with sand changing at greater depth to silt and clay. On December

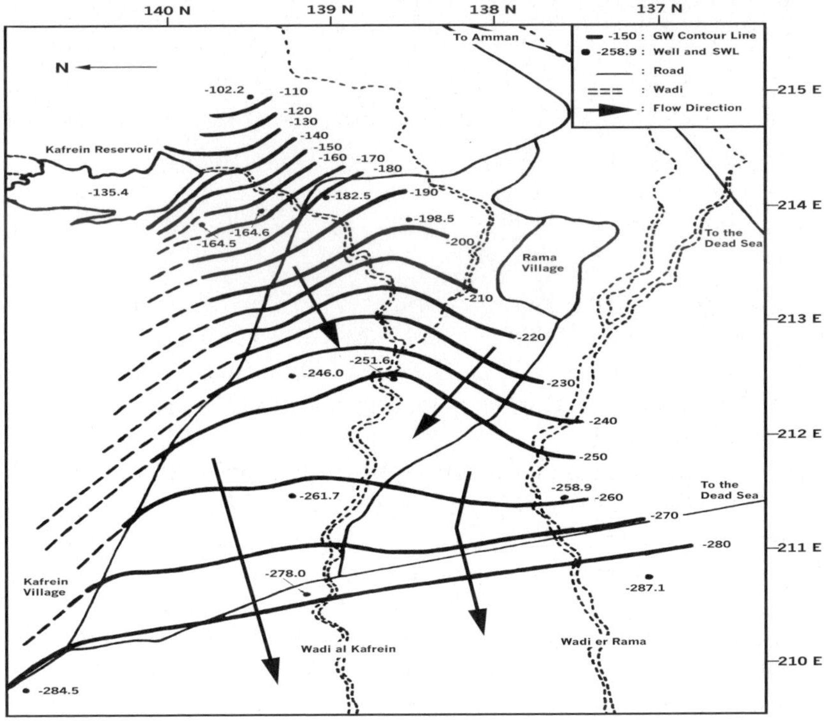

Fig. 3.4.3-13. Ground-water contour map of the shallow alluvial aquifer downstream of al Kafrein Reservoir. Water levels measured on the 8th of May 1999. Contour lines extrapolated after Thiessen Polygons (Lenz 1999)

13th 1998 about 50 g of Eosin dissolved in about 50 l of water were injected in zone I. In zone II, 50 g of Uranin dissolved in about 50 l of water were injected. In both locations, the injection was influenced by the small infiltration rate in the sinkholes. Over a four-month period of observation, 80 samples from the wells AN 1025 and Wadi 2 were analyzed for Uranin and Eosin. In well Wadi 2 no tracer substance was identified. Eosin from Zone I was first found on April 10th with 0.038 µg/l. Uranin from Zone II appeared in three samples between April 20th and 26th with concentrations of 0.124 µg/l, 0.0072 µg/l and 0.0017 µg/l (Fig. 3.4.3-15). The details of the calculated parameters from the breakthrough of the tracers are given in Table 3.4.3-2.

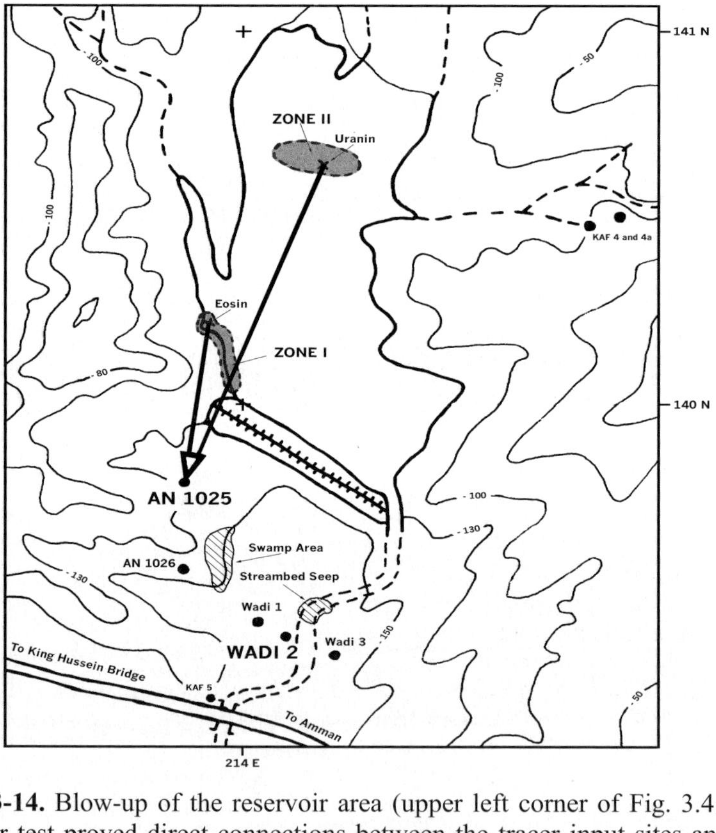

Fig. 3.4.3-14. Blow-up of the reservoir area (upper left corner of Fig. 3.4.3-13). The tracer test proved direct connections between the tracer input sites and well AN 1025. The zipper-like line represents the dam.

The tracer experiment proved the connection between the sinkholes in the reservoir and the downstream groundwater flow. Obviously, the proven underground water flow follows mainly a highly permeable fault zone of limited discharge capacity causing a long flow time. The infiltration conditions in the whole reservoir area are rather restricted, which may be due to (i) the absence of a thick and high permeable gravel layer and (ii) the high content of silty and clayey material deposited in the reservoir by the dammed water. In case of an increased artificial groundwater recharge, additional technical measures have to be implemented to avoid the sealing effect by the silty and clayey sediments of the reservoir.

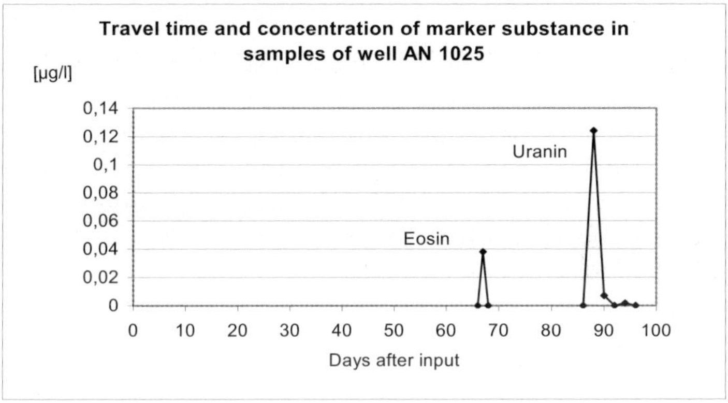

Fig. 3.4.3-15. Breakthrough curves of the tracers in well AN 1025 in days after the injection

Table 3.4.3-2. Results of the tracer test (SWL = Static Water Level; DWL = Dynamic Water Level)

Tracer	EOSIN	URANIN
Input Site (Fig. 6-6)	ZONE I	ZONE II
Altitude	~ -136 m msl	~ -136.5 m msl
Distance to well AN 1025	~ 400 m	~ 900 m
Gradient to well AN 1025 → SWL (~ -160 m) msl)	6%	4.80%
Gradient to well AN 1025 → DWL (~ -194 m) msl)	14.50%	6.40%
Distance to well Wadi 2	~ 800 m	~ 1200 m
Gradient to well Wadi 2 → SWL (~ -165 m) msl)	3.60%	2.40%
Gradient to well Wadi 2 → DWL (~ -175 m) msl)	4.90%	3.20%
Input Date	13.12.1998	13.12.1998
Site certainly covered with water from	02.02.1999	22.01.1999
to	15.05.1999	21.05.1999
Appearance in well AN 1025	10.04.1999	20.- 26.04.1999
Approximate Travel Time to well AN 1025	67 days	88 days
Distance Velocity v_d	6.0 m/day	10.2 m/day

3.4.3.4 Hydrochemistry

Hydrochemical and isotope analyses were carried out in order to support regional hydrogeologic interpretations. The sampling campaigns were restricted to the contact zone between the Kurnub and the Ajlun Group in the southern part and to the Kafrein reservoir with its downstream groundwater flow.

Suweimeh

Numerous water samples from the contact zone between the Kurnub and the Ajlun Group southeast of Suweimeh were collected from springs in order to gain an impression of the local variety within one aquifer (Table 3.4.3-3). Fig. 3.4.3-16 shows a map of the investigated area, which is mainly built of rock-units from the Kurnub and Ajlun Groups.

Fifteen springs were investigated in the Suweimeh area (Fig. 3.4.3-16). The hydrochemical analyses show a sharp increase of salinization towards the Dead Sea and very different chemical compositions. The 15 sampled springs in the Suweimeh area can be subdivided into three groups of springs according the content of TDS:

- Two springs (S 01, S 04) show an extremely high content of TDS with maximal value of 40,670 mg/l which is about 5 to 10 times higher than in the springs S 02 and S 03 emerging at a distance of about 100 m. The S 01 and S 04 springs may be in contact with the Amman-Halibat Fault. The high salinity suggests a long flow path of the groundwater. The anion composition differs drastically from that in all other springs.
- Three springs (S 02, S 03, S 15) discharge from of the alluvial beds and emerge below an elevation of -300 m msl. They also show a very high content of TDS of about 4,000 to 4,500 mg/l. These springs are mainly influenced by dissolution of soluble salts in the beds along the shores of the Dead Sea below the elevation of -180 m msl. These beds were permeated by Dead Sea waters at its previous and higher stands.
- In springs S 05, S 06, S 07, S 08, S 09, S 10, S 11, S 12, S 13, S 14), the content of TDS is at about 1,200 up to 3,400 mg/l. Most of these springs are located on the Kurnub sandstone Group, especially in the transition zone between the Kurnub and Zarqa sandstones. Only two springs (S 12, S 14) are related to the Zarqa sandstone Group. Some of these springs (S 07, S 09, S 13) are heavily exploited for drinking and irrigation purposes. Springs S 07 and S 13 manifest the lowest TDS contents of all sampled spring waters.

During the sampling period from November 2001 until March 2002, the discharge of the springs has not changed significantly despite several precipitation events. The discharge ranges from seepage to only 0.75 l/s. In all springs a fluctuation of cation distribution is observed even within a time-span of two weeks. Ionic ratios show different origins and influences of the spring waters and are the results of different lithologies. The electric conductivities range from 1,730 µS/cm up to 58,800 µS/cm and the temperatures vary from 17.6–29.4°C. The distribution of the main ionic species is shown in the Piper diagram (Fig. 3.4.3-17).

The samples analyzed at the Hydrochemistry Laboratory of the University of Jordan, always show a Mg dominance over Ca. The water composition of the shallow aquifer on a regional scale is inhomogeneous regarding the major cations and anions. This is due to varying sediment components, evaporite distribution, and the changing influence of different mineralized water.

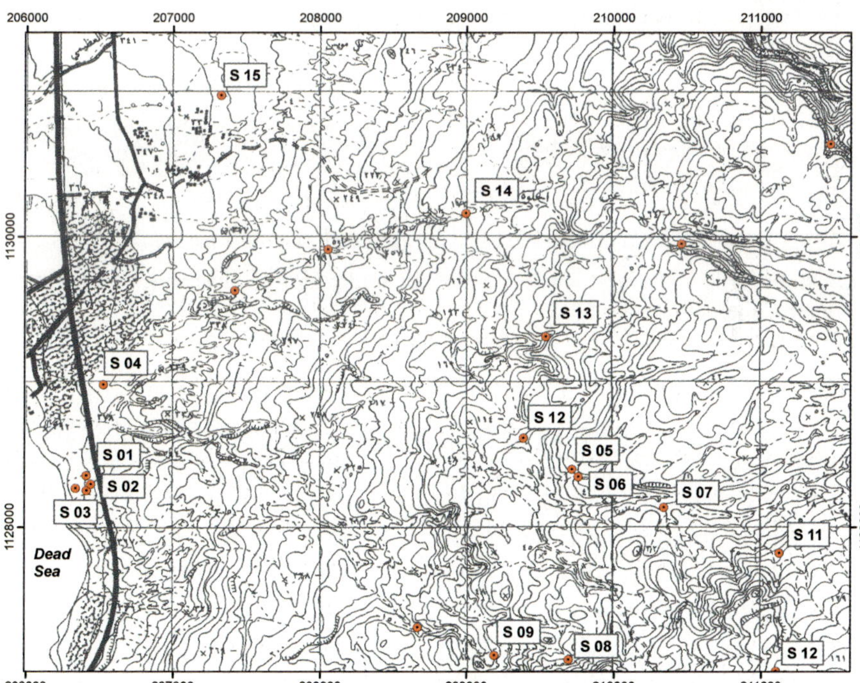

Fig. 3.4.3-16. Location of the springs in the Suweimeh study area

The stable isotopes ^{18}O and ^{2}H confirm in alleviated form (Fig. 3.4.3-18) the differences in the spring water composition (Table 3.4.3-3). This could be due to precipitation conditions of the infiltrating water but should not reflect large environmental differences. Isotope distribution in the samples collected during the spring season are located mainly at - or close to the global meteoric water line (GMWL) $\delta D=8\delta^{18}O+10$ or between GMWL and the Eastern Mediterranean Water Line $\delta D = 8\delta^{18}O+22$. Spring water samples of S 08, S 09, S 10 and S 11 trend with the thermal groundwater line $\delta D = 3.97\delta^{18}O-16.9$.

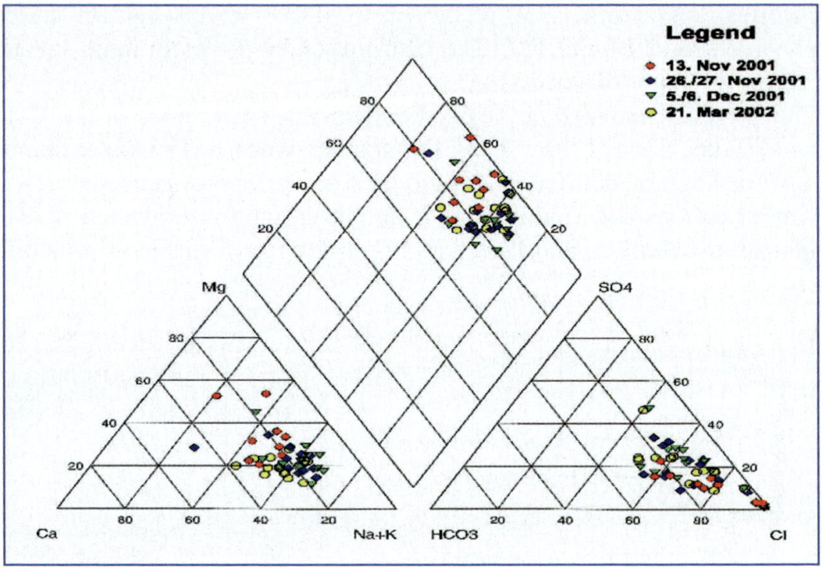

Fig.3.4.3-17 Piper diagram of the chemical composition of spring waters in the Suweimeh area

Kafrein

In the Kafrein area the water sampling program included 10 wells and the Kafrein reservoir. The sampling campaign in the study area was carried out in early May 1999 to determine the hydrochemical characteristics of the different surface- and well waters (Table 3.4.3-4). Water types have been classified after Furtak and Langguth (1967). Three major water types can be distinguished:

- Water from the shallow aquifer (wells 1-7) together with the reservoir water
- Thermal waters from wells KAF 4 and 10
- The thermal brackish water from well KAF 7

The results of the chemical analyses are shown on the Piper diagram (Fig. 3.4.3-19) and in Table 3.4.3-4. Reservoir water and water from the

Table 3.4.3-3. Analyses of the major anions and cations, $\delta^{18}O$ and $\delta^{2}H$ of spring water in the Suweimeh area (modified after Glaser 2002)

ID	North	East	Date	Temp	EC	pH	Ca^{2+}	Mg^{2+}	Na^{+}	K^{+}	Cl^{-}	F^{-}	Br^{-}	SO_4^{2-}	HCO_3^{-}	NO_3^{-}	TDS	$\delta^{18}O$	$\delta^{2}H$
				°C	mS/cm	mg/l	mg/l	mg/l	mg/l	mg/l	mg/l	mg/l	mg/l	mg/l	mg/l	mg/l	mg/l	o/oo	o/oo
S 01	31°45'00.6"	35°35'36.8"	2001-11-13	29.0	58.1	6.01	2156	2638	7246	751	22731		288	682	309	88	40670	-3.7	-16.3
S 01	31°45'00.6"	35°35'36.8"	2001-11-26	28.5	58.10	6.31	3252	1529	7996	802	22899		390	389	309	128	40670	-3.72	-18.2
S 01	31°45'00.6"	35°35'36.8"	2001-12-05	28.3	58.30	6.12	2938	1685	7826	1001	22801	2.5	417	551	290	102	40810	-3.66	-18.4
S 01	31°45'00.6"	35°35'36.8"	2002-03-21	28.4	54.80	6.84	2056	2001	6997	700	20382		494	681	317	104	38360		
S 02	31°44'49.2"	35°35'37.3"	2001-11-13	28.6	6.50	6.92	185	238	533	90	1471	2.5	15	388	309		4550	-4.4	-26.2
S 02	31°44'49.2"	35°35'37.3"	2001-11-26	27.7	6.40	7.12	171	200	900	102	1631	2.6	16	743	305	14	4480	-4.44	-26.5
S 02	31°44'49.2"	35°35'37.3"	2001-12-05	27.7	6.64	7.13	125	198	899	100	1666	1.6	15	566	290	11	4648	-4.48	-27
S 02	31°44'49.2"	35°35'37.3"	2002-03-21	25.3	6.57	7.68	198	172	835	110	1650		34	540	264	7	4599		
S 03	31°44'48.3"	35°35'36.2"	2001-11-13	28.3	6.70	6.88	216	191	577	110	1542	2.5	16	308	307		4690	-4.6	-26.5
S 03	31°44'48.3"	35°35'36.2"	2001-11-26	28.0	6.60	7.01	185	145	900	102	1716	2.5	18	411	299	11	4520	-4.6	-26.5
S 03	31°44'48.3"	35°35'36.2"	2001-12-05	28.3	6.75	6.90	187	160	1000	100	1716	2.9	19	764	303	10	4725	-4.5	-25.5
S 03	31°44'48.3"	35°35'36.2"	2002-03-21	26.6	6.81	7.56	202	181	900	130	1766			553	290	11	4767		
S 04	31°45'11.2"	35°35'41.9"	2001-11-13	29.6	24.1	6.35	801	1033	2768	350	8558		150	390	344	22	16870	-4.91	-23.6
S 04	31°45'11.2"	35°35'41.9"	2001-11-26	29.4	24.1	6.45	770	794	3332	266	8116		143	993	355	47	16870	-4.91	-23.6
S 04	31°45'11.2"	35°35'41.9"	2001-12-05	29.2	24.3	6.35	880	732	3586	199	8334	2.2	140	1350	343	42	17010	-4.87	-23.7
S 04	31°45'11.2"	35°35'41.9"	2002-03-21	29.4	24.1	6.97	776	800	3452	243	7994		263	1227	343	45	16870		
S 05	31°44'49.6"	35°37'44.2"	2001-11-13	28.2	2.47	7.29	138	168	100	45	526	2.5	2.2	283	321		1729	-5.24	-30.4
S 05	31°44'49.6"	35°37'44.2"	2001-11-26	25.0	2.40	7.25	105	53	260	30	483	2.6	2.2	178	287	12	1680	-5.29	-29.5
S 05	31°44'49.6"	35°37'44.2"	2001-12-05	24.2	2.76	7.35	70	109	375	50	592	2.6	2.6	402	277	16	1932	-5.33	-29.4
S 05	31°44'49.6"	35°37'44.2"	2002-03-21	21.6	2.51	6.19	124	58	300	42	483			288	317		1757		
S 06	31°44'48.4"	35°37'46.0"	2001-11-13	19.9	3.70	7.64	169	72	70	350	585	2.6	2.3	201	332		2590	-5.32	-33.2
S 06	31°44'48.4"	35°37'46.0"	2001-11-26	15.8	3.30	7.73	115	121	409	40	676	2.4	2.5	507	344	7	2310	-4.95	31.7
S 06	31°44'48.4"	35°37'46.0"	2002-03-21	15.3	3.36	8.82	194	61	350	65	655			358	317		2351		
S 07	31°44'42.6"	35°38'08.6"	2001-11-13	21.8	2.20	7.04	123	52	57	300	429	2.5	2.0	153	321		1540	-4.94	-27.3
S 07	31°44'42.6"	35°38'08.6"	2001-11-26	17.9	2.20	7.44	99	69	300	20	510	2.6	2.2	173	321	6	1540	-4.98	-27.2
S 07	31°44'42.6"	35°38'08.6"	2001-12-05	24.3	2.44	7.21	86	57	299	60	471	2.5	1.9	197	343	1	1708	-5.1	-29.5
S 07	31°44'42.6"	35°38'08.6"	2002-03-21	24.7	2.52	7.73	120	49	220	65	378			235	343	4	1764		
S 08	31°44'07.1"	35°37'42.5"	2001-11-13	26.9	5.35	6.63	246	177	120	551	1248	2.5	5.5	220	398		3745	-4.22	-33.1
S 08	31°44'07.1"	35°37'42.5"	2001-11-26	25.4	5.40	6.73	177	145	746	70	1328	2.6	5.1	546	311	9	3780	-4.44	-33.5
S 08	31°44'07.1"	35°37'42.5"	2001-12-05	24.6	5.45	6.68	179	146	754	100	1280	2.5	5.6	589	448	6	3815	-4.15	-32.4
S 08	31°44'07.1"	35°37'42.5"	2002-03-21	22.1	5.38	7.42	337	144	874	130	1766			606	422	12	3766		
S 09	31°44'08.8"	35°37'26.6"	2001-11-13	28.2	4.86	6.67	185	134	136	694	1111	2.7	4.7	246	390		3402	-4.13	-32.4
S 09	31°44'08.8"	35°37'26.6"	2001-11-26	28.1	4.85	7.34	162	89	756	70	1135	2.6	4.1	539	355	4	3395	-4.17	-31.9
S 09	31°44'08.8"	35°37'26.6"	2001-12-05	27.8	4.96	7.28	132	120	694	150	1088	2.6	4.1	677	343	5	3472	-4.22	-32
S 09	31°44'08.8"	35°37'26.6"	2002-03-21	27.1	4.85	8.00	192	67	650	85	1121			376	343		3395		
S 10	31°44'04.6"	35°37'36.2"	2001-11-27	24.0	3.97	6.27	166	86	477	51	913	2.4	3.1	158	482	4	2779	-4.31	-33.4
S 10	31°44'04.6"	35°37'36.2"	2001-12-06	21.5	4.05	6.77	140	52	531	100	858	2.4	3.4	267	435		2835	-4.39	-33.9
S 11	31°44'32.0"	35°37'39.6"	2001-11-27	23.6	4.09	7.45	153	76	508	60	951	2.7	3.5	203	367	9	2863	-4.29	-31.1
S 11	31°44'32.0"	35°37'39.6"	2001-12-06	23.0	3.84	7.70	101	90	600	100	844	2.7	3.2	572	317	42	2688	-4.22	-29.3
S 11	31°44'32.0"	35°37'39.6"	2002-03-21	19.8	4.02	8.37	175	61	475	40	878			200	369	9	2814		
S 12	31°44'59.0"	35°37'32.1"	2001-11-27	18.2	4.12	7.27	161	97	623	51	870	2.8	2.8	678	367	6	2884	-4.4	-28.2
S 12	31°44'59.0"	35°37'32.1"	2001-12-06	18.2	4.62	7.22	117	137	570	100	989	2.8	2.8	505	343	12	3234	-4.59	-27.9
S 12	31°44'59.0"	35°37'32.1"	2002-03-21	17.6	3.95	8.01	192	78	475	90	825	0		464	369	3	2765		
S 13	31°45'20.2"	35°37'39.7"	2001-11-27	26.5	1.73	7.07	71	74	175	15	338	2.7	1.5	135	299	13	1211	-5.23	-26.1
S 13	31°45'20.2"	35°37'39.7"	2001-12-06	25.7	1.85	7.09	93	48	177	20	307	2.6	1.4	161	277		1295	-5.2	-25.8
S 13	31°45'20.2"	35°37'39.7"	2002-03-21	26.4	1.87	8.01	144	47	170	35	320			210	290		1309		
S 14	31°45'50.2"	35°37'26.4"	2001-11-27	26.5	2.29	7.50	279	106	182	10	378	3.0	2.2	344	220	16	1603	-4.71	-22.5
S 14	31°45'50.2"	35°37'26.4"	2001-12-06	24.9	2.28	7.74	86	128	180	20	338	2.9	1.9	528	158	12	1596	-4.8	-23.6
S 14	31°45'50.2"	35°37'26.4"	2002-03-21	24.8	2.38	8.07	153	72	244	45	369			577	211	6	1666		
S 15	31°46'14.5"	35°36'13.9"	2002-03-21	26.7	5.93	7.33	279	163	600	100	1567		29	237	395	16	4151		

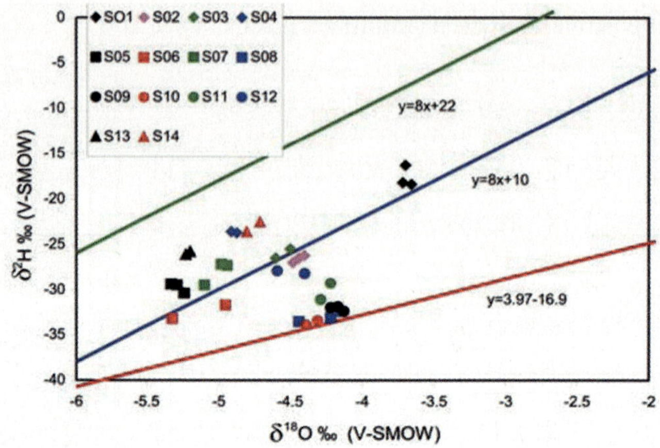

Fig. 3.4.3-18. Results of stable isotopes analyses of water samples from Suweimeh area

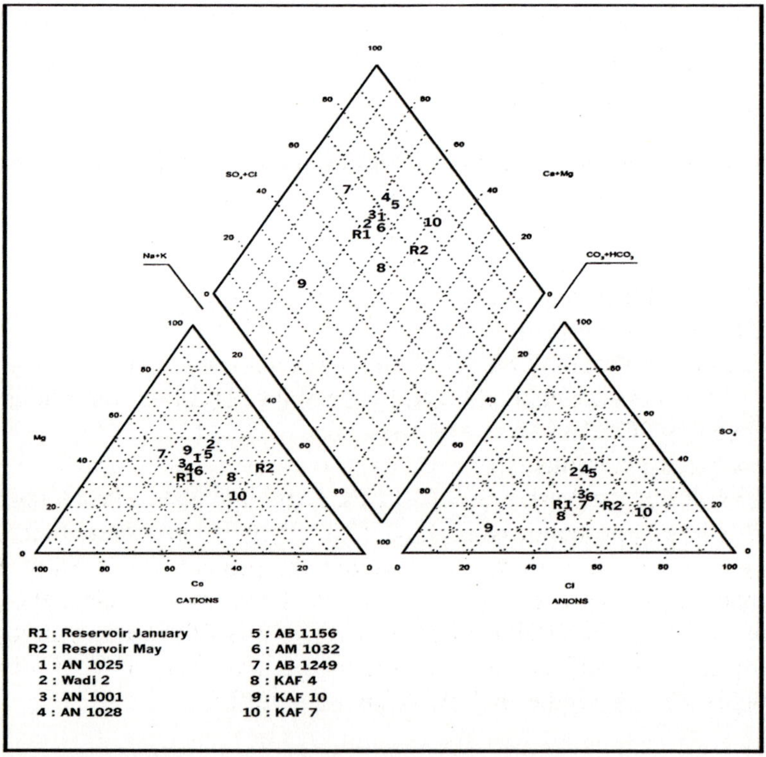

Fig. 3.4.3-19. Piper diagram of the results of samples from well waters and water from the reservoir. Samples were taken on May 18th 1999

Table 3.4.3-4. Chemical analysis of the major anions and cations of the Kafrein area. In brackets: numeration of wells (modified after Lenz 1999)

Source	Date	EC µs/cm	TDS mg/l	pH	Ca mg/l	Mg mg/l	Na mg/l	K mg/l	Cl mg/l	SO_4^{2-} mg/l	HCO_3^- mg/l	NO_3^- mg/l	Data source
Kafrein Dam	May 1995	736	478	8.14	29.1	15.2	42.3	5.9	86.9	34.2	152	10	Subah (1998)
	May 1996	660	429	9.07	16.2	16.2	46.6	5.9	88.6	36.2	74	8.1	Subah (1998)
	May 1997	833	542	8.20	19.8	20.5	65.0	7.4	127	53.5	98	2.9	Subah (1998)
	Jan 1999	1022	664	8.00	65.3	52.0	73.5	21.1	155	101	233	2.1	Lenz (1999)
	May 1999	803	522	9.14	20.2	35.2	91.9	7.8	152	62.7	107	2.1	Lenz (1999)
AN 1025 (1)	Jan 1999	1194	776	7.20	83.8	44.6	80.0	10.6	175	112	201	6.6	Lenz (1999)
	May 1999	1154	750	8.10	81.8	49.5	87.3	7.0	200	103	212	4.8	Lenz (1999)
Wadi 2 (2)	Jan 1999	1405	683	7.40	67.5	49.6	70.5	6.6	153	79.0	191	8.1	Lenz (1999)
	Feb 1999	1028	668	7.38	57.1	49.6	76.1	6.6	141	86.3	212	7.6	Lenz (1999)
	Mar 1999	1266	676	7.47	77.6	44.5	71.7	6.3	142	90.1	212	7.6	Lenz (1999)
	May 1999	957	622	7.28	44.9	57.0	62.0	16.0	117	140	173	8.9	Lenz (1999)
AN 1001 (3)	May 1996	907	590	7.87	36.3	17.4	58.1	7.0	107	50.6	190	7.9	Subah (1998)
	Mar 1997	910	592	7.88	36.1	17.5	58.1	7.0	108	50.1	189	7.9	Subah (1998)
	May 1999	942	612	7.26	65.3	42.0	63.4	6.6	142	106	160	7.9	Lenz (1999)
AN 1028 (4)	May 1999	1350	878	7.07	90.6	57.2	94.2	2.7	195	195	184	9.1	Lenz (1999)
AB 1156 (6)	May 1999	1288	837	7.32	47.7	22.0	97.9	12.9	178	72.8	233	8.3	Subah (1998)
	Mar 1997	1301	846	7.33	49.3	24.3	115	14.1	175	72.8	210	8.5	Subah (1998)
	May 1999	1037	674	7.14	52.3	51.0	77.9	5.1	145	140	150	8.9	Lenz (1999)
AM 1032	May 1996	1990	1294	6.90	76.6	41.4	148	10.9	267	105	426	22	Subah (1998)
	Mar 1997	2130	1385	7.30	78.6	42.0	156	10.6	301	111	390	23	Subah (1998)
	May 1999	1600	1040	7.50	110	69.3	130	8.2	268	135	281	23	Lenz (1999)
AB 1249 (7)	May 1999	1262	820	7.04	101	63.3	50.6	5.1	185	111	232	5.6	Lenz (1999)
KAF 4 (8)	Dec 1994	907	542	7.40	80.2	27.9	54.9	9.0	96.1	94.5	232	2.1	Jica (1995)
	Jun 1994	981	596	7.08	79.4	29.2	61.1	5.9	97.8	94.5	232	8.1	Jica (1995)
	Jan 1999	1078	700	7.50	48.9	49.6	85.5	22.7	130	89.7	239	10.1	Lenz (1999)
	May 1999	1012	657	6.50	44.1	40.1	96.5	14.9	142	67.5	235	1.4	Lenz (1999)
KAF 10 (9)	May 1999	602	391	6.88	36.1	34.0	33.3	2.7	45.0	24.1	222	1.4	Lenz (1999)
KAF 7 (10)	Dec 1994	8610	5508	6.40	397	158	1163	215	2124	506	1175	2.1	Jica (1995)
	Jun 1994	8350	5590	6.92	307	193	1367	166	2197	600	1013	48	Jica (1995)
	May 1999	4420	2872	6.78	245	121	418	144	978	390	423	0.6	Lenz (1999)

shallow aquifer (wells 1–7) are of predominantly chloridic/sulphatic earth-alkaline composition with enhanced alkali content showing the relation Mg > Ca > Na > K and HCO_3^- > Cl^- > SO_4^{2-}. The TDS range from 430–650 mg/l. The water from wells 1–6 is characterized by electrical conductivity (EC) of 960–2100 µS/cm and TDS exceeding 590 mg/kg. The anion pattern is Cl^- > HCO_3^- > SO_4^{2-}.

The water from wells KAF (8, 9 and 10) represents a special group in the area. All waters are characterized by a smell of H_2S and water temperatures exceeding 30 °C.

All analyzed constituents of the shallow aquifer waters are below the limiting values of the Jordan Standards for Drinking Water. Problems may arise because of the careless dealing with oil, fuel and other toxic substances at locations just a few meters above the groundwater surface. Pesticides and herbicides occur in low concentrations always below the permissible limits of the WHO guidelines (Al Kuisi 1998; Lenz 2000). If results of different studies are compared, no general trend of Mg/Ca ratios is visible in water from the shallow Alluvium aquifer.

3.4.3.7 Conclusions

The studied escarpment area represents specific drainage systems from the high mountain region extending from Amman westwards to the Jordan Valley. This region seems to be characteristic for all other parts of the Rift escarpment on the Jordanian site, except for those which are drained by the large wadi systems cutting through the escarpment and draining also parts of the eastern plateau.

Four major aquifer systems are relevant in the area. They are either exposed in the catchment area or are tapped by wells. The Cambrian to Ordovician Ram Group aquifer is composed of sandstones and not outcropping in the study area. It is a confined aquifer, tapped in adjacent areas by wells, and is producing mainly brackish water. It is overlain by the lower aquifer complex comprising the Zarqa Group (which except for sandy intercalations is an aquitard) and by the Kurnub Group. The aquifer system is predominantly composed of sandstones with marl and clay intercalations. In the study area, the Kurnub Group is the relevant aquifer. There are, however, only few, rather small springs emerging from its outcrops. Due to scarce precipitation in the lower parts of the escarpment, the direct recharge is limited and predominantly replenished by indirect vertical recharge from the overlying aquifer system. Whenever the Kurnub Group is tapped by deep wells, the groundwater is under confined artesian conditions and contains water of low mineralization but under reducing condition and a noticeable H_2S content.

The Upper Cretaceous aquifer complex includes the Ajlun and Belqa Groups. It is a carbonate aquifer system consisting of an alternation of several aquifers and aquitards. The local importance of this system is dominated by the lithological composition of the tapped formations. The main recharge takes place in the high rainfall areas, where the upper aquifer complex is exposed. Along this mountain ridge a water divide developed separating the groundwater flow into eastward and westward directions. The underground water divide does not differ largely from the surface divide. Therefore, in the study area only limited resources of groundwater can be taken in consideration. The catchment covers more or less only the westward sloping escarpment. This water discharges via numerous springs emerging at the base of the aquifer at elevations between +500 and +800 m msl, where it is used for local water supply and irrigation. The remaining portion generates surface flow in the wadis. In smaller channels most of the water completely disappears after short flow distances mainly due to evaporation or by seepage into the wadi fill, forming an underground flow towards the large alluvial fans which mouth into the wide Jordan Valley.

In the Jordan Valley, the alluvial aquifer complex is the main source for water production, especially from the marginal areas of gravel and sand. The large alluvial fans of the wadis and in the Jordan Valley receive their recharge through wadi flows or from the adjacent and interfingering stratigraphically deeper aquifer units. There is a high potential for artificial groundwater recharge by using flash floods and treated waste water. Compared with the dams and reservoirs, the artificial recharge in the underground favors the storage of water without the tremendous water losses by evaporation in the reservoirs.

References

Becker K (2000) Hydrological and Hydrogeological Investigations along the Wadi Shueib, Jordan; Master Thesis (unpubl.), Department of Applied Geology, University of Karlsruhe, pp 1–123

Furtak H, Langguth H R (1967) Zur hydrochemischen Kennzeichnung von Gewässern und Grundwassertypen mittels Kennzahlen. Mem IAH Congress, 1965, Hannover, VII pp 86–96

Glaser J (2002) Hydrochemical Investigations of the Suweimeh area, North-Eastern end of the Dead Sea; Master Thesis (unpubl.), Department of Applied Geology, University of Karlsruhe, pp 1–119

Horowitz A (2001) The Jordan Rift Valley, Tokyo: A.A. Balkema Publishers/ Lisse/Abingdon/Exton (Pa), pp 1–729

JICA Japan International Cooperation Agency (1995) The study on Brackish Groundwater Desalinisation in Jordan- Draft Final Report, Summary Report, Main, Supporting Report, Data and Drawings.- Yachiyo Engineering Co. Ltd & Mitsui Mineral Development Engineering Co. Ltd Tokyo, Japan; Ministry of Water and Irrigation MWI, Amman

Lenz S (1999) Hydrogeological Investigation along the Wadi al Kafrein and the Kafrein Reservoir. Unpubl. Masters Thesis (unpubl.), Department of Applied Geology University of Karlsruhe, pp 1- 108

Margane A, Hobler M, Almonani M, Subah A (2002) Contributions to the Hydrogeology of Northern and Central Jordan, Geologisches Jahrbuch –Reihe C, Heft 68, Bundesanstalt für Geowissenschaften und Rohstoffe, Hannover, pp1–52

Margane A, Hobler M, Subah A (1997) Groundwater Resources of Northern Jordan. - Special Report No. 3: Mapping of Groundwater Vulnerability and Hazards to Groundwater in the Irbid Area, prepared by BGR & WAJ. Bundesanstalt für Geowissenschaften und Rohstoffe, Hannover, Archive no. 118708 (unpubl.), pp 1–50

Parker DH (1970) The Hydrogeology of the Mesozoic, Cainozoic Aquifers of theWestern Highlands and Plateau of East Jordan, Technical Report Vol. 2: Hydrogeology; United Nations Development Program, Food and Agriculture Organisation of the United Nations; Rome, pp 1–79

Salameh E (2006) Structural control of hydrogeology and hydrochemistry in Jordan. Hydrogeol Umwelt (in press)

Salameh E, Bannayan H, (1993) Water Resources of Jordan. Present Status and Future Potentials. Friedrich Ebert Stiftung, Amman; pp 1–183

Salameh E and Udluft P (1985) The hydrodynamic pattern of the Central part of Jordan. Geol Jahrb 38, pp 39–53

Salzgitter Consult GmbH & Jordanian Consulting Engineer (JCE) Co. (1992) Studies on Raising Kafrein Dam- Final Report on Technical, Economic and Fiancial Feasibility and Preliminary Design; Part II-Main Report. Ministry of Water and Irrigation MWI and Jordan Valley Authority JVA, Amman, pp 1–40

Salzgitter Consult GmbH & Jordanian Consulting Engineer (JCE) Co. (1993) Raising Kafrein Dam Project – Site investigations reports. Ministry of Water and Irrigation MWI and Jordan Valley Authority JVA, Amman, pp 1–58

Sawarieh A (2004) Hydrogeology of thermal water from Zara-Zarqa Ma'in and Jiza areas, Jordan. Proc 5th Int Symp Eastern Mediterranean Geology, Extended abstracts. Thessaloniki, Greece. 14–20 April 2004. pp 1564–1567

Shawabkeh K F (2001) Geological map of Al Karama - 3153-IV, 1:50.000, The Hashemite Kingdom of Jordan, Natural Resources Authority, Geology Directorate

Thiel M (2002) Hydrogeological Investigations in the North-Eastern Dead Sea Area, Suweimeh, Jordan; Master Thesis (unpubl.), Department of Applied Geology, University of Karlsruhe, pp 1–142

3.4.4 Protective potential of soils, ground-water vulnerability and hazards in the Wadi Shueib, Jordan

H. Werz, H. Hötzl, D. Kuntz, S. Storz

3.4.4.1 Introduction

The study area is located on the western Rift slopes of the Jordanian highlands about 20 km northeast of the Dead Sea (Fig. 3.4.4-1). It is bordered by the transition zone from the Jordan Rift Valley to the eastern highlands in the south-west. The geographical border of the study area coincides with the catchment of about 185 km^2 of the Wadi Shueib. The area is characterized by a steep relief with elevations ranging from -200 m in the southwest up to 1240 m in the north-east. The climate is semi-arid, with low precipitation but high evaporation rates in the southern part and increasing precipitation in the northern parts. The majority of the springs in the study area are moderately to low contaminated, those however located close to Salt and other villages show higher levels of contamination (Ta'any 1992). The enhanced microbiological contamination and higher concentrations of Na^+, Cl^-, and NO_3^- are mainly caused by effluents from septic tanks, cesspits and discharge of sewage.

Land cover and land use of the area depends on variations of climate and relief (Figure 3.4.4.-2). Wherever possible, the middle and northern parts are mainly used for agriculture. Urbanization centers are only in the northern and north-eastern areas, whereas the southern parts are characterized by low-density of population. The agriculture in the southern parts is limited to the existence of soils along the wadi courses

Investigations in the Wadi Shueib catchment comprise the following topics:

- General hydrogeological and geological aspects such as karstification, location and characteristics of aquifers forming the base of groundwater vulnerability mapping
- Soil investigations, including the protective potential of soils
- Acquisition of remote sensing data as additional information on groundwater vulnerability and hazard assessment
- Investigations of regional ground-water vulnerability and hazards in order to assess long-term planning of protective measures for groundwater in an operational and cost-efficient way

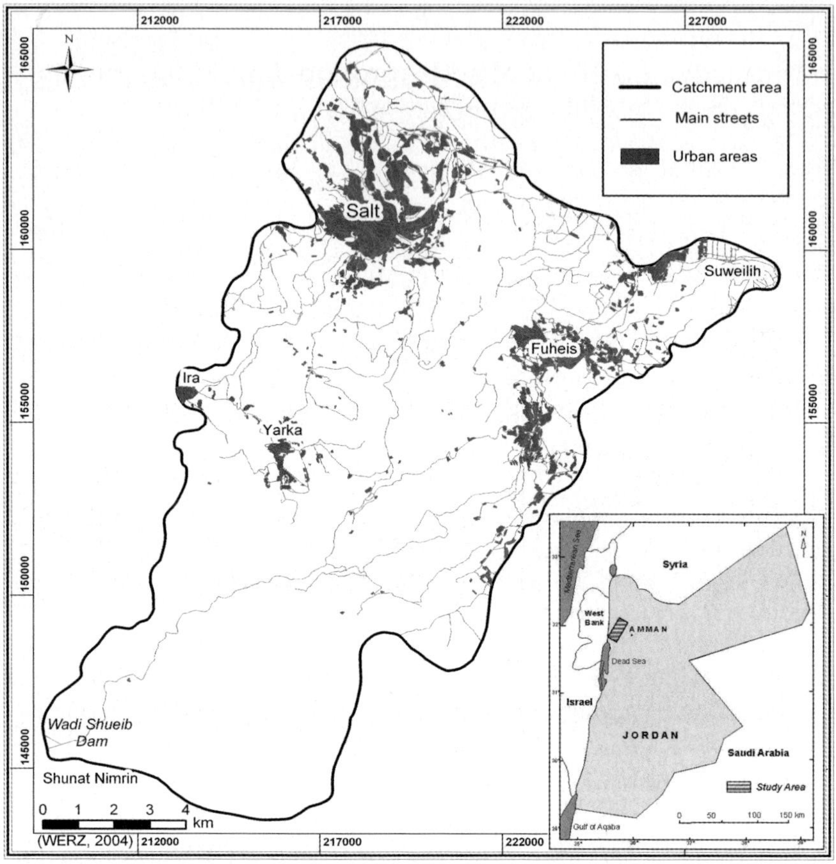

Fig. 3.4.4-1. Study area in Jordan with major towns and roads

Regional ground-water vulnerability maps contribute to the long-term planning of protective measures for the groundwater in an operational and cost-efficient way. The problems addressed in this study apply to the protection of freshwater resources in areas that suffer from a high degree of uncertainty due to a limited availability of observed and measured data.

3.4.4.2 Geology

The uplift of the Jordan Valley shoulders affected all previous structures of the bordering highlands (Horowitz 2001). Thus, all structures in the area are related to the Jordan Rift Valley fault system. The study area is one of the rift side catchments, which is extensively crossed by normal fault of variable throws and trends and by several anticlines and synclines. The prevailing fault directions are NW-SE and NNE-SSW. The general trend

Fig. 3.4.4-2. Characteristic land cover types: (a) In the southwestern part (below sea level) agriculture and dense vegetation is limited to the existence of soil cover on the valley floors. (b) In the northern part pasture and agricultural areas are dominant features (photo: H. Werz, April 2002).

of anticlines and synclines axes is NE-SW (Fig. 3.4.4-3). The dominating structural element is the Wadi Shueib structure, a fold belt of approximately 25 km length and 1–4 km width which extends from the south near Shunat Nimrin to northeast direction (Mikbel and Zacher 1981) and continues northwards into the Suweilih structure (Salameh 1980). Within the Wadi Shueib structure there are several tightly folded synclines and anticlines, overturned to the west, mainly on the western limb of the main flexure (Mikbel and Zacher 1981).

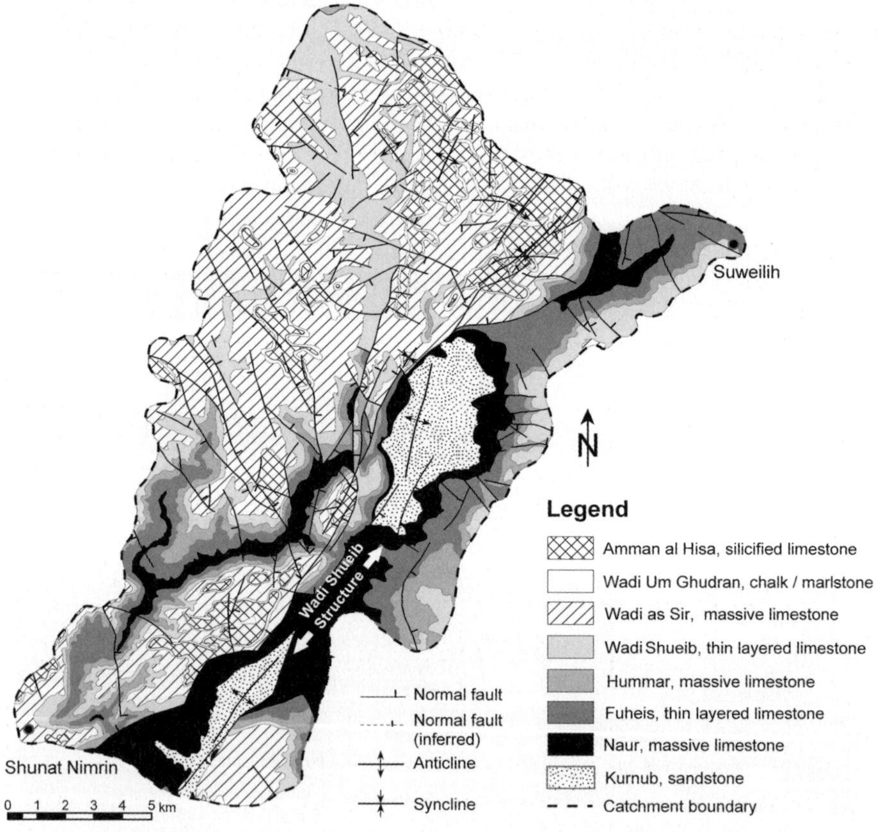

Fig. 3.4.4-3. Geological map of the study area (Werz and Hötzl 2007)

The area is dominated by Upper Cretaceous (Albian to Maastrichtian) sedimentary rocks such as limestones, dolomites and marls of the Ajlun and Belqa Groups which are underlain by Kurnub sandstones of the Lower Cretaceous age, with outcrops in fault zones of the Wadi Shueib structure (Mikbel and Zacher 1981; Salameh 1980; Bender 1974).

3.4.4.3 Hydrogeology

Precipitation varies between 800 mm/yr in the surroundings of Salt to less than 150 mm/yr near Shunat Nimrin (Orthofer 2001). Due to the high evaporation rate only approximately 12% of the precipitation contributes to groundwater recharge (Taa'ny 1992). Two main aquifer complexes can be discerned: the Lower and the Upper Cretaceous aquifer complex, the latter being more important and it includes the complete sequence of the Ajlun group (Table 3.4.4-1).

Table 3.4.4-1. Stratigraphy in the Wadi Shueib catchment area (modified after Powell 1989). The numbers (Z1, B2) indicate the commonly used symbols for the geological units. The Ajlun Group consists of approximately 350–400 m thick interbedded limestones, marly limestones and dolomitic limestones. The Belqa Group includes a sequence of mainly limestones intercalated by thick-bedded chert, chalk, marls, silicified limestones and phosphatic horizons

Period	Epoch	Quennel Group (1951)	Masri (1963)	Bender (1974)		Brief description	Thickness
Upper Cretaceous	Campanian Santonian	Belqa	Amman al Hisa	Phosphorite	B2	Phosphorite & silicified limestone	~??m
Upper Cretaceous	Coniasian	Belqa	Wadi Um Ghudran	Silicified limestone	B1	Massive chalk & marlstone, fossiliferous	~10m
Upper Cretaceous	Turonian	Ajlun	Wadi as Sir	Massive limestone	A7	Thinlayered dolomite or massive grey limestone, fossiliferous. To the top more Chertlayers.	~130m
Upper Cretaceous	Turonian / Cenomanian	Ajlun	Wadi Shu'ayb	Echinoidal limestone	A5,6	White crystalline limestone and Ammonites (A6), and thinlayered limestone and marlstone (A5)	~30m
Upper Cretaceous	Cenomanian	Ajlun	Hummar	Echinoidal limestone	A4	Massive grey, sometimes yellowish limestone, often crystalline and cavernous	~45m
Upper Cretaceous	Cenomanian	Ajlun	Fuheis	Nodular limestone	A3	Thinlayered, soft white limestone	~70m
Upper Cretaceous	Cenomanian	Ajlun	Na'ur	Nodular limestone	A1,2	Massive hard grey limestone + chertlayers (A2), and grey marlstone (A1)	~110m
Lower Cretaceous	Albian Abtian	Kurnub	Subeihi	Varicoloured sandstone	K2	Multicoloured sandstone	~250m
Lower Cretaceous	Neocomian	Kurnub	Aardo	White massive sandstone	K1	White, yellow massive sandstone (does not occur in study area)	~250m
Angular unconformity							
Jurassic Triassic		Zarqa	Azab	Zarqa	Z2	Does not occur in study area	~??m
Permian		Zarqa	Ma'in	Zarqa	Z1	Does not occur in study area	~??m

Three aquifers within the Ajlun Group, consisting of mainly massive and partly intensively karstified limestones, are relevant for the area: the Naur aquifer (A1/2); the Hummar aquifer (A4) and the Wadi es Sir aquifer (A7). They are drained by several springs that are partly used for drinking water supply. These aquifers are separated by aquitards of the Fuheis (A3) and the Shueib (A5/6) Formations, which are composed of thinly layered marls, dolomitic limestones, and siltstones.

In compressive tectonic areas within the study area, the aquifers are connected by fractures and broad fault zones. The lower aquifer complex is is only represented by a small outcrop of the colored Kurnub Group sandstones along the Wadi Shueib structure (Fig. 3.4.4-3). Springs are absent.

3.4.4.4 Soil Investigations

Overview

Detailed investigations of soil with the aim of quantifying their protective potential for groundwater against contamination were performed in the Wadi Shueib area (Kuntz, 2003). A scale soil map (1:50,000) was drafted based on the Jordanian Soil Atlas (National Soil Map Project, Ministry of Agriculture 1994) and field investigations. Soil samples were taken at representative locations attended by infiltration tests and soil corings. At 22 sites (Fig. 3.4.4-4) soil cores were taken from depths of 1 to 2 meters and sampled for pH and moisture analyses. The soil samples were analyzed for pH, cation exchange capacity, particle size distribution, bulk density, pore volume, and natural water content (Table 3.4.4-2).

As a result of this study a soil thickness map and soil cover class map are constructed presenting the protective potential of the soils with respect to potential contaminants released at the surface and migrating into the groundwater.

Table 3.4.4-2. Scheme of soil analyses: field and laboratory analysis

Test / Parameter	Intention
Infiltrometer test, bulk density / pore volume, particle size distribution	Estimation of hydraulic conductivity, permeability and Infiltration rates
Soil coring	Information about soil thickness
Soil-pH, natural water content cation exchange capacity	Result of these analyses show the general chemical state of the soil. Estimation of the retention potential for harmful pollutants
X-Ray Diffractometry	To get a quick overview of clay mineral content to choose samples for CEC analysis
Suction test	For the calculation of the field capacity (eFC) for some soils, which is utilized by certain groundwater vulnerability mapping methods

Soil Map

The soil map of the study area is shown in Fig. 3.4.4-4. The employed soil map units correspond with those of the Jordan Soil Atlas with soil map units range from 1–81. In order to avoid confusion with the Jordanian Soil Atlas, new soil map units are introduced (82–84) for parts of the southern study area which were not mapped so far. However, these soil types may have been defined in other areas. The boundaries of the soil types were compared with those of geological units in the study area in order to delineate any geological influence on soil development. Due to the high relief of the area, the thickness and type of soil occur vary on small distances. This limits detailed mapping to a 1:50,000 scale.

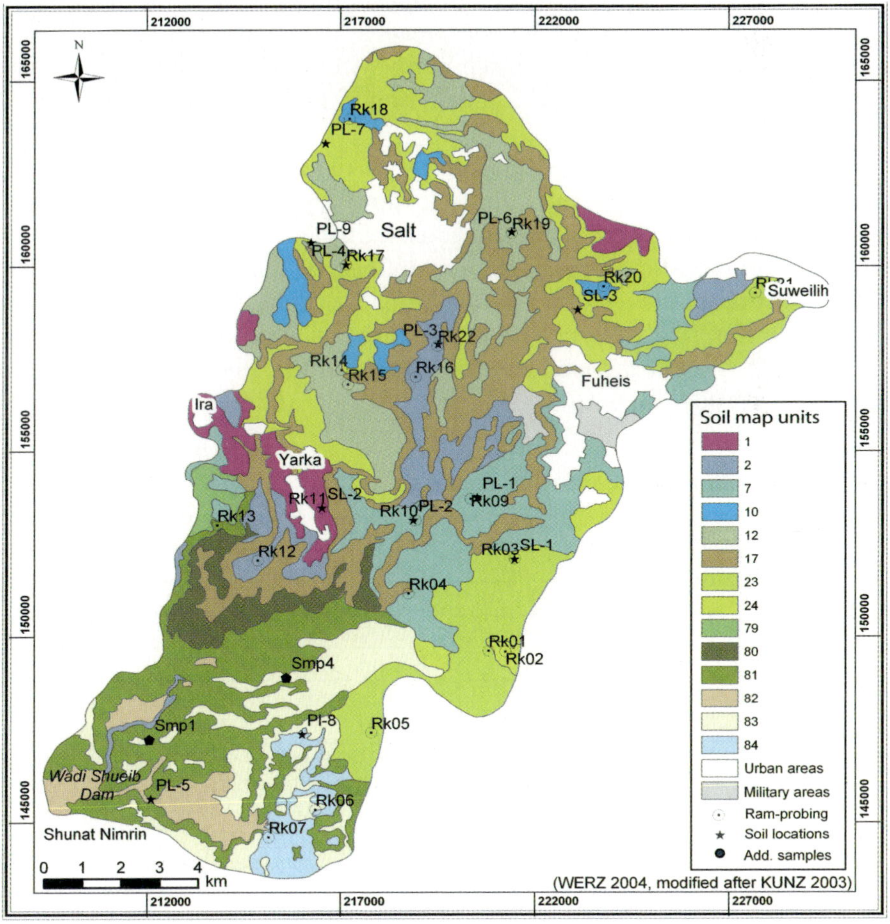

Fig. 3.4.4-4. Soil map of the study area localizing sites of soil coring, soil sampling, soil profile, and other sample sites

Basically, the soil map outlines areas where a specific soil taxon or a combination of certain soil types prevails. The soil map contains areas covered by soils not fitting into the applied scheme (inclusions). These areas are given as % error and their soil-type (Table 3.4.4-3).

Factors influencing soil development and their protective efficiency

In addition to surface geology, climate is a major factor for soil development. Unlike the area of western Jordan, where the climate determines the type of soil developed, in the study area the climate variations are too small to produce significant differences in soil types. However, due to the changes in moisture regime, slightly different types of soils of varying thicknesses developed in the northern and southern parts of the study area. Calcaric soils or calcisoils are widespread in the south with less then 150 mm/yr precipitation, whereas in the north with more then 800 mm/yr precipitation cambisoils developed. In addition to the moisture regime, the relief is also an important factor.

Furthermore, the protective efficiency of the soils is influenced by the thickness that acts as a major factor controlling the transit time of contaminants. Because the soil nomenclature of the Jordanian Soil Atlas relies mainly on diagnostic horizons that are defined by combinations of characteristic soil properties and/or soil materials (and not on the thickness of a specific soil unit), our soil cover classes are based on the thickness of the various soil types, e.g., moderate thickness on the floor of valleys or as thin cover on steep slopes.

Table 3.4.4-3 Soil map units in the study area

Soil map unit 1

Area [km^2]	5.8	
Area [%]	2.9	
Elevation [m]	300-670	
Slope [%]	0–8	
Thickness [cm]	100-170	
Predominant Soils	Chromic Vertisol	55%
	Vertic Cambisol	30%
Error	Rendzic Leptosol	15%
Sample Location	SL-2	

Soil map unit 2

Area [km^2]	9.5	
Area [%]	4.8	
Elevation[m]	300-960	
Slope [%]e	5–40	
Thickness [cm]	100-175	
Predominant Soils	Calcaric Cambisol	65 %
	Chromic Cambisol	35 %
Error	-	
Sample Location	PL-3 N	

Soil map unit 7

Area [km²]	17,3	
Area [%]	8,7	
Elevation [m]	400-1150	
Slope [%]	0-25	
Thickness [cm]	70-150	
Predominant Soils	Chromic Cambisol	60%
	Vertic Cambisol	20%
Error	Rendzic Leptosol	20%
Sample Location	PL-1 / PL-2	

Soil map unit 10

Area [km²]	3,1	
Area [%]	1,6	
Elevation [m]	450-1100	
Slope [%]	5-16	
Thickness [cm]	60-110	
Predominant Soils	Chromic Cambisol	70%
Error	Leptosols	30%
Sample Location	(Rks 20)	

Soil map unit 12

Area [km²]	17,9	
Area [%]	9,1	
Elevation [m]	700-1050	
Slope [%]	9-25	
Thickness [cm]	60-120	
Predominant Soils	Chromic Cambisol	50%
	Calcaric Cambisol	30%
Error	Dystric Leptosol	20%
Sample Location	PL-6 / PL-4	

Soil map unit 17

Area [km²]	35,6	
Area [%]	18,0	
Elevation [m]	400-1000	
Slope [%]	26-60	
Thickness [cm]	20-60	
Predominant Soils	Mollic Leptosol	30%
	Dystric Leptosol	30%
	Lithic Leptosol	20%
Error	Chromic Cambi-sols	20%
Sample,ocation	SL-3	

Soil map unit 23

Area [km²]	15,6	
Area [%]	7,9	
Elevation [m]	600-1100	
Slope [%]	9-16	
Thickness [cm]	30-70	
Predominant Soils	Dystric Leptosol	35%
	Rendzic Leptosol	35%
Error	Chromic Cambisol	30%
Sample Location	SL-1	

Soil map unit 24

Area [km²]	24,8	
Area [%]	12,5	
Elevation [m]	450-1100	
Slope [%]	9-16	
Thickness [cm]	10-40	
Predominant Soils	Lithic Leptosols	50%
	Mollic Leptosols	30%
Error	Chromic Cambisols	20%
Sample Location	PL-7 / PL-9	

Soil map unit 79

Area [km²]	1,3	
Area [%]	0,7	
Elevation [m]	400-600	
Slope [%]	0-40	
Thickness [cm]	30-80	
Predominant Soils	Aridic Cambisols	70%
Error	Lithic Leptosols	30%
Sample Location	-	

Soil map unit 80

Area [km²]	5,2	
Area [%]	2,6	
Elevation [m]	0-300	
Slope [%]	5-40	
Thickness [cm]	60-110	
Predominant Soils	Aridic Cambisols	40%
	Aridic Calcisols	60%
Error	-	
Sample Location	-	

Soil map unit 81

Area [km²]	22,7	
Area [%]	11,5	
Elevation [m]	-300-300	
Slope [%]	9-100	
Thickness [cm]	10-50	
Predominant Soils	Aridic Cambisol	50%
	Aridic Calcisol	50%
Error	-	
Sample Location	-	

Soil map unit 82

Area [km²]	5,4	
Area [%]	2,7	
Elevation [m]	-300-300	
Slope [%]	0-26	
Thickness [cm]	10-80	
Predominant Soils	Lithic Leptosol	30%
	Yermic Calcisol	40%
Error	Ochric Cambisol	30%
Sample Location	PL-5	

Soil map unit 83

Area [km²]	12,6	
Area [%]	6,4	
Elevation [m]	-100-500	
Slope [%]	0-18	
Thickness [cm]	10-80	
Predominant Soils	Yermic Calcisol	60%
	Lithic Leptosol	30%
Error	Calcaric Cambisol	10%
Sample Location	Smp1 / Smp4	

Soil map unit 84

Area [km²]	3,9	
Area [%]	2,0	
Elevation [m]	-300	
Slope [%]	6-20	
Thickness [cm]	40-120	
Predominant Soils	Takyric Calcisol	30%
	Aridic Calcisol	40%
Error	Leptosols	30%
Sample	PL-8	

Methods of soil investigation

Using a double-ring infiltrometer, the infiltration rate of specific soils were measured as the decrease of the water level in the inner ring. In accordance with DIN19682-7 (German Institute for Norms 1997) two metal cylinder rings, 63 cm and 31 cm in diameter and 25 cm high were buried at 5–10 cm depth. There should be no hydraulic connection between the inner and outer ring. The use of two nested rings ensures a vertical flow in the inner ring where the drop in water level is observed (Fig. 3.4.4-5). Both rings were filled with water up to a height of 10 cm. The decrease of water level in the inner ring is recorded at increasing time intervals and then re-filled. The water level in the outer ring is continuously adjusted to the level in the inner ring in order to minimize possible shortcuts below the rings. Infiltration rates are repeatedly recorder during one hour. Gradually, the infiltration rate converges towards the saturated conductivity, which is usually reached after approximately one hour. These resulting infiltration curves are then approximated by an exponential regression technique providing information on the infiltration development from a dry to a wet soil. All infiltration rates are converted to m/s.

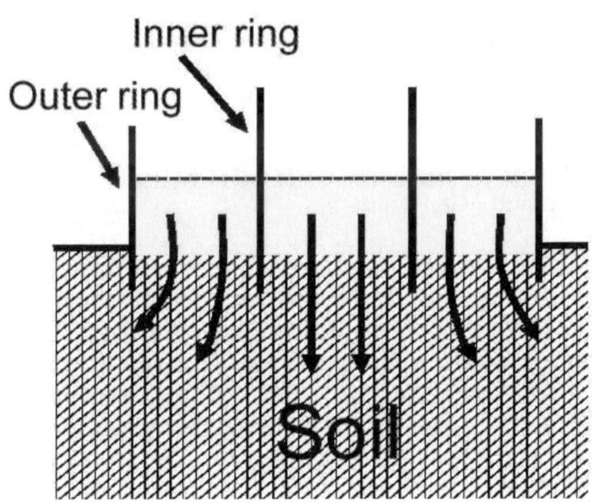

Fig. 3.4.4-5. Principle of double-ring infiltrometer

The water content of selective soil samples was determined according to German DIN 18121 (German Institute for Norms 1970). The bulk density/pore volume was specified using the DIN 18125. The particle size distribution was determined according to DIN 18123. The following mesh sizes were used for the sieving process: 37.5, 25, 9.5, 4, 2, 1, 0.5, 0.25, 0.125 mm. Grain sizes smaller then 0.125 mm were analyzed by sedimentation analysis. The retention potential of the soils for pollutants likely to contaminate groundwater was determined using the methylene-blue adsorption method. Fig. 3.4.4-6 summarizes granulometric graphs, densities, infiltration rates, retention potentials, cation exchange capacity (CEC), and soil thickness of typical soil profiles in the area.

Special attention was paid to the investigation of the effective field capacity (eFC) which is influencing the residence time of infiltrating water in the soils. In the unsaturated zone soil water retention curves define the relationship between volumetric water content Θ (cm^3/cm^3) and the soil matrix potential h (cm water column). The curves vary depending mainly on the particle size distribution of a soil. The soil matrix potential is zero when all pores are filled with water and the soil is saturated. In the unsaturated zone the matrix potential is negative, therefore also called suction power. It is expressed as logarithm of the suction power in (cm), so-called pF-value. The amount of water stored in the soil between pF-values of 1.8 (field capacity) and 4.2 (wilting point) is the plant available water; the range between these two pF-values is called eFC.

Applying the Van Genuchten's model on the soil-water retention curves the eFC can be calculated for known water contents. The Van Genuchten's model is a mathematical function describing the relationship between the water content and matrix potential (Eq.(3.4.4-1)).

$$\Theta = \Theta_R + \frac{\Theta_S - \Theta_R}{[1+(\alpha \cdot h)^n]^m} \quad (3.4.4\text{-}1)$$

Θ	Water content
Θ_S	Saturated water content
Θ_R	Residual water content
m,n	Empiric parameters, calculated experimental (with m = 1-1/n)
α	Air entry point, calculated from the experiment
h	Suction power (cm water column)

The soil samples were saturated first and then drained under specified suction power in several steps. Water content at each adjusted matrix potential was measured. From these measurements the unknown parameters of the Van Genuchten function are derived. For pF-values of 4.2 (wilting point) and higher the water content is calculated from the Van Genuchten function, since at pF-values of 4.2 and higher the suction power needed from the measuring devices to drain the sample is too high.

With known water content at pF 1.8 and 4.2 the effective field capacity is calculated and expressed as either (Vol%) or (cm^3/cm^3). However, for some soil locations, the eFC could not be measured and had to be estimated (AG Boden 1996). Fig. 3.4.4-6 shows the measured, estimated and final eFC values.

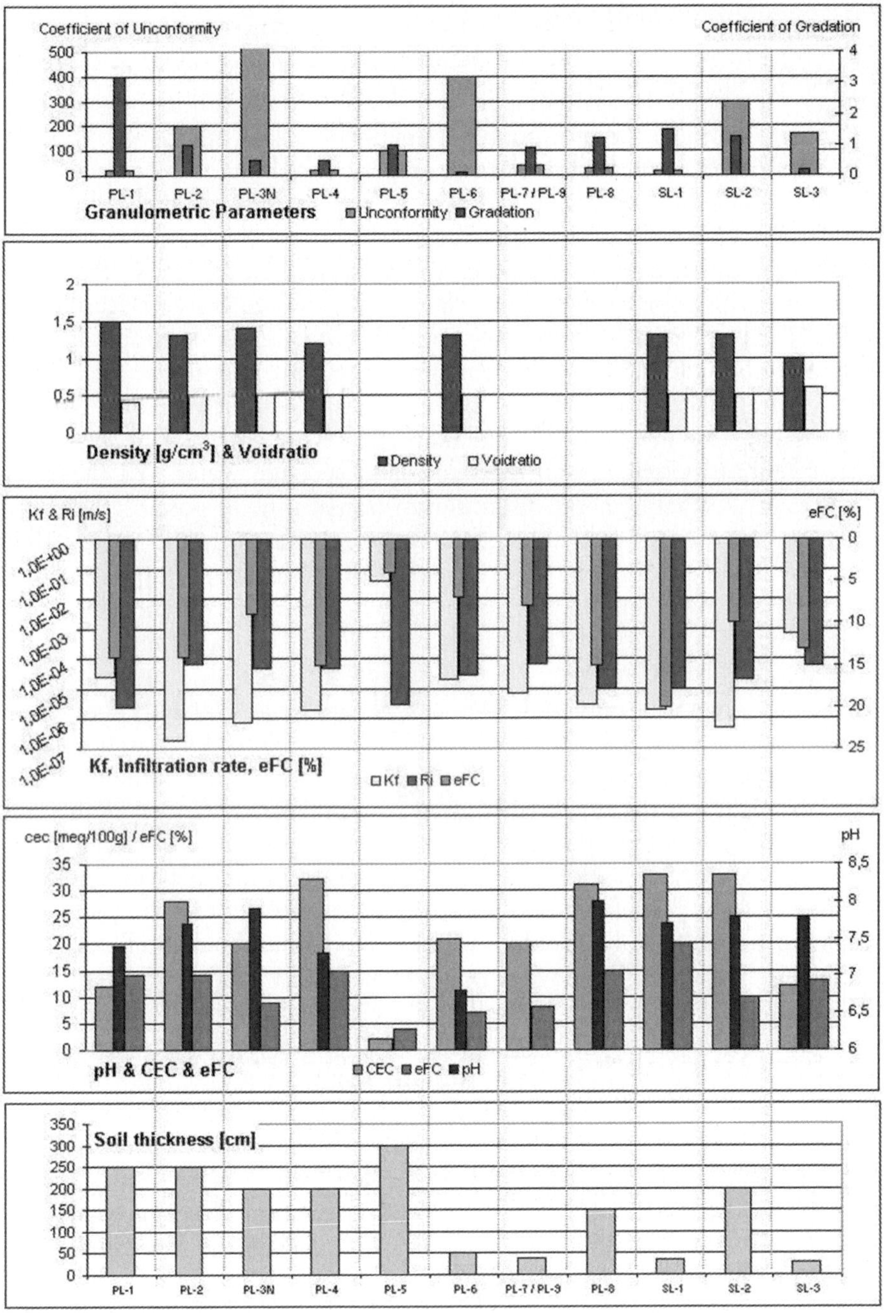

Fig. 3.4.4-6. Typical soil parameters at locations in the study area (Fig. 3.4.4-4)

Protective potential of soils

Microbiological, physico-chemical and mechanical processes in soils depend on the transit time and degree of saturation of the soil in the unsaturated zone. The transit time of contaminants depends on the thickness of the soils, the presence of macro pores (such as root channels or tubes burrowed by mice and other animals) and the permeability of the soil matrix. Evaluating the results of soil investigation, soil cover classes were determined reflecting the protective potential of the soils in the study area.

For the evaluation of the various factors for the protective effectiveness of the soils, a point system was applied taking into consideration the ground-water vulnerability concept introduced by Hölting et al. (1995). The scheme for estimating protection points is outlined in Fig. 3.4.4-7. This scheme includes several calculations of factors such as residence time, retention potential, soil thickness and seepage water. All these parameters are relevant for the definition of transit time of possible contaminants. In order to derive the point for the residence time (PP_{RT}) the effective field capacity (eFE) in (%) is squared and divided by 10, thus yielding values between 2.5 and 90 for ranges of eFC between 5% and 30%. This formula was first used by Hölting et al. (1995) in order to approximate a score system for eFC ranges.

For the retention potential, the cation exchange capacity (CEC) is merged with the soil-pH. The retention potential is considered to be mainly a function of the cation exchange capacity of soil, i.e., the content of clay minerals. CEC values are directly used to represent protection points and weigh slightly less than the residence time. The soil pH is considered because the mobility of cations is greatly enhanced at low pH-values. In order to transform pH-values into a numerical value characterizing soils, the measured pH-values are squared, divided by 2, and from the result 15 is subtracted. Thus, the final values range from -7 to 17 for pH values from 4 to 8, modifying the CEC points with a weighting factor of only 0.5. The pH-points are added to the measured CEC to yield the final protection points for the retention potential. Summarizing the points residence time (PPRT) and retention potential (PP_{RP}) yields the protection points (PP_s).

The soil thickness (ST) is taken from the soil map that was constructed by extrapolating the measured thickness of the soils. For the seepage water part, groundwater recharge values from Margane et al (2002) are used and converted into values between 0.75 and 1.75 based on the scheme of Hölting et al. (1995).

The protection points (PP_s), the soil thickness points (ST) and the groundwater recharge value (R) are multiplied to yield the final total protective point score (PP_T). The cover classes (according to the total protective point score) are listed in the cover class table in Fig. 3.4.4-7 ranging from very low to very high protective potential.

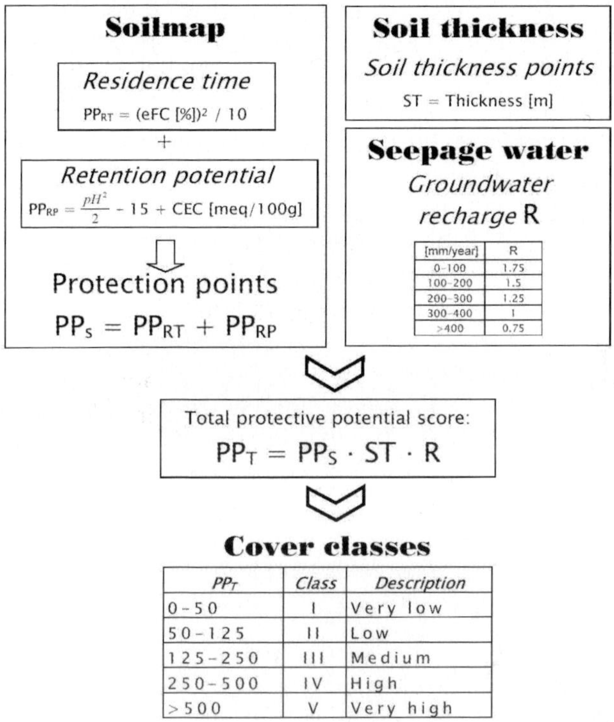

Fig. 3.4.4-7. Calculation scheme of protection points

Not for every soil map unit a representative location was analyzed. Several map units are represented by more than one sample. This inevitable inconvenience leads to an approximated extrapolation of the known parameters onto other map units. For example, CEC values measured at PL-8 and PL-5 were extrapolated to all map units in the southern study area. From the soil cover class map (Fig. 3.4.4-8) it is apparent that approximately half of the investigated area shows a very low groundwater protective potential.

It should be noted that in the southern part of the study area the surface rocks are exposed with only minor, local and patchy soil cover. Unlike exposed hilltops or steep slopes, relatively high protective soils occur on valley floors and terraces. However, many broad wadis show absolutely no soil cover. Nevertheless, the applied calculation scheme for protection points indicates the highest qualitative soil thickness in wadis. It is therefore suggested to delineate in a further step buffer zones around the wadi floors that can be subtracted from the final thickness map. A high protective potential of soils is reached in the surroundings of Yarka and Ira, where thick and mainly clayey soils with a high retention capacity dominate. Since other protective geological layers are absent, the soils are often the sole protective layer preventing contaminants from entering the unsaturated zone of the aquifer.

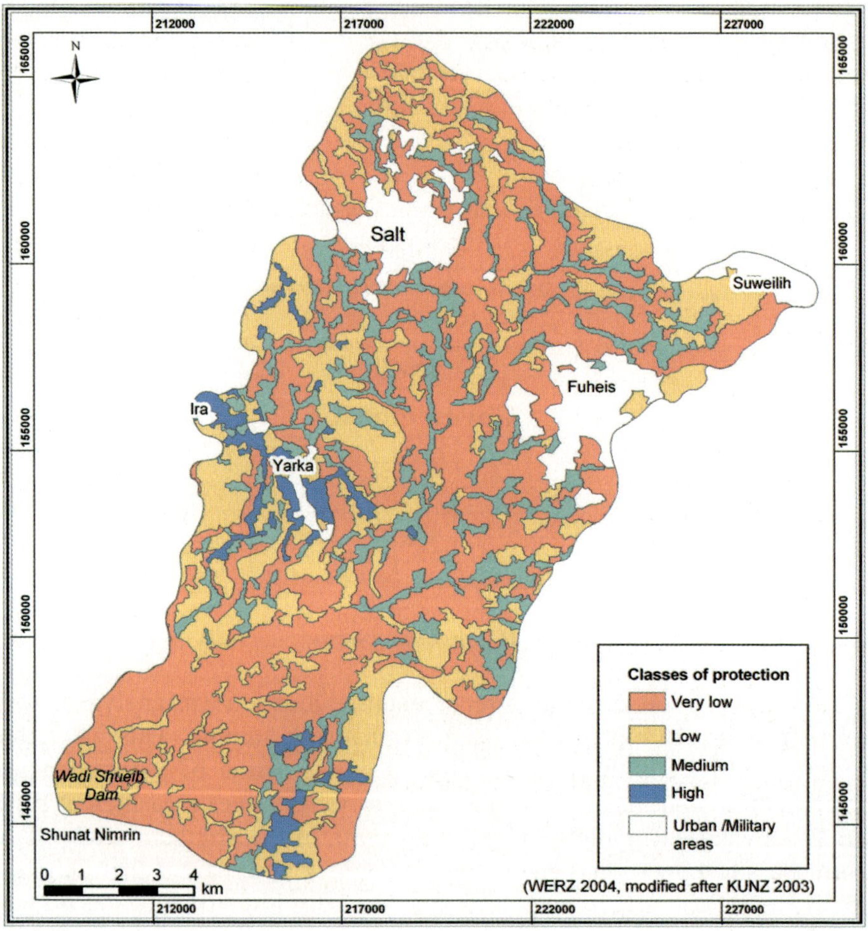

Fig. 3.4.4-8. Soil cover class map of the study area constructed according to their protective function for protecting groundwater from potential contaminations

3.4.4.5 Ground-water vulnerability and hazard mapping

Overview

The following techniques, surveys and existing data bases are used as supporting tools for ground-water vulnerability and hazard mapping in the study area: Landsat ETM^+ data, aerial photographs at different scales, age and quality, field survey, GPS data, topographic maps, geographic information systems, existing thematic data layers. Different thematic layers (agricultural areas, urban areas, infrastructure, etc.) are established by the analysis of remote sensing data (Werz 2006).

The main advantage of the application of remote sensing data compared to any conventional technique is that it allows fast, low-cost and complete collection of regional thematic data. This is particularly useful in inaccessible areas in which missing topographic data can be gained by land-cover classifications. Examples of thematic layers gained by on-screen digitizing of orthophotos within GIS and the section of Landsat ETM^+ data from 2001 including the study area are shown in Fig. 3.4.4-9.

An intrinsic ground-water vulnerability and hazard map (1:50,000) is constructed by applying the PI-method for vulnerability (Goldscheider et al. (2000) and the European concept of the COST Action 620 guidelines for hazards (Zwahlen 2004). This applied European concept focuses on an improved and consistent approach for the protection of karst groundwater based on the European Water Framework Directive, which provides a common framework for water resource policy and management (Zwahlen et al. 2004). COST is the acronym for "European Cooperation in the field of Scientific and Technical Research". The COST Action 620 "Vulnerability and risk mapping for the protection of carbonate (karst) aquifers" was established to develop an approach for the protection of karst aquifers. The COST Action 620 approach focuses on the assessment of the ground-water vulnerability, the potential ground-water hazards and the ground-water risk intensity of an aquifer as a basis for protection strategies. The intrinsic vulnerability of groundwater to contaminants is independent of the nature of the contaminants. It depends on the geological, hydrological and hydrogeological characteristics of an area.

The assessment of the ground-water vulnerability is based on the PI-method (Goldscheider et al. (2000), which was originally developed for European karst areas. The PI-method consists of two main factors: the P-factor and the I-factor. The P-factor considers the effectiveness of the overlying layers including the properties of the geological materials between the ground-water table and the earth surface. Factors influencing the effectiveness of the overlying layers comprise (a) the thickness and effective field capacity of the topsoil, (b) the recharge rate, (c) the thickness and

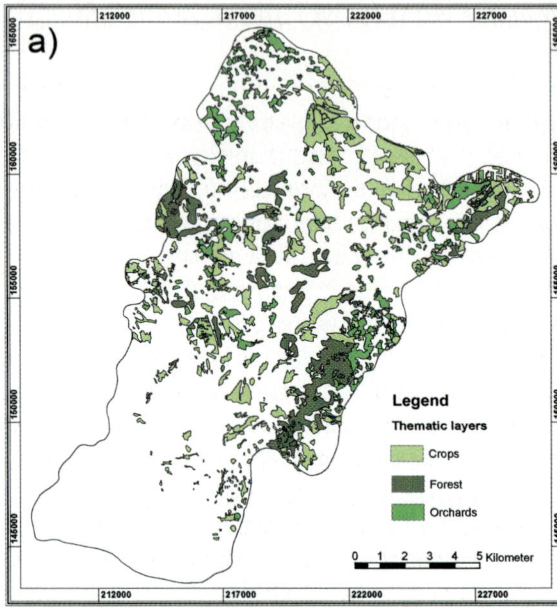

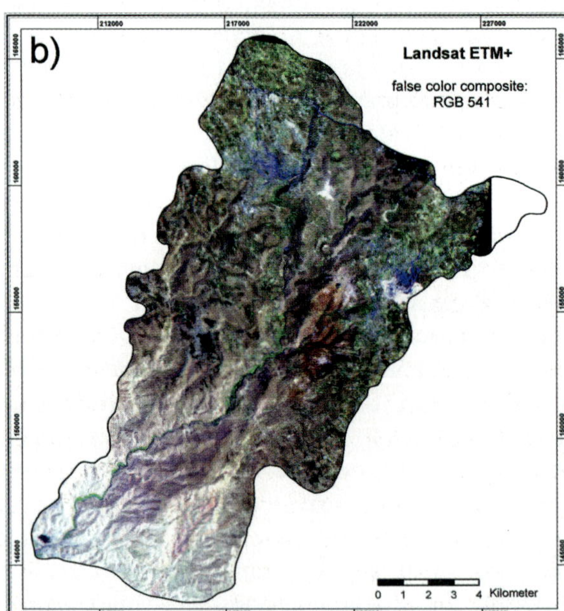

Fig. 3.4.4-9. Thematic maps derived by on-screen digitizing of orthophotos (a) and color composite of the Landsat ETM$^+$ image section (b) (band combination 5/4/1). Cleary visible are the urban areas (blue/white) and the vegetated areas (green).

grain size distribution of the subsoil, and (d) the lithology and fracturing of rocks.
The I-factor considers infiltration conditions, i.e., the flow and concentration of surface-water depending on the slope, the hydraulic conductivity, the depth to low-permeability layers, the land-cover conditions and the existence of sinking streams, sinkholes and discharge direction. The protective effectiveness of the rock cover above the aquifer is determined by a point-counting system.

Some aspects of the PI-method are modified in order to fit the climatic and land-cover conditions of the study area.

The COST Action 620 defines hazards as "a potential source of contamination resulting from human activities taking place mainly at the land surface". Numerous anthropogenic activities affect the environment and the groundwater in an adverse way, thus, it is impossible to include all these variants in a hazard map. Therefore, the hazard map can only incorporate the main impacts which are based on the production, transport, storage, utilization and disposal of harmful substances. It is assumed that the release of the contaminants takes place over a longer period of time, e.g., when fertilizers are applied, or in a singular event, e.g., via an accidental spill. Considering the transport of hazardous substances, fixed locations cannot be given.

For their evaluation and classification it is essential to define the quantity and the toxicity of the substances which can be released. Hazards can be classified by their duration of disposal, their geometry, type of human activity or the nature of the contaminants. According to their geometrical features, hazards can be differentiated in punctiform (e.g., septic tanks) linear (e.g., streets) and polygonal hazards (e.g., agricultural area), strongly depending on the mapping scale.

Three main categories of hazards are distinguishable which are in line with the type of land use: agriculture, infrastructure and industry (level I in Table 3.4.4-4 and then further distinguished in sub-categories (level II in Table 3.4.4-4). The sub-categories make use of additional criteria which distinguish between hazards according to the main source (solid or liquid contaminants) of possible ground-water contamination or refer to types of industrial or agricultural activities with their corresponding spectrum of possible pollutants. In a last step, the sub-categories are subdivided in level III giving all possible types of hazards (Table 3.4.4-4) resulting in weighting values for each hazard. The harmfulness of the hazard types is determined in a weighting and ranking scheme and displayed in an unclassified hazard map indicating the location, type, and size of the hazards and in a classified hazard map showing the degree of harmfulness for the groundwater.

Ground-water vulnerability in the study area

The ground-water vulnerability map using the PI method is based on the results of the thematic layers of the remote sensing data (Fig. 3.4.4-9), data from soil investigations, field survey data including hydrogeological investigations (degree of karstification, observed karst morphology etc), DEM data and geological maps (Werz and Hötzl 2004). Climatic conditions and vegetation in the area of the Wadi Shueib differ from those in most parts of central Europe where the PI-Method was developed. Therefore, some methodological modifications had to be applied in order to represent the hydrological conditions in the study area.

Table 3.4.4-4. Example of the first, second and third levels of hazard classification. The third level is derived for the municipal waste only. Excerption of the COST Action 620 hazard inventory list (modified after De Katelaere et al 2004)

Number	Level I Categories of hazards	Level II Subcategories of hazards	Level III hazard type	Weighting Value
1	Infrastructural development		…	
1.1		Waste water		
1.2		Municipal waste →	Garbage dump	40
1.3		Fuels	Waste loading station	40
1.4		Transport / traffic		
1.5		Recreational facilities	Sanitary landfill	50
			Spoils depository	35
1.6		Diverse hazards	Sludge from treatment plants	35
			…	
2	Industrial Activities		…	…
2.1		Mining		
2.2		Excavations sites		
2.3		Oil / gas exploration		
2.4		Industrial plants		
2.5		Power plants		
2.6		Industrial storage		
2.7		Diverting and treatment of waste water		
3	Livestock and Agriculture		…	…
3.1		Livestock		
3.2		agriculture		

For the assessment of ground-water vulnerability, only the uppermost aquifers are considered. The relevant aquifers within the Ajlun group are the Naur aquifer, the Hummar aquifer and the Wadi es Sir aquifer. According to the geological map (Shawabkeh 2001; Moh'd et al. 1998; Sawarieh and Barjous 1993) the Wadi es Sir aquifer layers are exposed over approximately 44% of the study area whereas the Hummar and Naur aquifers are both exposed only over approximately 24% of the study area. The Kurnub sandstone aquifer belongs to the lower aquifer system with an exposed area of approximately 7% and contributes little to the ground-water vulnerability.

For the topsoil layer of the PI-map, a modified version of the soil map and the effective field capacities (eFCs) is used. The eFC values of the soils are converted from vol% to mm ranging from 90 to 200 mm in order to fit the PI-method. Farming is considered as an indicator that soils extend area-wide and are not just patchy. For the main parts of the study area in which the soil cover is not widely and uniformly distributed and with bare-rock outcrops between thin soils, no protective points are assigned to the topsoil layer (Fig. 3.4.4.-10).

In order to estimate the recharge volume (R), the modeling results of Orthofer (2001) for the precipitation in the study area are used taking into account that only approximately 12% of the precipitation is assumed for groundwater recharge (Ta'any 1992). The estimated recharge in the study area is less than 100 mm/a and corresponds to an R factor of 1.75. Unfortunately the PI-method does not differentiate recharges below 100 mm/a.

Very limited and small subsoil deposits (wadi floor deposits) exist beneath the course of Wadi Shueib (Al-Daher 1989). These subsoil deposits are of no significance for the protective function of groundwater and therefore they are not considered as a special layer. The lithology was taken from the digitized geological maps based on 1:50,000 geological maps by Shawabkeh (2001), Moh'd et al. (1998) and Sawarieh and Barjous (1993). For assessment of fracturing in the study area, a moderately jointed or low-degree karstification is assumed. The thickness of the geological layers was taken from Ta'any (1992).

The thickness of the unsaturated zone of the karstified aquifers is assessed considering well data. Information about the Hummar aquifer is taken mainly from well loggings in the study area. Most of the wells are concentrated in two clusters in the area of Salt in the north (four wells) and near the Wadi Shueib dam in the south (four wells). In other parts of the area, the thickness of the unsaturated zones is assessed by considering topography and the elevation of the discharging springs. Due to the lack of information, both constant thickness and depth to the groundwater table within the aquifers was assumed for the unsaturated zones, even if local changes exist within the aquifers due to the complex tectonic situation.

Topsoil - T

eFC [mm] up to 1 m depth	T
> 250	750
> 200-250	500
> 140-200	250
> 90-140	125
> 50-90	50
< 50	0

Recharge - R

Recharge [mm/y]	R
0-100	1.75
>100-200	1.50
>200-300	1.25
>300-400	1.00
>400	0.75

Subsoil - S

Type of subsoil (grain size distribution)	S
clay	500
loamy clay, slightly silty clay	400
slightly sandy clay	350
silty clay, clayey silty loam	320
clayey loam	300
very silty clay, sandy clay	270
very loamy silt	250
slightly clayey loam, clayey silty loam	240
very clayey silt, silty loam	220
very sandy clay, sandy silty loam, slightly sandy loam, loamy silt, clayey silt	200
sandy loam, slightly loamy silt	180
slightly clayey silt, sandy loamy silt, silt, very sandy loam	160

Type of subsoil (grain size distribution)	S
very clayey sand, clayey sand, loamy silty sand	140
sandy silt, very loamy sand	120
loamy sand, very silty sand	90
slightly clayey sand, silty sand, sandy clayey gravel	75
slightly loamy sand, sandy silty gravel	60
slightly silty sand, slightly silty sand with gravel	50
sand	25
sand with gravel, sandy gravel	10
gravel, gravel with breccia	5
non-lithified volcanic material (pyroklastica)	200
peat	400
sapropel	300

Lithology - L

Lithology	L
claystone, slate, marl, siltstone	20
sandstone, quarzite, volcanic rock, plutonite, metamorphite	15
porous sandstone, porous volcanic rock (e.g. tuff)	10
conglomerate, breccia, limestone, dolomitic rock, gypsum rock	5

Fracturing - F

Fracturing	F
non-jointed	25.0
slightly jointed	4.0
moderately jointed, slightly karstified or karst features completely sealed	1.0
moderately karstic or karst features mostly sealed	0.5
strongly fractured or strongly karstified and not sealed	0.3
Epikarst strongly developed, not sealed	0.0
not known	1.0

Thickness of each stratum in [m] - M

Bedrock - B $B = L \cdot F$

Artesian pressure A 1500 points

Total protective function P_{TS}

$$P_{TS} = \left[T + \left(\sum_{i=1}^{m} S_i \cdot M_i + \sum_{j=1}^{n} B_j \cdot M_j \right) \right] \cdot R + A$$

score P_{TS}	effectiveness of protective cover	P-factor	example
0-10	very low	1	0-2 m gravel
>10-100	low	2	1-10 m sand with gravel
>100-1000	medium	3	2-20 m slightly silty sand
>1000-10000	high	4	2-20 m clay
>10000	very high	5	> 20 m clay

P-map

Fig. 3.4.4-10 Determination of the P-Factor (after Goldscheider et al. 2000)

Finally all these parameters are combined in order to derive the total protective functions (P_{TS}) of the layers that overlie the aquifer (Fig. 3.4.4-10).

The I-factor (infiltration conditions, degree of bypassing) comprises three steps of calculation (Fig. 3.4.4-11). The dominant flow process is assessed by combining the depth to the low permeability layer with the saturated hydraulic conductivity. For the hydraulic conductivity of the soils the k_f-values are derived from granulometric curves of the soil samples.

The depths to the low permeability layers are determined based on the soil map combined with field observations and extrapolations from locations with known soil thickness to locations with similar conditions. In a further step the dominant flow process map is combined with the slope and the vegetation cover leading to a map showing where surface and subsurface flow predominates.

For the I'-factor, the type of vegetation, the slope and the dominant flow process are combined. The PI-method distinguishes between two types of vegetation cover: forest and field/meadow/pasture areas (Fig. 3.4.4-11). Under moderate European climate conditions, woodlands infiltration is enhanced because of the characteristics of the forest soil (loose organic material, abundant macropores) and due to the reduction of the precipitation intensity through interception. Agricultural areas often enhance surface runoff due to the soil characteristics (clayey, loamy soils) and increased compaction of the not permanently vegetated surface. In the study area no woodland comparable to a typical European forest exists. Two forest types are mapped: oak tree forests and pine forests. Both tree types grow directly on more or less karstified limestone where hardly any soil cover is present. Therefore for the I'-factor in the study area, just the field/grassland/pasture area table was accepted for calculation and was combined with the slope information. For further vulnerability calculations applying the PI-method, it could be useful to distinguish between field, grassland and pasture areas taking into account the density and intensity of vegetation. Agricultural areas with dense vegetation could favor the infiltration process even on steep hills.

The slope thematic layers are derived from digitized elevation contour lines of 1:50.000 topographic maps (elevation interval of 20 m). During a further step a hydrologically correct elevation grid was constructed and was later converted from slope degree to slope percent and assembled to the three slope classes needed for the PI-method.

In a next step, the results of the I'-factor are combined with the surface catchment map to get the final I-factor. The surface catchment map includes the existence of relevant karst features that favor infiltration like sinking streams, swallow holes and their catchments. Furthermore, it indicates whether areas have surface runoff inside or outside karstified areas. During field work in the study area, no swallow holes or sinking streams were found. However, deep wadis can be considered as potentially more vulnerable zones because they are often deeply incised into the aquifer and their depth to the groundwater table is small. Furthermore many wadis follow the direction of faults and thus indicate zones of higher permeabilities.

Thus, it could be possible that contaminants can infiltrate relatively fast through the wadi bed into the aquifer. Further investigations should be done to assess whether zones under the wadi beds are more permeable and may possibly lead to direct and fast infiltration of contaminated water into the aquifer.

1st Step: Determination of the dominant flow process

		Depth to low permeability layer		
		< 30 cm	30-100 cm	> 100 cm
Saturated	> 10^{-4}	Type D	Type C	Type A
hydraulic	> 10^{-5}-10^{-4}		Type B	
conductivity	> 10^{-6}-10^{-5}	Type E		
[m/s]	< 10^{-6}	Type F		

2nd Step: Determination of the I'-factor

Forest				
dominant flow process		Slope < 3.5 %	Slope 3.5 - 27 %	Slope > 27 %
infiltration	Type A	1.0	1.0	1.0
subsurface flow	Type B	1.0	0.8	0.6
	Type C	1.0	0.6	0.6
surface flow	Type D	0.8	0.6	0.4
	Type E	1.0	0.6	0.4
	Type F	0.8	0.4	0.2

Field/Meadow/Pature				
dominant flow process		Slope < 3.5 %	Slope 3.5 - 27 %	Slope > 27 %
infiltration	Type A	1.0	1.0	0.8
subsurface flow	Type B	1.0	0.6	0.4
	Type C	1.0	0.4	0.2
surface flow	Type D	0.6	0.4	0.2
	Type E	0.8	0.4	0.2
	Type F	0.6	0.2	0.0

3d Step: Determination of the I-factor

Surface Catchment Map		I' factor 0.0	0.2	0.4	0.6	0.8	1.0
a	swallow hole, sinking stream and 10 m buffer	0.0	0.0	0.0	0.0	0.0	0.0
b	100 m buffer on both sides of sinking stream	0.0	0.2	0.4	0.6	0.8	1.0
c	catchment of sinking stream	0.2	0.4	0.6	0.8	1.0	1.0
d	area discharging inside karst area	0.4	0.6	0.8	1.0	1.0	1.0
e	area discharging out of the karst area	1.0	1.0	1.0	1.0	1.0	1.0

↓ **I-map**

Fig. 3.4.4-11 Determination of the I-Factor (after Goldscheider et al. 2000)

Interpretation of vulnerability mapping

The protective function map (P-map in Fig. 3.4.4-12a) suggests that two levels of the protective function dominate: a moderate (yellow color) and a low protective function (red color). The areas with low protective function match with the outcrops of mainly bare karstified limestones and dolomites with well developed karst features and with only patchy soil cover. A high protective function dominates along outcrops of the Kurnub sandstone. Enhanced protection points are generated due to the thickness of the unsaturated zone and to the type of lithology.

Generally, the contribution of the topsoil to the protective function of the aquifer is very small. As shown in Fig. 3.4.4-13, 82.9% of the area has no protection points for the topsoil layer because of the absence of a continuous soil cover. The class "very high protective function" of the overlying layers does not exist in the study area.

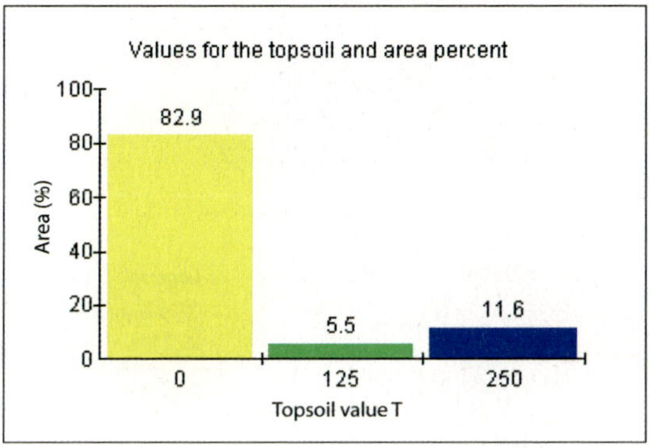

Fig. 3.4.4-13. Area percent of the three existing protection values for the topsoil

The I-map includes the whole range of bypass probabilities, from very low to very high (Fig. 3.4.4-12b). It is dominated by a very low degree of bypassing of the protective function (blue color), and by a low degree of bypassing (green color). The areas with a very low degree of bypassing are those with bare karst where the main flow process is infiltration. The small areas with a very high degree of bypassing (orange color) are limited to steep slopes favoring surface flow. They are also limited to areas of missing soil cover characterized by a fast surface- or subsurface flow.

For the final PI-factor and the degree of vulnerability, the P (protective cover) and I factor (infiltration conditions) are multiplied. By combining the P-map with the I-map, parts of the moderate protective function areas are reduced and are assigned in the final vulnerability map to highly vulnerable areas due to a high bypass probability. The class "very low vulnerability" (Fig. 3.4.4-14) is not present. In the study area a moderate vulnerability (45.6% of the area) dominates (Fig. 3.4.4-15). About one third of the map shows high vulnerability. This is due to different reasons such as large areas of barely or patchily covered, highly karstified rocks, or areas in which a continuous soil cover is missing, or steep slopes favoring bypassing of the protective functions.

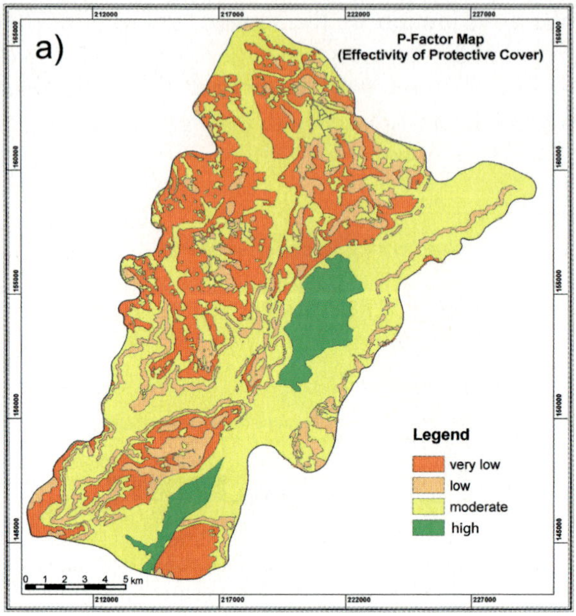

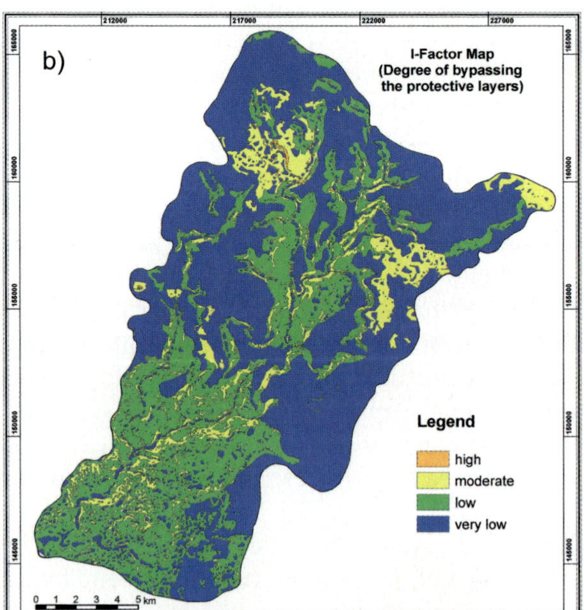

Fig. 3.4.4-12 P-Factor map (a) showing the effectivity of the protective cover (after Werz and Hötzl 2007) and I-factor map (b) showing the degree of bypassing of the protective layers

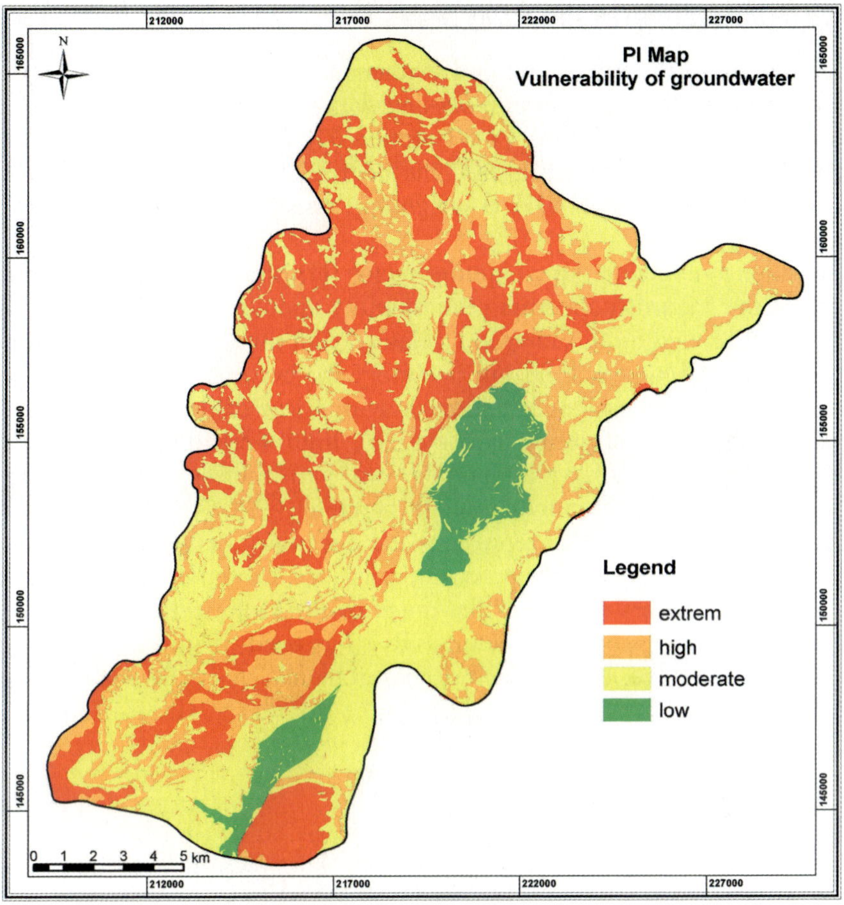

Fig. 3.4.4-14. PI-Map showing the groundwater-vulnerability (after Werz and Hötzl 2007)

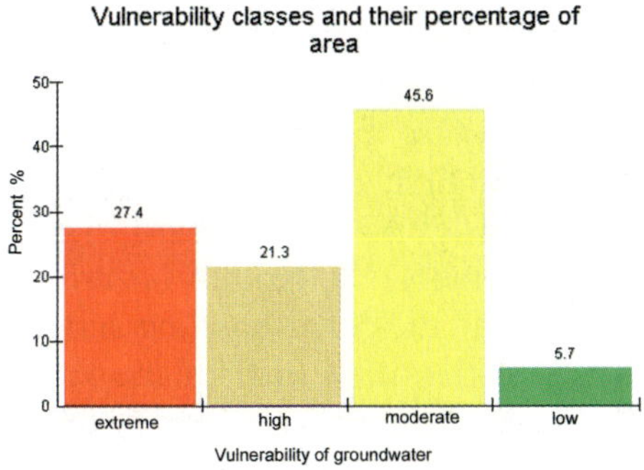

Fig. 3.4.4-15. Area percent of the existing vulnerability classes

3.4.4.6 Hazard map of the study area

The hazards in the study area are evaluated based on field work and on thematic maps based on remote sensing data (1:50,000 scale) and by employing the guidelines of the COST Action 620 (De Katelaere et al. 2004). In order to define for each hazard the degree of harmfulness (finally expressed as Hazard Index HI), a weighting, ranking, and reduction procedure was applied.

In order to evaluate the weighting factor (H), the toxicity and the properties of the hazards concerning the mobility and solubility, are taken into account. The weighting factor is a relative, non-measurable value defined by the COST Action 620 hazard inventory list between 0 and 100. A low weighting factor is assigned to hazards with a low degree of harmfulness, e.g., greenhouses (weighting factor = 20), high weighting values are assigned to hazards that have a high degree of harmfulness, e.g., oil refinery, or a waste water injection well (weighting factor = 85).

The ranking factor (Q_N) represents the difference between various hazard types of the same group, considering for example the size or the amount of fertilizers used. For the range of the ranking factor, 0.8 to 1.2 were chosen. The reduction factor $0<R_f<1$ is used, if information on the status, maintenance, age, security measurements etc. exist including the probability that a hazard release might take place. The Hazard Index (HI) is given by Eq.(3.4.4-2).

$$HI = H \cdot Q_N \cdot R_f \tag{3.4.4-2}$$

HI: Hazard Index
H: Weighting Factor
Q_N: Ranking factor
R_f: Reduction Factor

First, an unclassified hazard map (Fig. 3.4.4-16) is created showing the size, position and type of hazard according to the hazard inventory list. Based on this map, a classified hazard map is generated after application of the weighting, ranking and reduction factor. This map displays the degree of harmfulness of groundwater hazards (HI). Table 3.4.4-4 shows the scheme used to establish the first, second level and - by the example of the municipal waste - the third level of the hazard classification and the assigned weighting values. Finally, the type of hazard is assessed according to main hazard categories and the corresponding sub-categories.

Fig. 3.4.4-16 shows the unclassified map of the study area. In the study area twenty seven different hazard types are evaluated. The dominating hazard types in the study area are small industries. Agricultural and municipal hazard point-source contaminations that can be caused by manure heaps. Filling stations are also significant potential contamination sources. The hazards are concentrated in the larger villages like Salt, Fuheis and Suweilih. For streets, additional considerations according to the spreading of traffic related contaminants have to be considered. Harmful substances, related to traffic do not only affect the ground directly under streets. The generated contaminants are partly spread onto the adjacent roadsides, condense in the surrounding areas or accumulate in grounds beyond the unpaved streets. To represent this additional sphere of influence, a buffer zone of ten meters at each side is placed along streets.

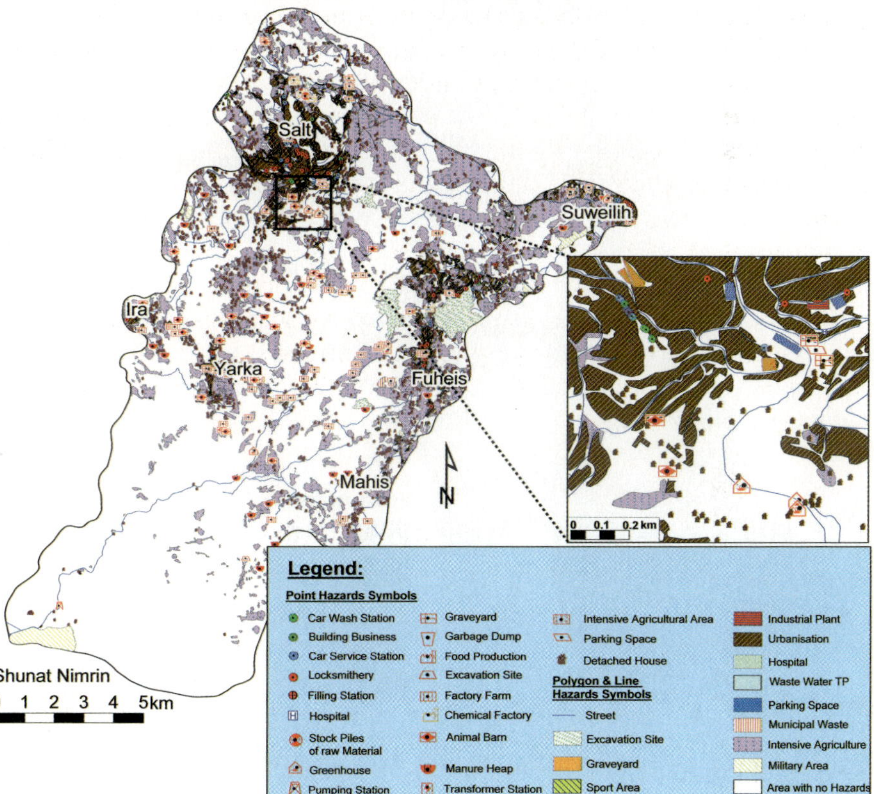

Fig. 3.4.4-16. Rescaled unclassified hazard map of the test site. The legend shows the point, line and polygonal hazard symbols of the unclassified hazards. For the city of Salt a section of the map is shown at larger scale. The hazards were originally mapped in the scale of 1:50.000. Due to the reduction of the original scale some details are not any more recognizable in this illustration. (Werz and Hötzl 2007)

Fig. 3.4.4-17 shows the classified map of the study area. The major part of the study area shows low- or no-hazard level (hazard index less than 24) which implies a very low degree of harmfulness. This is due to the fact that the major part in the south is not populated and is not used for agriculture. Hazards with HI less than 24 are not distinguishable from areas without any existing hazards due to the fact that they are placed in the same hazard level class. Table 3.4.4-5 shows the hazard index, the hazard index classes, the description and the area percent of each hazard index class in the study area.

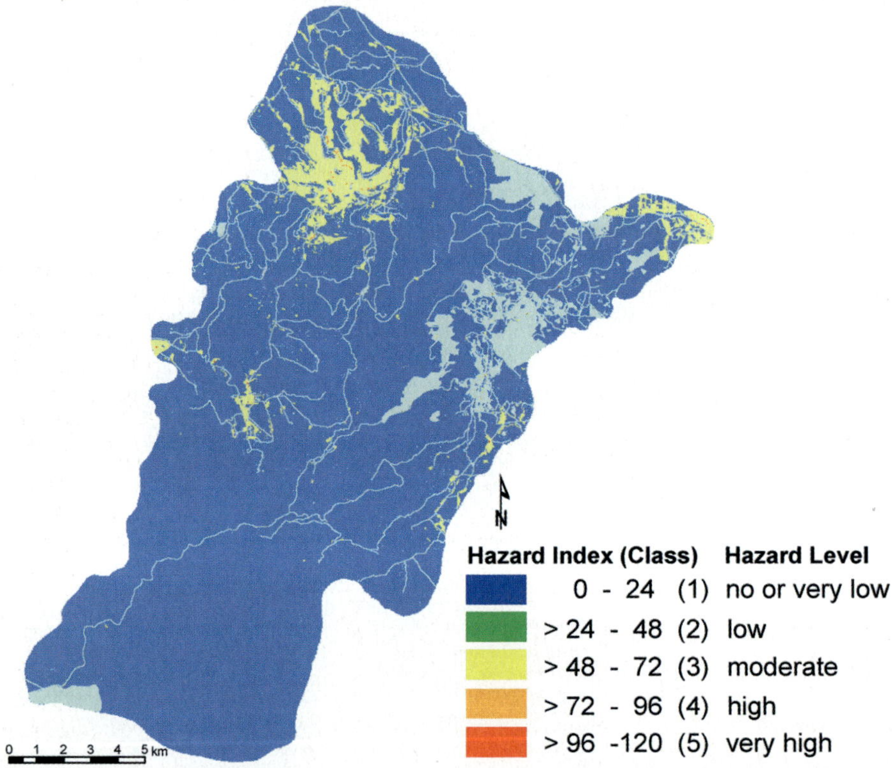

Fig. 3.4.4-17. Classified hazard map of the study area (after Werz and Hötzl 2007)

Interpretation of the hazard mapping and conclusions

The hazard index is calculated for every evaluated hazard type based on the three different parameters: weighting factor (depending on the hazard type), ranking factor and reduction factor, representing the degree of harmfulness for any ground-water hazard.

The study area is considered as an insufficiently developed area of Jordan, the major difficulty is based on the lack of information as well as on problems of data evaluation. Besides the inspection of the area, interviewing inhabitants and employees of companies and municipalities, the analysis of orthophotos and satellite images are important tools for data evaluation.

Because of the special situation and the character of the study area in Jordan, the approach of the COST Action 620 had to be modified in various aspects. Hazards such car service stations, car washing stations, hospitals, locksmiths, and pumping stations, not included in the hazard inventory list of the COST Action 620, had to be included.

One of the prevalent problems arose from lack of information concerning the connection of settlements with a communal waste-water system. It is solved by establishing a scale of weighting values between the values of connected and unconnected areas.

Generally, areas with predominant agriculture are affected only by small to no degree of harmfulness. Significant hazards are only located in the industrialized areas of Salt, Fuheis, Mahis and Suweilih. Regarding the study area in total, 82.4% of the investigated sites show no or very low degrees of harmfulness (Table 3.4.4-5).

Table 3.4.4-5. Distribution of the different hazard-index classes within the study area

Hazard Index	Hazard Index Class	Hazard Level	Color	Area percentage in the study area (%)
0 - 24	1	No or very low	Blue	82.432
25 - 48	2	Low	Green	12.574
49 - 72	3	Moderate	Yellow	4.939
73 - 96	4	High	Orange	0.042
97 - 120	5	Very high	Red	0.012
> 120	>5	Extremely high	Violet	0.00106

However, action is called upon the improvement of the environmental situation of abundant illegal waste dumps or the necessity to connect all households to waste-water treatment plants. The ground-water vulnerability and hazard map show clearly the vulnerable areas and the “hot spots” of contamination in the study area and supply information for planners and decision makers. To maximize the possibilities, the hazard map should be combined in a next step with a ground-water vulnerability map to create a risk intensity map. The risk intensity map reflects the probability of groundwater to be actually contaminated by human activities showing that the highest risk of contamination exists, where dangerous hazards are located in zones hosting highly vulnerable aquifers.

The ground-water vulnerability and hazard maps provide only relative answers and no quantitative or absolute estimates of the actual sensitivity of the groundwater to contamination or the probability that contamination will occur. For quantitative assessments, tracer tests should be carried out in the study area to prove the hydraulic connection between "hot spots" of contaminations at ground surface and in the aquifer. This helps to determine the underground flow paths, to delineate the catchment areas of springs used for drinking water, to get information about the mechanism of contamination transport, and to evaluate the risk potential resulting from existing hazards. With the results of the tracer tests, groundwater protection areas can be established following for example the German guidelines for the establishment of water protection areas including individual protection zones (DVGW 1992). The establishment of protection zones for the springs in the study area can help primarily to reduce the microbiological contamination existing in some of the springs.

3.4.4.7 References

AG BODEN (1996) Bodenkundliche Kartieranleitung, 4th Edition, Schweizerbart, Hannover, pp 1–392

Al-Daher N (1989) Soil erosion susceptibility survey and evaluation: the example of Wadi Shueib Basin, Jordan. Diploma thesis University of Amman, Department of Geology, (unpubl.)

Bender F (1974) Geology of Jordan. Gebrüder Borntraeger Berlin, pp 1–196

De Ketelaere D, Hötzl H, Neukum C, Civita M, and Sappa G (2004) Hazard Analysis and Mapping. In: Zwahlen F (ed) (2004): COST Action 620- Vulnerability and Risk Mapping for the Protection of Carbonate (Karst) Aquifers-Final Report Neufchatel, European Union, pp 1–297

DIN 18125 Deutscher Normenaussschuss (1970) Baugrund - Untersuchung von Bodenproben – Bestimmung der Rohdichte. Beuth-Verlag GmbH, Berlin, pp 1–18

DIN 19682-7 Deutsches Institut für Normung E.V. (1997) Felduntersuchungen – Teil 7: Bestimmung der Infiltrationsrate mit dem Doppelringinfiltrometer. Beuth Verlag GmbH, Berlin, pp 1–20

DVGW (1992) Richtlinien für Trinkwasserschutzgebiete. I. Teil, Schutzgebiete für Grundwasser, Stand: Oktober 1992. – Deutscher Verein des Gas- und Wasserfaches, Technische Regeln, Arbeitsblatt W 101: 38 ., 3 Anl.; Eschborn, pp 1–30

Goldscheider N, Klute M, Sturm S, Hötzl H (2000) The PI method-a GIS based approach to mapping groundwater vulnerability with special considerations of karst aquifers, Z angew Geol 46:157–166

Hölting B, Haebertle T, Hohberger KH, Nachtigall, KH, Villinger, E, Weinzierl, W, Wrobel, JP (1995) Konzept zur Ermittlung der Schutzfunktion der Grundwasserüberdeckung. Geologisches Jahrbuch, C63:5–24

Horowitz A (2001) The Jordan Rift Valley, Tokyo: Balkema Publishers/ Lisse/Abingdon/Exton (Pa), pp 1–729

Kuntz D (2003) Soils in the Wadi Shueib catchment area and their protective potential for the Groundwater- Salt area.- Master Thesis University of Karlsruhe, Karlsruhe (unpublished), pp 1–161

Margane A, Hobler M, Almonani M, Subah A (2002) Contributions to the Hydrogeology of Northern and Central Jordan, Bundesanstalt für Geowissenschaften und Rohstoffe, Hannover, Geol Jahrb C 68, pp 1–68

Mikbel S, Zacher W (1981) The Wadi Shu'eib structure in Jordan. N Jb Geol Paläont Mh 9:571–576

Ministry of Agriculture (1994) National Soil Map and Land Use Project – The Soils of Jordan, Level 2–Semi detailed studies, Hunting Technical Services Ltd., Amman, Vol 2 & 3, pp 1–105

Moh'd A, Basem K, Muneizel S (1998) The geology of Al Salt area – Map sheet. No. 3154III. National Resources Authority (NRA) Bulletin No.41, Geology Directorate, Subsurface Geology Division, Amman, pp 1–67

Orthofer R, Shuval H, Issac J, Daoud R, Kupferberger H (2001) Developing Sustainable Water Management in the Jordan Valley Final Report, INCO-DC Contract No ERBIC18CT970161, Seibersdorf, pp 1–64

Powell JH (1989) Stratigraphy and sedimentation of the phanerozoic rocks in central and south Jordan: Part B Kurnub, Ajlun and Belqa Groups. Geological Mapping Division Bulletin 11B, Amman, pp 1–127

Salameh E (1980) The Suweilih Structure.- N Jb Geolog Paläont Mh 1980:428–438

Shawabkeh KF (2001) Map sheet 3153 IV, NRA, Geology Directorate, Subsurface Geology Division, Amman

Sawarieh A, Barjous M (1993) Map sheet 3154 II, NRA, Geology Directorate, Subsurface Geology Division, Amman

Ta'any Rakad AA (1992) Hydrological and Hydrochemical Study of the Major Springs in the Wadi Shueib Catchment Area, Master Thesis, Yarmouk University, Irbid (Jordan), pp 1–300

Werz H (2006) The use of remote sensing imagery for groundwater risk intensity mapping in the Wadi Shueib, Jordan. PhD thesis, University of Karlsruhe; online available, http://www.ubka.uni-karlsruhe.de/eva/index.html, pp 1–148

Werz H, Hötzl H (2007) Groundwater risk intensity mapping in semi-arid regions using optical remote sensing data as an additional tool. Hydrogeol J DOI 10.1007/s10040-007-0202-0

Werz H, Hötzl H (2004) Groundwater-Vulnerability Mapping and Hazard-Mapping in Semi-Arid Regions using Remote sensing Data. – Proc Intern Conf Groundwater vulnerability Assessment and Mapping, Ustron, Poland, 16 to 20 June 2004, pp 1–22

Zwahlen F (ed) (2004) COST Action 620- Vulnerability and Risk Mapping for the Protection of Carbonate (Karst) Aquifers-Final Report, Neufchatel, European Union, pp 1–297

3.4.5. An integrated approach for the hydrogeological investigation of unconsolidated aquifers in the lower Jordan Valley

M. Toll, T. Heinrichs, M. Sauter, E. Salameh

3.4.5.1 Introduction

The lower Jordan Valley is part of the Jordan - Dead Sea - Wadi Araba Rift Valley, which extends from the Red Sea to Lake Tiberias and beyond with a major 107 km sinistral strike-slip fault between the Arabian plate to the east and the northeastern part of the African plate to the west. Due to extensional forces, a deep depression called - for historical reasons - the Jordan "graben", has formed (Horowitz 2001). As a result of arid environment during geological periods, it is filled with evaporites, lacustrine sediments, and clastic fluvial components (Niemi et al. 1997).

The low elevation of -300 and -415 m below mean sea level near the Dead Sea along with the long and warm dry season, lead to excellent farming conditions. Therefore the area is used intensively for agriculture. As a result, hundreds of shallow wells were drilled and large amounts of groundwater are abstracted. Consequently since the 1960's, groundwater quality decreased rapidly and signs of overpumping and increasing soil salinity could clearly be seen. Along with heavy groundwater abstraction, replenishment of the Jordan Valley aquifers is reduced by diversion and storage of large quantities of freshwater both in dams built across the major wadi outlets on the Jordanian site of the Valley and by diversion of most of the freshwater from Lake Tiberias.

3.4.5.2 Integrated approach

In order to achieve sustainable water resources, a clear picture regarding the availability and distribution of groundwater of good quality is desirable. Therefore, this study aims at a better hydrogeological understanding of the groundwater in the unconsolidated sediments of the lower Jordan Valley (Fig. 3.4.5-1). Special emphasis is given to the spatial and temporal distribution of salt water. In order to get a comprehensive picture of the aquifer, an integrated approach, combining geological, geophysical, hydrogeological, historical, and chemical methods are chosen. The area of interest extends from the Dead Sea in the south to the city of Karameh to the north, from the Jordan River in the west to the margin of the western hills of the East Bank in the east. Nevertheless, the consolidated strata to the east are included in this study because it is believed that the major part of recharge to the unconsolidated aquifer derives from this area (Fig. 3.4.5-2).

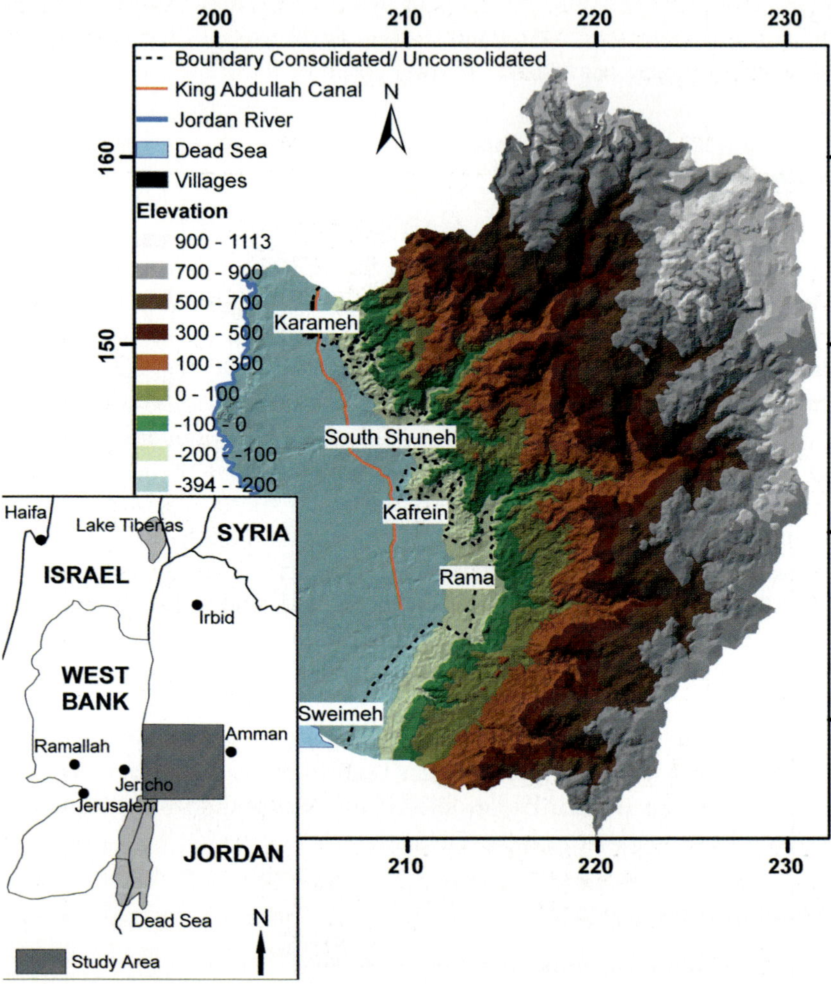

Fig. 3.4.5-1. Overview and digital elevation model of the study area (m msl)

3.4.5.3 Geology and hydrogeology

Consolidated-rock stratigraphy and aquifers

Along the lower Jordan Valley, NE of the Dead Sea foothill area, three consolidated sedimentary sequences occur, which are separated from each other and from the underlying crystalline basement by major unconformities. The Cambrian to Early Ordovician Ram Group represents a sequence of fluvial sandstone, transgressive siltstone/limestone and regressive-mature sandstone cycle overlying the crystalline basement. The thickness increases from an estimated 700 m near Wadi Zarqa in the north (Andrews 1991) to about 1 km near Madaba in the south. It contains the Disi aquifer, which is encountered in well JICA No. 5 south of Kafrein.

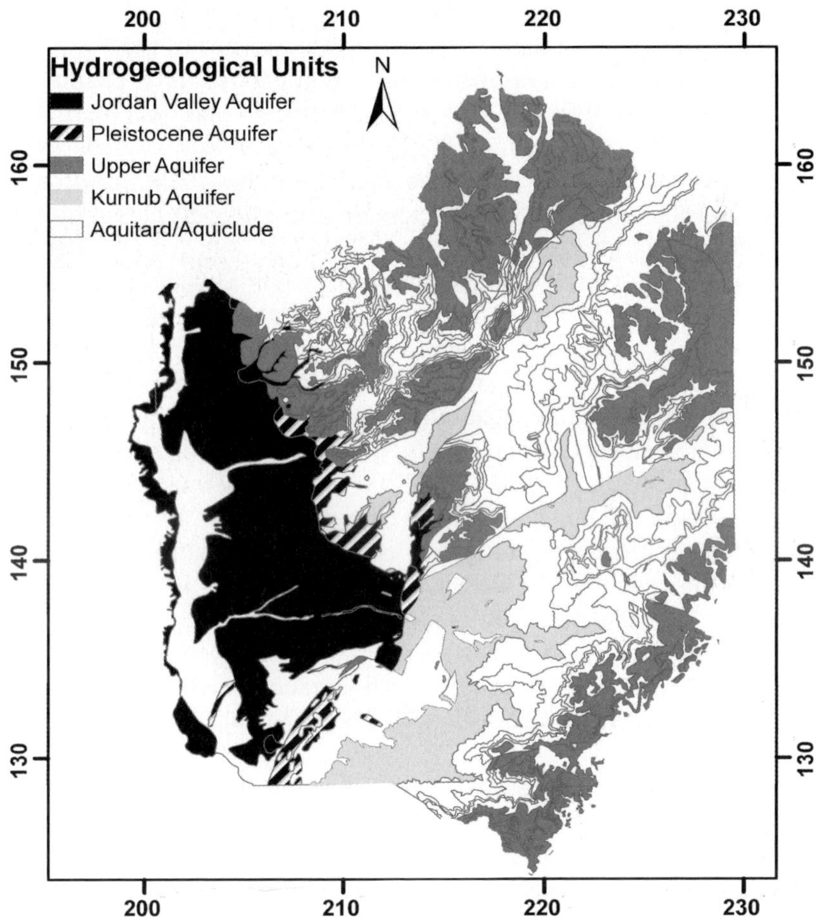

Fig. 3.4.5-2. Consolidated and unconsolidated aquifers within the study area

The following Permian to Jurassic Hudayb, Ramtha, Azab Groups (formerly agglomerated as Zarqa Group) consists of unconformity-bounded marginal marine sandstone-carbonate-shale associations most of which are considered an aquitard. Toward SE, all of them wedge out so that the total thickness of about 1000 m near Wadi Zarqa (Andrews 1992) is reduced to about 250–300 m south of Kafrein, where the Jurassic Azab Formation (Fig. 3.4.5-3) tapers out completely and only the lower part (Scythian to Anisian) of the Triassic Ramtha Group intervenes between the Ram Group and the overlying Cretaceous strata. SE of Wadi Naur-Wadi Kafrein, the gypsiferous Ladinian and Carnian Formations (Umm Tina, Abu Ruweis) are completely removed by pre-Cretaceous erosion (Bandel and Khoury 1981).

Above the Early Cretaceous transgressive unconformity, the sandy Kurnub Group (Neocomian), the limestone-marl sequences of the Ajlun Group (Late-Albian to end of Turonian) and the predominantly chalky Belqa Group (Coniacian to Eocene) follow. The fluvial to marginal marine sandstones of the Kurnub Group increase in thickness northwestwards attaining 180–200 m at the NE end of the Dead Sea and 250 to 300 m in the Wadi Zarqa area. They form the Kurnub aquifer (Fig. 3.4.5-3) containing water of low salinity.

The lower Ajlun Group (Naur, Fuheis, Hummar, Shueib Formations; A1 to A6) with 300 m of subtidal to peritidal marls, nodular- and massive limestones, reflect several transgressive to highstand cycles on a permanently submerged carbonate platform which, however, end with a late Middle Turonian lowstand generating the Upper Shueib Formation with shallow-water dolostones and marls grading southward into supratidal gypsiferous claystones (Schulze et al. 2003). Thus, because of the presence of the Fuheis and the Shueib marls and claystones, the Lower Ajlun (A1 to A6) forms a multi-layered aquitard, locally even an aquiclude, with embedded aquifers such as the massive limestones of the Upper Naur and of the Hummar Formations. Contrastingly, the overlying, ca 100 m thick Wadi Sir Formation (A7) consists in the area mainly of well bedded and massive limestone of a prograding and agrading carbonate platform. It forms together with the basal Belqa Group the important upper aquifer (A7–B2 aquifer).

After an episode of non-deposition, the pelagic chalks of the Wadi Umm Gudran and Amman al Hisa cherty limestone Formations (Belqa Group) were deposited and are preserved in synclines of the west-facing slopes. Only on the crest of the Transjordanian Mts. and beyond, they form a gently eastward sloping continuous layer overlain by bituminous marls and chalks of the B3 aquitard (Muwaqqar Formation).

Unconsolidated rocks: Jordan Valley Group and its aquifers

Two aquifers of significance are located within the study area. These aquifers are hydraulically interconnected and consist of terrigenous clastic components of Pleistocene to Recent age.

Period	Epoch	Group	Formation	Hydrogeol. Unit		Brief Description	Regional Thickness
Quaternary	Holocene	Jordan Valley	Alluvium	Aquifer	Qal	Clastic sediments	
	Pleistocene		Lisan	Aquitard	Jv3	Marl, Clay, Evaporites	0 - 300m
Teritary	Pliocene		Ghor el Katar	Pleistocene Aquifer	Jv2	Conglomerate with silicious cement, sand, gravel	100 - 350m
	Oligocene		Shagur Conglomerate		Jv1		
Upper Cretaceous	Campanian Santonian	Belqa	Amman al Hisa	Upper Aquifer	B2	Phosphorite, silicified limestone	100 - 350m
	Coniasian		Wadi Um Ghudran	Upper Aquifer	B1	Massive chalk & marlstone, fossiliferous	20 - 90m
	Turonian	Ajlun	Wadi as Sir	Upper Aquifer	A7	Thinlayered dolomite or massive gray limestone, fossiliferous. To the top more chertlayers	60 - 340m
	Cenomanian		Wadi Shu'eib	Aquitard	A5/6	White crystalline limestone and Ammonites (A6), and thinlayered limestone and marlstone (A5)	40 - 120m
			Hummar	Aquitard/ Aquifer	A4	Massive gray, sometimes yellowish limestone, often crystalline and cavernous	30 - 100m
			Fuheis	Aquiclude	A3	Thinlayered, marlstone, claystone	30 - 90m
			Na'ur	Aquitard	A1/2	Massive hard grey limestone + chertlayers (A2), and gray marlstone (A1)	90 - 220m
Lower Cretaceous	Albian Aptian	Kurnub	Subeihi	Kurnub Aquifer	K2	Multicoloured sandstone	120 - 350m
	Neocomian		Aardo	Kurnub Aquifer	K1	White, yellow massive sandstone	
Angular unconformity							
Jurassic Triassic Permian		Zarqa	Azab	Aquitard	Z2	Marine sandstone-carbonate-shale associations	0 - 600m
			Ma'in	Aquitard	Z1		0 - 1250m

Fig. 3.4.5-3. Stratigraphic table of rock units in the study area (modified after Shawabkeh 2001)

The Alluvial aquifer (Holocene or sub-recent)

This aquifer is built of sub-recent terrigenous deposits formed along the outlets of major wadis. As a result of large floods, these alluvial fans are still under accumulation. They consist of debris from all neighboring lithologies and are deposited according to their transport energy. The largest components are found close to the apex and the smallest, close to the fan margin. The transport occurs normally as sheet flow along alternating channels or after very heavy rainstorms. Thus within the deposits, permeable horizons alternate with impermeable ones. It is believed that the maximal thickness of the aquifer is near the Rift margins, thinning out towards the centre of the Rift basin. Well depth rarely exceeds several tens of meters. The alluvial aquifer often overlies directly the Pleistocene gravel aquifer and it is hydraulically interconnected with it.

The Pleistocene gravel aquifer

The coarse clastics, silt and Lisan sediments are a vertical and lateral facies succession from terrestrial/fluvial, to deltaic/limnic and limnic/brackish lake environments. They reflect the Plio-Pleistocene depositional conditions. The Lisan Lake (60,000–18,000 year b.p.) reached up to an elevation of -180 m MSL. The Lisan Formation containing marly, gypsiferous and silty lacustrine units, is generally considered an aquiclude, without any exploitable water. It is distributed mainly towards the middle of the graben. The Al Ghor Formation consists of two members: a silty member underlying or interfingering with the Lisan Formation and a coarse clastic member further to the east which consists predominantly of gravel, interbedded with clay, sand and marl horizons. The Al Ghor Formation is henceforth referred to as the "Pleistocene aquifer". However, exploitable water resources are principally restricted to the coarse clastic member.

Natural recharge by rain is small. Therefore, the aquifer is mainly fed by inflows of excess water from neighboring aquifers, namely the East Bank Mountain aquifers and from runoff in the wadis on the western slopes of the East Bank. Unlike the middle and northern part of the Jordan Valley, recharge due to infiltration of irrigation water can be neglected, since drip irrigation is the main irrigation practice.

The quality of water varies with low Cl^- concentrations near the areas of recharge and high ones in areas under the influence of the highly saline Lisan Formation, hypersaline brines and/or Dead Sea water intrusion. The hydraulic gradient adjacent to the Jordan valley escarpment is steep and becomes mild towards the west.

Geologic Structure

Pre-Cretaceous structural imprint on the subsurface rocks of the area is virtually unknown. Triassic and/or Jurassic phases of extension are indicated by a number of basaltoid dykes and sills in the Ramtha and Azab Groups, which did not intrude the Cretaceous cover (Bandel and Khoury 1981). The structures most pertinent to groundwater flow probably originated during the Late Cretaceous to Eocene Syrian Arc folding episode and they were modified during the following Miocene to present-day deformation associated with shear on the N-S-trending Dead Sea Transform Fault (DSTF) which in the area runs close to the River Jordan. The Syrian Arc folding produced the following fold- and fault-structures exposed along the eastern slopes of the southern Jordan Valley between Wadi Zarqa and the eastern shore of the Dead Sea. They are listed from N to S (Fig. 3.4.5-4):

- A dome-like structure around Salt where the base of the Cretaceous (bCr) rises to +400 m msl (Mueizel and Khalil 1993)
- The Wadi Shueib Syncline, where bCr drops to -500 m msl near Shuna
- The Wadi Shueib NW-facing monoclinal flexure plus associated normal faults; the Wadi Shueib composite anticline where bCr between Shuna and Kafrein rises to -250 m ms
- The NNE-trending Kafrein normal fault and associated faults bounding the Kafrein syncline to the W
- The Kafrein asymmetric syncline (bCr approx. -400 to -800 m msl)
- The NE to ENE trending Amman flexure and associated faults facing toward NW
- The gently warped dip slope SE of the Amman flexure where bCr rises again to about +300 m msl

These structures are modified or enhanced by sets of normal faults trending NNW and NW, which may be regarded as syn- and antithetic to the subsiding Jordan Valley depression.

The network of wadis flowing down the western slopes of the Transjordanian Mountains toward the Jordan Valley depression seem to be fairly well controlled by the above structures of the Cretaceous rocks. For example, wadis follow synclines (Wadi Shueib, Wadi Kafrein etc). They also follow a combination of structural dip with NW-trending antithetic faults on the dip slope SE of the Amman Flexure (Wadi Hisban/Ar Rama, Wadi al Muhtariqa, Wadi al Hiri E of Sweimeh). This may reflect some of the controls geologic structure exerts on subsurface flow paths of groundwater in the consolidated rock sequences.

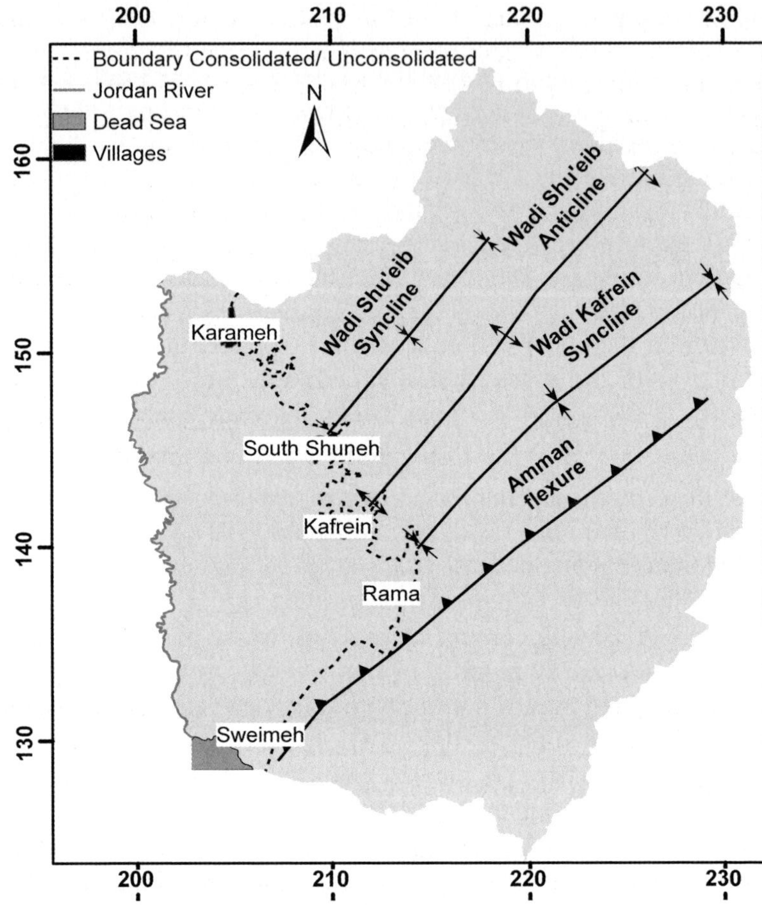

Fig. 3.4.5-4. Main structural features in the area of interest

Subsurface contacts between consolidated rocks and unconsolidated Jordan Valley Group sediments

In view of the multilayer nature of aquifers and aquitards in the consolidated rock sequence, it follows that many aquifers are confined. Therefore it seems likely that the unconsolidated sediments are infiltrated from beneath via consolidated layers dipping into the Jordan Valley.

At the foot of the Transjordanian Mts. near Al Karameh, Shuna, Kafrein, Sweimeh, the base of Jordan Valley Group sediments (bJVG) crops out at an averaged altitude of -200 m msl From a preliminary analysis of reflection seismics in combination with the deep wells JV1 and JV2, it appears that the Late Cretaceous structures continuously underlie the younger Jordan Valley Group sediments (Heinrichs et al. 2004). From the same data we may estimate that the base of the Jordan Valley Group sediments (bJVG) along the Wadi Shueib composite anticline drops from -200 m at the surface to about -500 m msl at 4–5 km distance WSW of the outcropping contact. This dip corresponds roughly to the dip of the Cretaceous strata east of the contact. The generalized conclusion is that up to this distance on this anticline, the Naur and Kurnub Formations are directly overlain by unconsolidated sediments. Further west and toward southwest, seismic data suggest that higher stratigraphic levels of the Cretaceous are preserved and in contact with the unconsolidated sediments in spite of the bJVG sinking rapidly westwards and southwards toward the Dead Sea. A similar inference can be made for the Wadi Shueib Syncline where the Wadi es Sir Formation should be continuous for some distance under the base of the unconsolidated sediments. At present insufficient seismic information prevents tracing the major faults into the subsurface of the Jordan Valley.

3.4.5.4 Historical review of land use

The lower Jordan Valley has a long history of human settlements and hence, of agriculture though periods of minor usage or even of abandonment occurred throughout history. The city of Jericho, founded in the 9th till 8th century BC, is known to be the oldest continuously populated city in the world and is located just 8 km west of the study area. Excavations in the Wadi Shueib dating back to the Neolithic Period show manifold human activities also within the study area.

The settlement and agricultural activities in former times were usually restricted to the availability of water resources and good farm land. Within the area of interest the Jordan River itself flows in 15–30 m deep gorge (in Arabic: Zor). The residues of former Jordan River flood plains can be considered as good farming land. Contrarily, the areas in which the Jordan River flows built of the formerly described Lisan Formation are unsuitable for farming due to the high salt content. As a consequence, only places near the outlets of major wadis, where a perennial flow of water from the eastern hills supplies enough water, were suitable for agricultural activities. Upon the arrival of the first European and American explorers in the late 19th century the study area was described as a hot and uninhabited area, unsuitable for human settlements (Burckhardt 1822).

The first comprehensive report describing water use in the Jordan Valley was by Ionides (1939) "The water resources of Transjordan and their development". According to this report only 9 wells were drilled prior to 1938. In 1938 a Haifa-based company drilled 18 modern wells. Ionides noted that farming is only possible with the help of irrigation. The irrigation water came largely from perennial streams in the valley and only a few progressive irrigation systems existed. Bedouins irrigated their farmland via earth channels diverting the water from the outlets of major wadis. He described these channels as highly inefficient, because most of the water infiltrates or evaporates on its way to the farmland. Wheat, barley, and corn were the dominant crops at that time.

A major transformation from an almost uninhabited area with only minor agriculture started after 1948. After the 1948 war, Palestinian refugees along with increased engagement of major landowning Jordanian families were instrumental for the agricultural development in the Jordan Valley (Khoury 1981). This increase in farming activities is reflected by major increase of well drilling. According to Tleel (1963), the major phase of new well drilling started around mid-1950 and clear signs of overpumping could be seen by the beginning of the 1960's. As a consequence first drilling restrictions were issued in 1961. Between 1953 and 1963 62 new wells were drilled in the Karameh area, 122 in the area of Shuna Nimreen, and 133 in the area of Sweimeh, Ghor el Rama, and Ghor el Kafrein. The drilling of new wells and hence abstraction of larger quantities of groundwater reached its first peak in 1967. The events of 1967 brought almost all agricultural activities to a halt or even a major step back. It was not before 1971 that agricultural activities began to increase again.

In the study area two earth-filled dams deliver excess water for irrigation. Theses are the Wadi Shueib dam with a capacity of around 2.3 Mm^3 and the Wadi Kafrein dam with a capacity of 3.8 Mm^3, both built in 1968 (Salameh and Bannayan 1993) located at the outlets of these major wadis (Fig. 3.4.5-5). Due to the ongoing deposition of sedimentary load transported mainly after heavy storm events, the capacity of both dams is diminishing. According to Lenz (1999) they also seem to contribute to local recharge by seepage into the ground.

The King Abdullah Canal (KAC) was known before 1986 as the East Ghor canal. After its third extension (by 14.3 km) completed between 1984 and 1989, it is the third allochthonous source of irrigation water in the study area (Fig. 3.4.5-1). The KAC carries water from as far as the Yarmouk River. On the way towards the south, additional water from side wadis is added and water for domestic as well as irrigation purposes is diverted from the canal.

At present, groundwater is not the major source of irrigation water within the study area and drilling of new wells requires a permit issued by the Ministry of Water and Irrigation. Nevertheless many illegal wells in fenced private farms still exist and extract unknown quantities of groundwater.

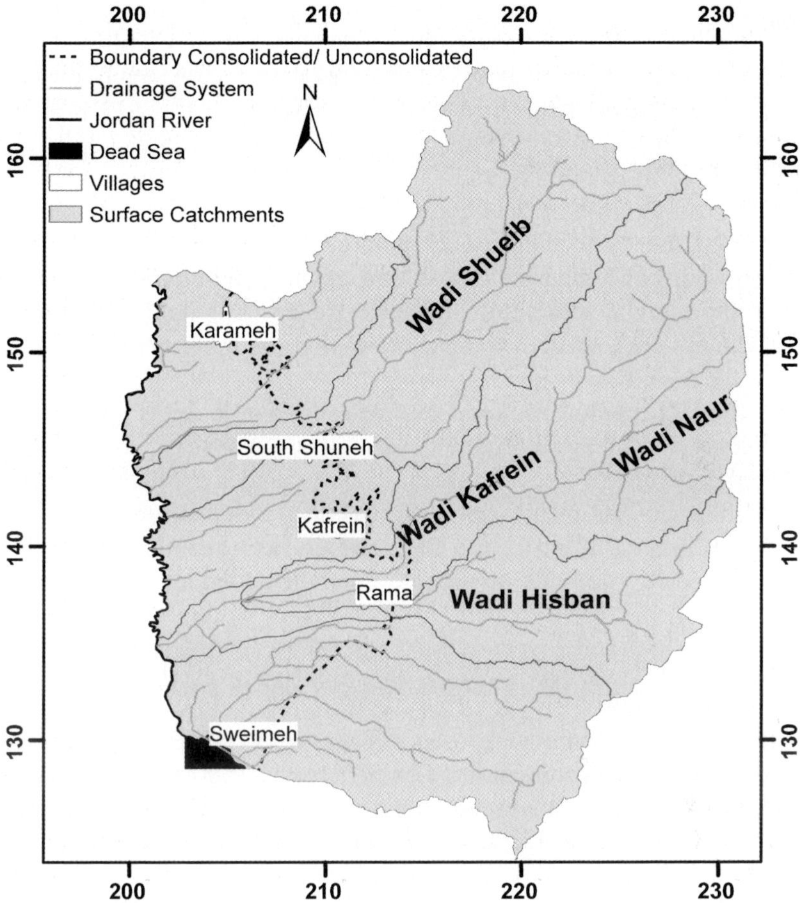

Fig. 3.4.5-5. Surface catchment areas and drainage lines of the studied unconsolidated aquifers

3.4.5.5 Distribution of salinity

Due to the regional semi-arid to arid conditions and to the input from the highly saline lacustrine sediments, different methods were applied in order to distinguish water of different qualities. The methods include surface geophysics (vertical electric Schlumberger depth soundings, frequency domain electromagnetic method MaxMin I-8) direct push measurements (Dipole and Wenner four point geoelectric probes, soil and water sampling, mostly with the help of the Geoprobe 66DT drilling apparatus), well sampling (electric conductivity, short term pumping test with the help of CTD DIVER), and soil sampling (distribution of soluble salt content with depth).

MaxMin

A steep resistivity contrast between the fresh- and saltwater bodies is expected in some parts of the Jordan Valley. Therefore, the APEX MaxMin I-8 equipment based on electromagnetic technique was applied for differentiation of saline water layers in the unconsolidated aquifer. One of the advantages of the MaxMin technique is the small inter-coil separation compared to the large electrode separations necessary in Schlumberger depth soundings and the time consuming placing and coupling of the electrodes to the alluvial sediments of Jordan Valley. Also lateral variations of layer thickness are expected to be more easily and more precisely detected by using the MaxMin.

Since the EM method is very sensitive to magnetic fields and the area is used intensively for agriculture, it was not always very easy to identify possible measurement sites. Sources of disturbances are power lines, fuel driven generators, or metal pipes. In addition, salinization of the Jordan Valley soils proved to be far too high and no deep response could be obtained. Therefore, methods applicable in highly conductive environments were chosen.

Vertical electric soundings (Schlumberger depth soundings)

Since the early sixties the "deeper" salinization and its development can be seen by vertical electric soundings (VES) conducted by various agencies and intermittently continued until today. Although VES has proved to be a tool for detecting a possible salt-freshwater interface, it is difficult to detect the exact depths due to large electrode separations that are needed to penetrate the highly conductive layers. Since it is believed that water flows along alternating channels or areas, more selective information would be desirable. Fig. 3.4.5-6 shows an example of a Schlumberger depth sounding six km to the north of the Dead Sea. The very high conductivities between the modeled 1.8 to 5.2 m depth can be clearly attributed to the accumulation of salt within the upper part of the wet soil. However, a clearer picture can be obtained by direct-push measurements as can be seen in the left part of Fig. 3.4.5-7. To avoid confusion between the interpretation of the Schlumberger depth soundings and the direct-push plot, it should be noted that the Schlumberger sounding diagram is given in resistivities [Ohm*m] per depth [m] and the direct-push diagram in electrical conductivity [mS/m] per depth [m].

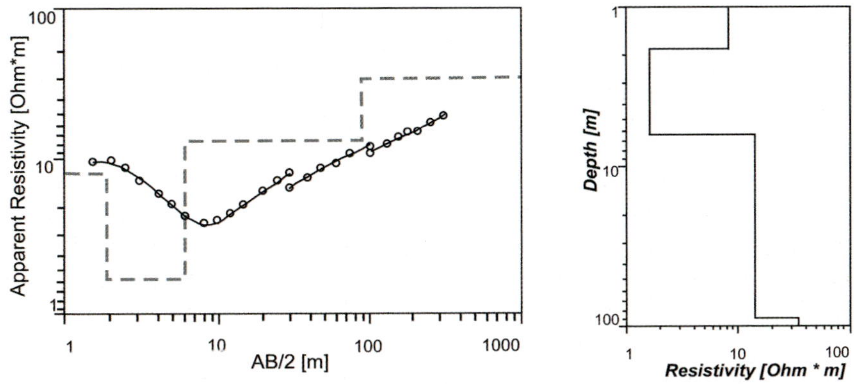

Fig. 3.4.5-6. Surface Schlumberger depth sounding; left: VES depth sounding curve; right: depth model of the VES. The soundings are made in the same location as the Geoprobe geoelectric direct-push measurement in Fig. 3.4.5-7, about 6 km north of the Dead Sea

Fig 3.4.5-7. Geoprobe direct- push depth log; left: Wenner geoelectrical conductivity depth log; right: Geoprobe equipment near the village of Sweimeh; the measurement was taken at the same location as the VES in Fig. 3.4.5-6

Direct-push geoprobe

As stated before, investigating the unconsolidated aquifers in the Jordan Valley is a challenging and an ambitious task. Hence direct measurements of geological composition and geoelectric resistivities are desirable. The employment of the Geoprobe direct-push technique for larger depths has proved to be a suitable tool. It allows an almost undisturbed quick determination of different parameters, such as resistivity, gamma-ray and of various geotechnical parameters and facilitates emplacement of sampling points. When compared to conventional drilling, this method is very quick and reasonably priced The method implies direct pushing (or hammering) of a probe into the ground for either analysis or coring and hence the name of “direct-push” technique. Since the different parameters are directly measured at the tip of the probe and different parameters can be determined at the same time, the problem of equivalence or of ambiguity can be minimized. A geoelectric 4-point “Pole-Pole” or "Wenner" arrangement located on top of a series of drill rods is pushed into the ground. Consequently the apparent resistivities of the material in the vicinity of the top of the rod are determinable and logging of electrical conductivities with depth is directly possible (Fig. 3.4.5-7). By using drilling rods of two different diameters (i.e., 2 1/8" and 1" inch rods), soil samples from any desired depth can be taken. Samples are not taken with the help of coring rods (as it was the case in the soil sampling campaign), but are taken with the help of disposable plastic tubes. The ground-water table is often encountered during the measurements. Since the degree of salinization in subsurface soil and in groundwater is of major interest, the depth of the water table and field parameters (pH, temperature, and water conductivity) are determined, and water is sampled from boreholes.

Chemical analysis of soil samples

To obtain more information on the degree and nature of the “upper” salinization and to correlate the direct-push measurements, depth-oriented soil samples are taken. Soil leachates are analysed by ion chromatography for major ion components according to DIN38414 part 4. For anions Dionex D320 KOH isocratic ASMHC is used, for cations Dionex DX500 MSA isocratic CS16. Fig. 3.4.5-8 shows an example for depth-oriented samples taken within the soil profile of the Lisan Formation, just 1 km north of the Dead Sea. Starting with highest salt content in the upper centimetre of the soil crust, maximum salt content within the soil profile was found at depth between 10 and 30 centimeters. This result is in very good agreement with the results of the geoelectric direct-push sounding measured at the same location. A direct-push log directly on the crust of the soil is not possible and would not coincide with the chemical analysis, since the upper crust of the soil is dried out. Consequently very low conductivities are expected.

"Direct-push" method for small depths

The degree of salinization of soils and its distribution with depth was determined with the help of a "direct push" method. The sounding locations were chosen in order to get a regional representative distribution of soil salinity. In addition, date and time were recorded and the site was described, in order to correlate salinization with different agricultural practices. For instance, soil salinity increases in farming areas of plants with high water demand.

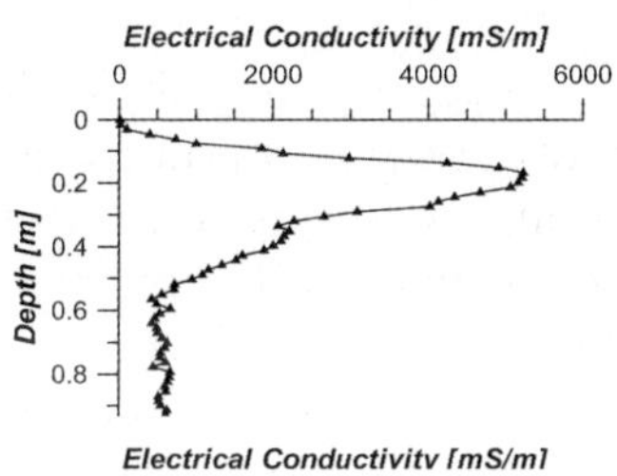

Depth	Sodium	Potassium	Magnesium	Calcium	Chloride	Sulfate	Bromide	Nitrate
[m]	g/kg	g/kg	g/kg	g/kg	g/kg	g/kg	g/kg	g/kg
surface	83,766	1,946	5,508	7,964	194,462	3,434	0,305	2,330
0,1	10,356	0,594	2,002	8,068	31,364	16,805	0,369	0,457
0,2	11,455	0,824	2,727	10,595	36,640	21,972	0,532	0,634
0,3	11,956	1,000	3,281	11,073	40,767	22,827	0,656	0,715
0,4	5,214	0,713	1,415	2,824	17,434	4,735	0,273	0,248
0,5	2,997	0,472	0,603	1,016	8,304	1,618	0,116	0,106
0,6	2,884	0,456	0,495	0,715	7,618	0,874	0,097	0,093
0,7	2,559	0,418	0,400	0,557	7,091	0,751	0,088	0,087
0,8	2,551	0,426	0,383	0,515	6,756	0,736	0,080	0,079
0,9	2,371	0,411	0,325	0,426	6,303	0,680	0,074	0,075
Depth	Sodium	Potassium	Magnesium	Calcium	Chloride	Sulfate	Bromide	Nitrate

Fig. 3.4.5-8. Direct-push geoelectric depth profile compared with results of chemical analysis of depth-oriented soil samples. The measurement of the geoelectric profile and collecting of samples took place at the same location, about 1 km north of the Dead Sea

Sources of salinization

Different sources of salinization are present in the soils of the lower Jordan Valley. The primary source for high salt content in groundwater is the presence of the Lisan Formation. Although the Lisan Formation is considered as an aquitard, the presence of the Lisan Formation in the vicinity of the gravel aquifer is often responsible for the degradation of groundwater quality.

Due to high evaporation rates, salt accumulates in the upper parts of the soils either from precipitation in the rainy season or from irrigation during the dry season. Since the grain size of most soils and sediments in the Jordan Valley is very small and the climate is arid, the upper parts of the soils and sediments dry out in the summer season. This leads to capillary flow which could carry saline water from the underlying Lisan Formation into the upper parts of the soil.

Another reason for high salt content in the upper soil layer is the local shallow groundwater table. In the area close to the Dead Sea and in certain areas to the west of the study area, the soils have a wet appearance. This makes them easily detectable. Due to the long absence of rainfall this phenomenon can be clearly seen during summer and autumn. Geoprobe direct-push soundings undertaken at the end of October 2004 showed within these wet soil areas a very shallow groundwater table often between 10–30 cm below surface. The Wenner four-point geoelectric direct-push log showed soil conductivities as high as 5,300 mS/m. The highest electrical conductivities within the soils are found in the upper decimeters of the soils. This can be explained by accumulation of salt as a result of groundwater evaporation. Electrical conductivity measurements of the groundwater resulted in conductivities as high as 16–20 mS/m. The locally high contents of salt within this soil make the land unusable for agriculture.

Since freshwater in the area is very scarce, water of high salinity is often used for irrigation and this further deteriorates the quality of the Jordan Valley soils. The overuse of fertilizer and the improper use of drip irrigation is another anthropogenic factor.

Spatial distribution of salinization

From east to west a general increase of salt content in the groundwater of the shallow aquifer complex can be observed. An exception is the area south of Tal ar Rama, there, the salt bearing Lisan Formation deposited next to some wells is responsible for the local high salt content. The lowest salinities are found at the outlets of the major wadis Shueib, Kafrein, and Hisban. This may be related to two major factors. The unconsolidated Jordan Valley Group in this area is in direct contact or it overlies parts of two local synclines: the Wadi Shueib syncline and the asymmetric Kafrein syncline. Consequently, not only surface freshwater may infiltrate into the Jordan Valley Group but also groundwater of good quality carried by the consolidated mountain aquifers. The second reason may be the composition of the unconsolidated material. Since this area is directly next to the outlets of the above named wadis, the area can also be described as the apex of alluvial fans deposited at the surroundings of these major wadi outlets. Due to the occurrence of relatively coarse materials, the hydraulic conductivity in these areas is higher. Though the enhanced flow of water from the wadis flushed out the accumulated salts, the relics of salt-bearing formations such as the Lisan Formation remain as sources of salt deteriorating the local water resources.

3.4.5.6 Summary

In order to get a better understanding of the study area an integrated approach of geological, geophysical, hydrogeological, historical, and chemical methods is applied.

In the study area occur aquifers of local importance which are made up of clastic material deposited at the outlets of major wadis. These unconsolidated aquifers interfinger with - or are underlain by - the highly saline lacustrine Lisan Formation. Groundwater quality varies from fresh to brackish in the area close to the outlets of major wadis and from saline to highly saline in the area west and southwest. Most recharge comes from the adjacent consolidated mountain aquifers extending in the east or during heavy rain-events, from infiltrating runoff water draining from the western hills of the East Bank.

The transformation from an almost abandoned area into a major food producer in Jordan started around mid 1950's. As a consequence, large quantities of groundwater were abstracted and first signs of overpumping were already observed at the beginning of the 1960's. The agricultural development of the area went along with the major development of the infrastructure, including two earth-filled dams (Wadi Shueib and Wadi Kafrein dam), a north-south trending canal (King Abdullah Canal), and the construction of an irrigation pipe network. Nevertheless, unknown quantities of groundwater are still extracted from the local aquifers.

Since the area suffers from severe problems of water salinization, several methods sensitive to changes in the electrical conductivity of soils and or water were applied. VES proved to be a tool for investigating a salt-freshwater interface. However, direct measurements such as the direct-push techniques for larger depths, yielded a better depth resolution. Due to the high salt-content within the soils of the area of interest, the frequency-dependent electromagnetic method (FDEM) MaxMin was not applicable.

As far as the sources of salinization are concerned, four different processes are identified: the high salt content of the lacustrine Lisan Formation in the vicinity of the Jordan Valley aquifers, the upward movement of saline water into the soils due to capillary forces, the accumulation of salt in soils due to evaporation of very shallow groundwater and the anthropogenic factor such as overuse of fertilizer, improper use of drip irrigation, and irrigation with brackish water.

References

Andrews IJ (1991) Palaeozoic lithostratigraphy in the subsurface of Jordan. Subsurface NRA Amman,Geol Bull 2:1–75

Andrews IJ (1992) Permian, Triassic and Jurassic lithostratigraphy in the subsurface of Jordan. NRA Amman, Subsurface Geol Bull 4:1–60

Bandel K, Khoury H (1981) Lithostratigraphy of the Triassic in Jordan. Facies, 4:1–26

Burckhardt JL (1822) Travels in Syria and the Holy Land, Association for Promoting the Discovery of the Interior Parts of Africa. London, pp 1–309

DIN 38414, Teil 4 (1983) Bestimmung der Eluierbarkeit mit Wasser, Beuth Verlag, pp 1–4

Heinrichs T, Al-Zoubi A, Sauter M, Salameh E, Qabbani I (2004) Structure of the Shuna (Eastern Jericho) Basin, Lower Jordan Valley - interpretation of reflection seismic data, preliminary results. 5th Int Symp Eastern Mediterranean Geol, Thessaloniki, Proc 3:1522–1525.

Horowitz A (2001) The Jordan Rift Valley, Balkema Publication, pp1–730

Ionides MG (1939) Report on the Water Resources of Transjordan and their Development, Amman, pp 1–169

Khoury RG (1981) The Jordan Valley. Life and society below Sea Level. Longman Group Limited. London, pp 1–238

Lenz S. (1999) Hydrogeological Investigation along the Wadi al Kafrein and the Kafrein Reservoir. Unpubl. Masters thesis, Dept Applied Geology Univ Karlsruhe, typescript, 108 p, 47 figs, 16 tables, 1 geol map 1:10.000.

Muneizel S, Khalil I (1993) Geological map of As Salt-3154 III, scale 1: 50000. Jordan, Natural Resources Authority.

Niemi TM, Ben-Avraham Z, Gat JR (1997) The Dead Sea: The Lake and Its Setting. Oxford University Press, New York, USA, pp1–286

Salameh E, Bannayan H (1993) Water Resources of Jordan. Present Status and Future Potentials. Friedrich Ebert Stiftung, Amman, pp 1–183

Schulze F, Levy Z, Kuss J, Gharaibeh A (2003) Cenomanian - Turonian carbonate platform deposits in west central Jordan. Int J Earth Sci (Geol Rundsch) 92:641-660.

Shawabkeh KF (2001) Geological map of Al Karama - 3153-IV, 1:50.000, The Hashemite Kingdom of Jordan, Natural Resources Authority, Geology Directorate

Tleel JW (1963) Inventory and Groundwater Evaluation Jordan Valley. The Hashemite Kingdom of Jordan, Central Water Authority, Amman, pp 1–80

4. Groundwater Management

4. 1 Overview of available water resources and uses in the Jordan Rift Valley

Marwan Haddad

4.1.1 Regional water scarcity

4.1.1.1 Population growth and water utilisation

Water resource management issues in the Middle East (ME) are different because of rapid population growth, changing ecosystems, political and military disputes and poverty (Gleick 2002). Water management practices either in domestic, industrial or agricultural sector in many of the ME countries in general are poor and governmental investment in developing the agricultural sector is not adequate. Furthermore, economic efficiency of agriculture in those countries is negative and governmental subsidies are practiced in some countries (Haddad and Lindner 1999).

Annual per capita water availability for the Southern Mediterranean region including ME countries has fallen from approximately 3,400 m^3 in 1960 to an average of about 1,200 m^3 in 2000 (WRI 1996; WWF 2003). It is projected to fall to just over 500 m^3 in 2025 due to a high population growth rate and increased urban and rural water use. In many parts of the region, dwindling water resources are threatening people's livelihood, the environment, and the economic growth. Intermittent supply is common in many cities and agricultural water supply faces the challenge of competing demands for water (WRI 1996; WWF 2003).

Countries bordering the Jordan Rift Valley (JRV) are also characterized by scarce water resources and rapid population growth. The water resources of the JRV are barely adequate to satisfy even current demand.

Consequently, the per capita water supply and food production is decreasing dramatically with time. The region is experiencing already moderate to high water stress and referring to the natural resources it is expected that by 2040, the whole population would be under high water stress conditions, if no economic replacement is possible. Accordingly water is the most important and binding constraint for any future economic development in this region including plans and activities for ensuring sustainable agriculture (Haddad and Lindner 2001).

Because of the usage of water for food production, food security is connected to water security and both pose national security challenges. Countries that suffer from water deficits invariably have food deficits and these conditions contribute to tensions between themselves and neighboring countries that can erupt into violent confrontation. Of all the uses of water in the ME, food and agricultural production are the most extensive and intensive. Per unit of outcome, their per unit input dwarfs any other use of water. It follows, therefore, that food security is particularly problematic in the ME as arid and semi-arid regions.

There is currently a widespread discussion of water as a human right which intersects with discussions of food security and minimal guaranteed food as a basic human right. Some advocates go so far as to campaign for free or nearly free water unlimited in scope. This is easily dismissed, because of resource constraints, uneven allocation of water resources across regions and countries, and the basic economic law that a low or no price for a commodity will encourage waste, over-consumption, and inefficiencies, all of which threaten the environment and do not conserve a natural resource (Wachtel 2002).

In a balanced assessment of these intersections, Gleick (2002) concluded that 50 liters per person per day of fresh water is a minimal requirement for domestic-human-purposes, broken down among five liters per day for drinking, 20 for sanitation, 15 for bathing, and 10 for food preparation. This does not include the heavy use of water for food production, which he estimates at 2,700 liters per day, ranging from current usage of about 5,700 liters in North America to 1,800 in Sub-Saharan Africa.

Alternative solutions to present and future water shortages in the region include (1) improving water demand management practices including technical, financial, economic, institutional, legal, and educational measures, (2) enhancing wastewater treatment and reuse, (3) enhancing runoff management and various water harvesting techniques, and (4) bringing water from outside the region through water transport by land or sea or producing new and additional water resources by desalinating brackish and seawater (Haddad and Mizyed 2004).

4.1.1.2 Water resources: limitations and future demands

If we consider the broader Middle East (ME) region from Turkey in the North to Yemen in the South, the following notes could be drawn concerning water supply and demand and its relation to available water resources:

- The countries of the ME are projected to be in a state of absolute water scarcity by 2025. Only Turkey and Lebanon are projected to be in relatively good water conditions.
- Jordan River area is experiencing and expected to face in the next fifty years mean annual and summer increase in ambient temperature leading to changes in agricultural water demand, decrease in winter and increase in spring precipitation leading to change in groundwater recharge conditions, and a decrease of precipitation intensities
- The riparians of the Jordan basin do not have sufficient annual water resources to meet reasonable per capita water needs for their rapidly expanding populations. Projected domestic water needs of most ME countries are frozen at present levels meaning that the per capita domestic water use will decrease with time leading to less human well-being conditions. Most ME countries will almost certainly have to reduce the amount of water used in irrigated agriculture and transfer it to the other sectors, importing more food instead.
- While importing more food is a viable option for countries with high earnings, it will impose additional burdens on others, leading to more, poverty, political and economic instability, and possibly conflicts on water (in the absence of no basin agreement such as the case of the Jordan River Basin).
- Many of ME countries will also have to increase their dependency on expensive and energy-consuming desalinization plants to meet domestic and as well as economic developmental needs.

The water resources of the ME are barely adequate to satisfy current demands. Annual average rainfall of over 500 mm is limited to the northern parts to Turkey and Lebanon and a small strip along the eastern Mediterranean. Most ME countries have annual average rainfalls ranging between bellow 250 to a maximum of 500 mm. Such rainfall intensity is either not suitable for any cultivation or suitable for limited crops, and heavy supplementary irrigation is needed.

In the coming decades, water demand is expected to rise and the situation will become more and more critical considering the quantity and quality of water needed for economic and social development. Various water supply and demand scenarios show that despite an optimal development of conventional water resources, the size of the regional net water gap will greatly depend on the anticipated future utilization pattern (GTZ 1998; ESCWA 2003).

The estimates of future water demand for the ME countries indicate that although some countries such as Jordan and Israel will substantially decrease their total per capita water withdrawal by the year 2025, their water demands are still much over the naturally available and renewable water resources (Table 4.1-1). Forecasted irrigation water demands for the same period were 24 to 50% of existing ones for most ME countries. If such reduction in forecasted irrigation demand do not go parallel with high investment in improving irrigation techniques and management drastic reductions in agricultural production and consequently food supply will take place in those countries leading to increasing poverty in countries such as Jordan and Palestine.

Table 4.1-1. Available water resources in the Jordan Rift Valley (GTZ 1997, 1998)

Sub-Region	**Surface Water** (Mm^3/yr)	**Ground Water** (Mm^3/yr)	**Treated Wastewater** (Mm^3/yr)	**Total Water** (Mm^3/yr)
West of Jordan Rift Valley	685	1234	285*	2204
East of Jordan Rift Valley	350	507	52	909
Total Region	1035	1741	337	3113

Note: Groundwater includes renewable and non-renewable groundwater (fresh and brackish water quality). Surface water includes fresh and brackish water quality. (*) = This treated wastewater volume was 213 Mm^3/yr in 1994, 285 Mm^3/yr in 2000 and forecasted to be 460 Mm^3/yr in the year 2010/11.

4.1.1.3 Water as an element for peace, stability and cooperation

The vital role of water in sustaining socio-economic, peace, and political stability and development in the ME calls for joint regional water planning to reach a regional strategy for water development and management and to narrow the gap in perceptions and experiences and promote cooperation in the medium- and long-term.

The water shortages of each of the countries of the ME region can be resolved separately and individually by each country of the region. However, for lesser costs, better affordability, enhancement of peace and stability in the region, and better quality of life, it is preferable that countries of the region co-operate and work jointly together to resolve those shortages. Cooperative paths are numerous beginning with joint water related studies and research, through joint monitoring and data collection, compilation, and analysis, to transfer of technology, know-how, and expertise, to implementing small joint water projects, to regional co-operative agreements and institutional structure to implement those agreements, and to more advanced forms of co-operation such as large scale regional water projects and joint regional basin water authority and so on.

Joint regional water cooperation is important not only in developing and managing existing as well as additional and new water resources but also resolving water conflicts and reducing the differences in perception among the regional parties.

4.1.2 Available water resources

Because of the limitation of the current project to the study area of the Jordan Rift System the following section on water use and future demands will be limited to JRV countries: Jordan, Israel and Palestine.

The consolidated regional data for 1994, representing estimated approximate figures in millions of cubic meters per year (Mm^3/yr), show that the sustainable average annual yield from all renewable resources is 2700 Mm^3/yr: 1300 Mm^3/yr surface water and 1400 Mm^3/yr groundwater (Haddad and Lindner 1999). The total water resources available in the region are summarized in Figure 4.1-1. A separate water balances are presented in Table 4.1-1 for the sub-regions "West of the Jordan Rift Valley" (Israel and West Bank/Gaza Strip), and "East of the Jordan Rift Valley" (Jordan).

4.1.2.1 Surface water

The surface water, of which a major portion originates in Syria and Lebanon, can be hydrographically divided between the Upper Jordan River, Lake Tiberias, the Yarmouk River, and the Lower Jordan River. About 1,000 Mm^3/yr of surface water have been "developed" and are mainly used by Israel and Jordan. Water use is so high that the Jordan River decreases significantly in volume as it flows downstream, with the result that (1) Dead Sea water level lowering with time reaching already critical levels, and (2) downstream users face shortages, and ecosystems suffer. The Jordan River is severely contaminated by human, industrial and agriculture wastes, affecting the freshwater quality.

4.1.2.1 Groundwater

About 1700 Mm^3/yr of groundwater (renewable and non-renewable aquifers of different qualities) are currently used in the region. The largest share of about 1200 Mm^3/yr is used in Israel and in the West Bank/Gaza

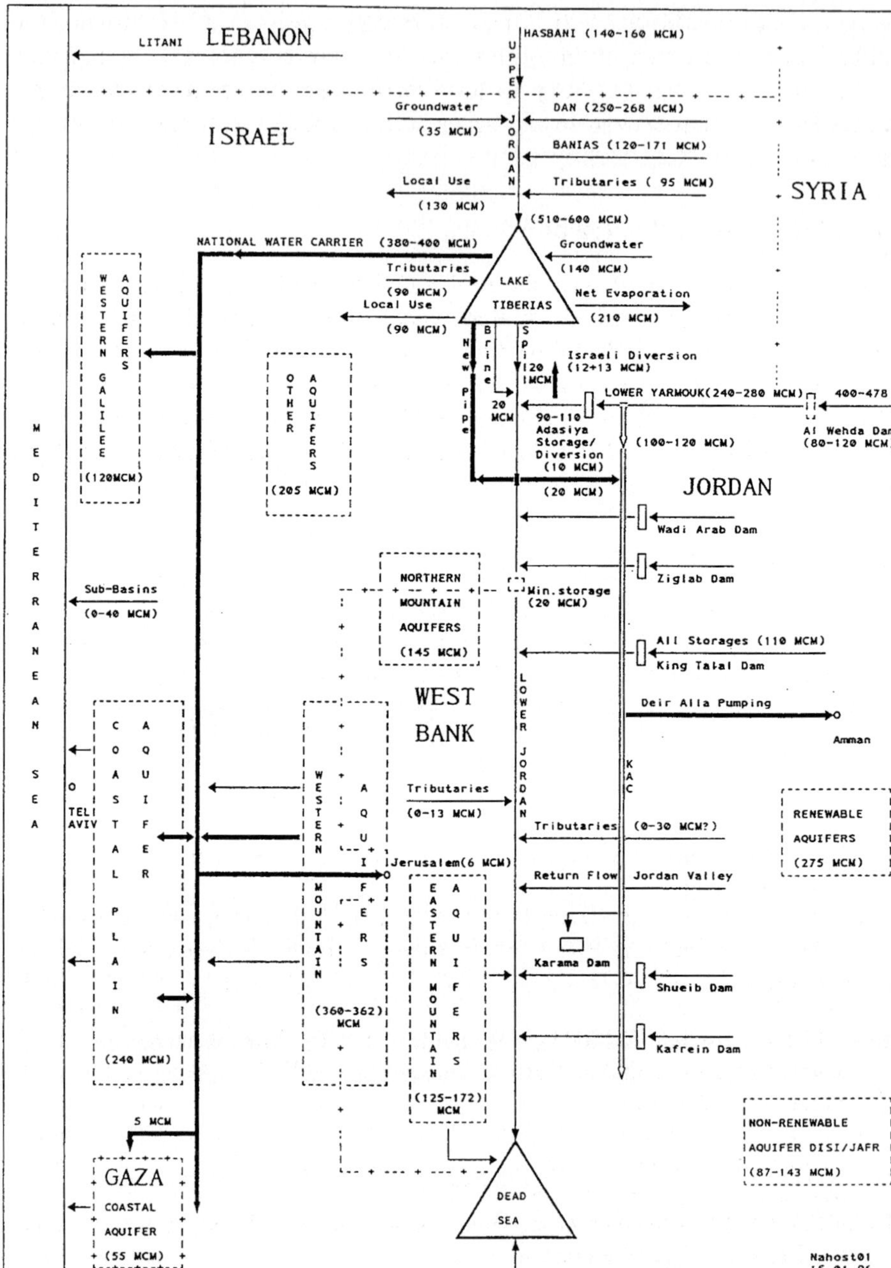

Figure 4.1-1. Water resources in the Jordan Valley (Source: GTZ 1998)

Strip, compared to about 500 Mm^3/yr currently used in Jordan. Most of the groundwater resources in the region are fully exploited and some aquifers are overexploited, particularly in Jordan. In addition, non-renewable resources are currently used in the order of 80 Mm^3/yr west and 90 Mm^3/yr east of the Jordan Rift Valley.

4.1.2.3 Reclaimed Wastewater

Total water use in 1994 was about 3050 Mm^3/yr (Table 4.1-1), which includes 265 Mm^3/yr reuse of treated wastewater for irrigation (the major part - 213 Mm^3/yr - in Israel). This amount of treated wastewater has increased in the year 2000 to 285 Mm^3/yr and planned to increase to 460 Mm^3/yr in the year 2010/11 (Shelef 2002). Of this total water use, irrigation accounts for about 2,000 Mm^3/yr (66%), domestic water use for about 900 Mm^3/yr (30%), and industrial water use for about 150 Mm^3/yr (4%).

References

ESCWA Economic and Social Commission Water in Asia (2003). Up-dating the Assessment of Water Resources in ESCWA Member Countries E/ESCWA/ENR/1999/13, United Nations, New York, pp 23-46

Gleick P (2002) The World's Water 2002-2003: The Biennial Report on Fresh Water resources. Island Press, Washington DC, 2002, pp 87-112

GTZ (1997) Deutsche Gesellschaft für Technische Zusammenarbeit GmbH and Consulting Engineers Salzgitter (CES) GmbH: Multilateral Working Group on Water Resources: Middle East Regional Study on Water Supply and Demand Development. Regional Overview Phase II, November 1997 (restricted distribution), pp 2-1–2-24

GTZ (1998) Deutsche Gesellschaft für Technische Zusammenarbeit GmbH: Multilateral Working Group on Water Resources: Middle East Regional Study on Water Supply and Demand Development. Concluding Report, February 1998. pp 1-24

Haddad M and Lindner K (2001) Sustainable Water Demand Management Versus Developing New and Additional Water in the Middle East: A Critical review. Water Policy J 3:143–163

Haddad M and Mizyed N (2004) Non-Conventional Options for Water Supply Augmentation in the Middle East: A Case Study. Water International Journal, International Water Resources 29:232–242

Shelef G (2002) Wastewater Treatment, Reclamation and Reuse in Israel, source: www.Biu.Ac.Il/Besa/Waterarticle3.html

Wachtel H (2002) The Intersection of Food and Water Security: A Case Study of the Near East. Source: www.american.edu/cas/econ/faculty/

WRI World Resources Institute (1996) World Resources 1996–1997 - A Guide to Global Environment. New York: Oxford University Press, pp 149-158

WWF World Water Forum (2003) 3rd World Water Forum: March 16-23, 2003

Kyoto, Shiga and Osaka, Japan, http://www.world.water.forum3.com2003/eng/secretariat/0322-3.hmtl

4.2 Sustainable management of groundwater resources

J. Guttman, E. Salameh, E. Rosenthal, AR. Tamimi, A. Flexer

4.2.1 Significance of aquifer systems and recharge conditions

Groundwater is the main water resource in the region. The morphology of the study area includes two mountain ridges separated by the deep Jordan Valley. As will be further shown, this pattern defines the distribution as well as the way of exploiting and utilizing the groundwater resources in the study area.

The study area is a multi-aquifer system that includes several regional aquifers which are located in the mountainous region and downstream. In the foothills they are laterally connected to alluvium aquifers extending in the subsurface of the Jordan Valley.

The most important regional aquifer that covers extensive areas in the highlands is the limestone-dolomite aquifer (the Judea and Ajlun Groups). Several case studies demonstrated that this aquifer is divided into the upper and the lower sub-aquifers (Guttman 2000; see also Chap. 3.2.2). The hydraulic behavior of the two sub-aquifers is different and bears on their ability to serve as long-term sources for the population in the region.

The natural replenishment to the regional aquifers occurs in the high mountain region above the areas in which the aquifer is exposed. The highlands receive an average amount of precipitation of 300–550 mm/yr on the Jordanian (eastern) side of the Rift and about 500–700 mm/yr on the western side (Israel and the Palestinian Authority). The outlets of these aquifers are always at much lower topographical levels in the Jordan Valley and along the shores of the Dead Sea. The outlets of groundwater are through springs and as lateral flow into the alluvial formations which are limited to the Jordan Valley (Guttman et al. 2004). This is in contrast to the area extending east of the Valley, where springs occur at medium elevations at the base of the Ajlun aquifer (Parker 1970; Salameh 1996). The outflow of the springs feeds the flow in the wadis. These aquifers are also replenished by leakage of runoff, irrigation return flow, and by small amounts of direct recharge.

Because of the irregular incidence of natural replenishment on both sides of the Rift, the storage capacity of the aquifers is of particular importance. The basic elements for any possibly sustainable exploitation of groundwater resources in the study area are the understanding of the flow patterns from the regional aquifers into the Rift Valley, the water balance, the storage capacity of each aquifer and the interaction of fresh and saline water bodies.

Because of their proximity to replenishment areas, the water in wells located at high topographic levels on both sides of the study area, are of low Cl^- concentrations. This natural distribution of water of potable quality luckily coincides with the dense concentration of the large populations residing in the highlands such as: Hebron, Jerusalem, Ramallah in the west and Amman, E-Salt, Zerqa in the east. Some wells which are located in these urban centers (and pump water from the upper unconfined aquifer) are polluted by anthropogenic activities. The storage of these aquifers is quite small because they are located close to the recharge area. Contrasting wells located far from recharge area, the water levels in wells located close to replenishment areas are sensitive to seasonal climatic changes.

The combination of limited storage, intensive exploitation and large variability of natural recharge caused a strong drop of water levels in several wells. Fig. 4.2-1 shows the drawdown in two wells, one on the Jordanian side (well AL1430) and the other on the western side of the Valley, near Bet Lehem (well Herodion 2). In well Herodion 2 the drop of the water level during the last 15 years was of the order of 20 m. The drop of the water level implied reduction of the hourly discharge of the well. In the Jordanian well AL1430 the water level dropped during the last 15 years by 15 m. The drop of groundwater levels influences negatively the discharge of some wells and of few adjacent springs that emerge from the same aquifers. In the highlands the deep confined aquifers contain water of good water quality. A major drop of water levels was also observed in these confined aquifers. It occurs that this happened as the result of reasons similar to those mentioned above and is mainly due to overexploitation (Fig. 4.2-1).

4.2.2 Seepage of surface water and ground water to deeper aquifers

The local water resources are not sufficient to satisfy the increasing water requirements of the populations in the urban centers. Therefore, water has to be imported from outside of the region. Flood water and surface drainage of major urban centers in Jordan such as Amman, Irbid, E-Salt and Madaba, flow westward through wadis towards the Jordan Valley and is captured in the Shueib, Kafrain and King Talal dams. This water is mixed with treated waste water and with spring water. Together with groundwater

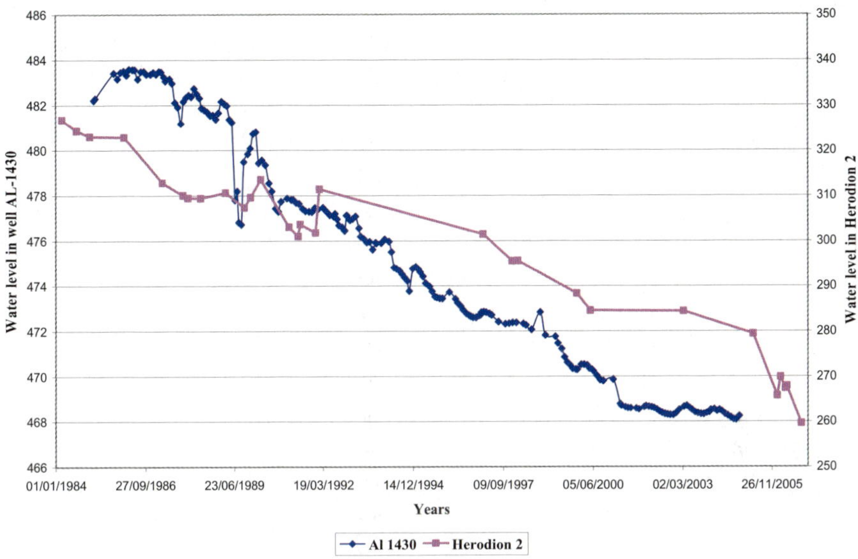

Fig. 4.2-1. Drop of water level in wells located on the topographic highlands (WAJ open files)

it serves the agriculture along the eastern wadis of the Rift (Salameh and Bannagan 1993; Margane et al. 2002). The remaining portions of water recharged along the highlands and flowing westward, discharge through springs or seepages along the western slopes of the Jordanian highlands infiltrating laterally into the recent sediments or discharging as submarine springs into the Dead Sea. The other portion of recharged groundwater flows within the Upper Cretaceous and Tertiary rocks eastwards and is either exploited there or percolates through the secondary porosity of the aquitards to reach the deep aquifer system of the Lower Cretaceous, Jurassic, Triassic and locally Ordovician sandstones and siltstones (Fig. 4.2-2). In this deep aquifer system, the water flows once again westwards towards the Jordan Valley and the Dead Sea (Salameh and Udluft 1985). The water in the deep aquifer system is thermal and mineralized (Sawarieh 2005; cf. Chap. 3.4.1).

West of the river Jordan, the flow mechanism of the Uja spring emerging north of Jericho, manifests the hydraulic features of the upper subaquifer. It is characterized by high discharge fluctuations and quick response to seasonal climatic changes (Guttman 2007). During summer and throughout drought periods (which are common in this arid zone) the spring dries up. During these periods, the farmers that use spring water to irrigate their cultivations suffer from water shortage and the only solution to help them is by importing water from wells that are pumping from the

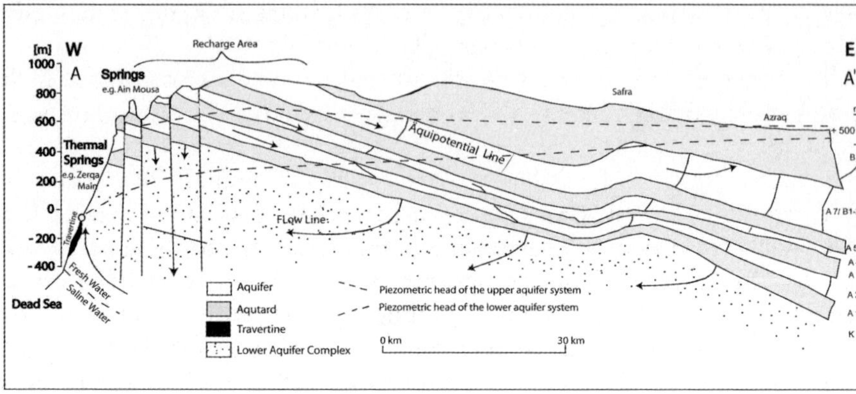

Fig. 4.2-2. Hydrodynamic pattern of the central part of Jordan (modified after Salameh and Udluft 1985)

lower sub-aquifer (which is the regional source of water). The storage of the lower sub-aquifer is much larger than that of the upper sub-aquifer. Its recharge area is close to the mountain axis and far from the wells that are located close to the foothills. Because of the large distance between the recharge area and the well fields and the low permeability, the impact of recharge (as during the winter of 1992) on the raise of the water levels, appears after a delay of 2–2.5 years after this particular rain-rich year. Because of the large storage capacity of the lower sub-aquifer, it should be regarded as the main water resource for drinking and agricultural purposes.

The understanding and especially the identification of the main flowpaths bear on the successful emplacement of wells for the exploitation of fresh water along the foothills.

4.2.3 Effects of tectonic structures on groundwater storage and quality

Some of the studies (see Chaps. 3.2.2 and 3.3.2) that were carried out west of the river Jordan during the current project, show that there is a strong relationship between structural elements, the thickness of the saturated aquifer and its hydraulic characteristics, the regional water level, flow directions and salinity. Groundwater flow from recharge areas to the Jordan Valley is influenced by the tectonic pattern which creates zones of low permeability ("cascades") that cause groundwater not to flow as a uniform flow-front but to bypass the low permeability zones and to flow along preferential paths. Along the western side of the rift, this flow mechanism can explain salinity fluctuations and differences in the chemical compositions of water occurring particularly in wells located along the eastern edge of the Fazael-En Samiya area (Fazael, Gitit and Uja wells).

In some wells Cl^- concentrations are similar to that measured in the natural replenishment zones (30–50 mg/l) and remained constant throughout the period of exploitation. From the aspect of exploitation, wells that are located along the foothills or close to the preferential flow paths, receive fresh water from the recharge area and their salinity remains low. Other wells located in the same region are characterized by salinity fluctuations of tens to hundreds of mg/l per season (Guttman 2000). Such behavior indicates that these wells are pumping water which circulates close to a source of salinity. Therefore, in order to avoid any salinization and to supply the farmers with water of constant quality, the discharge of these wells should be reduced. On the other hand, reduction of pumping rates implies that a certain quantity of fresh water (mixed with the saline component) will flow to the Rift Valley and will therefore be a lost resource.

4.2.4 Significance of operating procedure

The water supply regime is controlled by demand. In wells that supply potable water, the pumping regime is more or less constant throughout the year with small fluctuations between winter and summer (Fig. 4.2-3). Due to ongoing raise in life standard, there is also an increasing demand for potable water. In wells that supply water for agriculture, the pumping regime depends on the agricultural seasons. In Fig. 4.2-3, village 1 and the urban settlement are located in the Fazael area, whereas village 2 is located in the area north to the Dead Sea. In the Jordan Valley there are two agricultural seasons: a major one during April-May and a secondary season during August-October. During these seasons the overall demand increases dramatically and all farmers in the region need the water simultaneously. This situation forces the water suppliers to establish many production tools such as wells and reservoirs that enable them to supply large amounts of water over short periods.

Like on the Jordanian side, a similar situation occurs west of the river Jordan, in the Bet Lehem and Hebron regions. During the last drought period (1999–2002) the water level in the Herodion wellfield (which is part of the water supply system to the Bet Lehem and Hebron regions) dropped sharply. The drop of the water levels limits the pumping capability in several wells (Guttman and Percia 2004). The result was that the local water resources were not sufficient to satisfy the increasing water requirements of the populations in these urban centers. Additional water had to be imported from outside of the region, i.e., from the Israeli network.

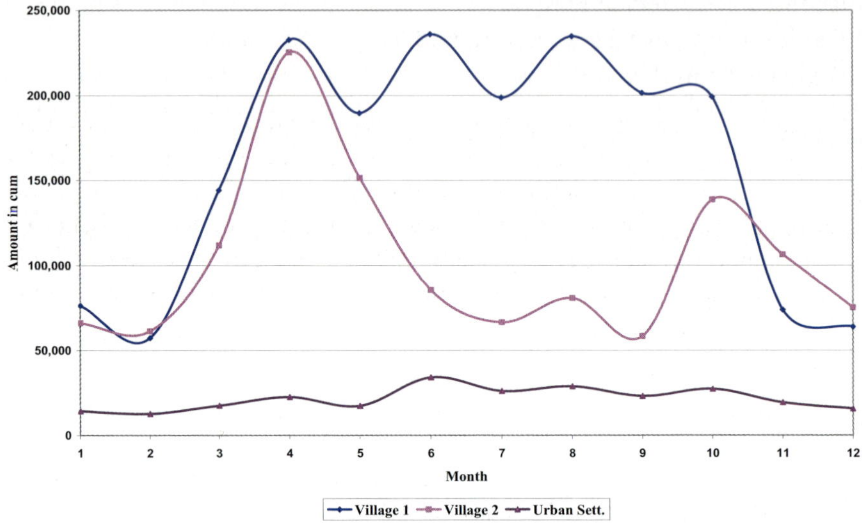

Fig. 4.2-3. Structure of water demand in Israeli villages and in an urban settlement located in the Rift Valley

4.2.5 Impacts by different land use

Although both agricultural and urban development requires the development of water resources, the growth of agriculture has a stronger negative impact on biodiversity than urban development. Agriculture, urban development and the infrastructure connecting them (e.g., roads and pipelines) adversely affect biodiversity and therefore changes in the agricultural land use are requested. The recent reduction in cotton production in Israel, not only saves water but also reduces the usage of pesticides. On the other hand, water drawn from wadis and from aquifers for agriculture contaminates aquifers and alters ecosystems. Because dry land ecosystems are limited not just by water but also by nutrients, the enrichment of fertilizers "escaping" from desert agriculture may dramatically change the functioning and structure of these ecosystems. Water quality in the Jordan Rift Valley is affected by the following activities:

- Discharge of inadequately treated effluent from municipal treatment plants,
- Discharge of untreated domestic and agricultural wastes,
- Extraction, use and disposal of low-quality groundwater,
- Leachates from solid waste landfills,

- Runoff from urban drainage,
- Fertilizer and pesticide residues, and
- Drainage of wetlands.

It should be stressed that some of the results from the case studies which were carried out during the current research project could serve as guidelines for the location of additional wells and for the development of local water for the people in the region. They can also help to achieve better water management of the existing wells and prevent further deterioration of the water quality as a result of over-pumping and inflow of water of lower quality.

References

Guttman J (2000) Hydrogeology of the Eastern Aquifer in the Judea Hills and Jordan Valley- Final Report. Mekorot no. 468, pp 1–87

Guttman, J. (2007) The karstic flow system in Uja Area-West Bank: An example of two separated systems in the same area. Water resources in the Middle East- Israel Palestinian water issues from conflict to cooperation. Vol. 2 Hexagon Series on Human and Environmental Security and Peace, pp 61–66.

Guttman J, Flexer A, Hoetzl H, Bensabat J, Ali W, Yellin-Dror A (2004) 3-D hydrogeological model in the arid zone of Marsaba-Feshkah region - Israel, 5th Intern Symp Eastern Mediterranean Geology, Thessaloniki, Greece, pp 1510–1513

Guttman, J. Percia, C. (2004) Educated water management under hydrological stress in the West Bank. 2nd Israel Palestinian international conference- Water for life in the Middle East, pp 937–944

Margane A, Hobler M, Almonani M, Subah A (2002) Contributions to the Hydrogeology of Northern and Central Jordan, Bundesanstalt für Geowissenschaften und Rohstoffe, Hannover, Geol Jahrb Reihe C, 68:1–125,

Parker D H (1970) The Hydrogeology of the Mesozoic, Cainozoic Aquifers of the Western Highlands and Plateau of East Jordan. Technical Report Vol. 2: Hydrogeology; United Nations Development Program, Food and Agriculture Organisation of the United Nations, Rome, 1–79

Salameh E (1996) Water quality degradation in Jordan. Impacts on environment, economy and future generations resources base. Royal Society for the Conservation of Nature and Friedrich Ebert Stiftung. Council of Science and Technology, Jordan, 1–120

Salameh E, Bannayan H (1993) Water resources of Jordan. Present status and future potentials. Friedrich Ebert Stiftung, Amman, 1–183

Salameh E, Udluft P (1985) The hydrodynamic pattern of the Central part of Jordan. Geol Jahrb 38:39–53

Sawarieh A (2005) Heat sources of groundwater in the Zara-Zarqa Ma'in-Jiza area, Central Jordan. PhD Thesis, University of Karlsruhe, Faculty of Civil Engineering, Germany, pp 1–165
WAJ: Water Authority of Jordan. Open Files.

4.3 State of water strategy and policy

H. Hötzl, J. Guttman, E. Salameh, AR. Tamimi

4.3.1 Strategic water situation in Jordan

The Kingdom of Jordan as the eastern riparian is facing a chronic imbalance between the available water resources and the demand of water. Shortage has become a permanent feature. Jordan has a per capita basis, which is one of the lowest levels of water resources in the world and this amount is still decreasing. Due to its scarcity, water is a precious resource and is of vital importance to the socio-economic development of the kingdom.

Jordan's renewable water resources are estimated at about 800 Mm^3/yr, consisting of more than 500 Mm^3/yr surface water and 275 Mm^3/yr groundwater (Table 4.3-1). Limited resources of non-renewable (fossil) groundwater are available in the Disi aquifer in the southern part of the country. Its estimated annual yield is of 125 Mm^3 and is expected to last for about 50 years. Currently, about 90% of treated wastewater (about 79 Mm^3/yr) is reused in irrigation. The main water consumers are agriculture (62%), municipalities and tourism (32%) and industry (4%). Only about 66% (825 Mm^3/yr) of the present total demand of about 1250 million m^3/yr can be supplied (Govern. Jordan 1989; 2003; JMWI 1997; Taha 2006). The mentioned demands result from a population of 4.6 million in 2004. An expected population growth till 2008 of about 8 million, and about 10 million in 2020 will lead to a dramatically increased water demand (Fig. 4.3-1).

Table 4.3-1. Available fresh water resources in Jordan

SOURCE	Quantity (Mm^3/yr)
Renewable groundwater	275
Surface water	505
Total renewable	**780**
Non Renewable Groundwater	125
Total renewable & non-renewable resources	**905**

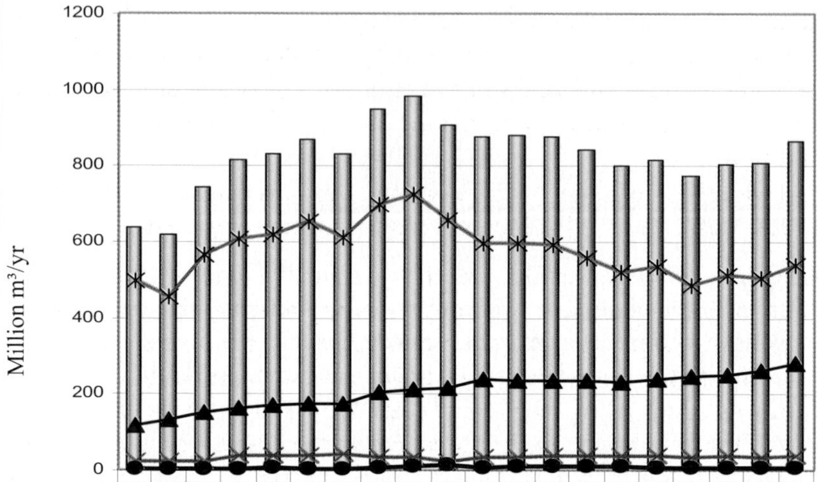

	1985	1986	1987	1988	1989	1990	1991	1992	1993	1994	1995	1996	1997	1998	1999	2000	2001	2002	2003	2004
Total	639	619	744	817	830	870	833	951	984	909	878	882	876	843	802	817	774	804	810	866
Municipal	116	135	150	165	170	176	173	207	214	216	240	236	236	236	232	239	246	249	262	281
Industrial	22	23	24	39	36	37	42	35	33	24	33	36	37	38	38	37	33	37	36	38
Irrigation	497	456	565	608	618	652	613	700	726	655	596	598	592	557	521	534	488	511	506	541
Livestock	4	5	5	5	6	5	5	9	10	13	9	12	11	12	11	7	8	7	7	7

Fig. 4.3-1. Water use in Jordan between 1985 and 2004 (after Taha 2006)

Current deficits are covered by non-sustainable overexploitation of aquifers and by non-renewable resources causing decline of groundwater tables and degradation of water quality (Taha 2006).

The Government of Jordan (1998) has adopted a comprehensive water strategy in order to mitigate the water situation. This is supported by a targeted policy and numerous supplemented measures to accelerate the implementation of the water regulations. The Ministry of Water and Irrigation (MWI) was nominated as the responsible organization in charge for the overall monitoring of the water sector, water supply and of wastewater systems. Directly incorporated are the Water Authority of Jordan (WAJ) taking care for water and sewage projects in Jordan as well as the Jordan Valley Authority (JVA) holding responsibility for the socio-economic development of the Jordan Rift Valley including water development and irrigation distribution (JMWI 2002).

The recent Jordan Water Strategy dates back to 1997 (JMWI 1997). It provided the base for developing policies addressing specific issues. Today four policies were outlined:

- Groundwater management: resource exploration and development, resource protection and sustainability, priority of allocation, legislation,
- Irrigation water: utilization of wastewater, sustainability of irrigated agriculture, irrigation water quality, pricing, regulation and controls,
- Water utility: reorganisation of water agencies, water pricing and cost recovery, water resources and quality management, service levels, public awareness,
- Wastewater management: development and management of the resources, wastewater collection and treatment, reuse of treated effluent.

The policies clearly emphasize the sustainable use of the country's scarce natural water resources aiming to improve the living conditions of the population. In the past, the main emphasis was put on the need to tape the full potential of surface and groundwater to a feasible extent. Special attention was given to the construction of dams in wadis and of irrigation projects in the Jordan Valley in order to maximize the utilization of these resources before their drainage to the Dead Sea (IDCR 1995; JMWI 1997, 2002, FAO 2002; World Bank 2001). These projects were, however, disadvantageous to the environment and contributed to the depletion of the Dead Sea (cf. Chaps. 3.1.3 and 3.2.4; Allan and Mallat 1995; Salameh 1996; Salameh and Naser 1999; JMWI 1998; McCornick et al. 2002; Faber et al. 2003). Within the new Water Strategy, priority in water resources allocation was given to allocation of domestic water supplies, where a modest daily per-capita share of 100 l was aspired. In the last years, additional 90 Mm^3/yr of fresh water could be gained by the construction of dams to store wadi and flood waters, by desalination of brackish water, and by developing groundwater resources (JMWI 2002; Taha 2006). However, according to the criteria of sustainability, most of the available renewable resources were already exhausted or even overexploited.

Upgrade and extension of existing water supply systems and reduction of water losses in the networks have received equal attention. The strategy also ensures that wastewater is collected and treated to standards that allow its reuse for unrestricted agriculture and non-domestic purposes (JMWI 1998; Govern. Jordan 1995; 2003; Sheikh 2001; Malkawi 2003). Treated wastewater is generated in about 20 wastewater treatment plants and is mostly discharged into the wadis or is used directly in irrigation. Due to the increasing population and in spite of all these measures, it will not be possible to bridge the gap between the available resources and the still very conservatively estimated consumption (Table 4.3-2) without additional water resources (Jamal and Mousa 2001; World Bank 2004).

Table 4.3-2. Development of resources and demands in Mm³/yr in Jordan for the period 2005-2020 (JMWI 2002)

Year	**2005**	**2010**	**2015**	**2020**
Renewable Groundwater	259	259	259	259
Surface Water (baseflows and reservoir safe yields)	382	404	419	433
Treated wastewater, not flowing into reservoirs	34	69	89	101
Additional Resources	344	511	454	456
Total Resources	**1019**	**1244**	**1220**	**1250**
Municipal, Industrial, Tourist demands	433	493	561	634
Agriculture incl. reuse schemes	1114	1120	1101	1052
Total Demands	**1546**	**1612**	**1661**	**1686**
Groundwater return flows from losses	66	63	62	63
Deficit in million m³/yr	**-461**	**-306**	**-379**	**-373**
Deficit in %	-45	25	-31	-30

In the last years the water legislation has empowered state organisation in the matters of national water resources. The new "Groundwater Control By-Law" of 2002 (JMWI 2002) supports the water rights in favour of the general public. According to the new law, the groundwater is owned and controlled by the state. Contrasting former tradition, ownership of the land does not anymore include ownership of the groundwater. Furthermore, a new by-law addressing pollution prevention and protection of water resources is currently in preparation.

4.3.2 Strategic water situation in Israel

As in other countries of the Jordan River Basin water-scarce region, Israel has chronic problems of water resources (Arlozoroff 1966; Brimberg et al. 1994; Exact 1998). The main sources of fresh water consist on one hand of surface water of Lake Tiberias which is fed by the Upper Jordan River and by run-offs from the Golan Heights and a part-diversion of the Yarmouk River. This surface water reservoir of the Lake conveys water directly to surrounding areas and to the Coastal Plain via the 130 km long National

Water Carrier. The available quantity of water in the Lake changes depending on the yearly precipitation rate which fluctuates between 400 and 700 Mm³/yr. The operational reserve of the Lake is of approximately 700 Mm³/yr.

The National Water Carrier (Fig. 4.3-2) was completed and became functional by 1964 (IMNI 2002; Mekorot 2007). The Carrier is a combination of underground pipelines, open canals, interim reservoirs and tunnels. Water from the Lake located some 210 m below sea level, is pumped to an elevation of about 152 m above sea level. From this height, the water flows by gravitation to the coastal region from where it is pumped to the Negev. The National Water Carrier functions not only as the main supplier of water but also as an outlet for surplus water from the north during winter and early spring. It is also a source of recharge to the aquifers in the coastal region (IMFA 2002, 2003). Most of the regional water systems are incorporated into the National Water Carrier to form a well-balanced network in which water can be shifted from one line to another according to conditions and needs.

The second main source is the groundwater from the mountain aquifer mainly built of karstic limestones and from the coastal aquifer of Quaternary sands and calcareous sandstones (cf. Chaps. 2.2, 3.3 and 4.2; Allan 1996; Exact 1998; IMFA 2003; IMNI 2002a). The first source occurs in the mountain ridge and extends from the Jordan Valley to the Coastal Plain. It is subdivided into the eastern-, northern- and western aquifer with renewable recharges of 170, 145 and 365 Mm³/yr, respectively. These water sources are supplemented by marginal sources including effluents from treatment plants, floodwater and saline water to meet consumptions needs.

The total average annual potential of renewable water amounts to approximately 1800 Mm³/yr (Table 4.3-3), of which about 95% is already exploited and used for domestic consumption and irrigation (Table 4.3-4 and Fig. 4.3-3). About 80% of the water potential is in the northern part of the country and only 20% in the south.

Table 4.3-3. Long-term potential of renewable water in Israel (after IMNI 2002)

Resource	**Replenishable quantities** (Mm³/yr)
Coastal aquifer	320
Mountain aquifer	370
Lake Tiberias	700
Additional regional resources	410
Total average	**1,800**

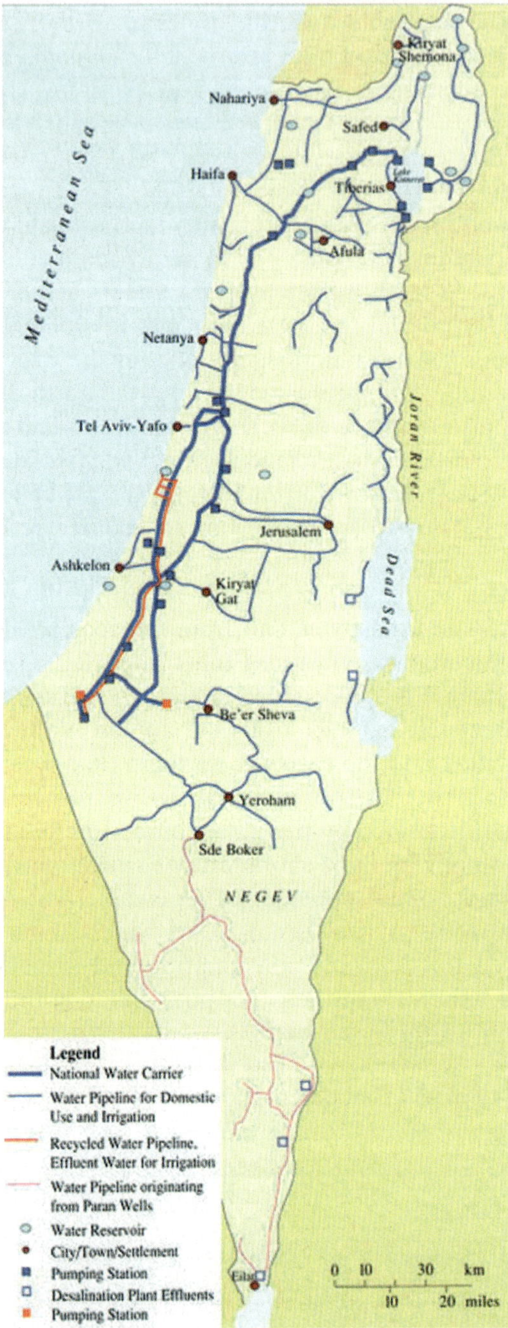

Fig. 4.3-2. The Israel National Water Carrier extending from the Lake Tiberias to the Negev with affiliated water network systems (after Mekorot1995)

Table 4.3-4. Water supply and demand - Israel 1998-2020 in Mm³/yr (Source: Israel Water Commission, IMNI 2002a)

WATER SUPPLY							
		Water Sources					
Year	**Population (Million)**	Surface Water	Ground Water	Brackish	Treated Effluents	Desalinsation	**Total**
1998	6.0	640	1050	140	260	10	2100
2005	6.9	645	1050	165	470	100	2430

WATER DEMAND						
Urban Sector		**Agricultural Sector**				**Total**
Year	Surface-, Groundwater, and Desalination	Natural	Brackish	Wastewater Effluents	Total	
1998	800	920	120	260	1300	2100
2005	980	750	95	380	1225	2430

Water is considered as a national resource of utmost importance. It is vital to ensure the population's well-being and to preserve the rural-agricultural sector (IMFA 2003; IMEP 2005; Murakami 1995; Allan 1996). The consumption has run always close to the renewable amounts. However, in the years 1999 to 2002 Israel has suffered from four consecutive years of drought creating a crisis so severe that water tables at the Lake Tiberias (Fig. 4.3-4) as well as in the main aquifers fell below the red lines due to overexploitation (IMNI 2002a,b; IMFA 2002, 2003). This led to a cumulative deficit in Israel's renewable water resources which reached approximately 2000 Mm^3, an amount equal to the annual consumption of the State including qualitative deterioration of potable aquifer water resources.

Water quality is an issue of equal importance to water scarcity and water quality degradation is a considerable issue in water management (Rosenthal 1987; Brimberg et al. 1994; Farber et al. 2003; IMEP 2005). Due to unbalanced exploitation and return flow from irrigation, groundwater salinity increased in many wells. In order to avoid further pollution advanced technologies and practices have been applied to protect water resources. Water conservation maps portraying restrictions of land use activities were produced to protect the underlying groundwater resources. Regular monitoring of water resources, including water recharge, water table levels, abstraction, salinity (chlorides) and pollution (nitrates) data, are regularly monitored and reported.

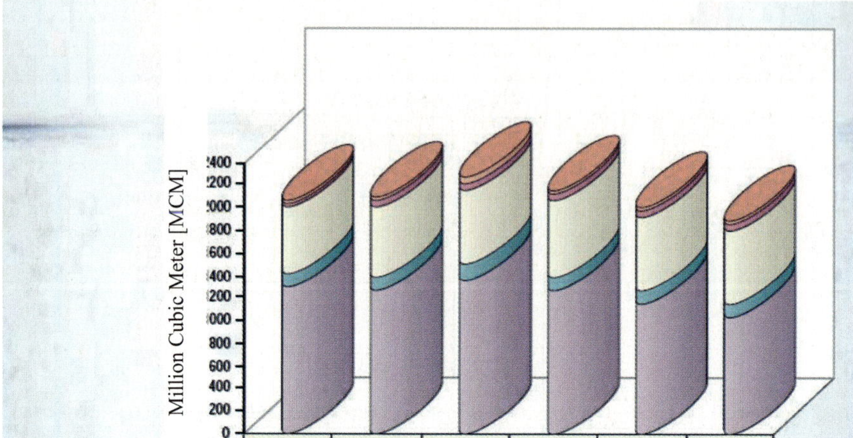

	1996	1997	1998	1999	2000	2001
Palestinian Authority	30.8	34.8	37.2	40.2	39.9	39.3
Kingdom of Jordan	31.6	49.0	59.0	44.5	54.2	46.1
Domestic consumption	604.0	621.2	671.7	681.8	662.1	658.4
Industrial consumption	124.4	122.8	129.2	126.5	124.2	120.1
Agricultural consumption	1284.3	1263.8	1364.9	1264.6	1137.4	1021.9

Fig. 4.3-3. Total Water Consumption in Israel in the years 1996–2001, including delivery to Jordan and Palestine according to special agreements (source: Israel Water Commission, in IMNI 2002a)

The causes of the mentioned crisis are both natural and man-made. The increase in demand for water for domestic uses caused by population growth and by the rising standard of living, together with the need to supply water pursuant to international agreements, led to over-exploitation of the renewable water sources. The agricultural sector has suffered most because of the crisis. Due to the shortage, water allocations to this sector had to be reduced drastically causing a reduction in the agricultural productivity. The crisis has led to the realization that a master plan for management policy implying institutional and operational changes, is required to stabilize the situation and to improve Israel's water balance with a long-term perspective (Plaut 2002; IMNI 2002a,b; IMEP 2005 a, b).

Water conservation is the most reliable and least expensive way to stretch the country's water resources and the challenge is being met in all sectors. Public water conservation campaigns coupled with technical and economic measures are being applied to reduce consumption and to increase awareness of water scarcity (IMFA 2002; IMNI 2002b). In the domestic and urban sectors, conservation efforts focus on improvements in

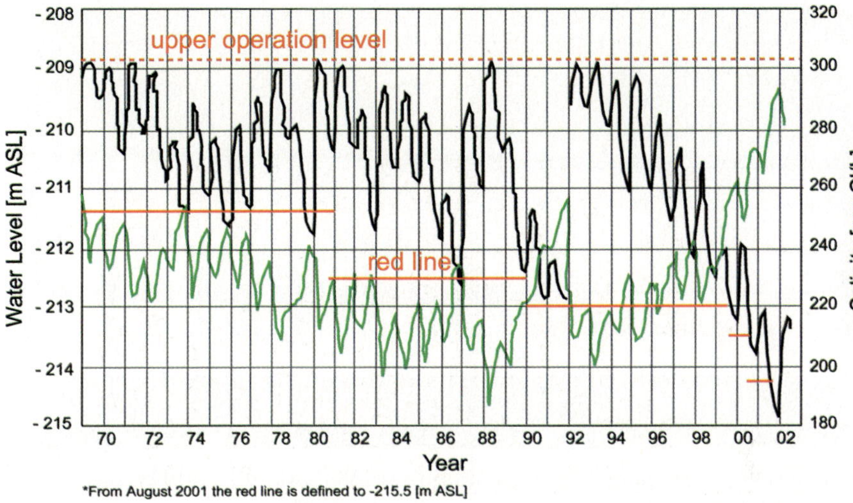

Fig. 4.3-4. Water level fluctuations in the Lake Tiberias from 1969 to 2002 with the strong depletion due to the drought period from 1999–2002. The green and red lines show the operating range between -208.8 and -213 m bsl. Due to the risk of strong increase of salinization the water level should not fall below -213 m bsl (after IMNI 2002a)

efficiency, resource management (Fig.4.3-5), repair, control and monitoring of municipal water systems. Water not required seasonally by consumers is recharged into the aquifer through spreading basins and dual-purpose wells. Recharging of aquifers helps to prevent evaporation losses and, in the coastal area, encroachment of seawater. In agriculture, the wide-scale adoption of low volume irrigation systems (e.g., drip, micro-sprinklers) and automation has increased the average efficiency to 90% as compared to 64% for furrow irrigation (Brimberg et al. 1994; Lonergan and Brooks 1995; Dinar and Wolf 1997; FAO 2002; IMNI 2002b). As a result, the average requirement of water per unit of land area has decreased from 8700 m^3/ha in 1975 to the current application rate of 5500 m^3/ha. At the same time agricultural output has increased twelve-fold, while total water consumption by the sector has remained almost constant.

After drawing on nearly all of its readily available water resources (Fig. 4.3-6–7.) and promoting vigorous conservation programs, Is- rael has long made it a national mission to stretch existing sources by developing non-conventional water sources (IMFA 2002; IMNI 2002b; World Bank 2004; Water Technology 2007; IMEP 2005; Shelef 2002; Haddad and Miszyed 2004). These efforts have focused on reclaimed wastewater effluents (cf. Fig.4.3-5), intercepted runoff and artificial recharge, artificially-induced rainfall by cloud seeding, and on desalination (Fig. 4.3-7). The crisis of the

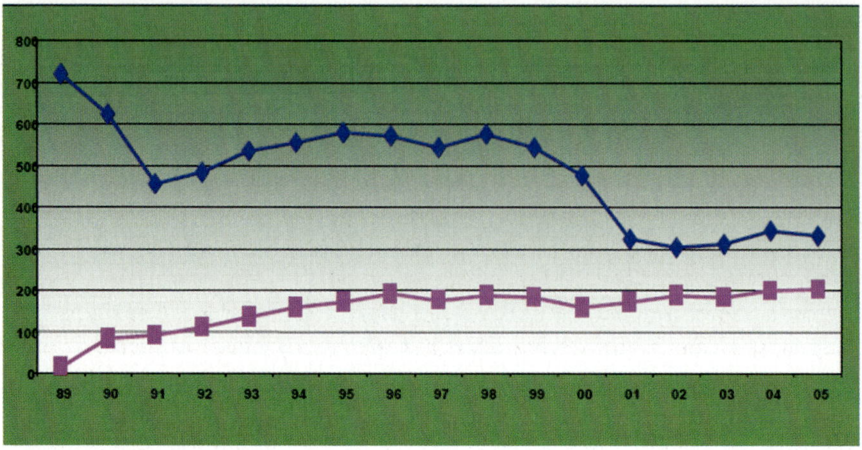

Figure 4.3-5. Water supply for agricultural exploitation by Israel National Water Company Mekorot in Mm³/yr for the period 1998-2005: replacement of tap-water (blue) by treated wastewater (pink). Mekorot covers around 60% of the water utilisation in agriculture in Israel (Mekorot 2007)

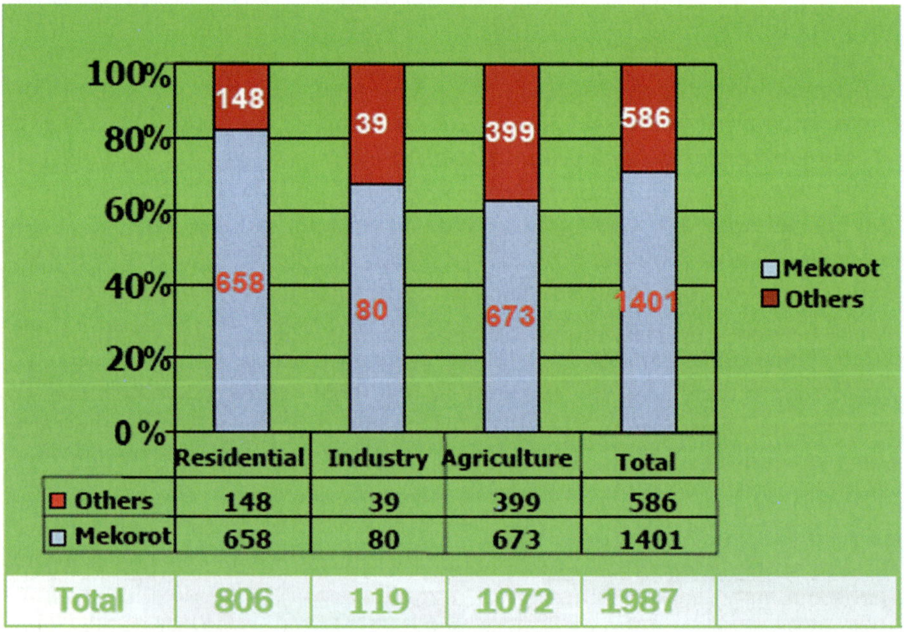

Figure 4.3-6. Annual water consumption in Israel by different branches, based on figures from 2004 (Mekorot 2007)

years 1999-2002 has led to the realization that a master plan for policy, institutional and operational changes is required to stabilize and improve Israel's water balance with a long-term perspective (IMNI 2002a, b). The Government of Israel decided on an integrated policy requesting:

- Preservation and protection of the existing reserves, their quality and quantity,
- Increase of the supply of potable water, mainly by seawater desalination and purification of water sources,
- Advancement of sewage treatment, its collection and purification, turning it into the main source of water for agriculture (replacing potable water),
- Improving of water saving in all sectors by economic measures as well as by legal means, enforcement, improved management and organization, and
- Investments in agro-technology in order to enable agriculture to use mainly low quality water, recycled wastewater, brackish and flood water.

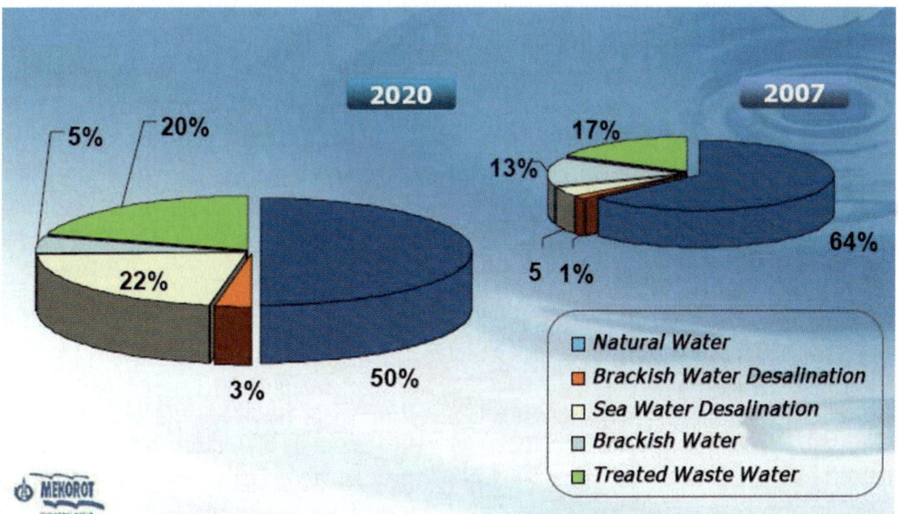

Fig. 4.3-7. Expected reallocation of the different water sources in Israel water supply between 2007 and 2020 (Mekorot 2007)

In early 2002 the government of Israel approved the installation of seawater desalination plants with a total output of 400 Mm^3/yr of potable water. The first plant in Ashkelon has already started production with 100 Mm^3/yr (Mekorot 2007; Water Technology 2007). All activities in the water sector will be based on a new water sector policy that incorporates a development plan and is founded upon three basic components:

- Ensuring water supply,
- Social and economic requirements, and
- Environmental and ecological needs.

These activities are based on the Water Law, 5719-1959 (State of Israel 1959). Its main intention is formulated in section 1: "The water resources of the State are public property; they are subject to the control of the State and are destined for the requirements of its inhabitants and for the development of the country". There is no private or governmental ownership of water. All water belongs to the public and is controlled by the State. The administrator is the Water Commissioner, who is appointed by the government. Ministerial responsibility rests with the Minister of National Infrastructure and in an amendment of the law by the Ministry of Environmental Protection. They are advised by the Water Council. The Israel National Water Company is Mekorot (2007) which is in charge of the wholesale supply of water to urban communities, industries and irrigation users. It produces and supplies about two-thirds of the total amount of water used in Israel. The company has developed Israel's water supply systems on a nationwide scale since 1937. In 1997, Mekorot supplied 1380 Mm^3/yr of water, of which 745 Mm^3/yr were supplied for irrigation, 540 Mm^3/yr for domestic use, 94 Mm^3/yr for industry, and 27 Mm^3/yr to replenish over-pumped aquifers. Water was also supplied to Jordan and the Palestinian Authority in accordance with the Peace Treaty and international agreements. Its crowning achievement, the National Water Carrier has been in operation since 1964.

4.3.3 Strategic water situation in Palestine

The subsequent depiction on the water situation in Palestine refers mainly to the West Bank, belonging with its larger part directly to the catchment area of the lower Jordan River and does not include the Gaza Strip. The access of Palestinians to fresh water for domestic, agricultural, industrial, tourism and other uses is extremely limited and constricted. There are many reasons for this particularly wrong situation (cf. Chap. 4.1; Haddad 1998 a; Isaac 1998; GTZ 1998). First reason is the limited natural water recharge from precipitation. Though the highlands of the West Bank (Fig.4.3-8.) belong to the rainiest area in the surrounding and the recharge there is very important for the mountain aquifers. The slopes towards the

Jordan Valley are extremely dry. Although Palestine borders to the Jordan River, due to political constriction it has neither access to the River nor to any other river bringing water from outside of the area. From the three regional portions of the mountain aquifers which are recharged mainly by the precipitation in the West Bank the western and northern are mainly exploited by Israel (Chap. 4.3.1.2). The Palestinian water installations are underdeveloped and of need for maintenance. New developments are very limited due to the control and restrictions by the Israeli since 1967 (Libiszewski 1995; Isaac 1998; Sofer 1998; Exact 1998; Husseini 2004). Another main source for the bad water situation is the natural and additionally artificially-enforced increase of the population.

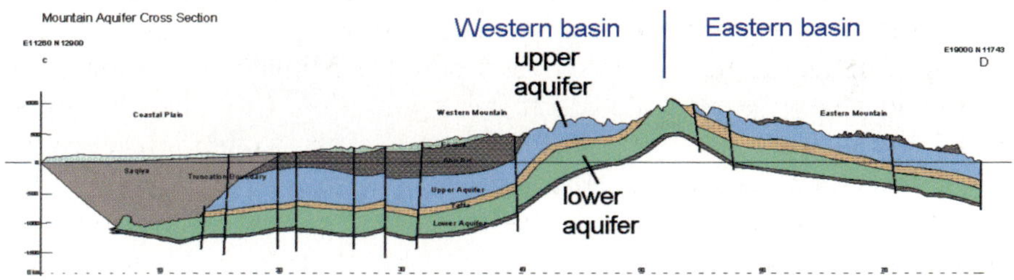

Fig. 4.3-8. Hydrogeological cross section from the Mediterranean to the Dead Sea (Messerschmidt 2003)

The total amount of rainwater varies between 2700 and 2900 Mm^3/yr. Between 65 to 70% of that evaporates, less than 100 Mm^3/yr are available for surface run-off to the wadis (mainly as flash floods during the winter seasons) and around 650 Mm^3/yr recharges into the underground. The most important aquifers in Palestine, which are used by Israel and Palestine, are the +700 m thick Upper Cretaceous limestones (Fig. 4.3-8 and Fig. 4.3-9) folded in a wide anticlinal structure (Chaps. 2.2 and 3.3.2). The eastern aquifer extends towards the Jordan Valley, and the Palestinians extract from it (via springs or wells; Fig. 4.4-10) about 65 Mm^3/yr (Murakami 1995; GTZ 1998; PHG 2003).

Additional water is extracted by Israeli settlers. An amount of about 50 Mm^3/yr might be still available, discharging mainly towards the Dead Sea as brackish water, after mixing with brines along the western margins of the Rift (Chap. 3.3.2). The slope of the anticline dipping towards the Mediterranean coastal plain forms the western aquifer. It was already mentioned that this part forms an important reservoir for Israel producing from this aquifer about 365 Mm^3/yr (IMNI 2002a; Isaac 1998; PNIC 1999a). The Palestinians produce from this aquifer about 22 Mm^3/yr. From the north-eastern aquifer Israel is pumping about 125 Mm^3/yr and Palestinians 25 Mm^3/yr (Table 4.3-5). The last two aquifers are already overexploited by

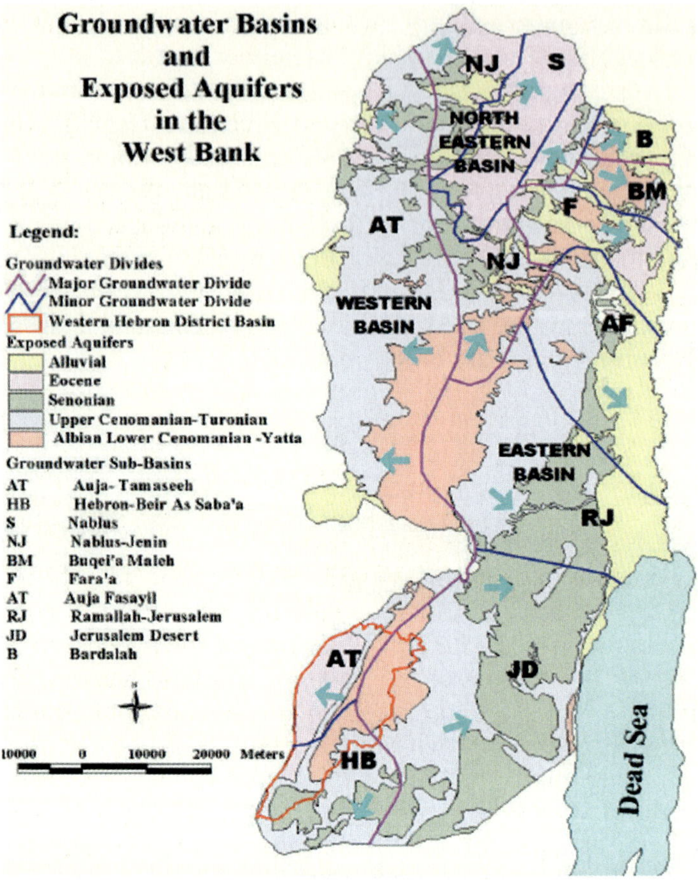

Fig. 4.3-9. Hydrogeological subdivision of the West Bank into the three main aquifers and into subbasins (after PHG 2003)

huge pumping rates. Additional water is drawn from the Quaternary sediments deposited in the deeply incised wadis and in the transition zones to the wide valleys or to the coastal plain.

The demand for water is strongly increasing. While the population was fifty years ago around 900,000, it has reached already 2.5 millions Palestinians in 2005 in addition to 300.000 Israeli settlers (Husseini 2004). In 2020 the population is expected to be approximately 5 million. The recent water consumption of 128 Mm^3/yr with additional 50 Mm^3/yr supplied by the Israeli system, covers only the most urgent demands. In many countryside regions, the per capita consumption is below the recommendations of the World Health Organisation of 65 l. Only 5% of the land is artificially irrigated as compared to 50% in Israel (GTZ 1998; Exact 1998; IDCR 1995; PNIC 1999a). The Palestinian agrarian consumption in the West

Bank amounts about 90 Mm^3/yr and needs much more for the production of food for the growing population.

Table 4.3-5. Average recharge in the mountain aquifers in Mm^3/yr. Figures after PNIC (1999a) and Messerschmid (2001, 2003).

MOUNTAIN AQUIFERS	Eastern aquifer basin	North eastern aquifer basin	Western aquifer basin
Israeli wells Israeli springs	40 --	103	340
Palestinian wells Palestinian springs	24 30	25 17	20 2
Total	**94**	**145**	**362**
Estimated potential	172	145	362
Remaining	78	0	0

Due to the historical and political development of the region, the legal framework governing water issues in Palestine has changed frequently during the last hundred years and still requests a final agreement in connection with the peace negotiations in the region. Under Ottoman rule, the Shari'a Law was the legal norm and provided in principle that water is God's property and therefore not owned by any person but owned jointly by all members of mankind. The British Mandatory Law in Palestine (1922–1948) was mainly referring to Ottomans Law and Order but added specific ordinances regarding for example safeguarding of public water supplies or water survey. The Jordanian legislation (1948-1967) kept in force mainly the old laws and introduced additional laws concerning water rights and management. After the occupation of the West Bank in 1967, Israel declared all water resources in the region as state property, which caused together with other military orders stringent restrictions on independent self-determination. These regulations are in force to the present days, so that the final decisions in water issues still belongs to the Israeli military governor (Shuval 1996; Sofer 1998; Husseini 2004).

In connection with the Oslo peace negotiations (1993) Israel and the Palestinian Liberation Organisation (PLO) signed a declaration of understanding, where water was one of the main issues. As a following step Israel recognized Palestinian water rights in the Taba Agreement (Oslo II, 1995). The details of the water issues were deferred to the final status of the peace negotiations which was not yet attained. The Joint Water Com mittee (JWC) was established in order to coordinate in the meantime the management and protection of water resources and sewage systems. The

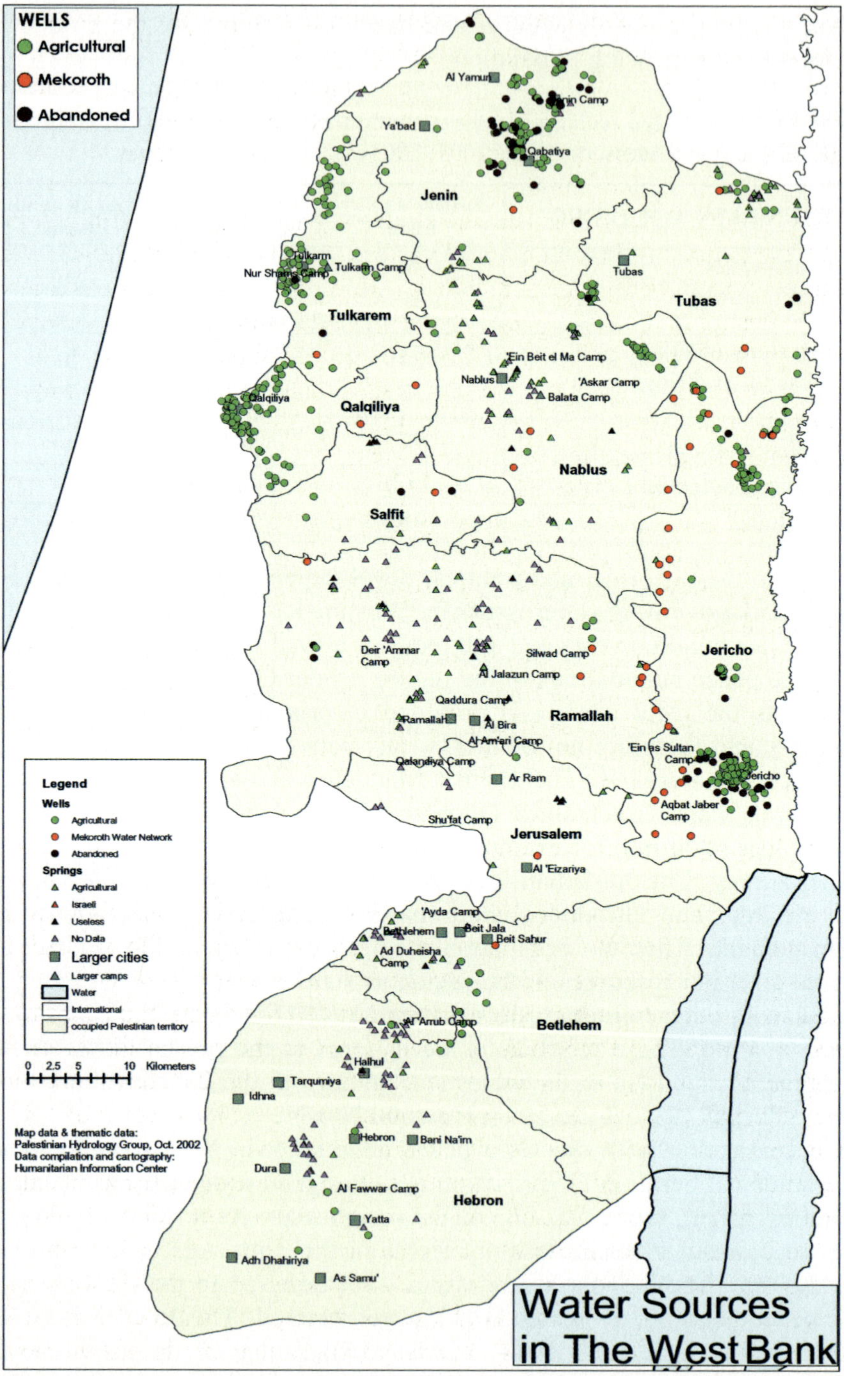

Fig.4.3-10. Location of springs and wells in the West Bank (PHG 2003)

appointment of the Palestinian Authority for the responsibility of parts of the West Bank in 1995 led to the establishing of the Palestinian Water Authority (PWA). By law it received the mandate to manage and supervise the water sector and resources as well as to execute a national water policy. One of the first main important achievements was the preparation of the National Water Plan of 2000.

The National Water Plan sets the strategic goals to the year 2020. It proposes the actions to be taken and describes the role of the Regional Water Utilities, which will be responsible for preliminary investigations and research, design and construction, operation and maintenance. The plan also emphasizes the responsibility of PWA to secure Palestinian water rights, to strengthen national policies, to build institutional capacity and to improve information services. With regard to these tasks PWA was substantially engaged in the preparation of the new water law.

This new Palestine Water Law No.3/2002 was enacted by the legislature in 2002 and encompasses the whole water sector (PNIC 1999b). It aims to develop and manage the water resources, to increase capacity, to improve quality, to preserve and protect against pollution and depletion. As leading principle the Law states that all water resources in Palestine are considered public property. The Law also regulates the position and tasks of the PWA at the same time, it also defines a new Water Council including the representatives of the mainly involved ministries, stakeholders and universities, who have to sanction water policies and to support PWA.

4.3.4 Transboundary water cooperation

Though the necessity of an Integrated Water Resources Management for the Jordan River Basin has become more and more urgent, the old tradition of delineating territories and insular policy as well as the recent political situation in the region has, so far, impeded the formulation of a comprehensive scheme for transboundary basin management. The divergent development of the region started at the beginning of the 20th century. At that time only parts of the fresh water resources were really used, most of the recharged fresh water flow freely towards the Mediterranean or the Dead Sea. With the enhanced Jewish immigration to Palestine, the demand for water increased proportionally with the necessary production of food. In the early stage of the State of Israel, the agricultural sector had absolute priority and developed very rapidly. By applying modern irrigating techniques, the area of cultivable land was extremely extended and the ambitious goal of Israel to include in this program also the northern Negev, could be realized with the National Water Carrier bringing the water from the Lake Tiberias to the coastal plain and further to the south. During the 1950' and 1960's – when these schemes were formulated, there was no contact whatsoever between Israel and the Palestinians, which were at that time under Jordanian rule. Against this geopolitical background the interests and demands of the other riparian of the Jordan River were not taken in consideration. Throughout the years the Palestinians were affected by a fast growth of their population therefore causing a high demand of water for domestic as well as for agricultural utilization. The water issues and particularly the unequal exploitation of the transboundary water resources became progressively part of the violent political dispute.

With the Oslo Agreement (1992) and the signing of the Declaration of Principles, Israelis and Palestinians agreed on the equitable utilization of water resources and the joint management of these resources. In the Oslo II Agreement (1995) there was also a joint understanding of the water demands of the West Bank and they agreed on the operation of a joint Water Commission. However, the translation of these principles into actual shares still requests successful peace negotiations.

Between Israel and Jordan a Peace Treaty has been signed in 1994. The water dispute was resolved based on mutual recognition of the rightful allocations of both parties to the Jordan and Yarmouk Rivers as well as to the groundwater in Wadi Araba. The agreement allows the use of Lake Tiberias for strong Jordanian surplus rain flows from the Yarmouk to be redrawn during the summer. It also maintained the right of Israeli farmers to draw water from the alluvial aquifers in Wadi Araba (Isaac 1998).

Agreements on water issues were not negotiated till now between Israel and Syria and between Israel and Lebanon.

After exploiting all of the renewable fresh water resources within their national boundaries, Israel, Palestine and Jordan have no choice except to develop transboundary waters and non-conventional waters. Water conservation is an important and essential issue in water management, but development of non-conventional water alternatives is becoming imperative to supply fresh potable water to the growing population in the region. Fortunately most of the responsible experts have realized that it makes no sense to concentrate just on a new partitioning of the existing overexploited resources. A reasonable solution can be only expected by integrated transboundary water management. The priorities and urgent countermeasures in water development have to focus on:

- Water conservation,
- Reuse of treated waste water,
- Desalination of brackish and sea water,
- Restoration of the overexploited aquifers, and
- Restoration of disequilibrated environment.

Fig.4.3-11. The cascaded shore line of the Dead Sea south of the Wadi Mujib, Jordan, due to the effects of the yearly depletion rate of the sea level (Photo Hötzl, 2003).

With regard to the last point, there is, beside many other cases, one particularly drastic example: the environmental degradation of the Dead Sea. The Dead Sea Basin is in grave danger because in less than 30 years, the water level declined from -395 m to -420 m below sea level. The current yearly rate of decline is about one meter. This decline created other environmental changes that are leading to rapid deterioration both on sea and shore (Fig. 4.3-11–12). Changes in the ecological balance of this fragile eco-system are causing deterioration of oases and wetlands. Sinkholes are formed and threaten the visitors to the Dead Sea.

The dramatic lowering of water level is largely attributed to the diversion of water sources upstream of its basin to meet domestic, agricultural, touristic and industrial demands and to the mineral extraction industry. About 800 Mm^3/yr additional inflows are necessary to keep the balance. There is an urgent need for concerted international action. Solutions must be large scale and comprehensive. Beside other proposals Jordan and Israel have jointly launched the project of the Red Sea-Dead Sea-Conduit. It addresses two issues of sustainable development: (1) sufficient quantities of water can be returned to the Dead Sea in order to restore it to historical levels and (2) significant quantities of desalinated water can be produced for domestic and agricultural use.

Fig.4.3-12. The flat coastline of the Dead Sea south of En Gedi. The brown mud plain documents the amount of the Sea regression during the last 30 years. In foreground the development of sinkholes due to solution of salt in the underground. (Photo: Hötzl, 2007).

References

Allan JA (ed.) (1996) Water, peace and the Middle East: Negotiating resources in the Jordan Basin. Tauris Academic Studies; St. Martin's [distributor], 1996. xix, Library of Modern Middle East Studies, London, pp 1–250

Allan JA, Mallat H, Chibli S (ed.) (1995) Water in the Middle East: Legal, political, and commercial implications. London: I.B. Tauris, 1995. ix,. Library of Modern Middle East Studies, 1, pp 1–358

Arlozoroff S (1966) Managing scarce water: Recent Israeli experience. In: Allan JA (ed) Water, Peace and the Middle East: Negotiating Resources in the Jordan Basin., Tauris Academic Studies; St. Martin's [distributor], 1996. xix, Library of Modern Middle East Studies, London, pp 21–48

Brimberg J, Mehrez A, Oron G (1994) Economic development of groundwater in arid zones with application to the Negev Desert, Israel. - Management Science, 40:353–363

Dinar A, Wolf A (1997) Economic and political considerations in regional cooperation models. Agricultural Resour Econ Rev 26:7–22

Exact (1998) Overview of Middle East Water Resources: Water resources of Palestinian, Jordanian and Israeli interest. Jordanian Ministry of Water and Irrigation, Palestinian Water Authority and Israeli Hydrological Service. Compiled by the U.S. Geological Survey for the Executive Action Team, Middle East Water Data Banks Project, pp 1–50

Farber E, Vengosh A, Gavrielli I, Marie A, Bullen TD, Mayer B, Holtzman R, Segal M and Shavit U (2004) The origin and mechanisms of salinization of the Lower Jordan River. Geochim Cosmochim Acta 68:1989–2006

FAO Food and Agriculture Organisation of the United Nations (2002) Water Resources, Development and Management Service (2002) AQUASTAT Information System on water in Agriculture: Rev Water Resour Statistics Country. Rome. Online at: http://www.fao.org/waicent/faoinfo/agricult/agl/aglw/aquastat/water_res/index.htm

Government of Jordan (1995) Jordanian Standards for Water Reuse. JS893/1995. Standard Publishers, Amman, Jordan, pp 1–20

Government of Jordan (1998) The Hashemite Kingdom of Jordan, Jordan's water shortage. http://www.kinghussein.gov.jo

Government of Jordan (2003) Technical Regulation for Reclaimed Domestic Wastewater. JS893/2002, Jordan Institution for Standards and Meteorology, Amman, Jordan, pp 1–25

GTZ (1998) Deutsche Gesellschaft für Technische Zusammenarbeit GmbH and Consulting Engineers Salzgitter (CES) GmbH: Multilateral Working Group on Water Resources: Middle East Regional Study on Water Supply and Demand Development. Regional Overview Phase II, November 1997 (restricted distribution).

Haddad M (1998) Planning Water Supply Under Complex and Changing Political Conditions: Palestine as a Case Study. Water Policy J 1:177–192

Haddad M and Mizyed N (2004) Non-Conventional Options for Water Supply Augmentation in the Middle East: A Case Study. Water Intern J, Intern Water Resour Ass 29:232–242

Husseini H (2004) The Palestinian Water Authority developments and challenges involving the legal framework and capacity of the PWA .- The International Israeli-Palestinian Water Conference, Antalia, Turkey, October 10-14, 2004

IDRC International Development Research Center (1995) More on water in Jordan. http://www.idrc.ca/en/ev-72165-201-1-DO_TOPIC.html

IDRC International Development Research Center (1995) Chapter 10 Palestinian water management in a transitional period. Document(s) 12 of 19, http://www.idrc.ca/en/ev-72165-201-1-DO_TOPIC.html

IMEP – Israel Ministry of Environment and Protection (2005) Environmental Topics, Water: Water Sources. http://www.sviva.gov.il

IMEP – Israel Ministry of Environment and Protection (2005) Environmental Topics, Water: Marginal water Sources. http://www.sviva.gov.il

IMEP – Israel Ministry of Environmental Protection (2007). Dead Sea, 09/09/2007- Environmental Topics, pp 1–2

IMFA – Israel Ministry of Foreign Affairs (2002) MFA Newsletter, MFA Library, Aug. 2002. Israel's Water Economy-Thinking of future generations

IMFA – Israel Ministry of Foreign Affairs (2002c) Israel and Jordan launch global campaign to save the Dead Sea, MFA Library 2000-2007, http://www.mfa.gov.il

IMFA – Israel Ministry of Foreign Affairs (2003) MFA Newsletter, MFA Library, Sept. 2003. Focus on Israel-Development of Limited Water Resources

IMNI - Israel Ministry of National Infrastructure, Water Commission (2002) Water in Israel consumption and production 2001, pp 1–52

IMNI - Israel Ministry of National Infrastructure, Water Commission (2002) Transitional Master Plan for Water Sector Development in the Period 2002-2010. pp 1–24

Isaac J (1998) A Palestinian perspective on the water crisis. Palestine-Israel journal of politics, economics and culture, Jerusalem, 5 (1998), 1, pp 54–58

Isaac J (1999) Water and Palestinian-Israeli Peace Negotiations. Policy Brief, 4, 19 August 1999, pp 1–2

JMWI (Jordan Ministry of Water and Irrigation) (1997) Water Resources Strategy. Ministry of Water and Irrigation (MWI), The Hashemite Kingdom of Jordan, Amman, Jordan, pp 1–7

JMWI -Jordan Ministry of Water and Irrigation (1998) Wastewater Management Policy Paper No.4. Ministry of Water and Irrigation (MWI), The Hashemite Kingdom of Jordan, Amman, Jordan, pp 1–10.

JMWI Jordan Ministry of Water and Irrigation (2002) Underground Water Control By-Law Underground Water Control By-Law By-Law No. (85) of 2002

JMWI Jordan Ministry of Water and Irrigation (2002) National Water Master Plan of Jordan, http://www.mwi.gov.jo/NWMP/INDEX.HTM

Lonergan SC, Brooks DB (1995) Watershed, The Role of Fresh Water in the Israeli-Palestinian Conflict. IDRC 1995, pp 1–220

Libiszewski S (1995) Water disputes in the Jordan Basin Region and their role in the resolution of the Arab Israeli Conflict. ETH-Zurich Swiss Federal Research Institute of Technology, Center for Security Studies and Conflict Research, Bern: Swiss Peace Foundation, 1995. Occasional paper, 13: 1–106

Malkawi SH (2003) Wastewater Management & Reuse in Jordan. First Regional Water Reuse Conference, Ministry of Water and Irrigation (7-9 Dec, 2003)

McCornick PG, Taha S and El Nasser H (2002) Planning for Reclaimed Water in the Amman-Zarqa Basin and Jordan Valley. American Society of Civil Engineering – Environmental and Water Resources Institute, Conf Symp Droughts and Floods, Roanoke, Virginia, pp 1–10

Mekorot (1995) Israel National Water Carrier, -from the Sea of Galilee to the frings of the desert. Mekorot brochure, Tel Aviv, pp 1–31

Mekorot (2007) Mekorot – Facts and figures. Mekorot brochure, Tel Aviv, pp 1–21

Messerschmid C (2001) Der westliche Bergaquifer. Berlin, INAMO, 27:1–15

Messerschmid C (2003) Till the last drop: the Palestinian water crisis in the West Bank, Hydrogeology and Hydropolitics of a regional conflict. http//www.alternative news.org

Murakami M (1995) Managing Water for Peace in the Middle East: Alternative Strategies. United Nations University Press; Tokyo - New York – Paris, pp 1–319

PHG Palestinian Hydrology Group (2003) Hydrogeology, hydrology and hydrochemistry of dug wells and springs in the western Hebron District Basin

PNIC Palestinian National Information Centre (1999a) Water resources, water consumption, balance. http://www.pnic.gov.ps/english/Water/Water.html

PNIC Palestinian National Information Centre (1999b) Laws and Legislation, Palest. Water Law. http://www.pnic.gov.ps/english/Water/Water_Law.html

Plaut S (2002) Water policy in Israel. Institute for Advanced Strategic and Political Studies, Policy Studies, 47, Jerusalem, pp 1–26 p

Rosenthal E (1987) Chemical composition of rainfall and groundwater in recharge areas of the Bet Shean-Harod multiple aquifer system, Israel. J Hydrol 155:329–352

Salameh E (1996) Water quality degradation in Jordan- Impacts on environment, economy and future generations resources base. Royal Society for the Conservation of Nature and Friedrich Ebert Stiftung. Council of Science and Technology, Jordan, pp 1–120

Salameh E, Naser H (1999) Does the actual drop in Dead Sea level reflect the development of water resources within its drainage basin. Hydrochim Hydrobiol Acta 27:5–11

Shelef G (2002) Wastewater Treatment, Reclamation and Reuse in Israel. Source: www.Biu.Ac.Il/Besa/Waterarticle3.html

Sofer A (1998) The Israeli-Palestinian conflict over water resources.- Palestine-Israel Journal of politics, economics and culture, 5 (1998), Vol.1, Jerusalem

State of Israel (1959): The Water Law 1959

Shuval H (1996) A water-for-peace plan. Palestine-Israel journal of politics, economics and culture, 3(1996), 3/4, Jerusalem, pp 74–83

Sheikh B (2001) Standards, Regulations and Legislation for Water Reuse in Jordan. Water Reuse Component, Water Policy Support Project, Ministry of Water and Irrigation, Amman, Jordan, pp 1–42

Taha S (2006) Jordan country paper. Conference of the Water directors of the Euro-Mediterranean & South Eastern European Countries. Athens, Greece 6-7 Nov. 2006

Water Technology (2007) Ashkelon Desalination Plant, Seawater Reverse Osmosis (SWRO), www.water-technology.net

WHO (World Health Organization) (1989) Health guidelines for the use of wastewater in agriculture and aquaculture: Report of a WHO Scientific Group. WHO Technical Report Series 778. World Health Organization, Geneva, Switzerland, pp1–74

World Bank, Rural Development, Water and Environment Group & Infrastructure Development Group, Middle East and North Africa Region (2001) The Hashemite Kingdom of Jordan Water Sector Review Update, Main Report, Report No. 21946-JO, February 1-5, 2001

World Bank (2004) Seawater and brackish water desalinisation in the Middle East. Final Report, Annex 3, Jordan, pp 1–80

4.4 Future water demands

Marwan Haddad

4.4.1 Scenarios

During the coming decades, water demand is expected to rise and the situation will become more and more critical considering the quantity and quality of water needed for economic and social development. Various water supply and demand scenarios show that despite an optimal development of conventional water resources, the size of the regional net water gap will greatly depend on the anticipated future utilization pattern (Haddad and Mizyed 2004).

Based on different population growth rates and separate demands for water, projections were developed for three planning horizons ending by the year 2040 and for three scenarios (low, base and high). The estimates of domestic and industrial demands were based on gross domestic per capita demand increasing (by 2040) from 270–400 l/d (for Israel), 100–240 l/d (for Jordan) and 100–230 l/d (for Palestine). Irrigation demands in the West Bank were projected to decrease, approaching low specific irrigation water use in Israel (500–550 m^3/donum) but to remain relatively high in Jordan (above 1000 m^3/donum) due to the assumed high intensity of leaching salt from the soil. This assumption on water use may be justified for the Jordan Valley irrigation schemes, but appear to be too high for the Jordanian highlands. Irrigation demand would be satisfied largely through increased reuse of treated wastewater, excepting only certain specific crops and areas (Fig. 4.3.1; GTZ 1997, 1998).

The predicted overall regional fresh water gaps for the period 2000–2040 range from 866 to 3381 to 3478 Mm^3/yr for the low to base or medium to high water demand scenarios (GTZ 1998). This outlook is alarming. The gap between supply and estimated demand, based on assumptions that were considered reasonable, shows no sign of closing, but widens considerably in the coming decades. If supply cannot be augmented to meet the projected demand in future, actual consumption will be lower than targeted and stress on both urban consumers and irrigation will result.

The comparison of future water demands in the Middle East (ME) demonstrates the following summary points:

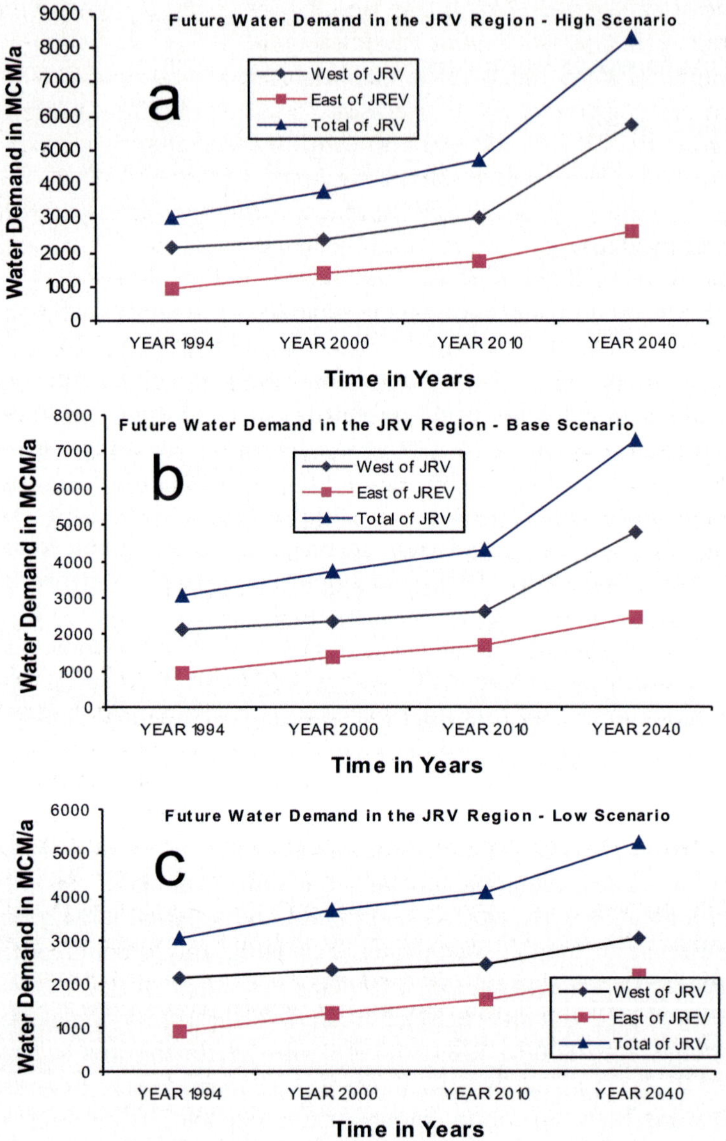

Fig. 4.4-1. Future water demand in the Jordan Rift Valley (JRV), high (a), base (b) and low (c) scenario

- The consolidated regional water demand scenarios (low-base-high) resulting from the cumulative individual demand projections of the three core parties show that substantial amounts of additional water will have to be mobilized during the coming decades. The amount of water currently disputed among the riparians is relatively small when measured against the long-term projected demand.

- The fresh water gap of the consolidated "base demand scenario" is based on long-term fresh water uses in irrigated agriculture in Israel and Jordan of 450 Mm^3/yr and 460 Mm^3/yr respectively, compared to 150 Mm^3/yr in West Bank. This estimate is considered not realistic since the future cost of water and more stringent water demand management would lead to a considerably lower demands.
- The water demand scenarios adopted by all three core parties reflect rather high demand planning assumptions, even for the "low demand scenarios". National interests and negotiation positions have apparently influenced the projections rather than using a realistic assessment of likely developments. Furthermore, the underlying assumptions on the overall macro-economic development were not clearly specified.
- The size of the future net water gap and thus the total amount of additional water needed, depends on future water sector policies and the utilization pattern in the region including
  implementation of bilateral agreements,
  stringency of demand management,
  recovery rate of treated wastewater,
  proportion of fresh water substitution in agriculture, and
  amount of inter-sectoral and intra-regional fresh water that is reallocated to higher value uses and future water quality management.

4.4.2 Future conventional water resources development

The data on available water resources presented in this chapter are based on available hydrological information and best estimates of groundwater conditions. Like all water data, they should be reviewed regularly and updated in view of the latest information provided by hydrological and hydrogeological services.

The reliability of water resources will be of increasing importance in the future, because of its implications for water management. This will require systematic contingency planning to ensure that responses to droughts and floods will minimise adverse effects. From the assessment of the local water resources development potential, the following conclusions can be drawn:

- Low-cost conventional resources in the region are already developed. A further potential exists, but marginal costs increase rapidly.

- The least-cost solution for short- and medium-term water development is water conservation through demand management which should be given high priority.
- Although brackish water desalination is relatively expensive and the potential is limited, it nevertheless might become increasingly important in some places in the region.
- Deteriorating water quality in some parts of the region is already a serious issue and more and more existing resources are in danger of becoming in the future unsuitable due to increasing pollution and salinization.
- Treatment and reuse of wastewater is expected to become the major local option to develop a large portion of water for irrigation which will require substantial investments in all core party areas.

4.4.3 Future non-conventional water development options

There are several non-conventional options to bridge these shortages including water imports by sea or land from outside the region and water desalination. Non-conventional options to bridge water shortages have many advantages. The decision on what option to choose and proceed is neither easy nor clear-cut and depends on many factors such as size, cost, environmental impact and others. However, experience proves that in the long-term, the integration of the various conventional and non-conventional water-demand management options must be considered.

In addition to sustainability, enhancing cooperation and improvement of socio-economic conditions of the peoples of these countries, such carefully implemented options will have positive environmental effects. Bringing large amounts of fresh water continuously to the region will improve the hydrologic cycle, enhance water reserves and improve green cover and possibly the climate in the region. Such environmental advantages will be with time cumulatively positive to the region.

The comparison of non-conventional water development options shows the following summary points:

- The least contentious option would be the coastal seawater desalination, as it depends much less on political and technical preconditions.
- Import of water from other countries cause strong political dependencies and requires development of improved transport feasibilities. All combined options which include water import from Lebanon are cheaper in the short- to medium-term. However, this option is assumed to be limited in quantity and time.

- Desalination scenarios including the intersea project (Med-Dead, Red Dead) are slightly cheaper than coastal desalination, only requiring an international agreement among the riparians of the Dead Sea.

4.4.4 Joint regional planning and cooperation

Genuine regional water resources planning and development in the ME requires the participation of all regional parties in good will and with high belief that regional solutions are of higher benefits to the region as a whole as well as to each of the regional parties

The vital role of water in sustaining economic and social development in the region calls for a regional strategy for its development and management, i.e., a set of medium- to long-term action programs conducive to achieving development goals and implementing water-related policies.

Joint regional water planning should start and continue, and recommended pilot projects should be implemented jointly in order to pave the way for possible joint decisions in the medium- and long-term. This is also important for reducing the differences in perception among the three core parties. Irrespective of the actual size of the future regional net water gap, regional co-operation in developing and managing additional water resources is indispensable and should be facilitated by regional institutional arrangements. External financial and technical support will have to be mobilized for the activities recommended jointly by the three parties.

Accepting and applying the concept of negotiation and dialogue to resolve differences and disagreements between the countries of the region is another way of co-operation and enhancement of peace and stability in the region.

References

GTZ (1997) Deutsche Gesellschaft für Technische Zusammenarbeit GmbH and Consulting Engineers Salzgitter (CES) GmbH: Multilateral Working Group on Water Resources: Middle East Regional Study on Water Supply and Demand Development. Regional Overview Phase II, November 1997 (restricted distribution), pp 2-1–2-24

GTZ (1998) Deutsche Gesellschaft für Technische Zusammenarbeit GmbH: Multilateral Working Group on Water Resources: Middle East Regional Study on Water Supply and Demand Development. Concluding Report, February 1998, pp 1–24

Haddad M and Mizyed N (2004) Non-Conventional Options for Water Supply Augmentation in the Middle East: A Case Study. Water Intern J, Intern Water Resour Assoc 29:232–242

5 Summary and conclusions

A. Flexer, J. Guttman, H. Hötzl, P. Möller, E. Rosenthal, E. Salameh, A-R. Tamimi

From a great variety of aspects, the Jordan Valley is an unique area and a very complicated one. The Central Jordan Valley is part of the major dislocation zone extending from Turkey to East Africa which is known as the Syrian African Rift system. The segment which includes the Gulf of Aqaba, the Arava Valley, the Dead Sea, the Jordan Valley Lake Tiberias and the Baqa'a in Lebanon is regarded as a transform fault with a left-lateral displacement of 105 km. The strike-slip movement along the transform fault is the major tectonic element of Lake Tiberias, the Jordan Valley, and the Dead Sea, all embraced by two mountain ridges. This tectonic system includes, beside the elongated "pull-apart-basins", which form the deep depressions along the main structure, also several secondary faults and series of asymmetric anticlines and synclines in the adjacent escarpments. The structural elements have great impact on the flow pattern, on the interconnections between the aquifers and on the locations of the natural outlets of the aquifers.

Nature endowed the area with an arid climate with scarce and irregular precipitations which could also be of torrential and destructive character. The study area is bordered in the west by the Judean and Samarian Hills with the cities of Hebron, Jerusalem, Ramallah and Nablus on the top of the ridges. In the east the natural borders are the highlands of Jordan extending over the entire length from north to south and hosting Jordan's main population centers, Amman, Salt, Zarqa and Irbid.

In the mountainous areas the climate is of Mediterranean type with precipitations of 500–700 mm/yr. The precipitation declines sharply towards the rift valley where it varies between 50–150 mm/yr. The natural replenishment to the regional Upper Cretaceous calcareous aquifers occurs in the high mountain regions over areas in which the formations are exposed. The outlets of these aquifers are located at much lower topographical levels in the Jordan Valley and at the Dead Sea and Lake Tiberias. The outlets of groundwater from these aquifers are through springs and as lateral flow into the alluvial formations of the Valley. The alluvial aquifer is limited to the Jordan Valley and is replenished from several sources such as lateral flow from the regional aquifers, leakage of runoff, irrigation return flow and small amount of direct recharge. The rough balance of renewable groundwater in the regional aquifers of the study area shows that out of the total of approximately 220 Mm^3/yr, around 160 Mm^3/yr accumulate on the western side of the Rift.

The difference in the water levels between the recharge areas and the aquifer outlets in the Jordan Valley and along the Dead Sea shore are between 500–850 m over a distance of only about 20–30 km implying a very steep flow gradient (assuming that the gradient is uniform). In order to maintain such a steep gradient, a huge amount of rainfall and recharge are required. However, rainfall data, size of the recharge areas and low recharge coefficients indicate that such huge volumes of recharge do not exist in this area. This means that the volume of groundwater in the aquifers cannot account for such a high and constant flow gradient. The only mechanism that yields a satisfactory hydrological explanation for such steep gradients requires a series of hydraulic barriers. These hydraulic barriers (cascades) have lower permeabilities and are connected to geological structures like low permeable layers in the core of anticlines or vertical displacements along faults. They constrain the groundwater to bypass them and to flow along individual paths towards the outlets in the Rift Valley and along the Dead Sea. The details of the flow paths will enable to position more effective and productive wells in the future. This is also confirmed by the springs that emerge in clusters along the western shore of the Dead Sea, indicating that groundwater flows along preferential paths and not in an uniform front.

Such a flow mechanism also explains fluctuations in salinity and the differences in the chemical compositions of water from wells located along the foothills. Low salinities occur near the recharge areas and in the foothills, particularly in sites close to the preferential flow-paths whereas high salinities occur in the areas close to salinization sources such as salt layers in the Lisan formation, salt domes, brines and irrigation return flow. Overpumping creates local water-level lows which cause flow of saline water into freshwater bodies enhancing their salinity.

Along the western side of the Jordan Valley, the recharge into the Quaternary alluvial aquifer of the valley is mainly from leakage of spring water, by lateral flow from the regional aquifers, by ascending saline water from deep-seated layers and by irrigation return flow. Direct recharge is negligible.

On the eastern side of the Valley, the Quaternary alluvial aquifer is also recharged by contribution of lateral flow from the regional aquifers. Recharge from surface water consists mainly of low salinity water. Nevertheless, most of the groundwater in the alluvial aquifer is brackish due to the occurrence of shallow and deep-seated brines and of evaporite bodies.

Wells drilled into the Quaternary aquifer pump water which is the result of mixing of fresh and saline water in different proportions.

A comprehensive chemical study of groundwater focused on two topics: Characterization of chemical composition presented as spider diagrams which show characteristic shapes depending on the presence of dissolved minerals. In the study area groundwaters derived from late and early diagenetic limestones, from sandstones and from basalts, are distinguishable by their spider patterns. Even spider patterns of water from zones of interaction between basalts and limestones show recognisable differences from waters originating from both host rocks.

Rare earth elements and yttrium (REY) and stable isotopes (^{2}H and ^{18}O) indicate two different aspects of the source of water. REY characterise in most cases the lithology of the recharge zone, whereas water isotopes record, among other parameters, the altitude of precipitation and the effect of evaporation.

Detailed flow paths of groundwater can be delineated by conjoint consideration of aquifer lithology and from information derived from REY and water isotope distribution. Knowledge of the source of water, which is not identical with the source of solutes, is mandatory for any reliable hydraulic modelling, land use planning and discharge of untreated and treated waste water.

Five types of brines causing salinization of freshwater are identified:

- Recent Dead Sea brine which is flushed out by freshwater from areas of former high stands of the Dea Sea,
- Remnants of the Neogene Sdom Sea brine which is manifested in Haon water on the eastern side of Lake Tiberias,
- Neogene brines of the Tiberias Hot spring type which is an alteration product of the Sdom Sea brines by albitization. This type is also present in the Arava Valley (Zofar 20),
- Neogene to recent ablation brines such as the Zemah type, and
- Deep-seated, thermal and pressurized brines reach the surface only in the surroundings of Mt. Sdom at the Dead Sea but are known from deep wells all over the area.

The salinization of groundwater in the Arava Valley is largely contributed by the lithologies of the hosting aquifers such as the Kurnub Group sandstones and Judea Group limestones.

The geology of the **Central Jordan Valley** in the study area has not been entirely elucidated. Interpretation of available seismic sections reveals several structural and stratigraphic features which are of considerable hydrogeological importance. For instance, in the central part of the western lower Jordan Valley between the Dead Sea in the south to the Fazael area in the north, the Judea Group aquifer is encountered at relatively shallow depth, which varies from a few hundred meters east of the Cretaceous outcrops to more then 1000 m close to the Jordan River. Several east-stepping fault blocks were encountered in the area. Some of these faults exhibit elements of "flower structure" at the Judea Group level, suggesting lateral displacement. A lenticular shaped sedimentary body was deposited within a structural depression located east of Auja. The analysis of seismic evidence suggests that this 10 km long and 700 m thick body is probably composed of interbedded evaporites and clays and thickens eastwards. Small-scale horizontal crescentic structures in several seismic cross-sections are interpreted as evaporitic bodies located in local basins within the subsurface of the Jordan Rift Valley. The Bouguer anomaly map of the Central Jordan Rift Valley reveals within the basin-fill, several well-established gravimetric lows which are interpreted as evaporite bodies. Two major results are important for further groundwater exploration: the upper Cenomanian-Turonian aquifer beds are rather shallow and enable economic exploitation, but the existence of evaporite bodies may cause salinization of freshwater.

A three-dimensional geohydrological, conceptual model and groundwater flow modeling of the **Marsaba-Feschka** region reveals that exploitation of freshwater in the study area can be extended to new locations and increased in amount. The results obtained are promising and indicate the possibility to produce additional significant amounts of freshwater in the next 20 years. This could clearly contribute to the development of this area. In order to better control the drawdown, it is suggested to spread the wells over a larger area in the Judea Desert well field.

In the **Wadi Qilt region** the uppermost karstified aquifer of the Shivta Formation, constitutes a less important water resource. This is due to its limited spatial extent within the study area and in its surroundings. The carbonate aquifer is highly conductive but of relatively low storage capacity. As the result of the large outcrop area and the relatively high precipitation rates in the highland, the morphologically deeply incised central hills, represent the location of highest recharge to the uppermost aquifer. The other areas are covered by more or less impermeable chalks. Discharge of the relatively low salinity groundwater occurs mainly through springs which can be categorized into two types: the topographically higher eastern slope springs with very quick responses to recharge events that dry up during the summer, and the perennial types of springs emerging along the Jordan Valley escarpment with dampened seasonal fluctuations. Not all groundwater is discharged through the springs. Numerical modeling reveals cross-aquifer flow between the Shivta Formation aquifer and the upper aquifer of the Judea Group. The more or less continuous flow pattern of the springs in the Jordan Valley suggests reversed flow from the Judea

Group to the Shivta Formation. Spring water is mainly used for irrigation purposes and watering of livestock. A major source of recharge is partly treated waste water infiltrating along the three wadis: Mukhmas, Muqalek and El Nar. However, with improved treatment, waste water can be regarded as a significant resource in this area.

Hydrogeological models of the **Jericho area** are based on recharge through flooding in wadis, return-flow from irrigation and groundwater abstraction. The quality of water of the eastern aquifer is very variable. In the Jericho area the $Ca\text{-}HCO_3$ water type was identified in sources located near the mountainous areas, whereas the Na-Cl and Ca-Cl types occur in the Jordan Valley. Some of the wells show pollution by NO_3^-, SO_4^{2-} and K^+ which are due to the use of fertilizers in the area. Chemical monitoring of groundwater and spring water is essential to show the origin and location of possible threats to water quality as well as effects of overpumping.

In the Jordan **Rift escarpment** specific hydrogeologic conditions occure being different from that on the western side. In the lower part of the slopes aquitards and aquifers with rather small yield predominate, whereas the main aquifer (Upper Cretaceous limestones) outcrops in the higher parts of the escarpment. Numerous springs at the base of the aquifer generate the main wadi discharge. It is important to store and conserve this groundwater discharge for further utilization. In the larger wadis dams and reservoirs are already constructed. Still there exists additional potential for water storage especially for flash floods in winter time as well as for the increasing amount of discharging effluents from waste water treatment plants. Artificial recharge of these waters in proper aquifers. are recommended in order to avoid large evaporation losses from surface reservoirs.

In the **Wadi Shueib-Kafrein-Suweimeh area**, groundwater quality varies from fresh (in locations close to the major wadi outlets) to brackish along the foothills and to saline in the more western and southwestern parts of the valley floor. Most recharge comes from infiltrating runoff water from the wadi discharges, particularly from flood waters after heavy rain events. A smaller amount of already brackish water recharges from the adjacent lower mountain aquifers discharging directly into the valley aquifer.

The agricultural development of the area started in the mid 1950's and went along with the major development of infrastructure. As a consequence, large quantities of groundwater were abstracted and first signs of overpumping were already observed at the beginning of the 1960's. As far as the sources of salinization are concerned, four different processes were identified: the high salt content of the lacustrine Lisan Formation in the vicinity of the Jordan Valley aquifers, upward movement of salt water into the soils due to capillary forces, salt accumulation in the soils due to evaporation of very shallow groundwater tables and the anthropogenic factor, i.e., overuse of fertilizers, improper use of drip irrigation and irrigation with brackish water.

The established groundwater vulnerability and hazard map of Wadi Shueib, clearly show the vulnerable areas and the "hot spots" of contamination in the study area and supply information for a variety of applications in the domain of urban management as well as for decision making and planning aspects. The general limitations of groundwater-vulnerability and -hazard maps are that they provide only relative answers and no quantitative or absolute estimates of the actual sensitivity of the groundwater to existing contamination or the probability that contamination will occur.

The areas of **Wadi Waleh** and **Wadi Zarqa Ma'in** belong to the western part of the central part of Jordan. They consist of upper and lower aquifer complexes and groundwater flow in both aquifers towards the Dead Sea in the west. The thermal water in the lower aquifer is the result of deep circulation within the Paleozoic sandstones receiving heat from a normal to slightly elevated geothermal gradient. Part of this thermal water is ascending along the Jiza, Daba'a and Masattarat faults into the upper aquifer as water anomalies at wellheads indicate. However, the main discharge of the lower aquifer flows to the Dead Sea as is documented by several thermal spring along and close to the shoreline. Chemical analyses show that groundwater in the upper aquifers is of the bicarbonate type and in the lower of chloride type. The stable isotopes of the thermal water from the two aquifers indicate meteoric origin.

The large freshwater reservoir of **Lake Tiberias** is endangered by input of saline water. Such processes are not uniform but depend on the particular hydraulic systems in various tectonic blocks and in areas adjoining the lake. The interactions between recharge, lake- and groundwater level and hydraulic head, vary regionally and seasonally (spring and autumn). Groundwater flowing to the Lake, represents mixtures of water bodies of different compositions and qualities (originating from spring clusters and probably from the Hermon Massif and Golan Heights) with saline waters residing in the sediments. Because of alternating flow systems, the processes of mixing of fresh and saline groundwaters vary seasonally.

Due to their topographical and geological position and to the seasonally varying hydrochemical composition, hydrochemical differences arise between the water from the different clusters of springs emerging along the shores of the Lake. Each spring cluster at the Lake displays typical compositions which depend on the lithology of the recharge area and their enclosed formation waters.

REY patterns of the water-sediment interface of Lake Tiberias are characteristically influenced by seasonally varying Ce- and Y-anomalies modified by infiltrating groundwater. Using the changes in Ce anomalies, minimum annual amounts of 44 Mm^3 of water with a load of 15.4 Gg Cl enter through the bottom of the Lake. From weighted calculations of diffusion and advection, an annual minimum of 22 Gg Cl is derived. Both results represent distinctly different approaches. These amounts are minima and correspond to an increase of 1.4–2 % of the average chlorinity in the lake which is just at the verge of the analytical limit of detection.

As a result of the **retreat of the Dead Sea**, since the early 1960's, the interface between the fresh groundwater and the Dead Sea water has steadily migrated seaward in order to reach a new hydrodynamic equilibrium. This has been achieved through the lowering of the groundwater levels by groundwater discharges into the Dead Sea itself. The riparians of the Dead Sea are currently loosing an average of 425–461 Mm^3 of fresh groundwater to the Dead Sea per meter drop in the sea level.

The induced and accelerated flow of groundwater towards the Dead Sea, resulted in the dissolution of salts and the erosion of friable, fine-grained, non-consolidated Lisan Marls and their removal to the Dead Sea, leaving behind cavities in the underground of the coastal areas which, when reaching a critical size, collapse and form huge sinkholes at the ground-surface. The shrinking Dead Sea resulted in lower total evaporation amounts from its surface and hence, lower humidity of the surrounding areas thus increasing regional aridity.

In the chapter on **groundwater management** an overview on available resources in the Rift Valley and their uses is given followed by a discussion of their sustainable management. The state of water strategy and policy is separately given for Jordan, Israel and the Palestinian Authority. This chapter finishes by outlining the future demands of water showing that improvements in reclaiming waste water and desalinization are the most urgent tasks. In summary, the water quality in the Jordan Rift is deteriorated by overpumpage causing upflow of saline water, inadequately treated sewage, untreated domestic and agricultural wastes, uses of low-quality groundwater in agriculture, leachates from solid waste landfills, runoff of urban drainage, and residues of fertilizers and pesticides. The growth of agriculture has a stronger impact on the biodiversity in the Rift.

The study area is governed by three political entities: Jordan, Israel and the Palestinian Authority. However, water does not respect fences or borders. It is a joint resource for all people living in the area. In each of these entities, the management of water resources is not only done according to their particular legislations but also considering ancient traditions, commitments and, above all, the limitations imposed by different socio-economic frameworks.

It is less than fifty years that life and environment in the Lower Jordan Valley was affected and animated by the clean and abundant water flow of the Jordan River and its tributaries. Nowadays about 90% of the river's natural flow, that is more than 1200 Mm^3/yr including discharges from the side wadis, has been diverted by Israel, Jordan and Syria for domestic and agricultural use with sewage flowing in its place, so that the Lower Jordan River today is a small, winding, dirty stream that makes its way to the Dead Sea.

Though everybody has to realize that the environment and ecosystems have been partly changed in an undesirable dramatic way, it also has to be accepted in a honest assessment that this was done in favor of the people living in the region. The growing population on both sides of the Jordan River, their increasing demand for potable water, their demand for enhanced food production as well as their demand for improved hygienic and living standards has set priorities against the original natural conditions. Nevertheless, the remaining salty rill in the Jordan River and the steady drop of the Dead Sea water level are clear indications that this natural resource of fresh surface water is overstrained and the strong withdrawals are not in compliance with a sustainable water management. But, how to satisfy the request of the further increasing population for more fresh water? It is natural that in addition to the surface water, more and more groundwater were explored by capturing springs and developing the aquifers with shallow and deep wells. But now even many of the aquifers reveal signs of overexploitation by dropping water levels and/or quality deterioration as it has been described in the previous chapters of this book. In addition, the deterioration of groundwater is also due to anthropogenic processes such as agriculture, inflow of waste water and of return flow from irrigation.

The results of the joint research project presented in the chapters of this book improve the understanding of the origin, the geologic connections, the quantitative and qualitative dependencies and the hydraulic relationships as well as the man-made interferences and impacts on the available water resources in the Jordan Valley. The deeper insight gained by these studies can support the realization of local solutions for new developments or restorations of aquifers up to integrated regional responses for water management. The studies also emphasized the extreme weight exerted by water consumption based on the natural resources of fresh water. The heavy inroad upon these resources is clearly indicated by the drop of the water level of the Dead Sea, where more than 800 Mm^3 water are missing yearly to keep the natural balance. In view of the expected further increase of the population in the whole region, it is imperative to consider not only the additional water demands which will have to be met for the well-being of the population, but also to find solutions which consider the natural equilibrium of the environment or strive to restore its original status. To reach these goals, a multitude of non-conservative schemes to develop water resources will have to be applied, starting from local brackish and saline water treatment, reuse of treated waste water up to joint transboundary supply from remote sea-water desalination plants. One such scheme could be by bringing additional water from the Red Sea to the Dead Sea through a proposed Red Sea-Dead Sea canal. This operation would restore the Dead Sea and its environment and as a result of the elevation difference between the two seas which is presently 418 m, to additionally produce vast amounts of hydroelectric energy to be used, among others, for seawater desalination.

Notwithstanding these differences and limitations, the only way to survive under the harsh natural conditions and to create life-supporting instruments for all people living in the area, is by regional cooperation, mutual help and support, free exchange of professional experience and by joint efforts. The present project which was generously financed by the government of the Federal Republic of Germany proved that all this is possible and a major step forward was achieved by all working together. We should be all inspired by Prince El Hassan bin Talal of Jordan who wrote recently (2006): "...I worked over the years to promote the formation of a new framework of cooperation in the greater Middle East and North Africa – a Community of Water and Energy similar to the Community of Coal and Steel that was established in Europe in 1951 and served as a precursor to the European Common Market. The ability of Europe to overcome its troubled past and achieve full-fledged union in only half a century, is an example to inspire us all..." The present project should be a first step to lead the Water Communities in the three countries to a future of close cooperation and mutual support.

Subject Index

B

C

T

U

V

W

Printing: Krips bv, Meppel, The Netherlands
Binding: Stürtz, Würzburg, Germany